# Gold Catalysis

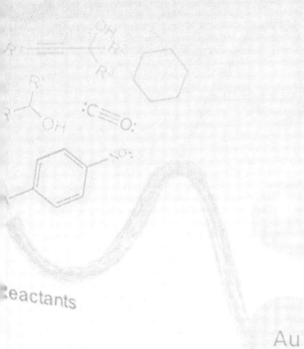

# Gold Catalysis

Preparation, Characterization, and Applications

edited by

**Laura Prati**
**Alberto Villa**

Pan Stanford Publishing

*Published by*

Pan Stanford Publishing Pte. Ltd.
Penthouse Level, Suntec Tower 3
8 Temasek Boulevard
Singapore 038988

Email: editorial@panstanford.com
Web: www.panstanford.com

**British Library Cataloguing-in-Publication Data**
A catalogue record for this book is available from the British Library.

**Gold Catalysis: Preparation, Characterization, and Applications**

Copyright © 2016 Pan Stanford Publishing Pte. Ltd.

*All rights reserved. This book, or parts thereof, may not be reproduced in any form or by any means, electronic or mechanical, including photocopying, recording or any information storage and retrieval system now known or to be invented, without written permission from the publisher.*

For photocopying of material in this volume, please pay a copying fee through the Copyright Clearance Center, Inc., 222 Rosewood Drive, Danvers, MA 01923, USA. In this case permission to photocopy is not required from the publisher.

ISBN 978-981-4669-28-3 (Hardcover)
ISBN 978-981-4669-29-0 (eBook)

Printed in the USA

# Contents

Preface     xv

**1 Deposition-Reduction**     1
*Catherine Louis*
1.1 Introduction     1
1.2 Synthesis Strategies     2
     1.2.1 Gold Precursor: Gold Speciation     2
         1.2.1.1 Warnings concerning the preparation of supported gold catalysts     3
1.3 Impregnation     4
     1.3.1 Principle     4
     1.3.2 Mere Impregnation     4
1.4 Anion Adsorption     6
     1.4.1 Principle of Ion Adsorption     6
     1.4.2 Gold Anion Adsorption     7
     1.4.3 Washing with Ammonia     8
1.5 Cation Adsorption     11
1.6 Deposition-Precipitation     13
     1.6.1 Principle of Deposition-Precipitation     13
     1.6.2 Deposition-Precipitation by Addition of a Base     14
         1.6.2.1 Deposition-precipitation at a fixed pH     14
         1.6.2.2 Influence of pH     16
         1.6.2.3 Mechanism of deposition-precipitation     18
         1.6.2.4 A base apart, $NH_4OH$, a precipitating agent, or a complexing agent?     20
     1.6.3 Deposition-Precipitation by Addition of a Delay Base     22

|  |  | 1.6.3.1 | Deposition-precipitation with urea | 22 |
|  |  | 1.6.3.2 | Mechanism of deposition-precipitation | 24 |
| 1.7 | Conclusion | | | 26 |

## 2 Immobilization of Preformed Gold Nanoparticles — 39
*Carine E. Chan-Thaw, Alberto Villa, and Laura Prati*

| 2.1 | Introduction | 39 |
| 2.2 | Nucleation and Growth Processes | 41 |
|  | 2.2.1 Homogeneous Nucleation: Growth | 41 |
|  | 2.2.2 Heterogeneous Nucleation: Growth | 44 |
| 2.3 | Stabilization of Gold Colloids | 45 |
|  | 2.3.1 Importance of Stabilization | 45 |
|  | 2.3.2 Electrostatic Stabilization | 48 |
|  | 2.3.3 Steric Stabilization | 49 |
|  | 2.3.4 Electrosteric Stabilization | 50 |
| 2.4 | Preparation of Gold Nanoparticles | 51 |
|  | 2.4.1 The Stabilizers | 51 |
|  | 2.4.1.1 Electrostatic stabilizers | 51 |
|  | 2.4.1.2 Steric stabilizers | 52 |
|  | 2.4.1.3 Electrosteric stabilizers | 52 |
|  | 2.4.2 Nature of the Reducing Agent | 53 |
|  | 2.4.2.1 Chemical reduction of metal salts | 53 |
|  | 2.4.2.2 Electrochemical reduction | 58 |
|  | 2.4.2.3 Other methods | 58 |
| 2.5 | Immobilization of Nanoparticles | 58 |
|  | 2.5.1 Adsorption | 59 |
|  | 2.5.2 Grafting | 63 |
| 2.6 | Conclusions | 65 |

## 3 Solvated Metal Atoms in the Preparation of Supported Gold Catalysts — 73
*Claudio Evangelisti, Eleonora Schiavi, Laura Antonella Aronica, Rinaldo Psaro, Antonella Balerna, and Gianmario Martra*

| 3.1 | Introduction | 73 |
| 3.2 | Synthetic Strategy | 74 |
| 3.3 | Metal Vapor Synthesis for Preparing Supported Gold Nanoparticles | 77 |

|   |     | 3.3.1 Preparation of Solvated Gold Atoms | 77 |
|---|-----|---|---|
|   |     | 3.3.2 Supported Catalysts from Solvated Gold Atoms | 82 |
|   | 3.4 | MVS-Derived Au–PD Bimetallic Catalysts | 90 |
|   | 3.5 | Concluding Remarks | 92 |

## 4 Microgels as Exotemplates in the Preparation of Au Nanoclusters — 99
*Andrea Biffis and Paolo Centomo*

|   |     |   |   |
|---|-----|---|---|
|   | 4.1 | Introduction | 99 |
|   | 4.2 | Microgel Preparation | 101 |
|   | 4.3 | Microgel-Stabilized Metal Nanoclusters | 103 |
|   | 4.4 | Advantages of Microgel-Stabilized Metal Nanoclusters | 109 |
|   | 4.5 | Microgel-Stabilized Gold Nanoclusters: Catalytic Applications | 112 |
|   | 4.6 | Conclusions | 119 |

## 5 Miscellaneous — 123
*Carine E. Chan-Thaw, Laura Prati, and Alberto Villa*

|   |     |   |   |
|---|-----|---|---|
|   | 5.1 | Introduction | 123 |
|   | 5.2 | Co-Precipitation | 124 |
|   | 5.3 | Chemical Vapor Deposition | 125 |
|   | 5.4 | Solid Grinding | 127 |
|   | 5.5 | Physical Vapor Deposition | 128 |
|   |     | 5.5.1 Thermal Evaporation | 128 |
|   |     | 5.5.2 Sputtering | 129 |
|   | 5.6 | Conclusions | 131 |

## 6 Transmission Electron Microscopy on Au-Based Catalysts — 135
*Di Wang*

|   |     |   |   |
|---|-----|---|---|
|   | 6.1 | Introduction | 135 |
|   | 6.2 | The Principle of Transmission Electron Microscopy | 137 |
|   |     | 6.2.1 Electron Diffraction | 137 |
|   |     | 6.2.2 HRTEM and HAADF STEM | 138 |
|   |     | 6.2.3 Analytic TEM | 140 |
|   |     | 6.2.4 Electron Tomography | 142 |
|   | 6.3 | Structures of Au-Based Catalysts | 144 |

|  |  | 6.3.1 | Crystal Structure of Au Nanoparticles | 144 |
|---|---|---|---|---|
|  |  | 6.3.2 | Interface and Surface Structures of Supported Au Nanoparticles | 147 |
|  |  | 6.3.3 | Structure of Au-Based Bimetallic Catalysts | 156 |
|  | 6.4 | Outlook |  | 160 |

**7  X-Ray Photoelectron Spectroscopy Characterization of Gold Catalysts** — **171**
*Gabriel M. Veith*

| | 7.1 | Introduction | 171 |
|---|---|---|---|
| | 7.2 | X-Ray Photoelectron Spectroscopy | 173 |
| | 7.3 | XPS of Gold Catalysts: Study of Gold Oxidation States | 175 |
| | | 7.3.1  Evolution of Gold Oxidation State with Synthesis | 176 |
| | | 7.3.2  Correlation of Oxidation State to Catalytic Activity | 177 |
| | | 7.3.3  Identification of Supported Gold Nanoparticle Oxidation States | 179 |
| | 7.4 | Postmortem Analysis | 184 |
| | | 7.4.1  Evolution of Gold Oxidation State with Reaction | 184 |
| | | 7.4.2  Changes in Support Oxygen Chemistry with Synthesis | 185 |
| | | 7.4.3  Changes in Support Metal Oxide Chemistry after Catalytic Reaction | 189 |
| | | 7.4.4  Quantifying Catalyst Coarsening | 191 |
| | | 7.4.5  Chemical Deactivation/Blocking of Gold Catalysts | 193 |
| | | 7.4.6  XPS to Understand the Nucleation and Growth of Gold Nanoparticles | 193 |
| | 7.5 | Frontiers in XPS Instrumentation | 194 |
| | 7.6 | Conclusions and Perspective | 195 |

**8  FTIR Techniques for the Characterization of Au(-Ceria)-Based Catalysts** — **205**
*Maela Manzoli and Floriana Vindigni*

| | 8.1 | An Overview of Gold/Ceria-Based Catalysts | 206 |
|---|---|---|---|
| | | 8.1.1  Preparation of the Samples | 209 |
| | 8.2 | Ex situ CO Adsorption at 100 K | 212 |

|  |  | 8.2.1 | Some Insights into CO Adsorption | 212 |
|---|---|---|---|---|
|  |  | 8.2.2 | As-Received Au/CeO$_2$ Catalyst | 214 |
|  |  | 8.2.3 | Effect of Pre-Oxidation at 473 K on the Exposed Sites | 215 |
|  |  | 8.2.4 | Au/CeO$_2$ Catalyst Reduced in H$_2$ | 217 |
|  | 8.3 | CO–O$_2$ Interaction at Low Temperature up to Room Temperature | | 220 |
|  |  | 8.3.1 | Effect of Doping: Au Supported on Zn-Modified Ceria | 221 |
|  |  | 8.3.2 | Effect of Doping by Other Elements (Sm, La) | 224 |
|  |  | 8.3.3 | On the Modification of the Support by Iron Oxide | 227 |
|  |  | 8.3.4 | Modification of Ceria by Other Oxides | 231 |
|  | 8.4 | Ex situ CO Adsorption at Room Temperature | | 234 |
|  | 8.5 | In situ FTIR Measurements at Increasing Temperature | | 238 |
|  | 8.6 | Operando Measurements of CO$_2$ Uptake | | 240 |
|  | 8.7 | Final Remarks | | 245 |
| 9 | **Determination of Dispersion of Gold-Based Catalysts by Selective Chemisorption** | | | **253** |
|  | *Michela Signoretto, Federica Menegazzo, Valentina Trevisan, and Francesco Pinna* | | | |
|  | 9.1 | On Gold Dispersion | | 253 |
|  | 9.2 | How to Measure Gold Dispersion | | 255 |
|  |  | 9.2.1 | Electron Microscopy | 255 |
|  |  | 9.2.2 | XRD | 256 |
|  |  | 9.2.3 | Selective Chemisorption | 257 |
|  |  | 9.2.4 | Other Techniques | 260 |
|  | 9.3 | Selective Chemisorption on Gold Catalysts | | 262 |
|  |  | 9.3.1 | O$_2$ Chemisorption | 262 |
|  |  | 9.3.2 | H$_2$ Chemisorption | 264 |
|  |  | 9.3.3 | Methyl Mercaptane Chemisorption | 264 |
|  |  | 9.3.4 | CO Chemisorption | 265 |
|  | 9.4 | Pulse Flow CO Chemisorption at Low Temperatures | | 266 |
|  |  | 9.4.1 | Apparatus for Pulse Flow CO Chemisorption | 267 |
|  |  | 9.4.2 | Pulse Flow CO Chemisorption Method | 269 |
|  |  |  | 9.4.2.1 Au/TiO$_2$ samples | 270 |
|  |  |  | 9.4.2.2 Au/ZrO$_2$ samples | 273 |

|          |                                                                      |     |
|----------|----------------------------------------------------------------------|-----|
|          | 9.4.2.3 Au/CeO$_2$ samples                                           | 276 |
|          | 9.4.2.4 Au/Fe$_2$O$_3$ sample                                        | 277 |
|          | 9.4.3 Consideration on the Chemisorption Stoichiometry               | 277 |
| 9.5      | Final Remarks                                                        | 279 |

## 10 New Findings in CO Oxidation — 285
*Yoshiro Shimojo and Masatake Haruta*

| 10.1 | Introduction | 285 |
|------|--------------|-----|
| 10.2 | An Overview of Catalytic CO Oxidation | 286 |
| 10.3 | Environmental TEM Observation under CO Oxidation | 291 |
| 10.4 | Stability of Nanoparticulate Gold Catalysts | 292 |
|      | 10.4.1 Al$_2$O$_3$ Support | 293 |
|      | 10.4.2 SiO$_2$ Support | 296 |
|      | 10.4.3 TiO$_2$ Support | 298 |
|      | 10.4.4 MnO$_x$ Support | 300 |
|      | 10.4.5 Fe$_2$O$_3$ Support | 300 |
|      | 10.4.6 CeO$_2$ Support | 303 |
| 10.5 | New Attempts in the Preparation of Gold Catalysts | 304 |
| 10.6 | Summary | 306 |

## 11 The Role of Gold Catalysts in C–H Bond Activation for the Selective Oxidation of Saturated Hydrocarbons — 311
*Sarwat Iqbal, Gemma L. Brett, and Graham J. Hutchings*

| 11.1 | Introduction | 311 |
|------|--------------|-----|
| 11.2 | Small Alkanes | 312 |
|      | 11.2.1 Methane and Ethane | 312 |
|      | 11.2.2 Propane | 315 |
| 11.3 | Propene | 316 |
|      | 11.3.1 Effect of Support | 317 |
|      | 11.3.2 Gold Particle Size and Shape Effects | 320 |
|      | 11.3.3 Promoters | 321 |
|      | 11.3.4 Oxidation of Propene with Oxygen | 321 |
| 11.4 | Effect of the Preparation Method | 323 |
|      | 11.4.1 Cyclohexane | 326 |
| 11.5 | Summary | 328 |

## 12 Liquid-Phase Oxidation Using Au-Based Catalysts — 341
*Nikolaos Dimitratos, Ceri Hammond, and Peter P. Wells*
- 12.1 Introduction — 341
- 12.2 Liquid-Phase Oxidation of Oxygen-Containing Organic Compounds — 342
  - 12.2.1 Supported Au Catalysts — 342
  - 12.2.2 Supported Au–Pd- and Au–Pt-Based Catalysts — 363
  - 12.2.3 Supported Au–Cu- and Au–Ag-Based Catalysts — 374
  - 12.2.4 Supported Au-Based Trimetallic Catalysts — 375
- 12.3 Conclusions and Future Perspectives — 376

## 13 Supported Gold Nanoparticles as Heterogeneous Catalysts for C–C Coupling Reactions — 389
*Ana Primo and Hermenegildo García*
- 13.1 Palladium as a Catalyst for Carbon–Carbon and Carbon–Heteroatom Cross-Coupling Reactions — 390
- 13.2 Gold vs. Palladium — 393
- 13.3 Homocoupling of Arylboronic Acids — 395
- 13.4 Suzuki–Miyaura Cross-Coupling Promoted by Supported Au NPs Assisted by Light — 398
- 13.5 Sonogashira Coupling — 402
- 13.6 Role of Pd Impurities on Au-Catalyzed Sonogashira Coupling — 407
- 13.7 Conclusions and Final Remarks — 410

## 14 Toward Chemoselectivity: The Case of Supported Au for Hydrogen-Mediated Reactions — 415
*Fernando Cárdenas-Lizana and Mark A. Keane*
- 14.1 Introduction/Scope — 415
- 14.2 Application of Gold in Hydrogen-Mediated Reactions — 417
  - 14.2.1 Hydrogen–Gold Interaction — 417
  - 14.2.2 Hydrogen-Mediated Reactions Catalyzed by Gold — 419
- 14.3 Case Study 1: Environmental Pollution Control; Hydrodechlorination of Chloroaromatics — 424

|  |  | 14.3.1 | Background | 424 |
|---|---|---|---|---|
|  |  | 14.3.2 | Gold-Promoted Gas-Phase Catalytic Hydrodechlorination of Chlorophenols | 425 |
|  | 14.4 | | Case Study 2: Production of Fine Chemicals; Hydrogenation of Nitroaromatics | 432 |
|  |  | 14.4.1 | Background | 432 |
|  |  | 14.4.2 | Gold-Promoted Gas-Phase Catalytic Hydrogenation of Nitrocompounds | 434 |
|  | 14.5 | | Concluding Remarks and a Look to the Future | 441 |

## 15 Homogenous Gold Catalysis  465
*David Zahner, Matthias Rudolph, and A. Stephen K. Hashmi*

| 15.1 | Introduction | 465 |
|---|---|---|
| 15.2 | The First Methodology: Asymmetric Gold Catalysis | 466 |
| 15.3 | The Most Basic Reactivity Pattern: Nucleophilic Attack on Carbon–Carbon Multiple Bonds | 467 |
| 15.3.1 | Nitrogen Nucleophiles | 467 |
| 15.3.2 | Oxygen Nucleophiles | 468 |
| 15.3.3 | Carbon Nucleophiles | 470 |
| 15.4 | Enyne Cyclizations | 471 |
| 15.5 | Gold Catalysis with Propargyl Esters and Related Compounds | 473 |
| 15.5.1 | 1,2-Migration | 474 |
| 15.5.2 | 1,3-Migration | 475 |
| 15.5.3 | Long-Range Migrations | 476 |
| 15.6 | Gold-Catalyzed Oxidations of Alkynes | 477 |
| 15.6.1 | Sulfoxides | 478 |
| 15.6.2 | Amine Oxides | 478 |
| 15.7 | Oxidative Couplings with Gold | 479 |
| 15.8 | Transmetalation/Cross-Coupling | 483 |
| 15.9 | Generation and Usage of Dipoles in Gold Catalysis | 484 |
| 15.10 | $A^3$-Couplings | 485 |
| 15.11 | Dual Activation | 485 |
| 15.12 | Gold Catalysis Combined with Organocatalysis | 486 |
| 15.13 | Functionalizing Deauration | 488 |
| 15.14 | Glycosylation via Gold Catalysis | 489 |

| | |
|---|---|
| 15.15 Ring Enlargements/Strained Substrates | 489 |
| 15.16 Dehydrative Gold Catalysis | 491 |
| 15.17 Conclusion | 492 |
| *Index* | 501 |

# Preface

Gold has always been recognized as a metal with special properties. Since ancient times gold has been used as a jewel, as a decorative material, or as a metal with therapeutic actions. More recently a new field of application of this extraordinary metal appears related to its characteristics as a catalytic material. Indeed, since the discovery in 1988 of its activity in two fundamental reactions—the oxidation of CO to $CO_2$ (Haruta) and the hydrochlorination of ethylene (Hutchings)—many studies and a constant growth of literature citations deal with this metal.

The discovery and the subsequent success of gold as a catalytically active metal were due to the discovery of suitable methods of obtaining finely dispersed nanoparticles. This is the main reason why the preparation for obtaining an active gold-based catalyst is so important. Gold is a metal with a relatively low melting point, especially if compared to the most used palladium and platinum. Therefore, it is difficult to disperse gold, especially at the nanoscale, the useful dimension in catalysis. The difficulties and trials devoted to obtaining highly dispersed gold catalysts are discussed in the first part of this book (Chapters 1–5) by experts of recognized reputation in this specific field. In particular Chapter 1 deals with the deposition-reduction method, which constitutes the historical method for gold catalyst synthesis. Chapter 2 presents the basis of one of the emerging techniques (sol immobilization) based on the use of preformed gold nanoparticles, whereas Chapter 3 deals with nascent nanoparticles trapped on a matrix (SMAD). A different approach for controlling the particle growth with the use of exotemplates is reported in Chapter 4. Many other methods that have led to excellent catalytic results are summarized in Chapter

5, which gives a comprehensive overview of the currently available methodologies.

The second part of the book is dedicated to characterization, where both surface and bulk techniques are presented. Actually there is not a single technique able to answer all the questions related to the possible correlation between structure and activity or disclose the real structure of gold catalysts under operative conditions. This section helps to understand the possibilities that different techniques offer from both a structural and a reactivity point of view. Structural problems solved by transmission electron microscopy (Chapter 6) and X-ray photoelectron spectroscopy (Chapter 7) related to gold catalysts are presented by experts in the field. Moreover, insights into the comprehension of real active sites are shown by the use of the interaction between gold active sites and molecular probes. It is the case of infrared studies (Chapter 8) as well as selective chemisorption (Chapter 9) that is able to provide information about the chemical activity of the systems.

The third part of the book is obviously devoted to the main gold catalytic applications, and we proudly present contributions of the founders of this chemistry, Prof. Masatake Haruta and Prof. Graham J. Hutchings, together with other leading exponents in the field. From Chapter 10 to Chapter 14 updated and still challenging applications of heterogeneous gold catalysts are shown: CO oxidation (Chapter 10), a well-studied reaction not yet completely understood; C–H activation (Chapter 11), a challenging application not only for gold but also where gold could be peculiar; oxidation reactions in the liquid phase (Chapter 12), where gold catalysts showed enhanced properties compared to classical oxidation catalysts; coupling reactions (Chapter 13), where gold catalysts represent one of the few examples of really heterogeneous catalysts; and hydrogenation reactions (Chapter 14), where gold is able to impart to the catalyst peculiarities in terms of selectivity.

The last chapter (Chapter 15) differs from the others because it presents the catalytic uses of gold in the homogeneous phase. The chapter is presented by one of the most important figures in the field, Prof. Hashmi, and it is a good source of information on the potentiality of gold, even for nonexperts in the field of homogeneous catalysis.

As the editors we believe that all the contents of this book constitute a valuable contribution in understanding not only why gold attracted so much interest in the catalysis scientific community but also why gold has become so popular in quite recent years. We would like to warmly thank all the authors for their excellent contributions, making this book useful for both researchers already involved in gold catalysis and the ones who would like to approach this field.

**Laura Prati**
**Alberto Villa**

# Chapter 1

# Deposition-Reduction

**Catherine Louis**
*Laboratoire de Réactivité de Surface, UPMC-CNRS, 4 Place Jussieu,*
*75005 Paris, France*
catherine.louis@upmc.fr

## 1.1 Introduction

It is well known that the size of gold particles in supported catalysts is a crucial parameter to obtain active catalysts, not only for the reaction of CO oxidation [1–5], but also for many other reactions [6–9]; the size of gold particles (average size and size distribution) drastically depends on the methods used for the catalyst preparation and support.

Most of the chemical preparations of supported gold catalysts are performed in aqueous solution and are based on the principle of a two-step procedure of deposition-reduction, that is, deposition of a gold precursor in the aqueous phase, followed by thermal reduction treatment to reduce it to metal. Most of the preparations involve tetrachloroauric acid (HAuCl$_4$) as a gold precursor. The emblematic method of preparation of supported gold catalysts is the so-called deposition-precipitation method (see Section 1.6). This method was first developed by Haruta's group [10, 11], who is at the origin of

*Gold Catalysis: Preparation, Characterization, and Applications*
Edited by Laura Prati and Alberto Villa
Copyright © 2016 Pan Stanford Publishing Pte. Ltd.
ISBN 978-981-4669-28-3 (Hardcover), 978-981-4669-29-0 (eBook)
www.panstanford.com

the discovery that gold can catalyse the reaction of CO oxidation at room temperature (RT); this major discovery was the starting point of widespread efforts to study catalysis by gold.

As thermal treatment to reduce the gold precursor into metallic particles, calcination in air is the most used because of the intrinsic instability of the Au$^{III}$ or Au$^{I}$ compounds, which are easily reduced to the metallic state even in an oxidising atmosphere; note that auric oxide (Au$_2$O$_3$) is the unique oxide that formed endothermically. One known exception is gold supported on ceria, which may remain unreduced after calcination, depending on the gold loading and ceria surface area [12–14]. Thermal reduction in hydrogen usually leads to the formation of slightly smaller gold particles.

The presence of chlorides in the samples promotes gold particle sintering during calcination [7, 15–19], and calcination under air even at 600°C does not lead to the total elimination of the chlorides [20, 21]. Gold sintering would arise from the formation of Au–Cl–Au bridges [22]. It can be avoided with a reduction treatment, but part of the chlorides also remains in the catalyst [23, 24] and may lead to poorly active catalysts (see Section 1.3.2).

The art of the preparation of supported gold catalysts is to obtain gold particles of small size and either to avoid the presence of chlorides or to get rid of them before thermal treatment.

The goal of this chapter is to describe the chemical methods of preparation of supported gold catalysts on the basis of the principle of deposition-reduction.

## 1.2 Synthesis Strategies

### 1.2.1 *Gold Precursor: Gold Speciation*

Only a small number of gold precursors are commercially available and soluble in water. These are tetrachloroauric acid (HAuCl$_4$.3H$_2$O with Au in the oxidation state III), its sodium and potassium salts, gold acetate, and gold nitrate, but the latter two are poorly soluble in water.

Tetrachloroauric acid is by far the most commonly used gold precursor for catalyst preparation. It is a hygroscopic solid, orange in colour; in aqueous solution, it acts as a strong acid, quite capable of dissolving oxide supports like alumina and magnesia. The speciation of gold strongly depends on the concentration, pH, and temperature of the solution. When tetrachloroauric acid is dissolved in water, chloroauric anions $[AuCl_4]^-$ hydrolyse and form hydroxychlorogold(III) complexes, $[Au(OH)_x Cl_{4-x}]^-$. The increase in pH induces a change of colour of the solution from yellow to colourless, indicating changes in gold speciation. This was attested by extended X-ray absorption fine structure (EXAFS) and Raman studies [25–27] that showed that increasing pH leads to increasing hydrolysis of the initial $[AuCl_4]^-$ complex and that the extent of hydrolysis also depends on the gold and chlorine concentrations, that is, on the ionic strength. Other studies showed that the extent of hydrolysis also depends on the aging time of the solution left at RT because speciation equilibration is slow [21, 28–30]; it also increases with temperature [28, 31, 32]. It is noteworthy that these studies did not reveal the presence of neutral gold species, in contrast with predictions given by thermodynamic calculations [33].

### 1.2.1.1 Warnings concerning the preparation of supported gold catalysts

The reader must be aware that each preparation method admits numerous variations, which may influence the composition and structure of the finished catalyst, and, as a consequence, induce problems of reproducibility. Thermal treatment to reduce gold must be performed under controlled conditions because parameters such as the nature of the gas, the flow rate, the heating rate, and the final temperature influence the final gold particle size.

There are reports mentioning that supported gold catalysts must never be prepared with solutions containing ammonia because *fulminating* gold, which is a family of ill-defined gold compounds containing nitrogen, may form, and these gold–nitrogen compounds contained in dried catalysts are extremely shock sensitive and may explode [34, 35]. However, in the academic literature reported in this chapter, there are no reports of any explosions during preparations of gold catalysts involving ammonia, cyanide, or urea. Of course, this does not mean that care must not be taken, since the chemistry of fulminating gold is not well established.

Some gold catalysts are very sensitive to ambient conditions (light and air), especially those supported on semi-conducting oxide supports like titania, zinc oxide, or ceria; in ambient conditions, uncontrolled gold reduction can occur if the catalysts are stored unreduced, and gold particle sintering can occur if they have been previously activated. To avoid these problems, it is advisable to dry the samples at RT under vacuum and not at around 100°C in air in order to avoid uncontrolled reduction, and to store them in the dark, either in a refrigerator or in a freezer [36, 37] or in a desiccator under vacuum and in the dark [38].

## 1.3 Impregnation

### 1.3.1 *Principle*

*Impregnation* is the most simple preparation method since it consists of wetting a powder support with an aqueous solution containing the metal precursor at natural pH. Afterward, the sample is dried and then thermal treatment is performed to reduce the precursor to metallic particles. The preparation can be performed with a volume of solution corresponding roughly to the pore volume of the support—in such a case, it is called *impregnation to incipient wetness*—or with an excess of solution and it is called *impregnation in excess of solution*; in the latter case, water is often removed using a rotating evaporator. During impregnation, and in fact it would be more correct to write during drying, both the metal ions and the counter-ions are deposited onto the support; in principle, any kind of support can be used.

### 1.3.2 *Mere Impregnation*

Historically, impregnation with tetrachloroauric acid was the first method applied to the preparation of supported gold catalysts because of its simplicity [39–41]. After thermal treatment in air or oxygen, most of the catalysts contained large gold particles

(10–35 nm) and exhibited poor activity in the very first reactions investigated, hydrogenation of alkenes [41, 42], reduction of NO by hydrogen [40], and CO oxidation [43, 44].

As mentioned in the introduction, large particles form because of the presence of chlorides that promote mobility and agglomeration of gold species during thermal treatment [7, 15–19]. In addition, part of the chlorides remains in the catalyst after calcination, even when it is performed at a temperature as high as 600°C [45]. Reduction under hydrogen leads to smaller particles, but part of the chlorides also remains in the samples [24]. Note that although chlorides act as a poison for catalytic reactions such as CO oxidation [20] or selective hydrogenation [24], this is not the case for all reactions. For instance, Baatz et al. [46] showed that impregnation of alumina with $HAuCl_4$ at several pHs led to small gold particles after reduction in $H_2$ at 250°C and that the most active catalyst in glucose oxidation was obtained for a sample prepared at pH < 1 after addition of HCl; the authors did not mention which amount of chlorides remained in the catalysts after reduction. They also confirmed former works [47–50] that gold reducibility under $H_2$ decreases as the amount of chlorides increases.

One can note that in spite of the fact that the aqueous solution of tetrachloroauric acid is acidic, some of the $AuCl_4^-$ ions (or $Au(OH)_xCl_{4-x}]^-$) can interact with the oxide support during impregnation. Indeed, they cannot be completely removed by washing with water. For example, 0.6 to 0.9 wt% gold remained on titania after water-washing for an initial gold loading of 4 wt% [51]. After calcination, the mean gold particle size in the washed sample was smaller (3 to 5 nm) than in an unwashed sample (>10 nm) containing the same gold loading (~1 wt%). The reason invoked was that chlorides were also removed during washing. However, washing impregnated samples with water is not a practicable method for obtaining small gold particles and low amount of chlorides, because of the gold loss in the washing solutions.

Impregnations with gold salts free of chlorides can be also performed. Impregnation with potassium aurocyanide ($KAu^I(CN)_2$) [40, 52] or gold acetate ($Au^{III}(O_2CCH_3)_3$) [17] led to smaller particles (~5 nm) than with $HAuCl_4$. However, $KAu(CN)_2$ leads also to the deposition of potassium onto the support, which may influence the catalytic properties, and gold acetate is hardly soluble in water. Regarding gold acetate, a very recent paper [53] reports attempts to improve the solubility of gold acetate. The powder of Au(OAc)$_3$ was first sonicated in water, resulting in a brown-coloured colloidal dispersion of Au(OAc)$_3$ that could be fully dissolved after addition of Na$_2$CO$_3$ at pH 10–11 and refluxing at the boiling temperature. The resultant solution was transparent and colourless and used to impregnate several metal oxide powders such as Al$_2$O$_3$, CeO$_2$, TiO$_2$, and SiO$_2$, as well as silicates such as a saponite clay and a Y-type zeolite. After calcination at 350°C, followed by washing with water to remove sodium and drying, the average size of the gold nanoparticles was smaller than 6 nm and the gold loading was close to the nominal one of 1 wt%. The smallest gold particles were obtained with silica (3.8 nm) and the saponite clay (2.3 nm).

## 1.4 Anion Adsorption

### 1.4.1 Principle of Ion Adsorption

A key for the formation of small metal particles is to establish, from the very beginning of the preparation, interactions between the metal precursor and the support in solution: these interactions can be electrostatic with the formation of outer sphere complexes or covalent with the formation of inner sphere complexes [54]. In both cases, the value of the point of zero charge (PZC) of the oxide support is crucial because the surface charge of the oxide is pH dependent. The PZC of an oxide is the pH at which the oxide surface is neutrally charged (Table 1.1) (PZC is also called isoelectric point

**Table 1.1** Points of zero charge (PZCs) of a selection of oxide supports

| Oxide support | PZC |
| --- | --- |
| SiO$_2$ | ~2 |
| TiO$_2$ | ~6 |
| Al$_2$O$_3$ | 7–9 |
| CeO$_2$ | 6–7 |
| $\alpha$-Fe$_2$O$_3$ | ~8 |
| ZnO | ~9 |

(IEP), although the definition is slightly different [55]). When the pH of the solution is lower than the oxide PZC, the overall charge of the oxide surface is positive ($OH_2^+$ groups are the main surface species) and anions from the solution can be adsorbed; conversely, when the pH is higher than the PZC, the overall charge of the oxide surface is negative ($O^-$ groups are the main surface species) and cations can be adsorbed. In both cases, the surface charge increases when the gap between the pH and the PZC increases.

Establishing an interaction between the gold precursor and the oxide support in aqueous solution therefore requires an appropriate choice of pH for the solution, precursor, and support. After gold deposition, the samples are washed to remove any potential excess of ions not interacting with the support, as well as the counter-ions.

### 1.4.2 Gold Anion Adsorption

Hence, adsorption of chloroauric anions on oxide supports is in principle possible if the pH of the solution is lower than the PZC of the support; this excludes supports with low PZCs, such as silica (PZC ≈ 2) (Table 1.1).

Practically, the oxide powder is immersed into a solution containing $HAuCl_4$ at a pH lower than the support PZC, and the suspension is stirred at RT or at 70°C–80°C, usually around 1 h because the equilibrium of adsorption is rapidly reached. Indeed, this method applied to the preparation of $Au/TiO_2$ at pH 2 showed that at either 25°C or 80°C, the equilibrium was reached within less than 15 min [51]. In contrast with impregnation in excess of solution, the sample is recovered after filtering or centrifugation, followed by thorough washing with water to eliminate the chlorides and any gold anionic species not interacting with the support. The amount of gold deposited is limited by the capacity of adsorption of the oxide support and also depends on the PZC and surface area of the support and on the pH of the solution. For usual supports like titania or alumina, the maximum gold loading is 1–2 wt%, but rather high levels of chloride remain on the support after washing (0.2–0.4 wt%) [51, 56]. The gold particles are reasonably small after calcination at 300°C (~4 nm) and smaller after reduction under hydrogen (~3 nm), but the chlorides are not totally eliminated either [56].

*Gold anion adsorption* has been extensively studied by Pitchon's group, who called it *direct anionic exchange (DAE)*. They performed a systematic study of the adsorption of chloroauric anions on alumina at 70°C using solutions at pHs from 1.5 to 4.5, which were obtained by varying the HAuCl$_4$ concentration (10$^{-2}$ to 10$^{-5}$ M) [21]. At the lowest pH, when AuCl$_4^-$ was the main species, and the alumina surface was the most positively charged, there was almost no adsorption. In contrast, at pH 4.5, when the main species in solution was AuCl$_2$(OH)$_2^-$, gold was adsorbed. After washing with water and calcination at 300°C, the gold loading was close to the nominal one (∼2 wt%), but the gold particles were large, 10–20 nm. This group also studied chloroauric anion adsorption on other oxides—titania, zirconia, and ceria—at pH 3.5 (10$^{-4}$ M) [57]. As shown in entry 1 of Table 1.2, the Au and Cl loadings varied with both the PZC and the surface area of the supports, and again, after calcination at 300°C, the gold particles were large.

### 1.4.3 Washing with Ammonia

As mentioned in Sections 1.3.2 and 1.4.2, whether impregnation or anion adsorption is carried out, the gold complexes deposited

**Table 1.2** Gold loadings and average particle sizes in gold catalysts prepared by anion adsorption and then washed either with water (entry 1) or ammonia (4 M) (entry 2) [57]. All samples were prepared as follows: 10$^{-4}$ M HAuCl$_4$, pH 3.5 (except for silica, 10$^{-2}$ M and pH 2), 2 wt% nominal Au loading and calcination at 300°C

| Support | Surface area (m$^2$·g$^{-1}$) | PZC | Entry 1: Washing with water Au loading (wt%) | Cl loading (ppm) | d Au$^0$ (nm) | Entry 2: Washing with ammonia Au loading (wt%) | Cl loading (ppm) | d Au$^0$ (nm) |
|---|---|---|---|---|---|---|---|---|
| TiO$_2$ | 40 | ∼6 | 1.53 | 480 | – | 1.19 | ≤150 | 3 |
| ZrO$_2$ | 44 | 6–7 | 1.63 | 2000 | – | 1.56 | 470 | 3 |
| MgO | 25 | ∼12 | 0.53 | 2000 | – | a | a | – |
| Al$_2$O$_3$ | 200 | 7–9 | 2.0 | 3300 | 16 | 1.38 | ≤150 | 1.9 |
| CeO$_2$ | 240 | 6–7 | 1.97 | 6600 | – | 1.98 | ≤150 | <4[b] |
| SiO$_2$ | 530 | ∼2 | 0.01 | ≤150 | – | a | a | – |

[a] Washing with ammonia not done because of the solubility of these oxides
[b] No contrast for transmission electron microscopy (TEM) but no visible X-ray diffraction XRD) peak of Au$^0$

on oxides contain chlorides in their coordination sphere, so the samples also contain chlorides, and the gold particles are rather large after calcination; they are smaller after reduction, but in both cases, chlorides remain present, which is in general detrimental for catalytic reactions. However, a solution has been suggested for eliminating chlorides after gold deposition; it consists of washing the samples with an ammonia solution after gold deposition. Xu et al. [58] and Pitchon et al. [21, 57, 59, 60] were the first groups to report that washing $Au/Al_2O_3$ samples prepared by DAE with an ammonia solution (4 and 25 M) led to chloride elimination without drastic decrease in the gold loading (~1.5 instead of 2 wt%) and that after calcination at 300°C, much smaller gold particles (4 nm) were obtained than after washing with water (10–20 nm) (entry 2, Table 1.2). Since then, this method has been used by many groups [61, 62] and applied to more complex oxides; goethite and Mn- and Co-substituted goethites [63]; ceria–zirconia mixed oxides [64]; ceria doped with Fe, La, or Zr [65]; and $La_2O_3$ [66]. Washing with ammonia has been also recently used to wash samples prepared by deposition-precipitation [67, 68].

Louis's group [69] applied this method of ammonia washing to a series of samples prepared by impregnation and containing 4 wt% Au on silica, alumina, and titania. After washing with an ammonia solution (1 M), they found that gold had not leached out into the solution, at least not significantly, and that small gold particles (~3 nm) were obtained after further calcination at 300°C, even on silica (Table 1.3). Note that because of the low PZC of silica, there were very few preparation methods at the time, in 2006, which allowed control of the gold loading and the formation of small gold particles on silica. Temperature-programmed desorption (TPD) coupled to mass spectrometry showed that the gold species formed during ammonia washing was an ammino-hydroxy or ammino-hydroxy-aquo gold cation complex $[Au(NH_3)_2(H_2O)_{2-x}(OH)_x]^{(3-x)+}$ and not $Au(OH)_3$ or $[Au(OH)_4]^-$, as proposed in former works [21, 58]. Louis et al. explained that most of the gold was retained on the supports because these cationic gold complexes could interact with the negatively charged sites of the support during washing at basic pH. They also used much more diluted $NH_3$ solutions (down to 0.01 M) [56] and showed that they were as efficient as more

**Table 1.3** Gold loadings after washing and average particle sizes after calcination at 300°C in oxide-supported gold samples prepared by impregnation with HAuCl$_4$ (4 wt% Au) and then washed with ammonia (1 M, pH = 11.5) [69]. The chlorine content was below 200 ppm (detection limit) in all the samples

| Support | Nominal Au loading (wt%) | Au loading (wt%) | Average gold particle size (nm) | Standard deviation (nm) |
| --- | --- | --- | --- | --- |
| SiO$_2$ | 1 | 0.94* | 4.0 | 1.67 |
| TiO$_2$ | 4 | 3.3 | 3.6 | 0.83 |
| TiO$_2$ | 1 | 0.97 | 3.2 | 0.6 |
| TiO$_2$ | 0.7 | 0.7 | 3.3 | 0.61 |
| Al$_2$O$_3$ | 4 | 3.4 | 3.8 | 0.82 |
| Al$_2$O$_3$ | 1 | 0.97 | 3.1 | 0.63 |

*Note that the same sample-washing with NaOH ($10^{-3}$ M, pH = 11) leads to the full removal of gold (0.1 wt%).

**Table 1.4** Gold and chlorine loadings after washing and average particle sizes after reduction in H$_2$ at 300°C in Au/TiO$_2$ samples prepared by impregnation with HAuCl$_4$ (4 wt% Au) and then washed with ammonia of various concentrations [56]

| Washing solution | Au loading (wt%) | Cl loading | Average gold particle size (nm) | Standard deviation (nm) |
| --- | --- | --- | --- | --- |
| – | 3.6 | 2.7 wt% | 7.9 | 2.4 |
| NH$_3$ (1M) | 3.6 | <200 ppm | 3.7 | 1.0 |
| NH$_3$ (0.1M) | 3.0 | <200 ppm | 2.8 | 0.8 |
| NH$_3$ (0.01M) | 3.4 | 300 ppm | 2.8 | 0.7 |
| H$_2$O | 0.9 | 0.2 wt% | 3.3 | 0.9 |

concentrated ones (Table 1.4) and had the advantage to prevent support dissolution, as was observed in the case of silica (Table 1.2).

Note that Li et al. [70] proposed a treatment with gaseous ammonia, followed by washing with water to remove the chlorides after impregnation and get small gold particles on titania. Also note that none of these groups cited above reported that the use of gold and ammonia led to the formation of the explosive fulminating gold mentioned in the introduction.

## 1.5 Cation Adsorption

The bis-ethylenediamine Au$^{III}$ cation, [Au(en)$_2$]$^{3+}$ (en = NH$_2$CH$_2$CH$_2$NH$_2$), was used to prepare gold catalysts with various types of supports, either by *cation exchange* of the protons of zeolites or by *cation adsorption* on oxide supports. [Au(en)$_2$]Cl$_3$ is not commercially available, but it is easy to synthesize [71].

[Au(en)$_2$]$^{3+}$ cation exchange with HY zeolite, followed by thermal treatment under He at 150°C, gave samples with 4 wt% gold and gold particles from 1 to 4 nm, mostly located within the zeolite framework and with only a few larger particles on the outside [72, 73]. Cation exchange of [Au(en)$_2$]$^{3+}$ was also used to insert gold nanoparticles into the layers of clay minerals, which also possess acidic properties [74]. Gold particles smaller than 5 nm were obtained in montmorillonite and sepiolite after calcination at 350°C–450°C. However, gold loadings higher than 2 wt% led to agglomeration of gold nanoparticles.

This gold cation was also used for depositing gold onto oxide supports by cation adsorption; the pH of the precursor solution must be higher than the PZC of the oxide support to make the support surface negatively charged. Cation adsorption is straightforward with [Au(en)$_2$]Cl$_3$ solution since it gives a basic solution, and its pH can be easily adjusted by addition of ethylenediamine. Adsorption must be performed at temperature close to ambient because [Au(en)$_2$]$^{3+}$ decomposes into metallic gold at ~60°C.

Au/TiO$_2$ samples were prepared at RT and pH 9; the gold loadings and particle sizes obtained after 1 and 16 h of contact time with the Au*en* solution, and then after washing with water, drying, and calcination at 300°C, were the same: 1.2 and 1.1 wt% Au and 2.5 and 2.7 nm in size, respectively [75]. This indicated that as for anion adsorption, the adsorption equilibrium was fast, and the adsorption capacity of the titania was also limited to ~1 wt% (the nominal loading was 8 wt% Au in solution).

This method is especially interesting for the silica supports because of their low PZC. Indeed, Zanella's, and then Dai's group, successfully used it for the deposition of gold on silicas, amorphous silicas [76–78], and ordered mesoporous silicas [79, 80]. The gold loading could reach 9 wt% in SBA-15, leading to gold particles of

**Table 1.5** Gold loadings and average particle sizes in silica-based gold catalysts prepared by [Au(en)$_2$]$^{3+}$ adsorption

| Silica | Surface area (m$^2$.g$^{-1}$) | pH | Au loading (wt%) | Thermal treatment | dAu$^0$ (nm) | Ref. |
|---|---|---|---|---|---|---|
| AD200 | 200 | 10.5 | 5.5 | H$_2$/200°C | 2.4 | [76] |
| Cab-O-Sil | – | 10 | 2.3 | H$_2$/200°C, then air/500°C | 3.8 | [77] |
| SBA-15 | 560 | 6.0 | 2.7 | H$_2$/150°C, then air /400°C | 5.4 | [79] |
|  |  | 7.4 | 7.4 |  | 5.2 |  |
|  |  | 8.5 | 6.9 |  | 4.9 |  |
|  |  | 9.6 | 9 |  | 4.9 |  |

∼5 nm after reduction under H$_2$ at 150°C, followed by calcination at 400°C [79] (Table 1.5). The chlorine content was less than 50 ppm, and transmission electron microscopy (TEM) showed that the gold particles were located inside the pores of SBA-15. The same paper also reported the evolution of the gold speciation when pH increased from 6 to 10, and showed in particular the formation of a gold complex with deprotonated ethylenediamine ligands at pH above 8. The gold loading increased with pH, but the gold particle size was rather constant, ∼5 nm. Smaller gold particles of ∼3 nm were obtained after reduction under H$_2$ at 150°C, but the samples retained residual organic compounds; to eliminate them completely, it was necessary to proceed to further calcination at 400°C [77]. An alternative proposed by Dai's group was to proceed to reduction in H$_2$ at 130°C, followed by washing with an aqueous solution of KMnO$_4$, then by calcination at 300°C, a more moderate temperature than above [78]. However, if the resulting gold particles were small (∼3 nm), MnO$_x$ remained present in the sample, so the support could not be considered as purely silicic any longer.

Cation adsorption with [Au(en)$_2$]$^{3+}$ has been recently applied to the preparation of gold on zirconia and tungstated zirconia [81]. With the increase in the tungsten loading (5 to 20 wt%), the gold loading increased because of the higher concentration of acidic hydroxyls generated on tungstated zirconia on which [Au(en)$_2$]$^{3+}$ could adsorb. However, the basic medium used for the [Au(en)$_2$]$^{3+}$ adsorption caused some leaching of WO$_x$ species, and both large (8–10 nm) and small (∼3 nm) gold particles were obtained. On tunstated zirconia, it appears that the gold particles occupy preferentially the WO$_x$-free zirconia surface [81].

Another cation precursor, gold tetrammine, [Au(NH$_3$)$_4$]$^{3+}$, was also used for cation adsorption. Like [Au(en)$_2$]Cl$_3$, gold tetrammine nitrate must be synthesized. [Au(NH$_3$)$_4$](NO$_3$)$_3$ is obtained by reacting HAuCl$_4$ with NH$_3$ in a saturated NH$_4$NO$_3$ solution and purified from Cl$^-$ by recrystallization [82, 83]. It was used for the deposition of gold in mesoporous carbon materials [84] and on ETS-10 titanosilicate supports [85]. In the first case, 1.3 wt% Au could be deposited with a yield of 65%, and the average gold particle size was 3.2 nm after reduction under H$_2$ at 400°C. However, a large proportion of particles was located on the outer surface of the carbon grains. The authors mentioned that a yield of 90% could be achieved by increasing the gold concentration and the duration of the adsorption. In the case of the ETS-10 support, 0.8 wt% Au with a deposition yield of 80% was obtained, and the gold particles were as small as 1–2 nm after calcination at 300°C.

## 1.6  Deposition-Precipitation

### 1.6.1  *Principle of Deposition-Precipitation*

Strictly speaking, the preparation method of *deposition-precipitation* involves a process whereby a hydroxide or a hydrated oxide precipitates onto the surface of a support as a result of a gradual rise of pH of the metal precursor solution in which the support is suspended. In other words, the method of deposition-precipitation consists of the conversion of a highly soluble metal precursor into another compound of lower solubility, which must specifically precipitate onto the support and not in solution. So, when properly performed, the precipitate nucleates onto the support surface, and in fine the whole precipitate interacts with the support (*homogeneous deposition-precipitation [HDP]*) with no heterogeneous precipitation in solution, and no metal precursor remains in the solution. This procedure was originally developed by Geus and co-workers in patents and papers [86, 87] for preparing non-noble-metal catalysts such as supported nickel or copper catalysts with high metal loadings. The key point for successful deposition-precipitation onto the support is the gradual addition of the precipitating agent to avoid local rise of concentration and precipitation in solution. Several reviews have been devoted to this preparation method [86, 88, 89], and the complex chemical mechanism involved was discussed in detail [86, 90].

In practice, the powder support is stirred into a solution containing the soluble metal precursor to which the precipitating agent, usually a base, is gradually added. Then, the solid sample is recovered, washed, dried, and activated.

This preparation procedure has been altered for supported noble metals [91], although the related papers still refer to deposition-precipitation. In these papers, the method used for adding the base to the solution to reach the final pH of the preparation is generally poorly described: either 'slowly', 'dropwise', or no information. The support is usually added before pH adjustment but sometimes after. The suspension is aged for a given time at the final pH, so it turns out that the preparation is performed at a fixed pH. The supported phase is sometimes assumed to be a hydroxide, but no characterization is usually performed at this stage of the preparation. Moreover, most often these papers do not report whether the whole metal precursor is deposited onto the support.

### 1.6.2 Deposition-Precipitation by Addition of a Base

#### 1.6.2.1 Deposition-precipitation at a fixed pH

This deposition-precipitation method is probably the most used for the preparation of gold catalysts since it readily leads to the formation of small gold particles (2–3 nm) and to a low level of chlorides, in spite of the fact that the precursor is $HAuCl_4$ or one of its salts. As already mentioned in the introduction, this method was first proposed by Haruta et al. [10, 11] and led these authors to discover the extraordinary catalytic performance of gold in the CO oxidation reaction at RT.

The main difference with the procedure used for anion adsorption (see Section 2.4) is that the pH of the solution containing both HAuCl4 and the oxide support is adjusted to a fixed pH higher than the support PZC by addition of a base, most often NaOH; the mixture is usually heated at 70°C–80°C and stirred for around 1 h. As for anion adsorption, the catalyst is washed with water to remove as much of sodium and chlorine as possible, dried between RT and 100°C, and usually calcined in air. Many variations of this procedure are found in the literature, with changes for example in the pH, the duration or temperature of the preparation, the washing method, or the nature of the base; indeed, in addition to NaOH, other bases such as $Na_2CO_3$ [28, 92, 93], KOH [94], and $NH_3$ [92, 95–97] have also been used.

Several preparation parameters have been explored for several oxide supports: (i) In some cases, no variation in the results was found whether sodium hydroxide, sodium carbonate, or ammonia was used as a base [92], but other cases reported differences between preparations performed with sodium hydroxide or sodium carbonate [98, 99]; moreover, the use of ammonia as a basifying agent possibly leads to the formation of the $[Au(NH_3)_4]^{3+}$ cation (see Section 1.6.2.4); (ii) changes in the aging period from 2 to 12 h or in temperature (25°C and 70°C) were found to have no influence on the results [92]; Moreau and Bond prepared their $Au/Al_2O_3$ catalysts at RT [100]; and (iii) sodium carbonate added to the gold solution before or after addition of titania did not produce significant changes in the results [92]; however, another study reported that the pH of the solution significantly decreased when titania was added and also during deposition-precipitation [101]. Note that problems of reproducibility in the preparations are sometimes explicitly reported [92], while other papers explicitly mentioned the absence of problems [29].

This method was found suitable for the deposition of gold on oxide supports of a PZC higher than 5, such as magnesia, titania, alumina, zirconia, and ceria [7, 75, 92]. It is not suitable for silica (PZC ≈ 2), silica-alumina (PZC ≈ 1), and tungsta (PZC ≈ 1) [7, 92] or for activated carbon [102] or zeolites [103] due to their high acidity. However, deposition-precipitation with ammonia has been successfully applied to these supports (see Section 1.6.2.4).

These last years, deposition-precipitation was applied to new kinds of supports, such as cobalt-modified silica [104], buckminsterfullerene(C60)-modified silica [105] to favour Au deposition on silica, hydroxyapatite ($Ca_{10}(PO_4)_6(OH)_2$) [106–110], and various metal phosphates $MPO_4$ (M = Ca, Y, La, Pr, Nd, Sm, Eu, Ho, Er, Ce) [94, 111].

### 1.6.2.2 Influence of pH

pH is certainly the most studied parameter for this type of preparation and for the specific case of $Au/TiO_2$. This is also the most important parameter if one wants to investigate the chemical mechanisms involved in this preparation method. pH has been varied within a broad pH range, below and above the PZC of the oxide support.

Among the studies performed by three different groups [10, 92, 100, 101, 112–114], there is a general agreement that the evolution of the gold loading versus pH looks like a volcano curve with a maximum at a pH roughly corresponding to the PZC of titania (Fig. 1.1).

Except for the preparations with the lowest nominal gold loading of 1 wt% (Fig. 1.1c), the yield of deposition does not exceed 60%, so not all of the gold is deposited onto the support. The resulting average gold particle size varies differently with pH (Fig. 1.2); the particles are large (around 10 nm) when the preparation is performed below the PZC, but the size decreases when pH increases above the PZC. Hence, there is a narrow range of pH where sufficient amount of gold is deposited (Fig. 1.1), and small gold particles are obtained after thermal treatment. The choice of a solution at pH 7–8 for titania supports recommended by several groups [10, 29, 101] therefore results from this compromise.

The results obtained in the case of the $Au/Al_2O_3$ system, which is the second-most studied system, are somewhat different. At least, three different studies report the preparation of $Au/Al_2O_3$ catalysts at various pH values by addition of sodium carbonate [28, 115, 116]. They show that the gold loading versus pH (Fig. 1.3) behaves differently from $Au/TiO_2$ (Fig. 1.1); instead of a volcano-shaped curve, the curve rather exhibits an S-type shape. Above the PZC of alumina at pH around 8, the gold loading drops more drastically, while it slightly increases when pH decreases below the PZC.

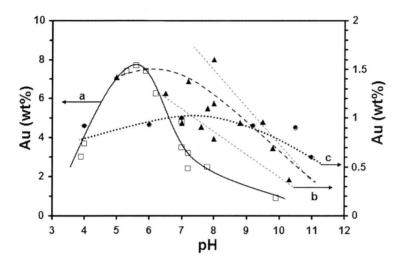

**Figure 1.1** Gold loadings and maximum yields of deposition-precipitation (DP) versus solution pH for Au/TiO$_2$ (P25 Degussa) catalyst prepared by DP (a) with NaOH added to the suspension, preparation at 70°C, nominal Au loading 13 wt%, and highest yield 60%; (b) with Na$_2$CO$_3$ (not mentioned when it is added), preparation at 20°C, nominal Au loading 2.4 wt%, and highest yield 60%; and (c) with NaOH, titania added to the NaOH solution, preparation at 70°C, nominal Au loading 1 wt%, and highest yield 100%.

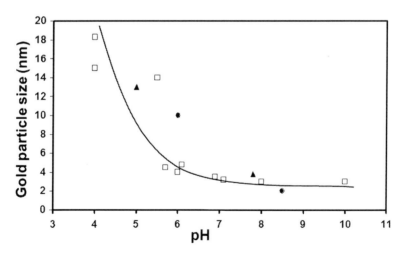

**Figure 1.2** Average gold particle size versus solution pH in Au/TiO$_2$ catalysts prepared by DP and after calcination at 400°C in air.

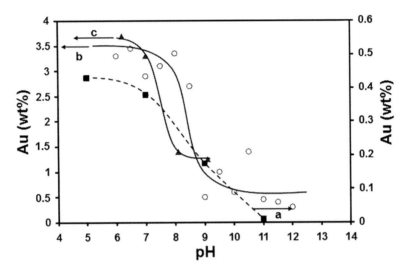

**Figure 1.3** Gold loadings and maximum yields of DP versus solution pH for Au/Al$_2$O$_3$ catalyst prepared by DP with Na$_2$CO$_3$. (a) Preparation at 20°C, nominal Au loading 1 wt%, and highest yield 45%; (b) preparation at 70°C, nominal Au loading 5 wt%, and highest yield 70%; and (c) preparation at 20°C, nominal Au loading 6 wt%, and highest yield 60%.

### 1.6.2.3 Mechanism of deposition-precipitation

Haruta's group [1, 10] was the first to propose a chemical mechanism to explain the evolution of the gold loading as a function of pH in the case of the Au/TiO$_2$ system (Fig. 1.1a). They proposed that at high pH, that is, above the support PZC, Au(OH)$_3$ precipitates on nucleation sites resulting from AuCl(OH)$_3^-$ adsorption on specific surface sites; the decreasing gold loading when pH increased from 6 to 10 (Fig. 1.1) was explained by the increasing the solubility of Au(OH)$_3$.

Several comments can be made: (i) This mechanism, which more or less directly derives from the principle of deposition-precipitation described in Section 1.6.1, does not explain why all of the gold in solution cannot be deposited onto the support when the nominal gold exceeds 1 wt%; (ii) according to several studies, there is no precipitation of gold hydroxide in the conditions of gold concentration and pH used [75, 101, 114, 117]; (iii) when the pH is above the PZC, both the titania surface and the gold species are negatively charged, so $AuCl(OH)_3^-$ adsorption cannot occur; and (iv) conversely, when the pH is below the PZC, the titania surface is positively charged, so increasing amount of adsorbed gold would be expected as pH decreases, which was not experimentally observed (Fig. 1.1).

Moreau and Bond [101, 114] proposed that an hypothetical neutral $Au(OH)_3(H_2O)$ species in equilibrium with $Au(OH)_4^-$ could react with the $TiO^-$ species present on the titania surface at pH >6, according to the following equation:

$$Au(OH)_3(H_2O) + 2TiO^- \rightarrow [Au(OH)_2(OTi)_2]^- + OH^- + H_2O \quad (1.1)$$

However, since the $TiO^-$ concentration is supposed to increase as pH increases above the PZC, increasing adsorbed Au species is expected, which was not experimentally observed (Fig. 1.1). The possible presence of the neutral species, $AuCl_3(H_2O)$, between pH 2 and 6 drawn from the diagram of gold speciation established by Nechayev [118] from thermodynamic data, could not explain either that the gold uptake increases as pH increases and becomes closer to the PZC since the maximum concentration of the $AuCl_3(H_2O)$ species is obtained at pH 4.

An alternative explanation was proposed: when pH is above or close to the PZC of titania, there is formation of a surface complex between the hydroxychloro-gold anion and OH of the support:

$$TiOH + [AuCl(OH)_3]^- \rightarrow Ti-[O-Au(OH)_3]^- + H^+ + Cl^- \quad (1.2)$$

This equation derived from an EXAFS study performed at the gold threshold after gold deposition-precipitation at pH 8 with NaOH and before washing; the results showed that the $Au^{III}$ complex with tetra-coordination did not contain any chlorides in its first coordination sphere but only oxygen atoms [75]. It is worth to note that the characterisation of the deposited gold phase is not straightforward, because it is amorphous, and EXAFS does not 'see' the second neighbours. The hypothesis of the formation of a gold surface complex was first proposed by geochemists; they also observed that (i) gold adsorption on alumina was maximum at a pH close to the PZC of $\gamma$-alumina ($\sim$8), that is, when the number of neutral hydroxyl groups was the highest [118], and (ii) the adsorption of an anionic hydroxychloro-gold(III) complex on goethite (FeO(OH); PZC $\approx$ 8) increased as pH increased from 4 to 7 [119], which is opposite to the typical behaviour for anion adsorption on positively charged oxide surfaces.

The formation of a surface complex is also consistent with the results of the three studies on Au/TiO$_2$ reported above, which showed that the gold loading was the highest when the pH was close to the PZC of titania and that as pH increased above the PZC or decreased below it, the amount of gold decreased, since the number of neutral hydroxyl groups of the support decreased (Fig. 1.1). The case of Au/Al$_2$O$_3$ appears different, since although the gold loading was the highest when the pH was close to the PZC, it did not seem to decrease below the PZC (Fig. 1.2). The reason could be that the number of surface hydroxyl groups is larger for alumina than for titania because of the larger surface area ($\sim$200 instead of $\sim$50 m$^2$/g) and/or that the surface charge does not vary as a function of the pH in the same way for alumina and titania; to fully elucidate this view, it would be interesting to compare the evolution of their surface charge as a function of the pH. On the other hand, the HCl released according to Eq. 1.2 could explain the decrease in pH during deposition-precipitation observed by Moreau and Bond [101]. Finally, the decrease in the gold particle size as pH increases above the PZC (Fig. 1.2) is easy to explain: when pH increases, the gold anion complexes are more and more hydrolysed, so the number of chlorides in the gold coordination sphere decreases; they are released in the solution and washed out during the washing step: this decreases the chlorine content in the samples and limits particle sintering during thermal treatment (see the introduction).

### 1.6.2.4 A base apart, NH$_4$OH, a precipitating agent, or a complexing agent?

As mentioned above, ammonium hydroxide was also used as a base for deposition-precipitation. One can wonder if the chemical process involved is really a deposition-precipitation mechanism, since ammonium hydroxide has been successfully used for depositing gold on silica or on acidic supports, both supports for which deposition-precipitation is supposed to fail.

Somodi et al. [96] mentioned that they applied the method of deposition-precipitation with ammonia (pH not reported) to a silica support, but they also mentioned that the intent was to form gold ammine cationic complexes that could electrostatically interact with the negatively charged silica surface. In other words, they prepared the $[Au(NH_3)_4]^{3+}$ cation in situ in the presence of the support after the introduction of a solution of $HAuCl_4$ into the pores of an amorphous silica, followed by the addition of ammonia (4 M). In the end, the preparation resembles cation adsorption (see Section 1.5) more than deposition-precipitation. The average gold particle size obtained in a sample containing 2.4 wt% Au was 3.6 nm after reduction under $H_2$ at 360°C, but the authors also reported that part of the silica dissolved (17%) during the preparation. In another work [120], 1.8 wt% Au was deposited (2 wt% nominal loading) at a pH between 9 and 10 and at 60°C, and particles smaller than 4.5 nm were obtained after reduction in $H_2$. The method was also applied to a hexagonal mesoporous silica (HMS) support modified with $Fe^{3+}$ ions (Si/Fe molar ratio of 40) at pH 9 [121]. After washing, drying, and calcination at 400°C, particles of 3.3–4.5 nm were obtained for gold loadings in the range of 0.4–5.1 wt%.

In the same way, Shimizu et al. [122] applied this preparation method with ammonia at pH 6.5 for depositing gold in HY zeolites, another type of support on which deposition-precipitation is supposed to fail. On the basis of EXAFS, the authors proposed that the gold precursor consisted of $Au(OH)_3$ deposited in the $NH_4^+$-exchanged Y zeolite; since it is not possible to distinguish Au-O from Au-N in EXAFS, the formation of a $[Au(NH_3)_4]^{3+}$ cation cannot be excluded. After calcination at 300°C in air, the average gold particle size was 1.1 nm in a sample containing 5 wt% Au.

The method was also used to deposit Au (0.5 to 1.5 wt%) on palygorskite, a clay mineral of fibre structure, at pH 10 and 60°C, leading to gold particles of 2–7 nm [123]. In this case also, probably an exchange of the $[Au(NH_3)_4]^{3+}$ cation with the acidic sites of the support took place.

Another interesting example is that of the preparation of $Au/ZrO_2$ catalysts (0.8 wt% Au) using ammonia as the basifying agent [124]. pH was adjusted to 9 by dropwise addition of ammonia (0.25 M). Playing with the conditions of preparation and of activation, the authors were able to decrease the final size of the gold particles from 1.8 to 0.8 nm, that is, to sub-nanometric sizes by lowering the HAuCl$_4$ concentration (0.25 mM instead of 1 mM), drying at RT under vacuum instead of 110°C in air after thorough water-washing in both cases, and reducing the sample under hydrogen up to 250°C instead of 350°C.

### 1.6.3 Deposition-Precipitation by Addition of a Delay Base

#### 1.6.3.1 Deposition-precipitation with urea

The gradual rise in pH required by the procedure of deposition-precipitation described in Section 1.6.1 can be easily performed thanks to the use of urea (CO(NH$_2$)$_2$), which is a delay base. It allows us to independently incorporate urea to the suspension and mix it homogeneously at RT and then start the basification only when the mixture is heated above 60°C, and urea is hydrolysed:

$$CO(NH_2)_2 + 3H_2O \rightarrow CO_2 + 2NH_4^+ + 2OH^- \quad (1.3)$$

In other words, it makes it possible to split the addition of the basifying agent and the effective basification into two steps.

Applied to the preparation of gold catalysts, the method consists of mixing the oxide support in an aqueous solution containing HAuCl$_4$ and urea and then heating the mixture to ~80°C. Initially, the pH is ~2, and upon heating, it gradually increases to ~7 after around 4 h in the case of a titania support.

An extensive study of this preparation method of *deposition-precipitation with urea* was performed by Louis's group [51, 75]. Table 1.6 shows that high gold loadings could be achieved, even on an oxide of rather low surface area, like titania of 50 m$^2$/g and with 100% yield of deposition, at variance with the method of deposition-precipitation at fixed pH (see Section 1.6.2). Moreover, small gold particles (~3 nm) were obtained on various oxide supports like TiO$_2$, Al$_2$O$_3$, and CeO$_2$ (entries 1 to 3), except for silica (entry 4), for which the yield of gold deposition was much lower, as in the case of the *deposition-precipitation at fixed pH*.

One can note from entries 1 to 3 of Table 1.6 that the deposition of the whole gold occurred within the first hour of preparation when the suspension pH was still acidic (~3), that is, when pH was

**Table 1.6** Au/oxide samples prepared by deposition-precipitation with urea at 80°C

| Entry | Oxide support | PZC | Nominal Au loading (wt%) | DP time (h) | Final pH | Au loading (wt%) | Cl loading (wt%) | Temperature treatment (°C) | Average gold particle size (nm) | Ref. |
|---|---|---|---|---|---|---|---|---|---|---|
| 1 | TiO$_2$ (~50 m$^2$/g) | ~6 | 8 | 1 | 3.0 | 7.8 | 0.041 | Air/300 | 5.6 | [51] |
|   |   |   |   | 2 | 6.3 | 6.5 | 0.122 | Air/300 | 5.2 |   |
|   |   |   |   | 4 | 7.0 | 7.7 | <0.03 | Air/300 | 2.7 |   |
|   |   |   |   | 16 | 7.3 | 6.8 | <0.03 | Air/300 | 2.5 |   |
|   |   |   |   | 90 | 7.8 | 7.4 | <0.03 | Air/300 | 2.4 |   |
| 2 | CeO$_2$ (250 m$^2$/g) | ~6 |   | 1 | 4.3 | 7.9 | – | Air/300 | 8.1 | [75] |
|   |   |   |   | 16 | 6.6 | 8.2 | – | Air/300 | <5* |   |
| 3 | δ–Al$_2$O$_3$ (100 m$^2$/g) | 7–8 |   | 1 | 4.3 | 6.9 | – | Air/300 | 6.9 |   |
|   |   |   |   | 16 | 7.1 | 7.2 | – | Air/300 | 2.3 |   |
| 4 | SiO$_2$ (250 m$^2$/g) | ~2 |   | 1 | 5.2 | 2.9 | – | Air/300 | >20 |   |
|   |   |   |   | 16 | 7.0 | 3.7 | – | Air/300 | >20 |   |
| 5 | γ–Al$_2$O$_3$ | 7–8 | 10 | 20 | 8 | 10.3 | – | Air/200 | 2–10** | [125] |
| 6 | Fe$_2$O$_3$ | ~8.5 | 3 | 8 | 7.3 | 3.0 | – | Air/350 | 3–7** | [126] |
| 7 | MgO | 10–12 | ? | 6 | 7.7 | 7.5 | – | Air/400 | 7.9 | [127] |
| 8 | CaO | ~12 | ? | 6 | 8.3 | 4.7 | – |   | 5.7 |   |
| 9 | TiO$_2$ (~50 m$^2$/g) | ~6 | 1 | 1 | 2.7 | 1.1 | <0.02 | H$_2$/300 | 2.2 | [24] |
|   |   |   |   | 16 | 7.4 | 1.1 | <0.02 | H$_2$/300 | 3.0 |   |
| 10 | δ–Al$_2$O$_3$ (100 m$^2$/g) | 7–8 | 1 | 1 | 4.5 | 0.9 | <0.02 | H$_2$/300 | 2.2 |   |
|   |   |   |   | 16 | 7.6 | 0.9 | <0.02 | H$_2$/300 | 2.2 |   |

*Estimated by XRD (poor contrast between gold particles and CeO$_2$ by TEM)
**Size range

still below the PZC of the oxide support. Then, as the pH of the suspension increases and reaches a plateau, the sample 'matures'. The consequence is that the size of the resulting metal gold particles decreases with the maturation time (entry 1), and the chloride loading decreases as well. These observations were confirmed by Baatz et al. [125] with the study of the Au/Al$_2$O$_3$ system (entry 5). They confirmed that up to 10 wt% Au could be deposited on alumina, with 95% of gold deposited after 2 min, and all gold within less than 20 min when pH was still acidic (pH 3).

The preparation method of gold deposition-precipitation with urea was applied to several other oxide supports. Au 3 wt% was deposited on ferric oxide with a yield close to 100%, and the gold particles were quite small (3–7 nm) after calcination at 350°C [126] (entry 6). Another study reports the use of this method to deposit gold on MgO and CaO supports [127] (entries 7 and 8). After calcination at 400°C, gold particles of moderate size were obtained on magnesium oxide (8 nm for a sample with 7.5 wt% Au) and on calcium oxide (6 nm for a sample with 4.7 wt% Au).

Recently, gold deposition-precipitation with urea has been also applied to more complex oxides like ceria-alumina [128, 129], ceria-zirconia [130], ceria-gallia [131], Ce$_x$Tb$_y$Zr$_z$O$_{2-v}$ [132], and Cu-Cr oxide spinel [133] and also to more exotic supports like titano-silicalite [134], $\beta$-MnO$_2$ [135], hydroxyapatite [136, 137], carbides [138], and nitrides [139].

### 1.6.3.2 Mechanism of deposition-precipitation

Attempts to elucidate the mechanism of gold deposition-precipitation with urea were made in the case of the preparation of Au/TiO$_2$ [75]. With the help of Raman spectroscopy, the precipitation of gold onto the support was found to result from a reaction between gold complexes in solution and products of hydrolysis of urea since there is no deposition as long as the suspension is not heated, that is, as long as urea is not decomposed [75].

The mechanism of gold deposition-precipitation with urea proposed was the following [75]: Some of the hydroxychloro-gold(III) anions present in the solution at a pH between ~2 (initial pH) and ∼3 (pH of precipitation) adsorb on the support, the surface of which is positively charged at this pH. This would explain why silica, which has a PZC ≈ 2, is not a suitable support for this preparation method (Table 1.6, entry 4). These adsorbed gold species would act as nucleation sites for the growth of particles arising from the precipitation of the gold–urea compound. The fact that the gold particle size decreases when the duration of deposition-precipitation increases (Table 1.6, entry 1) was attributed to a phenomenon of peptisation, that is, of redispersion of the precipitate when pH increases. The increase of pH induces an increase in the density of negative charges at the surface of the precipitate that would induce an increase in repulsive forces and lead to the fragmentation of the particles of precipitate. Baatz et al. [125], whose paper is mentioned above, agreed with this mechanism as they found the same experimental trends.

When lower gold loadings are needed (∼1 wt% Au), it is possible to drastically reduce the preparation time to ∼1 h and get small gold particles [24]. Whether the contact time is 1 or 16 h, the resulting gold particles are of the same size (Table 1.6, entries 9 and 10). In this case, the chemical mechanism involved is not that of deposition-precipitation described above, since the gold loading is lower than the adsorption capacity of the supports (see Section 1.4.2). As a consequence, all the hydroxychloro-gold anions can adsorb onto the support, but at variance with anion adsorption, they are transformed into gold–urea complexes without requiring a precipitation step. The advantage compared to anion adsorption is that the particles are smaller and the chlorine content is much lower (see Table 1.2 for comparison).

As already mentioned, deposition-precipitation with urea is applicable to the same type of supports as for deposition-precipitation at fixed pH. It also leads to small gold particles, but the advantage is that all of the gold in solution is deposited onto the support, so the Au loading can be easily controlled a priori. Moreover, it is possible to prepare a set of samples with the same gold loading but with different particle sizes, playing with the deposition-precipitation time (as in entry 1) and not with the temperature of the subsequent thermal treatment of gold reduction.

It is noteworthy that the fast deposition of the gold–urea compound compared to the slow precipitation of a hydroxide and the decrease of gold particle size versus the deposition-precipitation time indicate that the mechanism of gold deposition during deposition-precipitation with urea is different from that occurring during the preparation of catalysts with non-noble metals described in Section 1.6.1.

A recent study [132] showed that Au/Ce$_{0.50}$Tb$_{0.12}$Zr$_{0.38}$O$_{2-x}$ catalyst prepared by deposition–precipitation with urea exhibited a very thin carbonaceous layer when the sample was reduced under H$_2$ that did not exist for samples activated by calcination. Even though one cannot tell that this occurs with any type of support—the author of the present chapter never observed any inhibiting catalytic effect when the samples were reduced in hydrogen—this result must deserve attention.

## 1.7 Conclusion

By way of conclusion, one can tell that small gold particles (<5 nm) can be synthesized on almost any type of support using preparation methods based on the principle of deposition-reduction in aqueous solution, provided that the preparation method is well chosen. With some of these methods, it is also possible to control a priori the gold loading and to avoid the loss of gold in solution. Methods that do not involve gold chloride precursors or allow chloride elimination are definitively desirable because of their detrimental effect on the gold particle size and on the catalytic performance. Again, it must be stressed that it is important to pay attention to the conditions of preparation and storage of the samples so as to avoid uncontrolled reduction and gold particle sintering.

## References

1. Haruta, M. (1997). Size- and support-dependency in the catalysis of gold, *Catal. Today*, **36**, pp. 153–166.

2. Haruta, M. (2003). When gold is not noble: catalysis by nanopaticles, *Chem. Rec.*, **3**, pp. 75–87.
3. Mavrikakis, M., Stoltze, P., and Norskov, J.K. (2000). Making gold less noble, *Catal. Lett.*, **64** pp. 101–106.
4. Valden, M., Lai, X., and Goodman, D.W. (1998). Onset of catalytic activity of gold clusters on titania with the appearance of nonmetallic properties, *Science*, **281**, pp. 1647–1650.
5. Janssens, T.V.W., Clausen, B.S., Hvolbæk, B., Falsig, H., Christensen, C.H., Bligaard, T., and Nørskov, J.K. (2007). Insights into the reactivity of supported Au nanoparticles: combining theory and experiments, *Top. Catal.*, **44**, pp. 15–26.
6. Bond, G.C., and Thompson, D.T. (1999). Catalysis by gold, *Catal. Rev.: Sci. Eng.*, **41**, pp. 319–388.
7. Haruta, M. (2002). Catalysis of gold nanoparticles deposited on metal oxides, *Cattech*, **6**, pp. 102–115.
8. Hashmi, A.S.K., and Hutchings, G.J. (2006). Gold catalysis, *Angew. Chem., Int. Ed.*, **45**, pp. 7896–7936.
9. Thompson, D.T. (2006). An overview of gold-catalysed oxidation process, *Top. Catal.*, **38**, pp. 231–240.
10. Tsubota, S., Cunningham, D.A.H., Bando, Y., and Haruta, M. (1995). Preparation of nanometer gold strongly interacted with $TiO_2$ and the structure sensitivity in low temperature oxidation of CO, *Stud. Surf. Sci. Catal.*, **91**, pp. 227–235.
11. Bamwenda, G.R., Tsubota, S., Nakamura, T., and Haruta, M. (1997). The influence of the preparation methods on the catalytic activity of platinum and gold supported on $TiO_2$ for CO oxidation, *Catal. Lett.*, **44**, pp. 83–97.
12. Weiher, N., Bus, E., Prins, R., Delannoy, L., Louis, C., Ramaker, D.E., Miller, J.T., and van Bokhoven, J.A. (2006). Structure and oxidation state of gold on different supports under various CO oxidation conditions, *J. Catal.*, **240**, pp. 100–107.
13. Karpenko, A., Denkwitz, Y., Plzak, V., Cai, J., Leppelt, R., Schumacher, B., and Behm, R.J. (2007). Low-temperature water-gas shift reaction on $Au/CeO2$ catalysts: the influence of catalyst pre-treatment on the activity and deactivation in idealized reformate, *Catal. Lett.*, **116**, pp. 105–115.
14. Karpenko, A., Leppelt, R., Cai, J., Plzak, V., Chuvilin, A., Kaiser, U., and Behm, R.J. (2007). Deactivation of a $Au/CeO_2$ catalyst during the low-temperature water–gas shift reaction and its reactivation: a combined

TEM, XRD, XPS, DRIFTS, and activity study, *J. Catal.*, **250**, pp. 139–150.
15. Park, E.D., and Lee, J.S. (1999). Effects of pretreatment conditions on CO oxidation over supported Au catalysts, *J. Catal.*, **186**, pp. 1–11.
16. Kozlova, A.P., Kozlov, A.I., Sugiyama, S., Matsui, Y., Asakura, K., and Iwasawa, Y. (1999). Study of gold species in iron oxide supported gold catalysts derived from golg-phosphine complex Au(PPh$_3$)(NO$_3$) and as-precipitated wet Fe(OH)3, *J. Catal.*, **181**, pp. 37–48.
17. Oh, H.S., Yang, J.H., Costello, C.K., Wang, Y.M., Bare, S.R., Kung, H.H., and Kung, M.C. (2002). Selective catalytic oxidation of CO: effect of chloride on supported Au catalysts, *J. Catal.*, **210**, pp. 375–386.
18. Kung, H.H., Kung, M.C., and Costello, C.K. (2003). Supported Au catalysts for low temperature CO oxidation, *J. Catal.*, **216**, pp. 425–432.
19. Hutchings, G.J. (2005). Catalysis by gold, *Catal. Today*, **100**, pp. 55–61.
20. M. Haruta, (1999). The abilities and potential of gold as a catalyst, *Rep. Osaka Natl. Res. Inst.*, **393**, pp. 1–93.
21. Ivanova, S., Petit, C., and Pitchon, V. (2004). A new preparation method for the formation of gold nanoparticles on an oxide support, *Appl. Catal. A*, **267**, pp. 191–201.
22. Schulz, A., and Hargittai, M. (2001). Structural variations and bonding in gold halides: a quantum chemical study of monomeric and dimeric gold monohalide and gold trihalide molecules, AuX, Au$_2$X$_2$, AuX$_3$, and Au$_2$X$_6$ (X=F, Cl, Br, I), *Chem. Eur. J.*, **7**, pp. 3657–3670.
23. Bond, G.C., Louis, C., and Thompson, D. (2006). *Catalysis by Gold* (Imperial College Press, London).
24. Hugon, A., El Kolli, N., and Louis, C. (2010) Advances in the preparation of supported gold catalysts: mechanism of deposition, simplification of the procedures and relevance of the elimination of chlorine, *J. Catal.*, **274**, pp. 239–250.
25. Peck, J.A., and Brown, G.E. (1991). Speciation of aqueous gold(III) chlorides from ultraviolet/visible absorption and Raman/resonance Raman spectroscopies, *Geochim. Cosmochim. Acta*, **55**, pp. 671–676.
26. Farges, F., Sharps, J.A., and Brown, G.E. (1993) Local environment aroung gold(III) in aqueous chloride solutions: an EXAFS spectroscopy study, *Geochim. Cosmochim. Acta*, **57**, pp. 1243–1252.
27. Murphy, P.J., and LaGrange, M.S. (1998). Raman spectroscopy of gold chloro-hydroxy speciation in fluids at ambient temperature and pressure: a revaluation of the effects of pH and chloride concentration, *Geochim. Cosmochim. Acta*, **62**, pp. 3515–3526.

28. Lee, S.-J., and Gavriilidis, A. (2002) Supported Au catalyst for low-temperature CO oxidation prepared by impregnation, *J. Catal.*, **206**, pp. 305–313.
29. Moreau, F., Bond, G.C., and Taylor, A.O. (2004). The influence of metal loading and pH during preparation on the CO oxidation activity of Au/TiO$_2$ catalysts, *Chem. Commun.*, pp. 1642–1643.
30. Ivanova, S., Pitchon, V., Petit, C., Herschbach, H., Dorsselaer, A.V., and Leize, E. (2006). Preparation of alumina supported gold catalysts: gold complexes genesis, identification and speciation by mass spectrometry, *Appl. Catal. A*, **298**, pp. 203–210.
31. Belevantsev, V.I., Kolonin, G.R., and Ryakhovskaya, S.K.R. (1972). Hydrolysis of the AuCl$_4^-$ ion in the Range 21–90°C, *Russ. J. Inorg. Chem.*, **17**, pp. 1303–1306.
32. Murphy, P.L., Stevens, G., and LaGrange, M.S. (2000). The effects of temperature and pressure on gold-chloride speciation in hydrothermal fluids: a Raman spectroscopic study, *Geochim. Cosmochim. Acta*, **64**, pp. 479–494.
33. Nechayev, Y.A., and Zvonareva, G.V. (1983). Adsorption of gold (III) chloride complexes on hematite, *Geokhimiya*, **6**, pp. 919–924.
34. Cusumano, J.A. (1974). Safety in the preparation of multimetallic catalysts, *Nature*, **247**, pp. 456–456.
35. Fisher, J.M. (2003). Fulminating gold, *Gold Bull.*, **36**, p. 155.
36. Daté, M., Ichihashi, Y., Yamashita, T., Chiorino, A., Boccuzzi, F., and Haruta, M. (2002). Performance of Au/TiO$_2$ catalyst under ambient conditions, *Catal. Today*, **72**, pp. 89–94.
37. Schumacher, B., Plzak, V., Kinne, K., and Behm, R.J. (2003). Highly active Au/TiO$_2$ catalysts for low-temperature CO oxidation: preparation, conditioning and stability, *Catal. Lett.*, **89**, pp. 109–114.
38. Zanella, R., and Louis, C. (2005). Influence of the conditions of thermal treatments and of storage on the size of the gold particles in Au/TiO$_2$, *Catal. Today*, 107-**108**, pp. 768–777.
39. Cant, N.W., and Hall, W.K. (1971). Catalytic oxidation, IV. Ethylene and propylene oxidation over gold, *J. Phys. Chem.*, **75**, pp. 2914–2921.
40. Galvagno, S., and Parravano, G. (1978). Chemical reactivity of supported gold. IV. Reduction of NO by H$_2$, *J. Catal.*, **55**, pp. 178–190.
41. Sermon, P.A., Bond, G.C., and Wells, P.B. (1979). Hydrogenation of alkenes over supported gold, *J. Chem. Soc. Faraday Trans. I*, **75**, pp. 385–395.

42. Bond, G.C., Sermon, P.A., Webb, G., Buchanan, D.A., and Wells, P.B. (1973). Hydrogenation over supported gold catalysts, *Chem. Commun.*, pp. 444–445.
43. Lin, S., and Vannice, M.A. (1991). Gold dispersed on $TiO_2$ and $SiO_2$: Adsorption properties and catalytic behavior in hydrogenation reactions, *Catal. Lett.*, **10**, pp. 47–62.
44. Haruta, M., Tsubota, S., Kobayashi, T., Kageyama, H., Genet, M.J., and Delmon, B. (1993). Low-temperature oxidation of CO over gold supported on $TiO_2$, $\alpha$-$Fe_2O_3$ and $Co_3O_4$, *J. Catal.*, **144**, pp. 175–192.
45. Haruta, M. (1999). The abilities and potential of gold as a catalyst, *Rep. Osaka Natl.Res. Inst.*, **393**, pp. 1–93.
46. Baatz, C., Decker, N., and Prüße, U. (2008). New innovative gold catalysts prepared by an improved incipient wetness method, *J. Catal.*, **258**, pp. 165–169.
47. Bus, E., Prins, R., and van Bokhoven, J.A. (2007). Time-resolved in situ XAS study of the preparation of supported gold clusters, *Phys. Chem. Chem. Phys.*, 9, pp. 3312–3320.
48. Oxford, S.M., Henao, J.D., Yang, J.H., Kung, M.C., and Kung, H.H. (2008). Understanding the effect of halide poisoning in CO oxidation over $Au/TiO_2$, *Appl. Catal. A: Gen.*, **339**, pp. 180–186.
49. Hugon, A., Delannoy, L., and Louis, C. (2008). Supported gold catalysts for selective hydrogenation of 1,3-butadiene in the presence of an excess of alkenes, *Gold Bull.*, **41**, pp. 127–138.
50. Cárdenas-Lizana, F., Gómez-Quero, S., Hugon, A., Delannoy, L., Louis, C., and Keane, M.A. (2009). Pd-promoted selective gas phase hydrogenation of p-chloronitrobenzene over alumina supported Au, *J. Catal.*, **262**, pp. 235–243.
51. Zanella, R., Giorgio, S., Henry, C.R., and Louis, C. (2002). Alternative methods for the preparation of gold nanoparticles supported on $TiO_2$, *J. Phys. Chem. B*, **106**, pp. 7634–7642.
52. Delgass, W.N., Boudart, M., and Parravano, G. (1968). Moessbauer spectroscopy of supported gold catalysts, *J. Phys. Chem.*, **72**, pp. 3563–3567.
53. Sakurai, H., Koga, K., Iizuka, Y., and Kiuchi, M. (2013). Colorless alkaline solution of chloride-free gold acetate for impregnation: an innovative method for preparing highly active Au nanoparticles catalyst, *Appl. Catal. A*, **462– 463**, pp. 236– 246.

54. Boujday, S., Lehman, J., Lambert, J.-F., and Che, M. (2003). Evolution of transition metals speciation in the synthesis of supported catalysts: halogeno-platinates (IV) on silica, *Catal. Lett.*, **88**, pp. 23–30.

55. http://en.wikipedia.org/wiki/Isoelectric_point-Isoelectric_point_versus_point_of_zero_charge, 2012.

56. Hugon, A., Delannoy, L., Krafft, J.-M., and Louis, C. (2010). Supported gold-palladium catalysts for selective hydrogenation of 1,3 butadiene in an excess of propene, *J. Phys. Chem. C*, **114**, pp. 10823–10835.

57. Ivanova, S., Pitchon, V., and Petit, C. (2006). Application of the direct exchange method in the preparation of gold catalysts supported on different oxide materials, *J. Mol. Catal. A*, **256**, pp. 278–283.

58. Xu, Q., Kharas, K.C.C., and Datye, A.K. (2003). The preparation of highly dispersed Au/Al$_2$O$_3$ by aqueous impregnation, *Catal. Lett.*, **85**, pp. 229–235.

59. Ivanova, S., Pitchon, V., Zimmermann, Y., and Petit, C. (2006). Preparation of alumina supported gold catalysts: influence of washing procedures, mechanism of particles size growth, *Appl. Catal. A*, **298**, pp. 57–64.

60. Azizi, Y., Pitchon, V., and Petit, C. (2010). Effect of support parameters on activity of gold catalysts: studies of ZrO$_2$, TiO$_2$ and mixture, *Appl. Catal. A: Gen.*, **385**, pp. 170–177.

61. Milt, V.G., Ivanova, S., Sanz, O., Domínguez, M.I., Corrales, A., Odriozola, A., and Centeno, M.A. (2013). Au/TiO$_2$ supported on ferritic stainless steel monoliths as CO oxidation catalysts, *Appl. Surf. Sci.*, **270**, pp. 169–177.

62. Sobolev, V.I., and Pirutko, L.V. (2012). Room temperature reduction of N$_2$O by CO over Au/TiO$_2$, *Catal. Commun.*, **18**, pp. 147–150.

63. Campo, B.C., Rosseler, O., Alvarez, M., Rueda, E.H., and Volpe, M.A. (2008). On the nature of goethite, Mn-goethite and Co-goethite as supports for gold nanoparticles, *Mater. Chem. Phys.*, **109**, pp. 448–454.

64. Dobrosz-Gomez, I., Kocemba, I., and Rynkowski, J.M. (2009). Factors influencing structure and catalytic activity of Au/Ce$_{1-x}$Zr$_x$O$_2$ catalysts in CO oxidation, *Appl. Catal. B: Environ.*, **88**, pp. 83–97.

65. Sudarsanam, P., Mallesham, B., Reddy, P.S., Großmann, D., Grunert, W., and Reddy, B.M. (2014). Nano-Au/CeO$_2$ catalysts for CO oxidation: influence of dopants (Fe, Laand Zr) on the physicochemical properties and catalytic activity, *Appl. Catal. B*, **144**, pp. 900–908.

66. Lessard, J.D., Valsamakisz, I., and Flytzani-Stephanopoulos, M. (2012). Novel Au/La$_2$O$_3$ and Au/La$_2$O$_2$SO$_4$ catalysts for the water–gas shift

reaction prepared via an anion adsorption method, *Chem. Commun.*, **48**, pp. 4857–4859.

67. Pilasombat, R., Daly, H., Goguet, A., Breen, J.P., Burch, R., Hardacre, C., and Thompsett, D. (2012). Investigation of the effect of the preparation method on the activity and stability of Au/CeZrO$_4$ catalysts for the low temperature water gas shift reaction, *Catal. Today*, **180**, pp. 131–138.

68. Costa, V.V., Estrada, M., Demidova, Y., Prosvirin, I., Kriventsov, V., Cotta, R.F., Fuentes, S., Simakov, A., and Gusevskaya, E.V. (2012). Gold nanoparticles supported on magnesium oxide as catalysts for the aerobic oxidation of alcohols under alkali-free conditions, *J. Catal.*, **292**, pp. 148–156.

69. Delannoy, L., El Hassan, N., Musi, A., Nguyen Le To, N., Krafft, J.-M., and Louis, C. (2006). Preparation of supported gold nanoparticles by a modified incipient wetness impregnation method, *J. Phys. Chem. B*, **110**, pp. 22471–22478.

70. Li, W.-C., Comotti, M., and Schüth, F. (2006). Highly reproducible syntheses of active Au/TiO$_2$ catalysts for CO oxidation by deposition–precipitation or impregnation, *J. Catal.*, **237**, pp. 190–196.

71. Block, B.P., and Bailar, J.J. C. (1951). The reaction of gold(III) with some bidentate coordinating groups, *J. Am. Chem. Soc.*, **73**, pp. 4722–4725.

72. Guillemot, D., Borovskov, V.Y., Kazansky, V.B., Polisset-Thfoin, M., and Fraissard, J. (1997). Surface characterization of Au/HY by $^{129}$Xe NMR and diffuse reflectance IR spectroscopy of adsorbed CO. Formation of electron-deficient gold particles inside HY cavities, *J. Chem. Soc., Faraday Trans.*, **93**, pp. 3587–3591.

73. Guillemot, D., Polisset-Thfoin, M., and Fraissard, J. (1996). Preparation of nanometric gold particles on NaHY, *Catal. Lett.*, **41**, pp. 143–148.

74. Zhu, L., Letaief, S., Liu, Y., Gervais, F., and Detellier, C. (2009). Clay mineral-supported gold nanoparticles, *Appl. Clay Sci.*, **43**, pp. 439–446.

75. Zanella, R., Delannoy, L., and Louis, C. (2005). Mechanism of deposition of gold precursors onto TiO$_2$ during the preparation by deposition-precipitation with NaOH and with urea and by cation adsorption, *Appl. Catal. A*, **291**, pp. 62–72.

76. Zanella, R., Sandoval, A., Santiago, P., Basiuk, V.A., and Saniger, J.M. (2006). New preparation method of gold nanoparticles on SiO$_2$, *J. Phys. Chem. B*, **110**, pp. 8559–8565.

77. Zhu, H., Ma, Z., Clark, J.C., Pan, Z., Overbury, S.H., and Dai, S. (2007). Low-temperature CO oxidation on Au/fumed SiO$_2$-based catalysts prepared from Au(en)$_2$Cl$_3$ precursor, *Appl. Catal. A*, **326**, pp. 89–99.

78. Yin, H., Ma, Z., Overbury, S.H., and Dai, S. (2008). Promotion of Au(en)$_2$Cl$_3$-derived Au/fumed SiO$_2$ by treatment with KMnO$_4$, *J. Phys. Chem. C*, **112**, pp. 8349–8358.
79. Zhu, H., Liang, C., Yan, W., Overbury, S.H., and Dai, S. (2006). Preparation of highly active silica-supported Au catalysts for CO oxidation by a solution-based technique, *J. Phys. Chem. B*, **110**, pp. 10842–10848.
80. Guan, Y., and Hensen, E.J.M. (2009). Ethanol dehydrogenation by gold catalysts: the effect of the gold particle size and the presence of oxygen, *Appl. Catal. A*, **361**, pp. 49–56.
81. Kantcheva, M., Milanova, M., Avramova, I., and Mametsheripov, S. (2012). Spectroscopic characterization of gold supported on tungstated zirconia, *Catal. Today*, **187**, pp. 39–47.
82. Mason, W.R., and Gray, H.B. (1968). Electronic structures of square planar complexes, *J. Am. Chem. Soc.*, **90**, p. 5721.
83. Skibsted, L.H., and Bjerrum, J. (1974). Studies of gold complexes. I. Robutness, stability and acid dissociation of tetramminegold(III) ion, *Acta Chem. Scand. A 28*, pp. 740–746.
84. Pyryaev, P.A., Moroz, B.L., Zyuzin, D.A., Nartova, A.V., and Bukhtiyarov, V.I. (2010). Nanosized Au/C catalyst obtained from a tetraamminegold(III) precursor: synthesis, characterization, and catalytic activity in low-temperature CO oxidation, *Kinet. Catal.*, **51**, pp. 885–892.
85. Xu, J., Yueming, L., Wu, H., Li, X., He, M., and Wu, P. (2011). ETS-10 supported Au nanoparticles for solvent-free oxidation of 1-phenylethanol with oxygen, *Catal. Lett.*, **141**, pp. 860–865.
86. Hermans, L.A.M., and Geus, J.W. (1979). Interaction of nickel ions with silica supports during deposition-precipitation, *Stud. Surf. Sci. Catal.*, **3**, pp. 113–130.
87. van Dillen, J.A., Geus, J.W., Hermans, L.A.M., and van der Meijden, J. (1977). Production of supported copper and nickel catalysts by deposition-precipitation, *Proc. 6th Int. Congr. Catal., London*, pp. 677–685.
88. Geus, J.W. (1983). Production and thermal pretreatment of supported catalysts, *Stud. Surf. Sci. Catal.*, **16**, pp. 1–33.
89. Geus, J.W., and van Dillen, J.A. (1997). *Handbook on Heterogeneous Catalysis-Preparation of Solid Catalysts*, eds. Ertl, G., Knözinger, H., and Weitkamp, J. (VCH, Weinheim), pp. 460–486
90. Burattin, P., Che, M., and Louis, C. (1998). Molecular aproach to the mechanism of deposition-precipitation of the Ni(II) phase on silica, *J. Phys Chem. B*, **102**, pp. 2722–2732.

91. Louis, C. (2006). Deposition-precipitation synthesis of supported metal catalysts, in *Catalyst Preparation: Science and Engineering*, ed. Regalbuto, J.R. (Taylor & Francis/CRC Press), pp. 319–339.
92. Wolf, A., and Schüth, F. (2002). A systematic study of the synthesis conditions for the preparation of highly active gold catalysts, *Appl. Catal. A*, **226**, pp. 1–13.
93. Beck, A., Horvath, A., Stefler, G., Katona, R., Geszti, O., Tolnai, G., Liotta, L.F., and Guczi, L. (2008). Formation and structure of Au/TiO$_2$ and Au/CeO$_2$ nanostructures in mesoporous SBA-15, *Catal. Today*, **139**, pp. 180–187.
94. Ma, Z., Yin, H., Overbury, S.H., and Dai, S. (2008). Metal phosphates as a new class of supports for gold nanocatalysts, *Catal. Lett.*, **126**, pp. 20–30.
95. Nijhuis, T.A., Huizinga, B.J., Makkee, M., and Moulijn, J.A. (1999). Direct epoxidation of propene using gold dispersed on TS-1 and other titanium-containing supports, *Ind. Eng. Chem. Res.*, **38**, pp. 884–891.
96. Somodi, F., Borbath, I., Hegedus, M., Tompos, A., Sajo, I.E., Szegedi, A., Rojas, S., Fierro, J.L.G., and Margitfalvi, J.L. (2008). Modified preparation method for highly active Au/SiO$_2$ catalysts used in CO oxidation, *Appl. Catal. A: Gen.*, **347**, pp. 216–222.
97. Zwijnenburg, A., Saleh, M., Makkee, M., and Moulijn, J.A. (2002). Direct gas-phase epoxidation of propene over bimetallic Au catalysts, *Catal. Today*, **72**, pp. 59–62.
98. Sakwarathorn, T., Luengnaruemitchai, A., and Pongstabodee, S. (2011). Preferential CO oxidation in H$_2$-rich stream over Au/CeO$_2$ catalysts prepared via modified deposition–precipitation, *J. Ind. Eng. Chem.*, **17**, pp. 747–754.
99. Xua, H., Chu, W., Luo, J., and Zhang, T. (2011). Impacts of MgO promoter and preparation procedure on meso-silica supported nano gold catalysts for carbon monoxide total oxidation at low temperature, *Chem. Eng. J.*, **170**, pp. 419–423.
100. Moreau, F., and Bond, G.C. (2006). Gold on titania catalysts, influence of some physicochemical parameters on the activity and stability for the oxidation of carbon monoxide, *Appl. Catal. A*, **302**, pp. 110–117.
101. Moreau, F., Bond, G.C., and Taylor, A.O. (2005). Gold on titania for the oxidation of carbon monoxide: control of pH during preparation with various gold content, *J. Catal.*, **231**, pp. 105–114.
102. Prati, L., and Martra, G. (1999). New gold catalysts for liquid phase oxidation, *Gold Bull.*, **32**, pp. 96–101.

103. Lin, J.-N., and Wan, B.-Z. (2003). Effect of preparation conditions on gold /Y-type zeolite for CO oxidation, *Appl. Catal. B*, **41**, pp. 83–95.
104. Solsona, B., Perez-Cabero, M., Vazquez, I., Dejoz, A., García, T., Alvarez-Rodríguez, J., El-Haskourib, J., Beltran, D., and Amoros, P. (2012). Total oxidation of VOCs on Au nanoparticles anchored on Co doped mesoporous UVM-7 silica, *Chem. Eng. J.*, **187**, pp. 391–400.
105. Qian, K., Luo, L., Chen, C., Yang, S., and Huang, W. (2011). Finely dispersed Au nanoparticles on $SiO_2$ achieved by the $C_{60}$ additive and their catalytic activity, *ChemCatChem*, **3**, pp. 161–166.
106. Phonthammachai, N., Ziyi, Z., Jun, G., Han, F.Y., and White, T.J. (2008). Synthesis of high performance hydroxyapatite-gold catalysts for CO oxidation, *Gold Bull.*, **41**, pp. 42–50.
107. Dominguez, M.I., Romero-Sarria, F., Centeno, M.A., and Odriozola, J.A. (2009). Gold/hydroxyapatite catalysts. Synthesis, characterization and catalytic activity to CO oxidation, *Appl. Catal. B*, **87**, pp. 245–251.
108. Reddy, E.L., Prabhakharn, A., and Karuppiah, J. (2012). Gold supported calcium deficient hydroxyapatite for room temperature CO oxidation, *Int. J. Nanosci.*, **11**, pp. 1240004-1240001-1240007.
109. Zhao, K., Qiao, B., Wang, J., Zhang, Y., and Zhang, T. (2011). A highly active and sintering-resistant Au/FeOx–hydroxyapatite catalyst for CO oxidationw, *Chem. Commun.*, **47**, pp. 1779–1781.
110. Han, Y.-F., Phonthamamchai, N., Ramesh, K., Hong, Z., and White, T. (2008). Removing organic compounds from aqueous medium via wet peroxidation by gold catalysts, *Environ. Sci. Technol.*, **42**, pp. 908–912.
111. Yan, W., Brown, S., Pan, Z., Mahurin, S.M., Overbury, S.H., and Dai, S. (2006). Ultrastable gold nanocatalyst supported by nanosized non-oxide substrate, *Angew. Chem., Int. Ed.*, **45**, pp. 3614–3618.
112. Moreau, F., and Bond, G.C. (2006). CO oxidation activity of gold catalysts supported on various oxides and their improvement by inclusion of an iron component, *Catal. Today*, **114**, pp. 362–368.
113. Moreau, F., and Bond, G.C. (2007). Influence of the surface area of the support on the activity of gold catalysts for CO oxidation, *Catal. Today*, **122**, pp. 215–221.
114. Moreau, F., and Bond, G.C. (2007). Preparation and reactivation of $Au/TiO_2$ catalysts, *Catal. Today*, **122**, pp. 260–265.
115. Grisel, R.J.H., Kooyman, P.J., and Nieuwenhuys, B.E. (2000). Influence of the preparation of $Au/Al_2O_3$ on $CH_4$ oxidation activity, *J. Catal.*, **191**, pp. 430–437.

116. Chang, C.-K., Chen, Y.-J., and Yeh, C.-T. (1998). Characterizations of alumina- supported gold with temperature-programmed reduction, *Appl. Catal. A*, **174**, pp. 13–23.
117. Cellier, C., Lambert, S., Gaigneaux, E.M., Poleunis, C., Ruaux, V., Eloy, P., Lahousse, C., Bertrand, P., Pirard, J.-P., and Grange, P. (2007). Investigation of the preparation and activity of gold catalysts in the total oxidation of n-hexane, *Appl. Catal. B: Environ.*, **70**, pp. 406–416.
118. Nechayev, Y.A., and Nikolenko, N.V. (1985). Adsorption of gold(III) chloride complexes on alumina, silica and kaolin, *Geochem. Int.*, **11**, pp. 1656–1661.
119. Machesky, M.L., Andrade, W.O., and Rose, A.W. (1991). Adsoprtion of gold(III)-chloride and gold(I)-thiosulfate anions by goethite, *Geochim. Cosmochim. Acta*, **5**, pp. 769–776.
120. Qian, K., Luo, L., Bao, H., Hua, Q., Jiang, Z., and Huang, W. (2013). Catalytically active structures of $SiO_2$-supported Au nanoparticles in low-temperature CO oxidation, *Catal. Sci. Technol.*, **3**, pp. 679–687.
121. Ramirez-Garza, R.E., and Pawelec, B., Zepeda, T.A., and Martinez-Hernandez, A. (2011). Total CO oxidation over Fe-containing Au/HMS catalysts: effects of gold loadingand catalyst pretreatment, *Catal. Today*, **172**, pp. 95–102.
122. Shimizu, K.-I., Yamamoto, T., Tai, Y., Okumura, K. And Satsuma, A. (2011). Addition of olefins to acetylacetone catalyzed by cooperation of Brønsted acid site of zeolite and gold cluster, *Appl.Catal. A: Gen.*, **400**, pp. 171–175.
123. He, X., and Yang, H. (2013). Au nanoparticles assembled on palygorskite: Enhanced catalyticproperty and Au–Au2O3 coexistence, *J. Mol. Catal. A: Chem.*, **379**, pp. 219– 224.
124. Bi, Q.-Y., Du, X.-L., Liu, Y.-M., Cao, Y., He, H.-Y., and Fan, K.-N. (2012). Efficient subnanometric gold-catalyzed hydrogen generation via formic acid decomposition under ambient conditions, *J. Am. Chem. Soc.*, **134**, pp. 8926–8933.
125. Baatz, C., Thielecke, N., and Prüße, U. (2007). Influence of the preparation conditions on the properties of gold catalysts for the oxidation of glucose, *Appl. Catal. B: Environ.*, **70**, pp. 653–660.
126. Khoudiakov, M., Gupta, M.-C., and Deevi, S. (2005). Au/Fe2O3 nanocatalysts for CO oxidation: a comparative study of deposition–precipitation and coprecipitation techniques, *Appl. Catal. A*, **291**, pp. 151–161.

127. Patil, N.S., Uphade, B.S., Jana, P., Bharagava, S.K., and Choudhary, V.R. (2004). Epoxidation of styrene by anhydrous t-butyl hydroperoxide over reusable gold supported on MgO and other alkaline earth oxides, *J. Catal.*, **223**, pp. 236–239.
128. Lakshmanan, P., Delannoy, L., Richard, V., Méthivier, C., Potvin, C., and Louis, C. (2010). Total oxidation of propene over Au/xCeO2-Al2O3 catalysts: influence of the CeO2 loading and the activation treatment, *Appl. Catal. B*, **96**, pp. 117–125.
129. Smolentseva, E., Simakov, A., Beloshapkin, S., Estrada, M., Vargas, E., Sobolev, V., Kenzhin, R., and Fuentes, S. (2012). Gold catalysts supported on nanostructured Ce-Al-O mixed oxides prepared by organic sol-gel, *Appl. Catal. B*, **115–116**, pp. 117–128.
130. del Rio, E., Blanco, G., Collins, S., Lopez-Haro, M., Chen, X., Delgado, J.J., Calvino, J.J., and Bernal, S. (2011). CO oxidation activity of a Au/ceria-zirconia catalyst prepared by deposition-precipitation with urea, *Top. Catal.*, **54**, pp. 931–940.
131. Vecchietti, J., Collins, S., Delgado, J.J., Malecka, M., del Rio, E., Chen, X., Bernal, S., and Bonivardi, A. (2011). Gold catalysts supported on cerium-gallium mixed oxide for the carbon monoxide oxidation and water gas shift reaction, *Top. Catal.*, **54**, pp. 201–209.
132. del Rio, E., Lopez-Haro, M., Cies, J.M., Delgado, J.J., Calvino, J.J., Trasobares, S., Blanco, G., Cauqui, M.A., and Bernal, S. (2013). Dramatic effect of redox pre-treatments on the CO oxidation activity of Au/Ce0.50Tb0.12Zr0.38O2-x catalysts prepared by deposition-precipitation with urea: a nano-analytical and nano-structural study, *Chem. Commun.*, **49**, pp. 6722–6724.
133. Sobczak, I., Szrama, K., Wojcieszak, R., Gaigneaux, E., and Ziolek, M. (2012). CuxCryOz mixed oxide as a promising support for gold: the effect of Au loading method on the effectiveness in oxidation reactions, *Catal. Today*, **187**, pp. 48–55.
134. Lu, X., Zhao, G., and Lu, Y. (2013). Propylene epoxidation with $O_2$ and $H_2$: a high-performance Au/TS-1 catalyst prepared via a deposition-precipitation method using urea, *Catal. Sci. Technol.*, **3**, pp. 2906–2909.
135. Ye, Q., Zhao, J., Huo, F., Wang, D., Cheng, S., Kang, T., and Dai, H. (2013). Nanosized Au supported on three-dimensionally ordered mesoporous β-MnO2: Highly active catalysts for the low-temperature oxidation of carbon monoxide, benzene, and toluene, *Microporous Mesoporous Mater.*, **172**, pp. 20–29.

136. Huang, J., Wang, L.-C., Liu, Y.-M., Cao, Y., He, H.-Y., and Fan, K.-N. (2011). Gold nanoparticles supported on hydroxylapatite as high performance catalysts for low temperature CO oxidation, *Appl. Catal. B*, **101**, pp. 560–569.
137. Sun, H., Su, F.-Z., Ni, J., Cao, Y., He, H.-Y., and Fan, K.-N. (2009). Gold supported on hydroxyapatite as a versatile multifunctional catalyst for the direct tandem synthesis of imines and oximes, *Angew. Chem., Int. Ed.*, **48**, pp. 4390–4393.
138. Perret, N., Wang, X., Delannoy, L., Potvin, C., Louis, C., and Keane, M.A. (2012). Enhanced selective nitroarene hydrogenation over Au supported on $\beta$-Mo$_2$C and $\beta$-Mo$_2$C/Al$_2$O$_3$, *J. Catal.*, **286**, pp. 172–183.
139. Perret, N., Cardenas-Lizana, F., Lamey, D., Laporte, V., Kiwi-Minsker, L., and Keane, M.A. (2012). Effect of crystallographic phase ($\beta$ vs. $\gamma$) and surface area on gas phase nitroarene hydrogenation over Mo$_2$N and Au/Mo$_2$N, *Top. Catal.*, **55**, pp. 955–968.

# Chapter 2

# Immobilization of Preformed Gold Nanoparticles

### Carine E. Chan-Thaw, Alberto Villa, and Laura Prati

*Dipartimento di Chimica, Università degli Studi di Milano,*
*Via Golgi, 19-20133 Milano, Italy*
laura.prati@unimi.it

## 2.1 Introduction

Since the pioneer work of M. Haruta and G. J. Hutchings on gold as a catalytically active metal [1–3], enthusiasm to improve the understanding of new catalytic processes based on gold was observed. High metal dispersion is of importance to obtain an interesting catalytic activity. In the case of gold due to its low melting point, preparing a heterogeneous catalyst with high metal dispersion by the common procedures was not an easy task [4]. However, because of the peculiarity of the gold electronic states [5] leading to astonishing colors of the Lycurgus Cup, gold nanoparticles (GNPs) have attracted attention and not only from a catalytic point of view.

---

*Gold Catalysis: Preparation, Characterization, and Applications*
Edited by Laura Prati and Alberto Villa
Copyright © 2016 Pan Stanford Publishing Pte. Ltd.
ISBN 978-981-4669-28-3 (Hardcover), 978-981-4669-29-0 (eBook)
www.panstanford.com

The so-called immobilization technique, which has demonstrated its efficiency in catalysts preparation, implies the formation of GNPs before being immobilized on the support [4, 6]. Indeed, the advantage of using this technique principally is based on the applicability regardless of the type of support employed. Moreover, the use of solutions of molecular precursors as starting materials and mild chemical treatment for the generation of metal particles in principle leads to a careful control of composition, size, and morphology of the resulting particles [6, 7]. Common procedures include reduction of metal salts, photochemical or thermal decomposition, and reduction of organometallic complexes [7, 8].

The preparation of GNPs dispersed in a liquid phase, also called gold sol generation, is a method based on the reduction of a gold precursor in the presence of a stabilizing agent (polymer, surfactant, polar molecule, etc.). The subsequent immobilization of these GNPs on a support produces heterogeneous catalysts. The added molecules, namely the stabilizing agent, have the role to prevent any aggregation of the formed colloids into larger particles. Moreover, the temperature at which the reduction of the gold precursor occurs is an important parameter in the successful formation of stable GNPs (Section 2.2).

Section 2.3 will described in detail the theory behind the stabilization of the nanoparticles. Strategies for the preparation of small precious metal nanoparticles in the presence of a protective agent (Section 2.4) have been greatly developed in the last decade [9]. However, even stabilizer-free nanosized gold sols were claimed to be prepared [10]. The term "unprotected" metal colloids do not mean that the gold particles are truly bare; they are stabilized by solvents or simple anions adsorbed on them or by both. Moreover, unsupported gold particles have been studied in aerobic oxidation of glucose and it has been found that they behave as an efficient catalyst, showing a similar activity to enzymatic systems. In particular the catalytic activity has been found inversely correlated to the particle diameter [11].

However, by far, catalytic applications of gold sols were mainly as heterogeneous catalysts produced by immobilization of the preformed GNPs on a suitable support. Thus the immobilization of nanoparticles on the support, the consecutive step after sol generation, will be described (Section 2.5).

## 2.2  Nucleation and Growth Processes

The nature of gold colloids is highly dependent on the used reagents and preparation conditions. Indeed, particle shape, particle size, and particle size distribution may vary, not just according to the stabilizer and reducing agents used. Following LaMer theory, the formation of particles can be described by a two-step process, namely nucleation and growth. Turkevich [12] defined nucleation as the process where a particle of a new phase is formed in a previously single-phase system (homogeneous solution). Furthermore, growth is defined as addition of material deposit on the previously formed particle, which is thus increasing in size.

To achieve the synthesis of monodisperse GNPs, it is primordial to clearly distinguish the nucleation step to the growth one to avoid that new nuclei are formed together with the growth procedure. Formation of nuclei is such a fast step (milliseconds to a few seconds) that it has been difficult for the researchers to record and follow their formation.

### 2.2.1  *Homogeneous Nucleation: Growth*

Colloidal Au nanoparticles are synthesized through either reduction of $Au^{3+}$ ions with reducing reagents (most reported sodium borohydride, $NaBH_4$, ascorbic acid, $C_6H_8O_6$, sodium citrate) or thermal decomposition of organometallic compounds in the presence of surfactant molecules that can attach to the nanoparticles' surfaces to stabilize them.

The conversion of the ionic gold precursor into metallic gold nuclei occurs in a lapse time of milliseconds to a few seconds. LaMer and Dinegar [13] were the first to describe the formation of colloidal nanocrystals in a solution phase in 1950. They considered that the controlled growth of many nuclei, which are simultaneously generated, results in the formation of monodisperse nanoparticles. They named this concept "burst nucleation." From a homogeneous

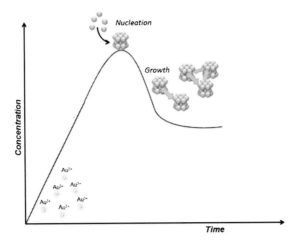

**Figure 2.1** Representative scheme of homogeneous nucleation and growth of nanoparticles by particle migration and coalescence.

supersaturated medium, a rapid nucleation of particles would initially occur (short burst nucleation). The nucleation process is directly followed by an initial fast rate of growth of these formed nuclei to reduce rapidly the concentration below the nucleation concentration (Fig. 2.1).

An eventual slow rate of growth can also occur and could lead to a long growth period compared to the nucleation period. These authors are clearly distinguishing the nucleation step to the growth one by working in homogeneous phase. Note that in this homogeneous nucleation process, nucleation occurs at a high-energy barrier as the system spontaneously changes from the homogeneous phase to the heterogeneous phase with the generation and accumulation of stable nuclei. Once this high-energy barrier is overcome, the monomer concentration initially provided starts to decrease with time concomitantly with the increase of the formed nuclei. There arrives a time where no nucleus is anymore formed, and only the growth stage takes place as long as the solution is supersaturated. This strategy is particularly important for shape control: to obtain a highly shape-monodisperse yield of nanocrystals, nucleation must occur rapidly and instantaneously.

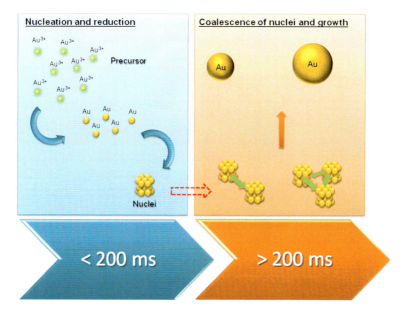

**Figure 2.2** Representative scheme of homogeneous nucleation and growth of nanoparticles by the mechanism of particle migration and coalescence.

As nucleation occurs rapidly, it is not an easy task to follow the formation of the nuclei. Recently, Polte et al. [14] reported that they could follow via in situ small-angle X-ray scattering at millisecond resolution time the nucleation and growth of GNPs.

Growth that consists of the deposition of materials on the surface of a particle is a heterogeneous reaction. The growth process can be explained by the Ostwald ripening mechanism. Dissolution of small crystals or sol particles and the redeposition of the dissolved species on the surfaces of larger crystals or sol particles were first described by Ostwald in 1896 [15]. Indeed, small crystallites or sol particles have a higher surface free energy than larger crystallites, so a small diffusion of material from small to larger particles occurs.

Moreover, the growth of nanoparticles can be also described by the mechanism of particle migration and coalescence (Fig. 2.2). When particles come in close proximity to each other because of the particles' motion, coalescence occurs.

Tao, Habas, and Yang [16] reviewed general strategies on controlling the shape of the nanoparticle with the purpose to increase functionality and selectivity during the process.

### 2.2.2 *Heterogeneous Nucleation: Growth*

An alternative to homogeneous nucleation to form monodisperse nanoparticles is the heterogeneous nucleation growth process. The reaction conditions for shape control are less stringent, given that seed particles are preformed in a separate synthetic step. Indeed, preformed nuclei (also called seed nuclei) are introduced into the reaction solution and then the monomers are supplied to precipitate on the surface of the existing nuclei.

Thus, utilizing heterogeneous nucleation allows a wider range of growth conditions that employ milder reducing agents, lower temperatures, or aqueous solutions (Fig. 2.3).

Heterogeneous growth is occurring at preferential sites where the surface energy is lower. This means that the free-energy barrier is diminished and that the growing step is facilitating. Moreover,

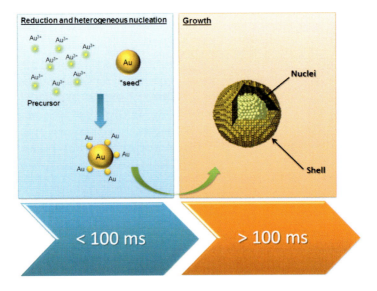

**Figure 2.3** Heterogeneous nucleation: growth.

the monomer concentration is kept low during growth to suppress homogeneous nucleation. As an example, small Au seeds with a diameter of about 3–5 nm are added to a Au precursor solution containing ascorbic acid that is only capable of reducing the Au precursor in the presence of seeds. The resulting structures, rods, and wires, as well as a variety of shapes [17], can be synthesized by carefully controlling the growth stage.

Heterogeneous nucleation can also result in the generation of core/shell nanocrystals. The surface structure of the seed determines the overgrowth morphology. The shape of the introduced Au seed has a significant impact on the resulting binary crystal structure if the added monomer solution is not composed of gold precursors. For example, the adsorption and dissociation of $H_2$ on catalytically active Pt and Pd seeds has been demonstrated to catalyze the growth of Au, Pt, and Pd shells [18]. Such seeded overgrowth of metal nanocrystals is used to produce bimetallic heterostructures.

## 2.3 Stabilization of Gold Colloids

### 2.3.1 *Importance of Stabilization*

GNPs have to show good stability under operative conditions. Good stability implies no aggregation of the nanoparticles as coagulation. In a liquid, the surface of a particle is charged either by dissociation of surface groups or by adsorption of charged molecules (e.g., polyelectrolytes) from the surrounding solution. In the case of colloidal gold particles, the surface is negatively charged and shows a surface potential $\Psi_0$. Thanks to the Coulomb attraction, the charged surface pulls the counterions back toward the surface. However, the osmotic pressure presents in the media forces the counterions away from the interface. This phenomenon results in the creation of a diffuse double layer.

The double layer is divided into three main parts consisting of the surface charged, the Stern layer, and the Gouy–Chapman layer. This Gouy–Chapman–Stern model is also including specifically adsorbed anions (Grahame model). The compact layer of counterions, very close in our case to the negatively charged surface, is known as the Stern layer in honor of Otto Stern. The potential corresponding to the Stern layer at a distance $d$ from the charged surface is called $\Psi_d$. The Stern layer is itself composed of the inner Helmholtz plane (IHP) at a distance $x_1$ and the outer Helmholtz plane (OHP) at a distance $x_2$ (equivalent to $d$). The potential at the IHP is named $\Psi_i$.

The counterions specifically adsorb tightly on the interface in the inner part of the Stern layer, namely the IHP. The potential that is dependent on the occupancy of the ions drops drastically. The OHP is located on the plane of the centers of the next layer of nonspecifically adsorbed ions. Normally the ions adsorbed electrostatically at the Stern layer will not completely neutralize the surface charge. However, if there is a chemical driving energy for adsorption, the adsorbed countercharge may exceed that of the "true" surface charge, and the overall charge on the particle may be reversed. Therefore, the potential changes sign before decaying slowly to zero in the bulk solution.

The diffuse layer begins at the distance $d$ from the interface, that is, at the end of the Stern layer or at the OHP and is known as the Gouy–Chapman layer. The potential varies in an approximately exponential manner from the Stern plane into the solution, through the diffuse layer. The schematic model of the electrical double layer and the change in the electric potential within the double layer are illustrated in Fig. 2.4. The thickness of the diffuse layer is called *Debye length* ($1/\kappa$), where $\kappa$ corresponds to the *Debye–Hückel parameter*. Moreover, this length indicates the distance from the end of the Stern layer up to the distance where the effect of the surface is felt by the ions.

When the distance $h$ between two particles is small enough, van der Waals forces are created (Fig. 2.5). If no forces are opposed to the attractive van der Waals ones, aggregation of GNPs occurs. To overcome such a problem, repulsive forces that can be induced by the use of a stabilizing agent are necessary. Such a balance between double-layer repulsion forces, which increase exponentially with decreasing distance $h$, and van der Waals attraction forces is described in Derjaguin, Landau, Verway, and Overbeek (DLVO) theory. DLVO theory [19, 20] gives a better understanding of the behavior of colloidal particles, that is, a quantitative aggregation

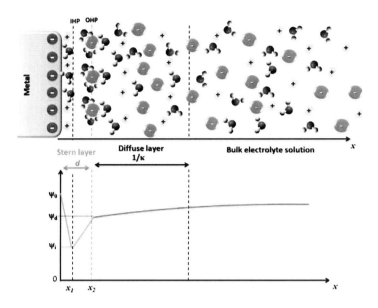

**Figure 2.4** Schematic model of the electrical diffuse double layer following the Gouy–Chapman–Stern–Grahame model and the corresponding electric potential.

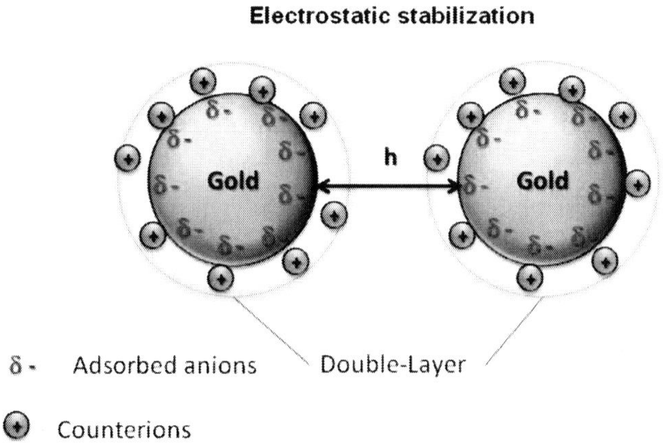

**Figure 2.5** Electrostatic stabilization of metal colloid particles.

in aqueous dispersions. It describes the force between charged surfaces interacting through a liquid medium. This theory combines the effects of the van der Waals attraction forces and the electrostatic repulsion due to overlapping of the diffuse layer.

In the present chapter, a distinction of three stabilization procedures dependent on the stabilizing agents that will be used will be described: (a) Electrostatic stabilization by the surface-adsorbed ions, (b) steric stabilization by the presence of a bulky group, (c) and the combination of these two procedures of stabilization with electrosteric stabilization, such as surfactants.

### 2.3.2 Electrostatic Stabilization

Electrostatic stabilization is based on the mutual repulsion of similar electrical charges. Most colloidal particles acquire a charge either from surface-charged groups or by specific ion adsorption from the solution.

A gold colloidal particle that is surrounded by a cloud of adsorbed anions/molecules (mainly $AuCl_4^-$ and $AuCl_2^-$ in the case of $NaBH_4$ reduction) [21] carries a negative charge in water.

When two charged surfaces approach each other in an electrolyte medium, an effective interaction that is determined by the Debye screening length $1/\kappa$ becomes appreciable. Decreasing the interparticle distance $h$, the counterion distributions start to overlap (Fig. 2.5).

The region near the surface of enhanced counterion concentration is also called the electrical double layer. For similar particles, that is, identically charged, a repulsive double-layer force is then generated. A competition between the van der Waals attraction forces and the double-layer repulsion, described by DLVO theory, determines the stability or instability of the colloid. Controlling any parameters of the electrostatic repulsion leads to an optimization of the electrostatic stabilization (Eq. 2.1 and Fig. 2.5) and the corresponding potential is

$$V(h) = V_A(h) + V_R(h) \qquad (2.1)$$

where $V_A(h)$ is the potential associated to attractive forces and $V_R(h)$ is the one associated to repulsive ones.

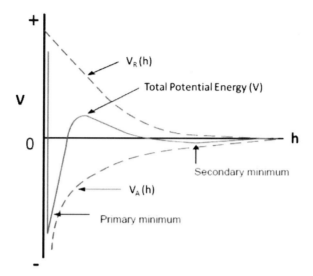

**Figure 2.6** Curve of the potential energy against the distance of the particle surfaces.

The curve of the potential energy of interaction against the distance of the particle surfaces is schematically represented in Fig. 2.6.

At a finite distance, where the surfaces do not come into molecular contact, equilibrium is reached between electrodynamic attractive and electrostatic repulsive forces (secondary minimum). At smaller distances there is a net energy barrier. Once overcome, the combination of strong short-range electrostatic repulsive forces and van der Waals attractive forces leads to a deep primary minimum. Both the height of the barrier and the secondary minimum depend on the ionic strength and electrostatic charges. The energy barrier is decreased in the presence of electrolytes (monovalent <divalent <trivalent) by compression of the double layer [8]. Thus, the stability of a colloid is a function of the energy of interaction between the particles.

### 2.3.3 Steric Stabilization

Steric stabilization of colloidal particles derives from covering the particles with polymers or oligomers (Fig. 2.7).

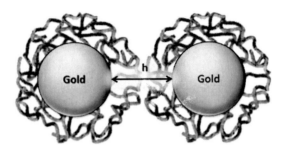

**Figure 2.7** Schematic representation of steric stabilization of gold colloid particles.

The absorption of flexible polymers of sufficiently high molecular weight leads to polymer chains protruding from the particle surfaces. The slight interpenetration of these chains keeps them at a distance too large to give a van der Waals interaction sufficient for coherence [22]. This type of stabilization was confirmed using gold particles as a test system and polyglycols as the absorptive; the stabilization was found to increase with polymer concentration and particularly with the molecular weight of the polymer [23]. Macromolecules, such as proteins and polysaccharides, were also used [24] and various synthetic polymers. The adsorbed layer of the stabilizer prevented cohesion of one particle to another and the lypophobic gold colloid became lypophilic.

### 2.3.4 Electrosteric Stabilization

Combination of both long-range electrostatic repulsion and short-range steric repulsion results as electrosteric stabilization.

Specifically, this type of stabilization features adsorbed polymers like in steric stabilization; however, the adsorbed polymers also bear electrostatic charge. Therefore, we get significant orientation of polymer chains deriving from the adsorption of charges on the metal surface and provide also electrostatic repulsion (Fig. 2.8).

For electrosteric stabilization, not only the electrostatic stabilization but also the steric stability imparted by the polymer depends on the pH, dielectric properties, and ionic strength of the solvent [25].

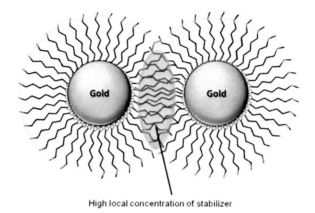

**Figure 2.8** Representation of electrosteric stabilization.

A common way to electrosterically stabilize a colloid is by adding a polymer (called a polyelectrolyte) with an ionizable group that is dissociated in the solvent to create charged polymers.

## 2.4 Preparation of Gold Nanoparticles

### 2.4.1 *The Stabilizers*

#### 2.4.1.1 Electrostatic stabilizers

An electrostatic stabilizer is an ionic compound such as a carboxylate or polyoxoanion that adsorbs on the metal surface. Together with its counterion it generates an electrical double layer around the gold particle.

Dextrin has been investigated as a rigid hydroxyl protective agent, affecting the attractive van der Waals forces. Small sizes of nanoparticles have been obtained with dextrin-stabilized sols (2–3 nm) [26].

Citrate is a common electrostatic stabilizing agent for GNPs through mutual electrostatic repulsion between neighboring GNPs; this occurs as a result of the negative surface charge of the citrate layer [12, 27].

Duff, Baiker, and Edwards reported for the first time the reduction of gold(III) ions by partial hydrolysis of tetrakis(hydroxymethyl)phosphonium chloride (THPC). The generated hydrosol of gold clusters shows a mean diameter of 1.5 nm (average nuclearity about $Au_{170}$) and even smaller [28].

### 2.4.1.2 Steric stabilizers

A steric stabilizer is a polymer or oligomer that adsorbs on the particle surface, providing a protecting layer. Among the polymeric stabilizers, Brij35 (polyoxyethylene(23)-laurylether, $MW_{av}$ 1200) [26]; poly(ethylene glycol) (PEG), reacted with bisphenol A diglycidyl ether, MW 15000–20000) [26]; and polyvinylpyrrolidone (PVP), $MW_{av}$ 55000) [29] have been reported to produce highly stable GNPs by steric stabilization, with a small particle size (2–6 nm).

### 2.4.1.3 Electrosteric stabilizers

An electrosteric stabilizer is characterized by a polar head group able to create an electric double layer and a lypophylic chain that provides the steric repulsion. This is, for example, the case of $N$-dodecil-$N$, $N$-dimethyl-3-amino-1-propansulphonate [30], which provides both sterical and electrostatic stabilization, a condition that has been demonstrated necessary to preserve particle dimension when carbon is used as the support [29]. Long-chain alcohols of 15,000–20,000 Dalton, like poly(vinyl alcohol) (PVA), MW 13000–23000, 98% hydrolyzed, are also suitable electrosteric stabilizers for nanoparticles of 3–8 nm mean diameter [7, 31].

Ammonium salts have been also used as stabilizers. The polycationic ammonium salt poly{ bis(2-chloroethyl)ether-*alt*-1,3-bis[3-(dimethyl-amino)propyl} urea] (PEU) can be chosen for increasing the ionic cloud in the electrical double layer, thus raising the electric repulsion between the gold particles in the sol. Small particles can be obtained (mean diameters between 3.1 and 4.2 nm) [26].

Typically, the stabilizer/Au weight ratio is maintained at 1, with a metal concentration ranging between $10^{-3}/10^{-4}$ M. The use of a higher ratio normally increases the stability of the GNPs and contributes to a decrease in the particle diameter. The same trend is followed by reducing the Au precursor concentration; the lesser the concentration, the smaller the diameter.

Ligands such as phosphine [32] and thiolates [33] also stabilized nanoparticles, providing direct ligand with metal nanoparticles. However, their use in catalysis is limited to a few examples, but they are very useful for other applications in different fields such as biosensors, nanomedicine, etc.

### 2.4.2 *Nature of the Reducing Agent*

Precursors of gold are most commonly based on Au(III) salts that can produce anionic species in aqueous solution such as $AuCl_4^-$. However, organometallic compounds are also used especially when nonaqueous media are required. Therefore, the nature of the reducing agent can be varied to study its influence on particle size and particle distribution. The size of the AuNPs is correlated to the rate of reduction. Generally, a fast reduction step induces an enhancement of the nucleation rate, which results in the production of small particles.

#### 2.4.2.1 Chemical reduction of metal salts

The chemical reduction method allows close control on the form (shape) and size of the nanoparticles by varying the reducing agent, the dispersing agent, the reaction time, and the temperature. The chemical reduction method carries out reduction of the metal ions to their 0 oxidation states (i.e., $M^{n+} \rightarrow M^0$). The main advantage of the salt reduction method in the liquid phase is that it is reproducible and allows the production of colloidal nanoparticles with narrow size distribution. However, chemical reduction obviously produces by-products that are not always just spectators in catalysis. Therefore, final catalysts are often washed thoroughly in order to remove any possible by-products or they undergo dialysis before being immobilized.

Diverse reducing agents have been reported to be effective in producing metal nanoparticles. Among all reducing agents, we can find sodium citrate, hydrogen, carbon monoxide, alcohols, hydrazine, and borohydrides.

**2.4.2.1.1 Sodium citrate**

Turkevich et al. reported for the first time a reproducible synthesis of the nucleation, growth, and agglomeration of GNPs reduced by sodium citrate. In this case, the sodium citrate plays both the role of a reducing agent but also an ionic stabilizer—a stable, deep-red dispersion of uniform 13 nm GNPs [12, 34, 35]. The synthesis consists of refluxing an aqueous mixture of gold salt and mono-, di-, or trisodium citrate till a color change in the solution is observed.

More recently, Plech et al. revisited this method [36]. They studied the growth of GNPs reducing Au(III) with a diverse concentration of citrate and ascorbic acid. A wide range of gold particles sizes, from 9 to 120 nm, with a defined size distribution, was obtained. The reaction was initiated thermally or by UV irradiation [36]. They indicated that a high concentration of citrate more rapidly stabilizes AuNPs of smaller sizes, whereas a low concentration of citrate leads to large-size AuNPs and even to the aggregation of AuNPs. It should be noted that citrate is a reagent (a reductant) as well as a stabilizer [12, 34, 35]. Therefore, citrate is a noninnocent ligand, becoming oxidized simultaneously to the intermediate ketone (acetone dicarboxylic acid), which in turn is an even better [12] reducing agent. An important aspect of this preparation lies in the intrinsic modest reducing ability of citrate. Indeed, citrate results in a low reducing rate and consequently the formation of large particles. This methodology produces always particles larger than 10 nm [27].

**2.4.2.1.2 Borohydrides**

Schmid et al. introduced the diborane reducing agent for the synthesis of Au55 nanoclusters stabilized by phosphine ligands [37].

$NaBH_4$ is the most widely used hydride for the reduction of transition metals salts. In this approach, $NaBH_4$ is directly mixed with the metal precursor in the presence of a stabilizing agent. The most reported capping agents are generally surfactants or water-soluble polymers. The surfactants can be cationic, anionic, or nonionic. Indeed, quaternary ammonium, sulfates, or PEG were used to stabilized Au nanoparticles by Nakao et al. [38]. PVA [27] and PVP are intensively used to form GNPs that will be later used for catalysis. Poly(vinylic ether) (PVE) [4, 39] or cyclodextrine [26] are also reported. The diameter obtained with hydride reduction is tunable in a large range (2–10 nm), depending on the stoichiometry used, the pH, the rate of addition, and obviously the type of hydride. The Au/BH$_4^-$ molar ratio is normally set to use an excess of hydride corresponding to 1:4 (mol:mol) with respect to the overall stoichiometry.

Bönnemann et al. [40] reported the combined use of hydride and NR$_4^+$ in the case of hydrotriorganoborates with tetraalkylammonium counterions. The resulting M(0) particles protected with long-chain quaternary ammonium salts are soluble in organic solvents and can be easily extracted during their synthesis (Eq. 2.2).

$$MX_n + NR_4(BEt_3H) \rightarrow M_{\text{colloid}} + nNR_4X + nBEt_3 + \frac{n}{2}H_2 \quad (2.2)$$

where M is a metal of group VI–XI, X is Cl or Br, n is 1, 2, or 3, and R is an alkyl, $C_6$–$C_{20}$.

They can be purified and isolated in solid form by suitable treatments of the colloid with solvents having different polarities. Another not negligible advantage is the possibility to redissolve the isolated particles in various solvents in high concentration of the metal (up to 1 molar solutions).

### 2.4.2.1.3 Tetrakis(hydroxymethyl)phosphonium chloride

The use of in situ hydrolysis of THPC as a reducing agent was introduced by Baiker et al., and facilitates the synthesis of small monodispersed GNPs (2–4 nm) [28]. The method uses in situ hydrolysis of THPC as a reducing agent and stabilizer [28]. The same method was also applied by Villa et al. [27, 41] for producing 3–8 nm sized GNP. THPC is an electrostatic stabilizer, so the negatively charged AuNPs are stabilized by the positive part of the THPC molecules. THPC is partially hydrolyzed with NaOH to P(CH$_2$OH)$_3$ and HCHO, the latter acting as the actual reducing agent [42] following the mechanism [28]

$$P(CH_2OH)_4^+ + OH^- \rightarrow P(CH_2OH)_3 + CH_2O + H_2O \quad (2.3)$$

**Figure 2.9** Polyol process.

### 2.4.2.1.4 Alcohols

Using alcohol to reduce metal salts to prepare metal particles is known as the polyol process (Fig. 2.9). It is a convenient, versatile, and low-cost method for the synthesis of metal nanostructures on a large scale. The process is normally carried out at the refluxing temperature of alcohol, and a wide range of particle sizes can be obtained; even larger particles are more accessible than the smaller.

The most effective alcohols that have been reported are methanol, ethanol, and propan-2-ol as they are bearing $\alpha$-hydrogen. Li and co-authors [43] obtained gold nanocrystals through the reduction of gold metal ions (HAuCl$_4$) by ethanol at a temperature varying from 20°C to 100°C under hydrothermal or atmospheric conditions for 10 hours. By altering the temperature or the mole ratio of the protecting reagents to noble metal ions, the diameter of the gold nanocrystals, usually in a round shape with a smooth surface, is about 4–15 nm.

The alcoholic reduction preparation, in the presence of PVP polymers, was reported by Nalawade, Mukherjee, and Kapoor [39a] and Ayyappan and co-authors [39c]. PVP is added at the start of the reaction or continuously during metal reduction. PVP acts first as a stabilizing agent, preventing aggregation of metal particles and retaining a uniform colloidal dispersion. In addition, PVP is used as a shape control agent or "crystal habit modifier," preventing reduction onto specific crystal faces. PVP is thus a useful protagonist in the growing step, leading to the formation of stable-shaped GNPs.

Using this method in the presence of PVP, Yang et al. [39b] obtained gold nanocrystals with a particle size of about 100–300 nm. Their approach consists of injecting simultaneously hydrogen tetrachloroaurate (HAuCl$_4$·3 H$_2$O) and PVP into a boiling ethylene glycol solution. In this case, ethylene glycol is playing the role of both solvent and reducing agent. PVP not only stabilized the particles but also controlled the shape of the particles. Gold particles are formed within minutes.

Ayyappan et al. [39c] reported to be able to control the size of the GNPs and their stability. More recently, Nalawade, Mukherjee, and Kapoor [39a] described the one-pot synthesis of uniform and stable PVP-protected GNPs (AuNPs) using glycerol as a reducing agent. When only glycerol is used, that is, glycerol plays the role of both a reducing and a protective agent, nonspherical and nonstable GNPs with a mean size of 30 nm were obtained.

### 2.4.2.1.5 Other reducing agents

To prepare GNPs, hydrogen is also used a reducing agent. Tan and co-authors [44] prepared different metal colloids in which gold is reduced by hydrogen in the presence of PVA. The metal salt was added to a PVA solution and heated to 100°C before hydrogen was passed through the suspension till it turned dark. The so-obtained suspension was then centrifuged to recover the GNPs.

Merga et al. [45] also describe the synthesis of GNPs, free of any foreign stabilizer, upon reduction of $Au_2O_3$ by molecular hydrogen. The $Au_2O_3$ powder was dissolved in de-ionized water bubbled with argon gas. The temperature is increased to 70°C and the mixture inside the flask is flushed with pure $H_2$ and pressurized at 1.5 atm. Gold particles of 20–100 nm size were obtained.

Ascorbic acid is also used as a reducing agent for the preparation of GNPs. The reduction of $HAuCl_4$ to $Au^0$ in the presence of cetyltrimethylammonium bromide (CTAB) by ascorbic acid was studied by Sun [46] and Khan [47]. Varying the concentration of $HAuCl_4$ at a fixed concentration of ascorbic acid and CTAB, Khan and co-authors [47] demonstrated the variation of the shape, size, and size distribution on GNPs.

Carbon monoxide reduction has also been used with gold salts to produce poly(vinyl sulfate) (PVS)-stabilized gold colloid nanoparticles (Au/PVS) [48]. The authors bubbled CO through the stirred mixture of $HAuCl_4$ and PVS potassium salt dissolved in water and at room temperature. The Au/PVS of a mean size of 20 nm was formed after boiling the solution for an hour.

### 2.4.2.2 Electrochemical reduction

This method for preparing nanostructured mono- and bimetallic colloids was developed by Reetz and his group since 1994 [49]. Indeed, they demonstrated that thanks to the electrochemical method, it is possible to obtain palladium colloid in the nanometer range. This method was later applied by Wang et al. to obtain gold nanorods [50]. A cationic surfactant, hexa-decyltrimethylammonium bromide ($C_{16}TAB$) was used not only as the electrolyte but also as the stabilizer for nanoparticles to prevent their further growth and agglomeration. During the synthesis, the bulk gold metal is converted from the anode to form GNPs most probably at the interfacial region of the cathodic surface and within the electrolytic solution that contains $C_{16}TAB$ [50]. The length of gold nanorods depends on the deposition time, whereas their thickness was linked to the pore diameter of the nanoporous alumina template [51]. The main advantage of using this approach is the possibility to control the particle size by varying the current intensity [49].

### 2.4.2.3 Other methods

It is worth mentioning that other methods, such as sonochemical [52], microwave assisted [53], laser ablation in solution [54], and photochemical [55], are also reported in the literature for Au precursor reduction. The sonochemical reduction of transition metal salts occurs in three steps: the generation of the active species, the reduction of the metal, and the growth of the colloids.

However, beside a lot of methods available for generating sols, the one consisting of chemical reduction in a solution of salts is the simplest and a widely applicable technology that provides suitable-sized particles with narrow distribution. Such a control of the particle size and distribution is an important factor in the preparation of heterogeneous catalysts.

## 2.5 Immobilization of Nanoparticles

Immobilization of GNPs on the support is a nonnegligible point. Very often this step was not considered as influencing the metal particle

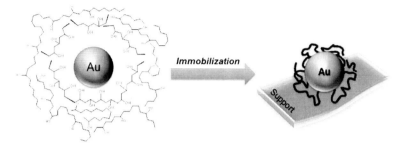

**Figure 2.10** Representative scheme of gold nanoparticles' immobilization on a support. For this scheme, PVA was chosen as the protective agent.

size. However, the morphology and the metal dispersion strictly depend on the surface properties and the morphology of the support [56, 57]. The immobilization (Fig. 2.10) is performed most often by simply dipping the support in the metal sol under vigorous stirring and different forces can be involved such as nonspecific electrostatic interaction (adsorption) or specific chemical bonding (grafting).

## 2.5.1 *Adsorption*

Charcoal, silica, alumina, or oxides such as $TiO_2$ and MgO are the most reported supports to adsorb colloidal GNPs. The main concern in the adsorption process is the size preservation of the GNPs during the transfer from the sol to the support surface. Indeed, GNP size is determined by the sol preparation, which is directly linked to the choice of the reducing agent and stabilizer. However, during their immobilization on the selected support, growing and aggregation of the nanoparticles can take place.

The kinetics of adsorption is correlated to the choice of the sol stabilizer and on the isoelectric point (IEP), also called the point of zero charge (PZC) of the support. One should remember that at a pH below the IEP, the surface is negatively charged, whilst at a pH over the IEP, the surface is positively charged. Therefore, by varying the pH of the metallic sol, one can modify the reciprocal affinity of metallic sol with the surface, varying thus the kinetic of adsorption. As an example, a PVA-stabilized gold sol is negatively charged within a large range of pH values (Fig. 2.11) and should be more easily

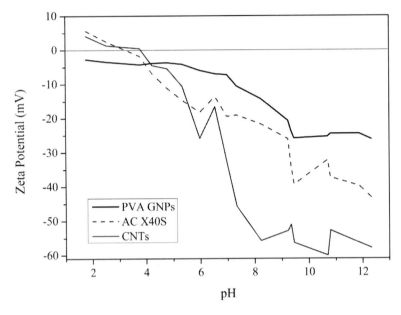

**Figure 2.11** Zeta potential measurement of PVA GNPs, AC X40S, and CNTs. Reprinted with permission from Ref. [58]. Copyright (2014) American Chemical Society.

adsorbed on a positively charged surface, if no other interactions occur. The IEP of the support should be investigated first, before the adsorption step. For example, the electrostatic interaction between PVA GNPs and activated carbon or carbon nanotubes (CNTs) is favored only at a pH below 3, where the two supports present a positively charged surface (Fig. 2.11).

Table 2.1 reports the variation of differently stabilized GNPs' mean size during the immobilization step. As noted above, the mean particle size is a function of the stabilizer/Au ratio (wt/wt). However, the maintenance of Au size during the immobilization appears dependent on both the amount and the nature of the protective agent related to a specific support. Indeed, it has been clearly shown that the stabilizer plays a role of mediator between the metal particle and the surface of the support, as shown in Table 2.1 [57].

**Table 2.1** Effect of the amount and nature of the protective agent on the growth of GNPs during the immobilization on different supports

| Stabilizer | Stabilizer/Au (wt/wt) | Particle size (sol) | Support | Particle size (nm) (supported) | Ref. |
|---|---|---|---|---|---|
| THPC | 0.8 | 4.0 | AC[a] | 8.6 | [31a] |
|  |  |  | TiO$_2$ | 4.1 | [31a] |
|  |  |  | Al$_2$O$_3$ | 3.9 | [31a] |
|  | 1 | 3.5 | AC | 8.2 | [31a] |
|  |  |  | TiO$_2$ | 3.7 | [31a] |
|  |  |  | Al$_2$O$_3$ | 3.8 | [31a] |
|  | 2 | 2.7 | AC | 4.2 | [31a] |
|  |  |  | TiO$_2$ | 4.2 | [31a] |
|  |  |  | Al$_2$O$_3$ | 4.3 | [31a] |
| PVA | 0.125 | 3.2 | AC | 4.1 | [58] |
|  | 0.25 | 3.1 | AC | 3.9 | [58] |
|  | 0.5 | 2.6 | AC | 3.6 | [58] |
|  |  |  | NiO | 3.8 | [59] |
|  |  |  | Nano-NiO | 3.6 | [59] |
|  |  |  | SiO$_2$ | 4.0 | [59] |
|  |  |  | MgO | 3.8 | [59] |
|  |  |  | TiO$_2$ | 4.0 | [59] |
|  |  |  | H-mordenite | 3.8 | [59] |
|  | 1 | 2.4 | AC | 3.5 | [58] |
|  |  |  | Graphite | 5.4 | [60] |
|  |  |  | CNTs | 4.6 | [60] |
|  |  |  | CNTs oxidized | 3.8 | [61] |
|  |  |  | CNFs PR24-LHT | 3.5 | [62] |
|  |  |  | CNFs PR24-PS | 3.8 | [62] |
|  |  |  | CNFs PR24-PS oxidized | 3.7 | [61] |
|  |  |  | N-CNFs PR24-PS 873K | 2.9 | [61] |
| PEG[b] | 0.3 | 5.3 | AC | 14.0 | [63] |
|  | 1 | 5.2 | AC | 12.0 | [63] |
| Dextrin | 0.3 | 3.5 | AC | 12.0 | [63] |
|  | 1 | 3.6 | AC | 8.4 | [63] |
|  | 1.8 | 2.8 | AC | 9.5 | [63] |
| PDDA[c] | 0.1 | 3.6 | AC | 27.0 | [63] |
|  | 0.3 | 2.8 | AC | 22.0 | [63] |
|  | 0.65 | 2.6 | AC | 15.0 | [63] |

[a] AC = activated carbon
[b] PEG = poly(ethylene glycol)
[c] PDDA = poly(diallyldimethylammonium bromide)
Preparation method for GNPs: Au precursor = NaAuCl$_4$ $10^{-4}$M, reducing agent NaBH$_4$ 0.1 M (Au/NaBH$_4$ 1/4 mol/mol)
Reprinted with permission from Ref. [58]. Copyright (2014) American Chemical Society

**Table 2.2** Immobilization step using carbon as the support

| Stabilizer | Stabilizer/Au (wt/wt) | Particle dimension (sol) | Particle dimension (supported) |
|---|---|---|---|
| THPC | 0.8 | 4.0 | 8.6 |
| | 1 | 3.5 | 8.16 |
| | 2 | 2.71 | 4.2 |
| PVA | 0.125 | 3.2 | 4.1 |
| | 0.25 | 3.1 | 3.9 |
| | 0.5 | 2.6 | 3.6 |
| PEG | 0.3 | 5.3 | 14 |
| | 1 | 5.2 | 12 |
| $C_{12}E_{23}^a$ | 0.3 | 5.3 | 8.2 |
| | 1 | 5.1 | 7.0 |
| | 1.8 | 4.3 | 7.0 |
| Dextrin | 0.3 | 3.5 | 12 |
| | 1 | 3.6 | 8.4 |
| | 1.8 | 2.8 | 9.5 |
| PEU | 0.3 | 3.1 | 4.7 |
| PDDA | 0.1 | 3.6 | 27 |
| | 0.3 | 2.8 | 22 |
| | 0.65 | 2.6 | 15 |

$^a C_{12}E_{23}$ = polyoxyethylene dodecyl ether
Reprinted with permission from Ref. [58]. Copyright (2014) American Chemical Society

Carbon as a support has been deeply studied and it has been noted that it shows a high tendency to make particles grow with respect to the original size. The use of PVA-protected particles appears the best choice, ensuring in a quite large range of the PVA/Au ratio, a good stability of the gold particles' size. Normally, the more the protective agent amount is, the more the GNP size of the sol is maintained. However, a direct correlation was not yet found. Table 2.2 presents selected results of different sols immobilized on active carbon.

It should be mentioned that recent advances have been made in improving the metal dispersion on a support by surface functionalization. Jiang et al. reported first on the introduction of nitrogen functionalities on the surface of CNTs to increase the dispersion and stability of gold NPs during catalytic reactions [65].

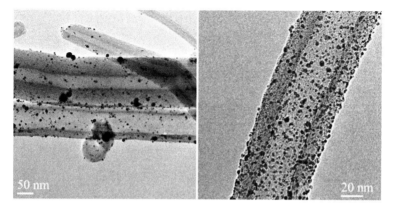

**Figure 2.12** TEM images of Au/CNFs (left) and Au/N-CNFs (right). Reprinted with permission from Ref. [58]. Copyright (2014) American Chemical Society.

More recently, Villa et al. also demonstrated that the introduction of nitrogen-containing groups by gas-phase amination favorably affected the dispersion on the support (Fig. 2.12 ) of PVA-protected GNPs [58, 62].

## 2.5.2 Grafting

Besides adsorption of GNPs on a support, grafting of colloids on a solid support is an alternative in immobilizing GNPs that normally produce a more ordered structure. This is explained by the creation of specific chemical bonds between the surface and the particle. Two different approaches can be followed. The first one consists of creating specific sites on the support that are able to react with functional groups on the protective layer. As an example, Fig. 2.13 reports the immobilization of gold particles bearing –COOH groups that can react with –NH$_2$ of the surface by creating an amide bond. The second approach was investigated by Wuelfing et al. [66].

The surface was functionalized with an appropriate molecule able to link by a coordinative bond the AuNPs. Typically, thiolates groups are employed for Au due to the high stability of the Au–S bond, as depicted in Fig. 2.14.

**64** | *Immobilization of Preformed Gold Nanoparticles*

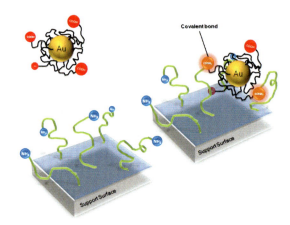

**Figure 2.13** Covalent immobilization of gold nanoparticles.

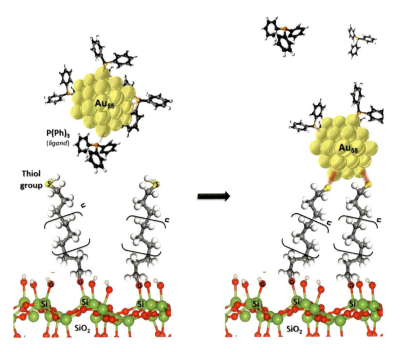

**Figure 2.14** Coordinated immobilization of gold colloid particles on a ligand-modified support.

## 2.6 Conclusions

Since the first report on the application of the sol immobilization method for preparing metal-supported catalysts, it has been claimed that the presence of a protective layer prolongs the catalyst life, preventing any metal particle coalescence [67]. Moreover, another advantage lies in the control that can be exerted on the particle size regardless (if protection is carefully checked) of the support. This aspect is of particular relevance for gold, which can often produce very large particles that which normally present very low catalytic activity.

In addition, it has been recently highlighted that there are other factors that should be taken into consideration. First, the quite obvious shielding effect that limits the catalytic activity and strictly depends on the amount and nature of the protective agent present on the surface [59]. Secondly but most importantly, the protective layer has been also shown to take part in the nanoparticles' location on the support surface [41] and in their shaping [18], actively affecting the activity and selectivity of the reaction [59]. Therefore, it should always be emphasized that the protective layer can actively participate in the reaction and therefore should be considered as an integral part of the catalyst. It has to be carefully checked in order to maximize not only the metal dispersion but also the selectivity of the reaction. Thermal treatment [68], washing procedure [59, 68, 69], or plasma oxygen [70] can be used to remove the capping agents; even these systems, at least partially, modify the metal nanoparticle structure.

## Acknowledgments

Sebastiano Campisi is gratefully acknowledged for graphical support.

## References

1. Haruta, M., Kobayashi, T., Sano, H., and Yamada, N. (1987). Novel gold catalysts for the oxidation of carbon-monoxide at a temperature far below 0°C, *Chem. Lett.*, pp. 405–408.

2. Haruta, M., Yamada, N., Kobayashi, T., and Iijima, S. (1989). Gold catalysts prepared by coprecipitation for low-temperature oxidation of hydrogen and of carbon monoxide, *J. Catal.*, **115**, pp. 301–309.
3. Hutchings, G.J. (1985). Vapor phase hydrochlorination of acetylene: correlation of catalytic activity of supported metal chloride catalysts, *J. Catal.*, **96**, pp. 292–295.
4. Prati, L., and Martra, G. (1999). New gold catalysts for liquid phase oxidation, *Gold Bull.*, **32**, pp. 96–101.
5. Roco, M.C. (1999). Nanoparticles and nanotechnology research, *J. Nanopart. Res.*, **1**, pp. 1–6.
6. Aiken III, J.D., Lin, Y., and Finke, R.G. (1996). A perspective on nanocluster catalysis: polyoxoanion and $(n\text{-}C_4H_9)_4N^+$ stabilized Ir(0)~300 nanocluster soluble heterogeneous catalysts, *J. Mol. Catal. A: Chem.*, **114**, pp. 29–51.
7. Porta, F., and Prati L. (2003). Pre-formed gold particle immobilised on supports: preparation and catalytic applications, *Recent Res. Dev. Vac. Sci. Technol.*, **4**, pp. 99–110.
8. Bradley, J.S. (1994). The chemistry of transition metal colloids, in *Cluster and Colloids: From Theory to Applications*, ed. Schmid, G. (Wiley-VCH, Weinheim), pp. 459–544.
9. Tan, Y., Li, Y., and Zhu, D. (2004). Noble metal nanoparticles, in *Encyclopedia of Nanoscience and Nanotechnology*, ed. Nalwa, H.S. (American Scientific, CA), pp. 9–40.
10. Andreescu, D., Sau, T.K., and Goia, D.V. (2006). Stabilizer-free nanosized gold sols, *J. Colloid Interface Sci.*, **298**, pp. 742–751.
11. Comotti, M., Della Pina, C., Matarrese, R., and Rossi, M. (2004). The catalytic activity of "naked" gold particles, *Angew. Chem., Int. Ed.*, **43**, pp. 5812–5815.
12. Turkevich, J., Stevenson, P.C., and Hillier J. (1951). A study of the nucleation and growth processes in the synthesis of colloidal gold, *Discuss. Faraday Soc.*, **11**, pp. 55–75.
13. LaMer, V.K., and Dinegar, R.H. (1950). Theory, production and mechanism of formation of monodispersed hydrosols, *J. Am. Chem. Soc.*, **72**, pp. 4847–4854.
14. Polte, J., Erler, R., Thunemann, A.F., Sokolov, S., Ahner, T.T., Rademann, K., Emmerling, F., and Kraehnert, R. (2010). Nucleation and growth of gold nanoparticles studied *via in situ* small angle X-ray scattering at millisecond time resolution, *ACS Nano*, **4**, pp. 1076–1082.

15. Ostwald, W. (1896). *Lehrbuch der Allgemeinen Chemie*, Vol. 2, Part 1 (Leipzig, Germany)
16. Tao, A.R., Habas, S., and Yang, P. (2008). Shape control of colloidal metal nanocrystals, *Small*, **4**, pp. 310–332.
17. Sau, T.K., and Murphy, C.J. (2004). Room temperature, high-yield synthesis of multiple shapes of gold nanoparticles in aqueous solution, *J. Am. Chem. Soc.*, **126**, pp. 8648–8649.
18. Njoki, P.N., Jacob, A., Khan, B., Luo, J., and Zhong, C.J. (2006). Formation of gold nanoparticles catalyzed by platinum nanoparticles, *J. Phys. Chem. B*, **110**, pp. 22503–22509.
19. Derjaguin, B.V., and Landau, L. (1941). Theory of the stability of strongly charged lyophobic sols and the adhesion of strongly charged particles in solutions of electrolytes, *Acta Physicochim. URSS*, **14**, pp. 633–662.
20. Verwey, E.J.W., and Overbeek, J.T.G. (1948). Theory of a single double layer, in *Theory of the Stability of Lyophobic Colloids. The Interaction of Sol Particles Having an Electric Double Layer* (Elsevier, New York), pp. 1–63.
21. Wieser, H.B. (1933). *Inorganic Colloid Chemistry* (John Wiley & Sons, New York); Kumar, A., Mandal, S., Selvakanna, P.R., Parischa, R., Mandal, L.E., and Sastry M. (2003). Investigation into the interaction between surface-bound alkylamines and gold nanoparticles, *Langmuir*, **19**, pp. 6277–6282; Leff, D.V., Brandt, L., and Heath, J.R. (1996). Synthesis and characterization of hydrophobic, organically-soluble gold nanocrystals functionalized with primary amines, *Langmuir*, **12**, pp. 4723–4730.
22. Evans, D.F., and Wennerstrom, H. (1994). *The Colloidal Domain* (VCH, New York).
23. Heller, W., and Pugh, T.L. (1960). Steric stabilization of colloidal solutions by adsorption of flexible macromolecule, *J. Polym. Sci.*, **47**, pp. 203–217.
24. Bontoux, J., Dauplan, A., and Marignan, R. (1969). Stabilisation des colloides mineraux. effet de la masse moleculaire du stabilisant, des dimensions micellaires et essais d'evaluation de la couche protectrice, *J. Chim. Phys. Phys. Chim. Biol.*, **66**, pp. 1259–1263.
25. Ortega-Vinuesa, J.L., Martín-Rodríguez, A., and Hidalgo-Älvarez, R. (1996). Colloidal stability of polymer colloids with different interfacial properties: mechanisms, *J. Colloid Interface Sci.*, **184**, pp. 259–267.
26. Porta, F., and Rossi, M. (2003). Gold nanostructured materials for the selective liquid phase catalytic oxidation, *J. Mol. Catal. A*, **204–205**, pp. 553–559.

27. Villa, A., Wang, D., Su, D.S., and Prati, L. (2009). Gold sols as catalysts for glycerol oxidation: the role of stabilizer, *ChemCatChem*, **1**, pp. 510–514.
28. Duff, D.G., Baiker, A., and Edwards, P.P. (1993). A new hydrosol of gold cluster, *J. Chem. Soc. Chem. Commun.*, pp. 96–98.
29. Porta, F., Prati, L., Rossi, M., Coluccia, S., and Martra G. (2000). Metal sols as a useful tool for heterogeneous gold catalyst preparation: reinvestigation of a liquid phase oxidation, *Catal. Today*, **61**, pp. 165–172.
30. Biella, S., Porta, F., Prati, L., and Rossi, M. (2003). Surfactant-protected gold particles: new challenge for gold-on-carbon catalysts, *Catal. Lett.*, **90**, pp. 23–29.
31. Bianchi, C.L., Porta, F., Prati, L., and Rossi, M. (2000). Selective liquid phase oxidation using gold catalysts, *Top. Catal.*, **13**, pp. 231–236; Porta, F., Prati, L., Rossi, M., and Scarì, G. (2002). New Au(0)sols as precursors for heterogeneous liquid-phase oxidation catalysts, *J. Catal.*, **211**, pp. 464–469.
32. Schmid, G., Pfeil, R., Boese, R., Bandermann, F., Meyers, S., Calis, G.H.M., and Van Der Velden, J.W.A. (1981). $Au_{55}[P(C_6H_5)_3]_{12}Cl_6$: ein Goldcluster ungewöhnlicher Größe, *Chem. Ber.*, **114**, pp. 3634–3642; Amiens, C., De Caro, D., Chaudret, B., Bradley, J.S., Mazel, R., and Roucau, C. (1993). Selective synthesis, characterization and spectroscopic properties of a novel class of reduced platinum and palladium particles stabilized by carbonyl and phosphine ligands, *J. Am. Chem. Soc.*, **115**, pp. 11638–11639.
33. Dassenoy, F., Philippot, K., Ould Ely, T., Amiens, C., Lecante, P., Snoeck, E., Mosset, A., Casanove, M.J., Chaudret, B. (1998). Platinum nanoparticles stabilized by CO and octanethiol ligands or polymers: FT-IR, NMR, HREM and WAXS studies, *New J. Chem.*, **22**, pp. 703–711.
34. Turkevich, J., and Kim, G. (1970). Palladium: preparation and catalytic properties of particles of uniform size, *Science*, **169**, pp. 873–879.
35. Turkevich, J. (1985). Colloidal gold. Part I, *Gold Bull.*, **18**, pp. 86–91.
36. Kimling, J., Maier, M., Okenve, B., Kotaidis, V., Ballot, H., and Plech, A. (2006). Turkevich method for gold nanoparticle synthesis revisited, *J. Phys. Chem. B*, **110**, pp. 15700–15707.
37. Schmid, G. (1992). Large clusters and colloids. Metals in the embryonic state, *Chem. Rev.*, **92**, pp. 1709–1727.
38. Nakao, Y., and Kaeriyama, K. (1986). Preparation of noble metal sols in the presence of surfactants and their properties, *J. Colloid Interface Sci.*, **110**, pp. 82–87.

39. Nalawade, P., Mukherjee, T., and Kapoor, S. (2013). Green synthesis of gold nanoparticles using glycerol as a reducing agent, *Adv. Nanopart.*, **2**, pp. 78–86; Kim, F., Connor, S., Song, H., Kuykendall T., and Yang, P. (2004). Platonic gold nanocrystals, *Angew. Chem., Int. Ed.,* **43,** pp. 3673–3677; Ayyappan, S., Srinivasa Gopalan, R., Subbanna, G.N., and Rao, C.N.R. (1997). Nanoparticles of Ag, Au, Pd, and Cu produced by alcohol reduction of the salts, *J. Mater. Res.*, **12**, pp. 398–401,

40. Bönnemann, H., Brijoux, W., Brinkmann, R., Joußen, T., Korall, B., and Dinjus, E. (1991). Formation of colloidal transition metals in organic phases and their application in catalysis, *Angew. Chem., Int. Ed. Engl.,* **30**, pp. 1312–1314; Bönnemann, H., Brijoux, W., Brinkmann, R., Dinjus, E., Fretzen, R., Joußen, T., and Korall, B. (1992). Highly dispersed metal clusters and colloids for the preparation of active liquid-phase hydrogenation catalysts, *J. Mol. Catal.*, **74**, pp. 323–333; Bönnemann, H., and Nagabhushana, K.S. (2004). Tunable synthetic approaches for the optimization of nanostructured fuel cell catalysts: an overview, *Chem. Ind.*, **58**, pp. 271–279.

41. Villa, A., Chan-Thaw, C.E., and Prati, L. (2010). Au NPs on anionic-exchange resin as catalyst for polyols oxidation in batch and fixed bed reactor, *Appl. Catal. B: Environ.,* **96**, pp. 541–547.

42. Handley, D.A. (1989). The development and application of colloidal gold as a microscopic probe, in*Colloidal Gold: Principles, Methods and Applications,* ed. Hayat, M.A. (Academic, San Diego, CA), pp. 1–11.

43. Wang, X., Zhuang, J., Peng, Q., and Li, Y. (2005). A general strategy for nanocrystal synthesis, *Nature*, **437**, pp. 121–124.

44. Tan, C.K., Newberry, V., Webb, T.R., and McAuliffe, C.A. (1987). Water photolysis. Part 2. An investigation of the relative advantages of various components of the sensitiser–electron relay–metal colloid system for the photoproduction of hydrogen from water, and the use of these systems in the photohydrogenation of unsaturated organic substrates, *J. Chem. Soc., Dalton Trans.*, pp. 1299–1303.

45. Merga, G., Saucedo, N., Cass, L.C., Puthussery, J., and Meisel, D. (2010). "Naked" gold nanoparticles: synthesis, characterization, catalytic hydrogen evolution, and SERS, *J. Phys. Chem. C*, **114**, pp. 14811–14818.

46. Sun, K., Qiu, J., Liu, J., and Miao, Y. (2009). Preparation and characterization of gold nanoparticles using ascorbic acid as reducing agent in reverse micelles, *J. Mater. Sci.*, **44**, pp. 754–758.

47. Khan, Z., Singh T., Hussain, J.I., and Hashmi, A.A. (2013). Au(III)-CTAB reduction by ascorbic acid: preparation and characterization of gold nanoparticles, *Colloids Surf. B: Biointerfaces*, **104**, pp. 11–17.

48. Kopple, K., Meyerstein, D., and Meisel, D. (1980). Mechanism of the catalytic hydrogen production by gold sols. H/D isotope effect studies, *J. Phys. Chem.*, **84**, pp. 870–875.
49. Reetz M.T., and Helbig, W. (1994). Size-selective synthesis of nanostructured transition metal clusters, *J. Am. Chem. Soc.*, **116**, pp. 7401–7402.
50. Yu, Y.Y., Chang, S.S., Lee, C.L., and Wang, C.R.C. (1997). Gold nanorods: electrochemical synthesis and optical properties, *J. Phys. Chem. B*, **101**, pp. 6661–6664.
51. van der Zande, B.M.I., Bohmer, M.R., Fokkink, L.G.J., and Schonenberger, C. (2000). Colloidal dispersions of gold rods:? synthesis and optical properties, *Langmuir*, **16**, pp. 451–458.
52. Okitsu, K., Ashokkumar, M., and Grieser, F. (2005). Sonochemical synthesis of gold nanoparticles: effects of ultrasound frequency, *J. Phys. Chem. B,* **109**, pp. 20673–20675.
53. Tsuji, M., Hashimoto, M., Nishizawa, Y., Kubokawa, M., and Tsuji, T. (2005). Microwave-assisted synthesis of metallic nanostructures in solution, *Chem. Eur. J.*, **11**, pp. 440–452.
54. Mafune, F., Kohno, J.Y., Takeda, Y., and Kondow, T. (2001). Formation of gold nanoparticles by laser ablation in aqueous solution of surfactant, *J. Phys. Chem. B*, **105**, pp. 5114–5120.
55. Han M.Y., and Quek, C.H. (2000). Photochemical synthesis in formamide and room-temperature Coulomb staircase behavior of size-controlled cold nanoparticles, *Langmuir,* **16**, pp. 362–367.
56. Prati, L., Villa, A., Lupini, A.R., and Veith, G.M. (**2012**). Gold on carbon. One billion catalysts under a single label, *Phys. Chem. Chem. Phys.*, **14**, pp. 2969–2978.
57. Villa, A., Schiavoni, M., Prati, L. (2012). Material science for the support design: a powerful challenge for catalysis, *Catal. Sci. Technol.*, **2**, pp. 673–682.
58. Prati, L., and Villa, A. (2014). Gold colloids: from quasi-homogeneous to heterogeneous catalytic systems, *Acc. Chem. Res.*, **47**, pp. 855–863.
59. Villa, A., Wang, D., Veith, G.M., Vindigni, F., and Prati, L. (2013). Sol immobilization technique: a delicate balance between activity, selectivity and stability for gold catalyst, *Catal. Sci. Technol.*, **3**, pp. 3036–3041.
60. Villa, A., Chan-Thaw, C.E., Veith, G.M., More, K.L., Ferri D., and Prati, L. (2011). Au on nanosized NiO: a cooperative effect between Au and nanosized NiO in the base-free alcohol oxidation, *ChemCatChem*, **3**, pp. 1612–1618.

61. Villa, A., Schiavoni, M., Campisi, S., Veith, G.M. Prati, L. (2013). Pd-modified Au on carbon as an effective and durable catalyst for the direct oxidation of HMF to 2,5-furandicarboxylic acid, *ChemSusChem*, **6**, pp. 609–612.
62. Prati, L., Villa, A., Chan-Thaw, C.E., Arrigo, R., Wang, D., and Su, D.S. (2011). Gold catalyzed liquid phase oxidation of alcohol: the issue of selectivity, *Faraday Discuss.*, **152**, pp. 353–365.
63. Wang, D., Villa, A., Su, D., Prati, L., and Schlögl, R. (2012). Carbon-supported gold nanocatalysts: shape effect in the selective glycerol oxidation, *ChemCatChem*, **9**, pp. 2717–2723.
64. Porta, F., and Prati, L. (2008). Gold colloidal nanoparticles sized to be suitable precursors for heterogeneous catalyst, in*Metal Nanocluster in Catalysis and Materials Science: The Issue of Size Control,* eds. Corain, B., Schmid, G., and Toshima, N. (Elsevier, Amsterdam), pp. 355–360.
65. Jiang, K., Eitan, A., Schadler, L.S., Ajayan, P.M., and Siegel, R.W. (2003). Selective attachment of gold nanoparticles to nitrogen-doped carbon nanotubes, *Nano Lett.*, **3**, pp. 275–277.
66. Wuelfing, W.P., Gross, S.M., Miles, D.T., and Murray, R.W. (1998). Nanometer gold clusters protected by surface-bound monolayers of thiolated poly(ethylene glycol) polymer electrolyte, *J. Am. Chem. Soc.*, **120**, pp. 12696–12697.
67. Bönnemann, H., and Richards, R.M. (2001). Nanoscopic metal particles: synthetic methods and potential applications, *Eur. J. Inorg. Chem.*, **1**, pp. 2455–2480.
68. Grunwaldt, J.D., Kiener, C., Wögerbauer C., and Baiker, A. (1999). Preparation of supported catalysts for low temperature CO oxidation via "size-controlled" gold colloids", *J. Catal.*, **181**, pp. 223–232.
69. Lopez-Sanchez, J.A., Dimitratos, N., Hammond, C., Brett, G.L., Kesavan, L., White, S., Miedziak, P., Tiruvalam, R., Jenkins, R.L., Carley, A.F., Knight, D., Kiely C.J., and Hutchings, G.J. (2011). Facile removal of stabilizer-ligands from supported gold nanoparticles, *Nat. Chem.*, **3**, pp. 551–556.
70. Gun, J., Rizkov, D., Lev, O., Abouzar, M.H., Poghossian, A., and Schöning, M.J. (2009). Oxygen plasma-treated gold nanoparticle-based field-effect devices as transducer structures for bio-chemical sensing, *Microchim. Acta*, **164**, pp. 395–404.

# Chapter 3

# Solvated Metal Atoms in the Preparation of Supported Gold Catalysts

**Claudio Evangelisti,[a] Eleonora Schiavi,[b] Laura Antonella Aronica,[b] Rinaldo Psaro,[a] Antonella Balerna,[c] and Gianmario Martra[d]**

[a]*CNR, Institute of Molecular Science and Technologies, Via G. Fantoli 16/15, 20138 Milano, Italy*
[b]*Department of Chemistry and Industrial Chemistry, University of Pisa, Via Risorgimento 35, 56126 Pisa, Italy*
[c]*INFN—Frascati National Laboratories, Via E. Fermi 40, 00044 Frascati, Roma, Italy*
[d]*Department of Chemistry IFM and NIS Interdipartimental Centre of Excellence, Via P. Giuria 7, 10125 Torino, Italy*
claudio.evangelisti@istm.cnr.it

## 3.1 Introduction

Several reviews appeared in the last years to emphasize the unique properties showed by gold metal when the particle size falls in the nanometer range [1–5]. Au nanoparticles (NPs) have found uses in ceramics, medicine, and other areas. Among them the most exciting and growing field of application is undoubtedly catalysis. Historically, gold was regarded to be catalytically inert, since the

---

*Gold Catalysis: Preparation, Characterization, and Applications*
Edited by Laura Prati and Alberto Villa
Copyright © 2016 Pan Stanford Publishing Pte. Ltd.
ISBN 978-981-4669-28-3 (Hardcover), 978-981-4669-29-0 (eBook)
www.panstanford.com

discoveries made by Haruta [6] and Hutchings [7] in the late 1980s. Surprisingly, they demonstrated simultaneously and independently that supported gold NPs are the best catalyst for low-temperature CO oxidation and ethyne chlorination to vinyl chloride. In recent years, it has been shown that gold becomes active for many novel reactions of synthetic interest when stabilized in the form of NPs deposited on several organic and inorganic supports [2, 4]. Supported Au NPs have found numerous applications as unique catalysts in aerobic oxidative processes [8–11], reduction of organic compounds [12, 13], and C–C coupling reactions [14]. Moreover, the ability of gold to coordinate with triple bonds has no parallel with other transition metals. Upon coordination and formation of the corresponding adduct, the alkyne becomes activated and more reactive toward nucleophiles such as alcohols [15], amines [16], and hydrosilanes [17].

It was clearly evidenced that particle size plays a crucial role in determining the catalytic activity of supported gold particles in CO oxidation, as well as in other reactions. On the other hand, it has been also demonstrated that the surface oxidation state of the metal, the nature of the support, the Au–support interface, and the particle morphology may strongly affect catalytic performance of Au NPs. All these findings have increased the efforts made by scientists to investigate how the different preparation methods can affects the above-mentioned factors.

In this chapter we intend to provide a synapsis of the synthesis of supported gold catalysts by the metal vapor synthesis (MVS) technique. In particular, the contribution includes the work undertaken till date in our and other research groups pointing out the key factors to control the size of Au NPs by this synthetic approach.

## 3.2 Synthetic Strategy

Among various preparative routes, the MVS technique provides a valuable tool to weakly stabilized nanostructured homo- and heterometallic particles [18, 19].

Metal vapors are co-condensed at low temperature (77 K) with vapors of weakly stabilizing organic ligands (such as acetone, toluene, tetrahydrofuran (THF), acetonitrile, etc.) using commercially available reactors [20, 21]. Upon warming the frozen matrix melts and nucleation and growth processes of the metal particles take place (Scheme 3.1), affording metal nanoclusters weakly

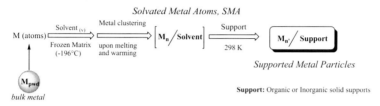

**Scheme 3.1**

stabilized by the solvent molecules, named solvated metal atoms (SMAs) [22]. The interaction of the metal vapors with the solvent matrix quenches very quickly the kinetic energy of metal atoms. As a result, during the further melting stage, metal NPs with unusual distorted geometries, thermodynamically not stable, are formed [23]. The final size of metal aggregates is greatly influenced by the solvent employed and amount used, allowing one to exert a good control on their size. SMAs are generally handled at low temperature (223–243 K) under an inert atmosphere and they are suitable precursors to prepare mono- and heterometallic supported NPs simply by mixing the SMAs with a solid support. The method is applied, as an alternative to the well-known traditional methods, to the deposition of metals on a different kind of supports, which include pristine inorganic materials in powder form (e.g., silica, alumina, carbon, organic polymers) [19, 24] or monolith supports, such as ceramic membranes [25].

The effectiveness of the deposition technique does not strongly depend on the material of the support or on the metal that can be employed, provided the choice of suitable operating conditions, like vaporization temperature or stabilizing solvent. The metal quickly separates from the solvent, by interacting with the support surface, affording solid catalytic systems without poisons (i.e., halide from metal salt precursors in catalysts prepared by reduction methods)

**Figure 3.1** Multi-electrode static glass reactor for simultaneous vaporization of two metals (A), example of condensation of solvent vapors (B), and co-condensation of metal vapors with solvent vapors (C).

and the presence of several defect sites (kinks, terraces, adatoms, etc.) due to the already mentioned distorted geometries.

The MVS approach allows us to prepare supported metal systems where the metal is deposited directly in its reduced form so that calcination and activation processes of the conventional wet deposition method are not required. The preparation of supported heterometal catalysts is affordable either by depositing two metals sequentially from two different metal atom solutions or by vaporizing two metals at the same time in a multi-electrode reactor (Fig. 3.1) [26], employing a suitable stabilizing solvent or a mixture of them.

The development of metal vapor routes to supported metal particles is often limited by the apparatus; this synthetic procedure can be conducted on lab scale (vaporization of 50/500 mg of metal) either in a rotating [27] or in a static glass reactor [28] (Fig. 3.1). Metal vapors are commonly produced by resistance heating or by electron beam vaporization. Recently stainless steel reactors developed by different research groups allow the preparation of

SMAs on a larger scale (vaporization of 1/10 g of metal) have been developed [29, 30].

## 3.3 Metal Vapor Synthesis for Preparing Supported Gold Nanoparticles

### 3.3.1 *Preparation of Solvated Gold Atoms*

According to the above-described method, gold in the form of a powder or pellets is evaporated under vacuum and deposited on the frozen walls (77 K) of a glass reactor simultaneously with a large excess of vapors of an organic solvent. In the vapor phase the gold is present mostly as a mono- or oligoatomic species and in this form is trapped in the solid matrix. During the warmup stage Au atoms can diffuse in the solid and particle nucleation and growth processes take place; this is clearly visible by changes in the color of the matrix. During further melting of the solid matrix Au atoms are very mobile, and reactions with the solvent can compete with the clustering process. As Au–Au bonds are formed, each resultant cluster may have a different reactivity with the solvent. Depending on different experimental parameters such as the kind of solvent used or the solvent/Au molar ratio, eventually a stage must be reached where it is energetically costly to move two clusters together and de-solvate them to allow further Au–Au bond formation. As a result, Au NPs remain solvated and suspended in solution by Brownian motion, leading to Au SMAs, which can be considered a kinetically stable dispersion of metal NPs.

Klabunde and co-workers reported extensive works on the use of this technique to prepare colloidal Au solutions [31]. They noted that polar organic solvents such as acetone, butanone, and Tetrahydrofuran (THF) allow us to obtain Au SMAs red-purple in color usually stable at room temperature, whereas nonpolar organics and water lead to large gold particles that precipitated after isolation.

Other solvents or solvent mixtures, such as acetone–water mixtures, toluene, and methylcyclohexane, were also reported to afford Au SMAs stable at low temperatures [32, 33]. In particular, the bis(toluene) gold (0) complex prepared by this way is reported to be extremely air sensitive and thermally unstable, decomposing into gold (0) and toluene at about 173 K. Polystyrene or poly(methyl methacrylate) polymers containing Au NPs ranging from 2 to 15 nm in diameter were obtained by co-condensation of Au vapors with the corresponding monomers followed by their further polymerization [34].

Ultraviolet-visible (UV-Vis) spectra of the colloidal Au SMAs show broad absorptions centered about at 520–560 nm, which is characteristic of gold nanocrystals less than 10 nm in size and due to the collective excitation of electrons confined in the metal NPs.

The influence of the gold content on the sizes of the Au NPs in SMA solutions has been deeply investigated [22, 35]. Transmission electron microscopy (TEM) micrographs obtained by evaporating acetone from Au/acetone SMAs containing low concentration (LC, 0.1/0.5 Au mg/mL, Scheme 2) of Au show small particles mainly in form of individual crystallites with very narrow size distribution (1.0/3.0 nm) (Fig. 3.2). On the other hand, initial high concentration (HCs, >1.0 Au mg/mL, Scheme 3.2) of Au afford larger particles with broader metal particle sizes (2.0/20.0 nm).

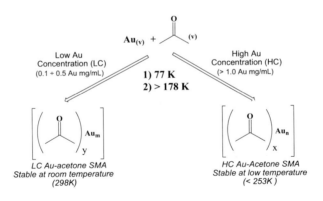

Scheme 3.2

Both HC and LC Au–acetone SMAs show Au NP morphologies that are predominantly comprised of highly defective spherical particles. Higher magnification of samples derived from HC Au–acetone SMAs (Fig. 3.3) show both the presence of individual

*Metal Vapor Synthesis for Preparing Supported Gold Nanoparticles* | 79

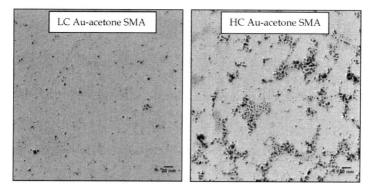

**Figure 3.2** Typical electron micrographs of gold nanoparticles dispersed on lacey carbon films obtained starting from (A) LC and (B) HC Au–acetone SMAs.

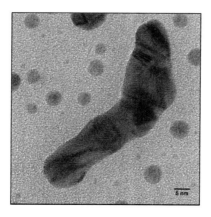

**Figure 3.3** High-resolution electron micrographs of gold nanoparticles dispersed on lacey carbon films obtained starting from HC Au–acetone SMAs.

crystallites, with a slightly broader particle size distribution than LC Au SMAs, and a predominant amount particles held together to form larger aggregates. The presence of these large aggregates most probably results from residual stresses accumulated during the chaotic nucleation and growth steps during the warm stage of the vaporization method.

The initial concentration of the Au–acetone SMAs greatly affects the color of the solution: passing from LC Au SMAs to the more concentrated ones the pink-brown color of the co-condensate becomes more intense until it results, in the case of HC Au SMAs, in retaining only a slight dark-purple color.

Mass spectrometry measurements carried out on an LC Au/acetone SMA using an ion source matrix-assisted laser desorption (MALD) with a time-of-flight (TOF) analyzer show a broad maximum centered at m/z 4857 upon which is superimposed a fine structure [36]. The observed spectrum is consistent with many possible combinations of Au and the ketone, suggesting the presence of very small gold clusters ($Au_{10-20}$), stabilized by acetone, having a particle diameter at the lower end of the size range determined by high-resolution transmission electron microscopy (HRTEM).

To study the effect of initial concentration of the solution on particle size, an Au–acetone SMA, containing different Au concentration ratios, was quenched by adding suitable stabilizing agents in order to prevent the aggregation of individual particles. Long-chain aliphatic amines, such as decylamine, appeared to be a useful ligand to quench the aggregation phenomenon, achieving a size-selective synthesis of Au NPs [37, 38]. The addition of large molar excesses of decylamine to the Au SMA gave a gray-purple powder of amine-capped gold NPs, which was conveniently isolated by adding ethanol. Dopamine (DA)-capped Au NPs are insoluble in acetone but can be easily redissolved in nonpolar solvents (toluene, benzene, *n*-pentane, etc.).

Starting from the LC Au–acetone SMA, DA-capped Au particles with a mean diameter of 2.8 nm were obtained. On the other hand, larger Au particles (mean diameter 5.8 nm) were achieved starting from HC SMA.

UV-Vis spectra of the two systems, named as A and B derived from LC and HC Au–acetone SMAs, respectively, show a broad absorption centered about at 530 nm, which is characteristic of the plasmon absorption resonance of gold nanocrystals (Fig. 3.4).

An increase in the intensity of the surface plasmon absorption resonance for the more concentrated solution (B) was detected accordingly with the TEM analysis in the presence of larger particles in an HC Au–acetone SMA.

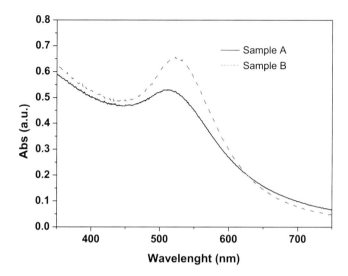

**Figure 3.4** UV-Vis absorbance assay of samples A and B prepared by addition of DA to Au–acetone SMAs with different Au concentrations: LC (sample A, straight line) and HC (sample B, dash line).

To gain a better understanding of the stability toward the aggregation at room temperature (298 K) of LC and HC Au–acetone SMAs, the DA ligand was added to portions of Au SMAs at different times spanning over 23 h and Au particle sizes were evaluated by TEM analysis. TEM micrographs on samples prepared from LC Au–acetone SMAs showed the presence of Au particles with mean diameters of 3.0 nm ± 1 nm that did not change significantly over time. On the other hand, TEM micrographs of samples obtained by the addition of DA to an HC Au–acetone SMA kept at 298 K, after $t = 3$ h and $t = 5$ h, evidenced the growth of gold NPs in solution ($d_\mathrm{m} = 6.5 \pm 1.3$ nm and $d_\mathrm{m} = 7.8 \pm 1.9$ nm, respectively). The increase in size over the aggregation time of Au–acetone SMAs at room temperature (298 K) was also accompanied by a significant widening of size distributions.

Recently, Klabunde and co-workers reported a reproducible gram-scale synthesis of ligand-stabilized gold NPs by thermal treatment of HC Au–acetone or Au–butanone SMAs with strong ligands such as thiols, amines, phosphines, and silanes [39, 40].

The combination of Au SMAs and the ligand dissolved in a toluene solution, at controlled times and temperatures, named *digestive ripening*, results in a great improvement of the size distribution of the Au particles and their subsequent organization in 2D and 3D superlattices. By this way particles of HC Au–acetone SMAs with sizes ranging from 1 to 20 nm are transformed into an almost monodisperse colloid with particle sizes of about 4.5 ± 0.4 nm.

### 3.3.2 Supported Catalysts from Solvated Gold Atoms

Solvated gold atoms obtained by the MVS technique can be employed as a source of gold NPs to obtain highly dispersed zero-valent supported systems. The method allows us to deposit Au NPs on a wide range of organic or inorganic supports with different sizes and metal loadings by simple impregnation of a suitable amount of support with Au SMAs at room temperature. The decolorization of the suspension is indicative of the complete deposition of the Au particles on the inorganic matrices. As already mentioned, in the case of gold-supported catalysts the dimensions of the particles are quite important since it is now well established that the catalytic activity of Au depends on a large extent on the size of the NPs, but other effects such as the nature of the support material, the Au/support interface, the particle shape, and the loading of the metal on the inorganic matrix may also be of great importance [41–43]. For these reasons the MVS approach for the synthesis of supported Au systems can be a valid alternative to traditional synthetic approaches, avoiding the use of oxidized gold chloride derivatives as precursor compounds and the eventual effect of residual chlorine ions on the catalytic behavior of the catalyst.

HRTEM analysis of samples prepared from LC Au SMAs deposited on carbon and $\gamma$-alumina supports, respectively (Fig. 3.5) revealed a homogeneous dispersion of the particles and a narrow distribution of sizes with an average particle diameter close to 4.5 nm.

The histograms of the metal particle distribution of supported samples containing different Au loadings show clearly that a large increase of the amount of the supported metal did not cause much variation of the Au particle sizes. As an example, in the case of alumina-supported Au samples the mean particle diameters of the

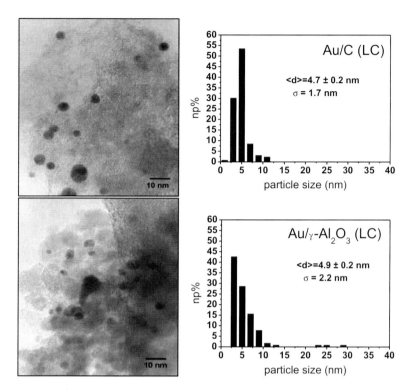

**Figure 3.5** HRTEM micrographs and particle size distributions of supported gold nanoparticles, 0.10 wt%, on carbon (top) and $\gamma$-alumina (bottom), starting from a concentration of 0.16 mg/ml (LC). Reprinted with permission from *J. Catal.* **266** (2009) 250–257. Copyright 2009 Elsevier.

particles increase from 4.2 to 4.8 nm, ranging from 0.10 Au wt% to 2.0 wt% (Fig. 3.6). This small increase in the mean particle sizes can be rationalized with the slight phenomenon of clustering during the Au deposition on the support, which is enhanced at higher metal loadings.

The microscopy analysis results confirmed that NPs essentially smaller than 10 nm in size and homogeneously dispersed on $\gamma$-$Al_2O_3$ and carbon can be obtained if LC Au–acetone SMAs are chosen as the starting material.

Extended X-ray absorption fine structure (EXAFS) spectroscopy [26] was also used to confirm the mean dimension of the 1.0 Au

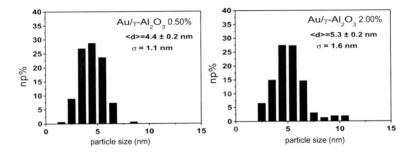

**Figure 3.6** Gold particle size distributions of supported gold nanoparticles, 0.50 wt% and 2.00 wt%, respectively, on γ-alumina, starting from an Au concentration of 0.16 mg/mL (LC).

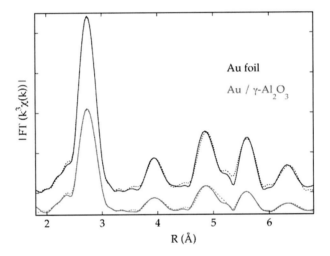

**Figure 3.7** Amplitudes of the Fourier transforms of the experimental EXAFS signals and best-fitting simulated signals (dotted lines).

wt% sample and determine its crystallographic structure. Using the results achieved by fitting the Au $L_3$ edge EXAFS spectra, it was determined that the gold NPs, like the Au reference sample, had a face-centered cubic (fcc) structure given also by the expected ratio between the second shell distance and the first one equal to 1.41. In Fig. 3.7 the Fourier transforms of the EXAFS spectra of the Au reference sample and of the 1.0 Au wt% one are reported. The reduction of the amplitude of the peaks of the 1.0 Au wt%

**Table 3.1** Hydrosilylation reaction of 1-hexyne

$$R_3SiH + nBu-\equiv \xrightarrow{[Au]} \underset{\beta(Z)}{\overset{nBu}{\underset{H}{\rightthreetimes}}\overset{H}{\underset{SiR_3}{}}} + \underset{\beta(E)}{\overset{nBu}{\underset{H}{\rightthreetimes}}\overset{H}{\underset{SiR_3}{}}} + \underset{\alpha}{\overset{nBu}{\underset{R_3Si}{\rightthreetimes}}\overset{H}{\underset{H}{}}}$$

| Catalyst | R₃SiH | β-(E) / α(%) | | SA[a] |
|---|---|---|---|---|
| Au/C | Cl₂MeSiH | / | / | / |
| | (EtO)₃SiH | 98 | 2 | 341 |
| | Et₃SiH | 97 | 3 | 1662 |
| | Me₂PhSiH | 98 | 2 | 7012 |
| Au/Al₂O₃ | Cl₂MeSiH | / | / | / |
| | (EtO)₃SiH | 92 | 8 | 1770 |
| | Et₃SiH | 93 | 4 | 6440 |
| | Me₂PhSiH | 99 | 1 | 8000 |
| Pt/C | Cl₂MeSiH | 95 | 5 | 4166 |
| | (EtO)₃SiH | 78 | 22 | 3571 |
| | Et₃SiH | 86 | 14 | 867 |
| | Me₂PhSiH | 79 | 21 | 2267 |

[a] Specific activity calculated as moles of silane converted/ moles of metal per hour.

sample is related to the reduction of the coordination numbers and was used to estimate the average NP diameter: assuming spherical NPs an average diameter of 4.6 nm was found, in agreement with the previously reported HRTEM results. Also the reduction in the interatomic distances and the increase in the Debye Waller factor given by the fitting procedure were compatible with the values reported for Au NPs of such mean size [44, 45].

Au/C and Au/γ-Al₂O₃ (0.10 Au wt%) samples prepared starting from LC Au–acetone SMA efficiently promote the highly regio- and stereoselective hydrosilylation of terminal acetylenes (Table 3.1).

The MVS Au samples clearly favor the formation of the *trans* isomer β-(E) that is almost the only product obtained. Moreover, the catalytic properties (expressed as specific activity) of the carbon-supported gold NPs are strictly related to the structure of the hydrosilane. Me₂PhSiH resulted as the most reactive compound, while very polarized silanes such as (EtO)₃SiH showed low activity and chloro-functionalized reagents yielded no products. Interestingly both Au/C and Au/γ-Al₂O₃ systems exhibited exactly the same order of reactivity, indicating that the peculiar catalytic behavior of the

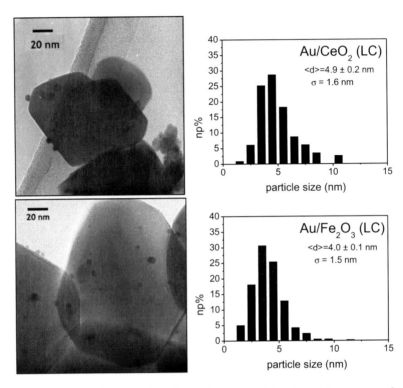

**Figure 3.8** HRTEM micrographs and gold particle size distributions of supported gold nanoparticles, 0.10 wt % on CeO$_2$ and Fe$_2$O$_3$, prepared from an LC Au SMA. Reprinted with permission from *J. Catal.* **266** (2009) 250–257. Copyright 2009 Elsevier.

Au catalyst is independent on the type of inorganic support. These data are in contrast with the well-known affinity between platinum catalysts and chlorosilane. Indeed using a MVS-derived Pt/C sample (0.10 wt%) a reverse order of specific activity has been observed and a considerable loss of regioselectivity detected.

Representative HRTEM images and particle size distributions of gold NPs supported on CeO$_2$ and Fe$_2$O$_3$, as well as on TiO$_2$ and ZrO$_2$, prepared starting from LC Au–acetone SMA, are reported in Fig. 3.8 and Fig. 3.9, respectively. The dispersion of Au on the support appeared highly homogeneous for all the samples considered, and the size of the particles lay mostly between 2.0 and 8.0 nm. The resulting size distributions exhibited some common features along

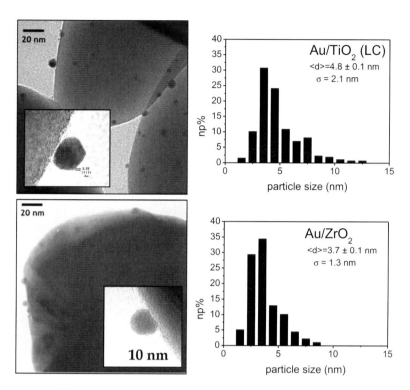

**Figure 3.9** HRTEM micrographs and gold particle size distributions of supported gold nanoparticles, 0.10 wt % on TiO₂ and ZrO₂, prepared from an LC Au SMA. Inset: Details showing the particle–support interaction. Reprinted with permission from *J. Catal.* **266** (2009) 250–257. Copyright 2009 Elsevier.

the series of the four materials: quite narrow (between 1.5 and 2.1 nm) around the mean value, which lay in the 3.7/4.9 nm range, and slightly asymmetric toward the larger sizes, especially for the Au/TiO₂ system. In this last case, general, HRTEM micrographs revealed a hemispheric/polyhedral shape for the gold NPs deposited on the metal oxides, with a quite wide area of interaction with the support (see inset in Fig. 3.9). The results indicate that the dimensions of the gold NPs are only slightly affected by the nature of the inorganic matrices chosen for the preparation of the supported catalysts. Moreover, as reported above for the $\gamma$-Al₂O₃ support, Fe₂O₃- and TiO₂-supported systems containing a higher gold weight percentage, showing only a slightly increase in particle size distribution, can be obtained.

Venezia et al. and Casaletto et al. reported the preparation of supported gold NPs by different synthetic approaches, including MVS, deposition-precipitation, and co-precipitation techniques [46, 47]. The influence of the different preparation techniques on metal–support interactions and structural and electronic properties of the gold-based catalysts was studied, with particular attention to their application in CO oxidation reactions. X-ray photoelectron spectroscopy (XPS) analysis of as-prepared Au/CeO$_2$ samples obtained from Au/acetone SMAs presented a main Au 4f$_{7/2}$ component at a binding energy of 84.4 eV, typical of metallic gold clusters. Interestingly, the signals of a just-dried sample and after calcination did not change. Further XPS, X-ray diffraction (XRD), and X-ray absorption near-edge structure (XANES) studies reported by the research group on Au/SiO$_2$ samples prepared by the same way gives evidence of the presence of small metallic gold clusters (ca. 2 nm) with a typical fcc structure highly dispersed on the silica matrix.

Recently Klabunde and co-workers reported the synthesis of Au NPs supported on different types of TiO$_2$ (anatase, rutile, and P25, a mixture of 75% anatase and 25% rutile) starting from Au–butanone SMAs [48, 49]. The systems were active as photocatalysts toward photocatalytic hydrogen production under UV-Vis and visible conditions. TEM bright-field images exhibited Au NPs of size ranging 1 to 8 nm. The small changes in gold average size were observed depending on the surface area of TiO$_2$ support: the authors conclude that gold NP growth is limited by the TiO$_2$ surface available.

TiO$_2$-supported Au-containing catalysts derived from Au–toluene SMAs were also reported [33]. By this way Au/TiO$_2$ systems containing small zero-valent Au particles ($d_{Au} = 1/4$ nm) were obtained that were active in CO oxidation at low temperature. These systems showed a good thermal stability below 303 K, whereas an increase in the average particle size, corresponding to a decrease in catalytic activity, was observed by increasing the temperature (333 K).

The comparison of the TEM results obtained for Au/C and Au/$\gamma$-Al$_2$O$_3$ samples prepared starting from HC Au/acetone SMAs (Fig.

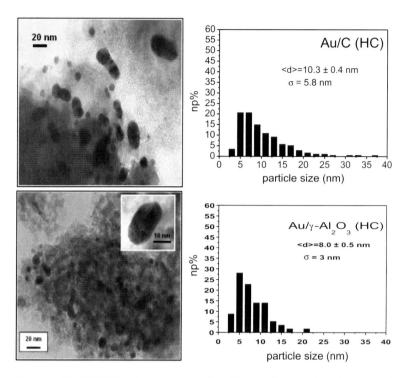

**Figure 3.10** HRTEM micrographs and gold particle size distributions of supported gold nanoparticles, 0.10 wt%, on carbon and $\gamma$-alumina, starting from a concentration of 1.42 mg/mL (HC). Reprinted with permission from *J. Catal.* **266** (2009) 250–257. Copyright 2009 Elsevier.

3.10) with those for the corresponding samples obtained from LC Au SMAs (Fig. 3.5) shows clearly a strong dependence of the deposition process and the morphology of the NPs from the initial Au concentration of SMAs. Samples derived from HC solutions exhibit a notably wider distribution of metal particle size (4.0–20 nm), with a significant fraction of particles larger than 10 nm in size. Furthermore, the large particles appeared to be polycrystalline in nature, as shown in the inset of Fig. 3.10 (bottom). The system Au/C (HC) 0.1% shows a more even distribution of the particles compared to the same system supported on alumina, but also in this case the size of particles are very high, because most of them have sizes ranging between 2 and 20 nm, and they are numerous with a

**Table 3.2** Hydrosilylation reaction of 1-hexyne

| Catalyst | β-(E) / α |  | SA[a] |
|---|---|---|---|
| Au/γ-Al$_2$O$_3$ (LC) | 96 | 4 | 6480 |
| Au/C (LC) | 98 | 2 | 5680 |
| Au/γ-Al$_2$O$_3$ (HC) | – | – | 0 |
| Au/C (HC) | – | – | 0 |

[a] Specific activity calculated as moles of silane converted/moles of metal per hour.

diameter between 20 and 40 nm, while those with less than 5 nm represent only 13% of the total.

The comparison of LC and HC Au/γ-Al$_2$O$_3$ and Au/C (0.10 Au wt%) systems in the hydrosilylation of 1-hexyne and triethylsilane, as a model reaction, indicated a strong influence of the metal particle size distribution on the catalytic activity of supported gold catalysts in this reaction (Table 3.2).

Indeed, while as already discussed supported Au samples derived from LC SMAs showed very high activity and regioselectivity toward the β-(E) isomer, the analogous samples derived from HC SMAs, containing particles of larger diameter (4.0–20.0 nm), are completely inactive, even after long reaction times. These results confirm the fundamental correlation between the NPs' sizes and the catalytic activity of gold, already observed by Haruta in his pioneer work on CO oxidation [6].

## 3.4 MVS-Derived Au–PD Bimetallic Catalysts

Recently the MVS technique has been reported as an alternative procedure for the preparation of supported Au–Pd bimetallic systems by using Pd and Au vapors as reagents [26]. Au and Pd vapors were co-condensed simultaneously with acetone vapors in order to obtain a close interaction between the two metals, leading to a bimetallic [Au–Pd]–acetone SMA. The [Au–Pd] SMAs, stable at low temperature (258 K), are used as the starting material to deposit Au–Pd bimetallic NPs on different supports (alumina, titania, carbon) (Scheme 3.3).

**Scheme 3.3**

HRTEM analysis of a sample prepared from [Au–Pd] SMAs containing a Pd/Au molar ratio of 1:1, deposited on a $\gamma$-alumina support, is reported in Fig. 3.11. The sample appeared densely populated by nanometric metal particles ranging from 1.0 nm to 6.0 nm in size and symmetrically distributed around a mean size of about $2.4 \pm 0.1$ nm.

Using EXAFS spectroscopy it was shown that, in the co-evaporated bimetallic [AuPd] samples, small bimetallic AuPd NPs were present, having an Au-rich core surrounded by an AuPd alloyed shell, while in the separately evaporated [Au][Pd] samples there was the presence of monometallic Au and Pd NPs showing some alloying only in the boundary regions. The EXAFS results were also qualitatively confirmed by the XANES spectra at the Au L3 edge reported in Fig. 3.12. The peak indicated by the arrow is known as the white line and its intensity is related to the presence of unoccupied $5d$ states ($d$-holes) [50]. In the NPs, the reduced number of Au–Au bonds results in an increase in the $5d$ level occupancy and hence in a decrease in its intensity. As shown by Liu et al. [50] in the presence of an AuPd alloyed shell, a further reduction in the intensity of the white line can be due to a charge transfer from Pd to Au. Therefore, the strong electronic effect observed in Fig. 3.12 comparing the [Au][Pd] sample to the [AuPd] one can be related to the increased Au–Pd coordination in the co-evaporated bimetallic samples.

The bimetallic Au–Pd/$\gamma$-Al$_2$O$_3$ systems obtained by the above-described procedure have reported to be active in the selective oxidation of benzyl alcohol to benzaldehyde with molecular oxygen (Table 3.3). Noteworthy, the Au–Pd/$\gamma$-Al$_2$O$_3$ sample resulted was largely more active and selective than an analogous sample obtained from separate Au SMA and Pd SMA precursors, the [Au][Pd]/$\gamma$-Al$_2$O$_3$ sample. Moreover, the [Au–Pd] sample showed remarkably higher efficiency than the corresponding monometallic Au and Pd catalysts.

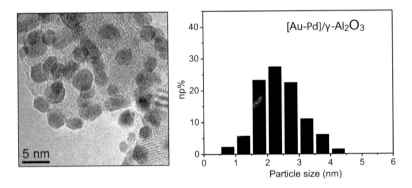

**Figure 3.11** HRTEM micrographs and particle size distributions of supported gold–palladium nanoparticles on $\gamma$-alumina (1 wt% of Au and 0.5 wt% of Pd). Reprinted with permission from *J. Catal.* **286** (2012) 224–236. Copyright 2012 Elsevier.

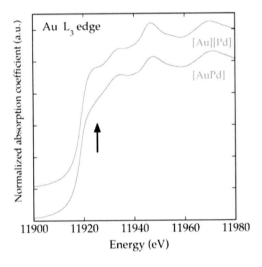

**Figure 3.12** XANES spectra co-evaporated [AuPd] and separately evaporated [Au][Pd] bimetallic samples supported on $\gamma$-Al$_2$O$_3$.

## 3.5 Concluding Remarks

In this chapter we have shown that solvated gold atoms, prepared according to the MVS technique, can be conveniently used to obtain highly dispersed zero-valent supported gold NPs. Among

**Table 3.3** Oxidation of benzyl alcohol over MVS-derived supported Au, Pd, and Au–Pd catalysts

$$R_3SiH + nBu{-}\!\!\equiv \xrightarrow{[Au]} \underset{\beta(Z)}{\overset{H}{\underset{nBu}{>\!\!=\!\!<}}\overset{H}{SiR_3}} + \underset{\beta(E)}{\overset{nBu}{\underset{H}{>\!\!=\!\!<}}\overset{H}{SiR_3}} + \underset{\alpha}{\overset{nBu}{\underset{R_3Si}{>\!\!=\!\!<}}\overset{H}{H}}$$

| Catalyst | Solvent | PhCHO selectivity | SA[a] |
|---|---|---|---|
| [Au–Pd]/γ-Al$_2$O$_3$ | Toluene | 84 | 25.4 |
| [Au][Pd]/γ-Al$_2$O$_3$ | Toluene | 81 | 5.5 |
| Pd/γ-Al$_2$O$_3$ | Toluene | 14 | 2.1 |
| Au/γ-Al$_2$O$_3$ | Toluene | 46 | 1.2 |
| [Au–Pd]/γ-Al$_2$O$_3$ | – | 84 | 250.6 |

[a] Specific activity calculated as moles of silane converted/moles of metal per hour.

the different experimental parameters that can be tuned using this approach, the kind of the co-evaporating solvent as well as the Au concentration of SMAs play a crucial role in defining the size of the gold particles and their interaction with the support. By using LC Au–acetone SMAs, a regime of preparative conditions to obtain supported small gold particles that show very good catalytic efficiency in the hydrosilylation reaction of acetylenes has been highlighted. In these conditions, the nature of the inorganic supports as well as the metal content (Au wt%) do not cause a marked difference in final gold particle sizes. Moreover, the synthesis of highly active supported Au–Pd systems using the simultaneous evaporation of Au and Pd vapors has been discussed. The preliminary results suggest this technique as a valuable preparative route to Au-based bimetallic metal particles that can be easily deposited in mild conditions on inorganic and organic supports.

## References

1. Gates, B.C. (2013). Supported gold catalysts: new properties offered by nanometer and sub-nanometer structures, *Chem. Commun.*, **49**, pp. 7876–7877.
2. Stratakis, M., and Garcia, H. (2012). Catalysis by supported gold nanoparticles: beyond aerobic oxidative processes, *Chem. Rev.*, **112**, pp. 4469–4506.

3. Haruta, M. (2008). Relevance of metal nanoclusters size control in gold(0) catalytic chemistry, in *Metal Nanoclusters in Catalysis and Materials Science: The Issue of Size Control*, eds. Corain, B., Schmid, G., and Toshima, N. (Elsevier, Amsterdam), pp. 183–199.
4. Hashmi, A.S.K., and Hutchings, G.J. (2006). Gold catalysis, *Angew. Chem., Int. Ed.*, **45**, pp. 7896–7936.
5. Schmid, G., and Corain, B. (2003). Nanoparticulated gold: syntheses, structures, electronics, and reactivities, *Eur. J. Inorg. Chem.*, pp. 3081–3098.
6. Haruta, M., Kobayashi, T., Sano, H., and Yamada, N. (1987). Gold catalysts for the oxidation of carbon monoxide at a temperature far below 0°C, *Chem. Lett.*, pp. 405–408.
7. Hutchings, G.J. (1985). Vapor phase hydrochlorination of acetylene: Correlation of catalytic activity of supported metal chloride catalysts, *J. Catal.*, **96**, pp. 292–295.
8. Della Pina, C., Falletta, E., Prati, L., and Rossi, M. (2008). Selective oxidation using gold, *Chem. Soc. Rev.*, **37**, pp. 2077–2095.
9. Hutchings, G.J. (2008). Nanocrystalline gold and gold–palladium alloy oxidation catalysts: a personal reflection on the nature of the active sites, *Dalton Trans.*, pp. 5523–5536.
10. Abad, A., Corma, A., and Garcia, H. (2008). Catalyst parameters determining activity and selectivity of supported gold nanoparticles for the aerobic oxidation of alcohols: the molecular reaction mechanism, *Chem. Eur. J.*, **14**, pp. 212–222.
11. Haruta, H., and Daté, M. (2001). Advances in the catalysis of Au nanoparticles, *Appl. Catal. A: Gen.*, **222**, pp. 427–437.
12. Corma, A., and Serna, P. (2006). Chemoselective hydrogenation of nitro compounds with supported gold catalysts, *Science*, **313**, pp. 332–333.
13. Okumura, M., Akita, T., and Haruta, M. (2002). Hydrogenation of 1,3-butadiene and of crotonaldehyde over highly dispersed Au catalysts, *Catal. Today*, **74**, pp. 265–269.
14. Carrettin, S., Guzman, J., and Corma, A. (2005). Supported gold catalyzes the homocoupling of phenylboronic acid with high conversion and selectivity, *Angew. Chem., Int. Ed.*, **44**, pp. 2242–2245.
15. Teles, J.H., Brode, S., and Chabanas, M. (1998). Cationic gold(I) complexes: highly efficient catalysts for the addition of alcohols to alkynes, *Angew. Chem., Int. Ed.*, **37**, pp. 1415–1418.
16. Muller, T.E., and Beller, M. (1998). Metal- initiated amination of alkenes and alkynes, *Chem. Rev.*, **98**, pp. 675–704.

17. Caporusso, A.M., Aronica, L.A., Schiavi, E., Martra, G., Vitulli, G., and Salvadori, P. (2005). Hydrosilylation of 1-hexyne promoted by acetone solvated gold atoms derived catalysts, *J. Organomet. Chem.*, **690**, pp. 1063–1066.
18. Klabunde, K.J. (1994). *Free Atoms, Clusters and Nanoscale Particles* (Academic Press, New York).
19. G. Vitulli, G., Evangelisti, C., Caporusso, A.M., Pertici, P, Panziera, N, Bertozzi, S., and Salvadori, P. (2008). Metal vapor-derived nanostructured catalysts in fine chemistry: the role played by particle size in the catalytic activity and selectivity, in *Metal Nanoclusters in Catalysis and Materials Science: The Issue of Size Control*, eds. Corain, B., Schmid, G., and Toshima, N. (Elsevier, Amsterdam), pp. 437–452.
20. Klabunde, K.J., and Cardenas-Trivino, G. (2007). Metal atom/vapor approaches to active metal clusters/particles, in*Active Metals*, ed. Fürstner, A. (VCH, Weinheim), pp. 237–278.
21. Bradley, J.S. (1994). The chemistry of transition metal colloids, in*Clusters and Colloids from Theory to the Applications,* ed. Schmid, G. (VCH, Weinheim), pp. 459–544.
22. Franklin, M., and Klabunde, K.J. (1987). Living colloidal metal particles from solvated metal atoms: clustering of metal atoms in organic media, in*Symposium on High Energy Processes in Organometallic Chemistry*, ed. Suslick, K. (ACS Symposium No. 333), pp. 246–259.
23. Stoeva, S.I., Prasad, B.L.V., Uma, S, Stoimenov, P.K., Zaikovski, V., Sorensen, C.M., and Klabunde, K.J. (2003). Face- centered cubic and hexagonal closed-packed nanocrystal superlattices of gold nanoparticles prepared by different methods, *J. Phys. Chem. B*, **107**, pp. 7441–7448.
24. Klabunde, K.J., Li, Y.X., and Tan, B.J. (1991). Solvated metal atom dispersed catalysts, *Chem. Mater.*, **3**, pp. 30–39.
25. Pitzalis, E., Evangelisti, C., Panziera, N. Basile, A., Capannelli, G., and Vitulli, G. (2011). Solvated metal atoms in the preparation of catalytic membranes, in *Membrane for Membrane Reactors*, eds. Basile, A., and Gallucci, F. (Wiley, UK), pp. 371–380.
26. Evangelisti, C., Schiavi, E., Aronica, L.A., Caporusso, A.M., Vitulli, G., Bertinetti, L., Martra, G., Balerna, A., and Mobilio, S. (2012). Bimetallic gold-palladium vapour derived catalysts: the role of structural features on their catalytic activity, *J. Catal.*, **286**, pp. 224–236.
27. Benfield, F. W. S., Green, M. L. H., Ogden, J. S., and Young, D. (1973). Synthesis of bis-$\pi$-benzene-titanium and -molybdenum using metal vapours, *J. Chem. Soc., Chem. Commun.*, pp. 866–873.

28. Klabunde, K. J., Timms, P., Skell, P. S., and Ittel, S. D. (1979). Metal atoms synthesis, *Inorg. Synth.*, **19**, 59–86.
29. Bradley, J.S. (1994). The chemistry of transition metal colloids, in*Clusters and Colloids*, ed. Schmid, G. (VCH, Weinheim), pp. 459–544.
30. http://www.advancedcatalysts.com.
31. Lin, S.-T, Franklin, M.T., and Klabunde, K.J. (1986). Nonaqueous colloidal gold. clustering of metal atoms in organic media. 12, *Langmuir*, **2**, pp. 259–260.
32. George, P.P., Gedanken, A., Perkas, N., and Zhong, Z. (2008). Selective oxidation of CO in the presence of air over gold-based catalysts Au/TiO2/C (sonochemistry) and Au/TiO2/C (microwave), *Ultrason. Sonochem.*, **15**, pp. 539–547.
33. Wu, S.-H, Zheng, X.-C., Wang, S.-R., Han, D.-Z., Huang, W.-P., and Zhang, S.-M. (2004). $TiO_2$ supported nano-Au catalysts prepared via solvated metal atom impregnation for low–temperature CO oxidation, *Catal. Lett.*, **96**, pp. 49–55.
34. Klabunde, K.J., Habdas, G., and Gardenas-Trivino, G. (1989). Colloidal metal particles dispersed in monomeric and polymeric styrene and methyl methacrylate, *Chem. Mater.*, **1**, pp. 481–483.
35. Aronica, L.A., Schiavi, L., Evangelisti, C., Caporusso, A.M., Salvadori, P., Vitulli, G., Bertinetti, L., and Martra, G. (2009). Solvated gold atoms in the preparation of efficient supported catalysts: correlation between morphological features and catalytic activity in the hydrosilylation of 1-hexyne, *J. Catal.*, **266**, 250–257.
36. Devenish, R.W., Goulding, T., Heaton, B.T., and Whyman, R. (1996). Preparation, characterization and properties of groups VIII and IB metal nanoparticles, *J. Chem. Soc., Dalton Trans.*, pp. 673–679.
37. Evangelisti, C., Raffa, P., Uccello-Barretta, G., Vitulli, G., Bertinetti, L., and Martra, G. (2011). A way to decylamine-stabilized gold nanoparticles of tailored sizes tuning their growth in solution, *J. Nanosci. Nanotechnol.*, **11**, pp. 2226–2231.
38. Tian, T., and Klabunde, K.J. (1998). Nonaqueous gold colloids. Investigations of deposition and growth on organically modified substrates and trapping of molecular gold clusters with an alkyl amine, *New J. Chem.*, pp. 1275–1283.
39. Stoeva, S., Klabunde, K.J., Sorensen, C.M., and Dragieva, I. (2002). Gram-scale synthesis of monodisperse gold colloids by the solvated metal atom dispersion method and digestive ripening and their organization into two- and three-dimensional structures, *J. Am. Chem. Soc.*, **124**, pp. 2305–2311.

40. Jose, D., Matthiesen, J.E., Parsons, C., Sorensen, C.M., and Klabunde, K.J. (2012). Size focusing of nanoparticles by thermodynamic control through ligand interactions. Molecular clusters compared with nanoparticles of metals, *Phys. Chem. Lett.*, **3**, 885–890.
41. Gates, B. (2013). Supported gold catalysts: new properties offered by nanometer and sub-nanometer structures, *Chem. Commun.*, **49**, pp. 7876–7877.
42. E. Bus, E., Prins, R., and van Bokhoven, J.A. (2007). Time-resolved in situ XAS study of the preparation of supported gold clusters, *Phys. Chem. Chem. Phys.*, **9**, pp. 3312–3320.
43. Hutchings, G.J., and Haruta, M. (2005). Catalysis by gold, *Appl. Catal. A: Gen.*, **291**, pp. 1–262.
44. Comaschi, T., Balerna, A., and Mobilio S. (2008). Temperature dependence of the structural parameters of gold nanoparticles investigated with EXAFS, *Phys. Rev. B*, **77**, pp. 075432–10.
45. Balerna, A., Bernieri, E., Picozzi, P., Reale, A., Santucci, S., Burattini, E., and Mobilio, S. (1985). Extended x-ray-absorption fine-structure and near-edge-structure studies on evaporated small clusters of Au, *Phys. Rev. B*, **31**, pp. 5058–5065.
46. Venezia, A.M., Pantaleo, G., Longo, A., Di Carlo, G., Casaletto, M.P. Liotta, F.L., and Deganello, G. (2005). Relationship between structure and CO oxidation activity of ceria-supported gold catalysts, *J. Phys. Chem. B*, **109**, pp. 2821–2827.
47. Casaletto, M.P., Longo, A., Venezia, A.M., Martorana A., and Prestianni, A. (2006). Metal-support and preparation influence on the structural and electronic properties of gold catalysts, *Appl. Catal. A: Gen.*, **302**, pp. 309–316.
48. Sorensen, C.M., Rayalu, S.S., Shrestha, K.M., and Klabunde, K.J. (2013). Au-TiO$_2$ nanocomposites and efficient photocatalytic hydrogen production under UV-visible and visible light illuminations: a comparison of different crystalline forms of TiO$_2$, *Int. J. Photoenergy*, pp 1–10.
49. Rayalu, S.S., Jose, D., Joshi, M.V., Mangrulkar, P.A., Shrestha, K., and Klabunde, K.J. (2013). Photocatalytic water splitting on Au/TiO$_2$ nanocomposites synthesized through various routes: enhancement in photocatalyticactivity due to SPR effect, *Appl. Catal. B: Environ.*, **142–143**, pp. 684–693.
50. Liu, F., and Zang, P. (2010). Tailoring the local structure and electronic property of AuPd nanoparticles by selecting capping molecules, *Appl. Phys. Lett.*, **96**, p. 043105.

# Chapter 4

# Microgels as Exotemplates in the Preparation of Au Nanoclusters

**Andrea Biffis and Paolo Centomo**

*Dipartimento di Scienze Chimiche, Università di Padova, Via Marzolo 1, Padova 35131, Italy*
andrea.biffis@unipd.it

## 4.1 Introduction

In the historical development of techniques for the preparation of metal nanoparticles, polymeric stabilizers have undoubtedly played a major role. Through the skilled use of commercial linear polymers as stabilizers, reliable synthetic routes for the facile preparation of significant amounts of stable, monodispersed metal nanoparticles in solution have been developed and have also been extensively reviewed [1]. More recently, investigations in this field have been extended to the use of more complex polymer architectures as stabilizing agents for metal nanoparticles, with the aim to ensure better control on the morphological properties of the nanoparticles (in particular their size), to tailor their solubility properties, and finally to prepare advanced composite nanomaterials that

---

*Gold Catalysis: Preparation, Characterization, and Applications*
Edited by Laura Prati and Alberto Villa
Copyright © 2016 Pan Stanford Publishing Pte. Ltd.
ISBN 978-981-4669-28-3 (Hardcover), 978-981-4669-29-0 (eBook)
www.panstanford.com

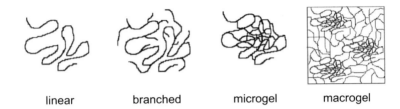

linear　　　branched　　　microgel　　　macrogel

**Figure 4.1** Different morphologies of polymer stabilizers for metal nanoparticles.

combine the properties of the metal nanoparticles with those of the polymeric stabilizer (e.g., reactivity, sensitivity to external stimuli, [bio]affinity). The resulting nanocomposite colloids currently find widespread application in different scientific fields, encompassing nanomedicine, sensorics, analytical chemistry, and also catalysis. Examples of the polymeric systems employed for this purpose include inter alia hyperbranched polymers or dendrimers [2], block co-polymer micelles [3], spherical polymer brushes [4], and especially soluble crosslinked polymers (*microgels*), which form the object of this chapter.

In polymer chemistry, the terms "microgel" and, more recently, "nanogel," define crosslinked polymer colloids that possess a size comparable to the statistical dimensions of uncrosslinked macromolecules ($10^1$–$10^2$ nm) and give rise to stable, low-viscosity solutions in appropriate solvents [5]. These colloids can be considered as an intermediate category of polymers, which combine characteristics of both linear macromolecules and 3D networks (Fig. 4.1).

In particular, the recognition of some peculiar properties of microgels (defined size, tailored morphology and porosity, variable nature and degree of chemical functionality, low viscosity, film-forming ability) has powered their application in widely different areas ranging from coatings to medicine and from molecular recognition to catalysis [5, 6]. This chapter will be devoted to the application of microgels as exotemplates and stabilizers in the preparation of metal nanoparticles, most notably gold nanoparticles, and to the use of the resulting nanocomposites for catalytic applications [7].

## 4.2 Microgel Preparation

Contrary to the soluble (linear) polymers that are commonly employed for the stabilization of Au nanoparticles in solution (e.g., polyvinylpyrrolidone [PVP], poly[ethylene glycol] [PEG], etc.), microgels for this specific use are not commercial and have to be manufactured on purpose.

Microgels can be prepared by virtually any conventional polymerization method, provided that measures are taken in order to promote intramolecular crosslinking of the growing polymer chains instead of intermolecular crosslinking, which would give rise to a macroscopic crosslinked polymer network. In practice, the most commonly employed techniques are emulsion, dispersion/precipitation, and solution polymerization.

Emulsion polymerization [8] exploits the compartimentalization of the monomer mixture by dispersing it inside a continuous phase made out of a monomer-immiscible solvent; the resulting dispersion is stabilized by a surfactant, which forms micelles around the monomer droplets. The presence of the surfactant allows one to reach a microgel size in the range of tens to hundreds of nanometers. Moreover, using the peculiar technique known as *miniemulsion polymerization* [8, 9], the size of the resulting microgel particles exactly matches that of the droplets in the starting emulsion, which allows one to predetermine the microgel particle size.

Dispersion/precipitation polymerization is another widely employed methodology for microgel synthesis [10]. In this approach, the key point is the use of a solvent where the monomers are soluble but the resulting microgel is not. This implies that in the course of the polymerization the growing polymer segregates from the solution while continuously capturing and consuming monomers and oligomers, eventually forming microgels in the submicrometer range (with size in the hundreds of nanometers) under proper conditions. In certain cases, the poorly soluble, typically hydrophobic growing polymer may become functionalized with polar heads derived from the initiator and thus form a sort of micelle in solution instead of precipitating out. This represents a technique known as surfactant-free emulsion polymerization.

Solution polymerization [10] exploits, instead, the fact that at high dilution intramolecular crosslinking becomes favored compared to intermolecular crosslinking for entropic reasons (Ziegler's dilution law); moreover, the growing microgels become stabilized toward macrogelation by the osmotic repulsion forces generated by the interaction of solvated polymer chains and loops at the periphery of the microgel particles (steric stabilization). This is sufficient to stabilize the growing microgels against macrogelation if the monomer concentration is reduced below a critical threshold value (critical monomer concentration, $C_m$). This value is dependent on many factors such as the polymerization methodology and conditions, the crosslinking monomer content, and of course the nature of the solvent, which can be conveniently rationalized in terms of the solubility parameter $\delta$ [11].

Clearly, to achieve steric stabilization the microgel chains must be efficiently solvated by the polymerization solvent so that the growing microgels can be considered to be in swelling equilibrium with the surrounding medium; as a consequence, the resulting crosslinked polymer particles are less rigid than particles prepared by compartimentalization: for example, the former tend to collapse when dried, forming films, whereas particles prepared by compartimentalization often retain their spherical shape, even in the dry state (for an illustrative example of such morphological differences as well as of the transition between the two kinds of particles with varying polymerization conditions, see Ref. [12]).

The key advantage of solution polymerization is that the onset of steric stabilization makes it possible to obtain particles of very small sizes without having to add stabilizers to prevent macrogelation. Steric stabilization is sufficient to stabilize the growing microgels against macrogelation if the monomer concentration is kept below, $C_m$. On the other hand, compared to polymerization in miniemulsion, solution polymerization suffers from a less precise control of the size and size distribution of the resulting microgels. However, the polymerization in solution offers the considerable advantage of allowing the variation of parameters, such as the type and degree of microgel functionalization (by utilizing suitable co-monomers), the nature of the nonfunctional co-monomers or the crosslinking degree, with less restrictions caused by the requirements of the polymerization procedure. As will be apparent in the following sections, these parameters play an important role in determining the solubility of the microgels in different solvents as well as the size of the metal nanoclusters generated inside them.

Polymerization in dilute solution also allows a very straightforward isolation and purification of the microgels. After polymerization, the resulting microgels can be conveniently precipitated from the reaction solution by using suitable nonsolvents for the microgel molecules. The resulting powders can be filtered off, dried, and redispersed in suitable solvents, when needed. On the other hand, microgels prepared by emulsion techniques need to be extensively purified after synthesis to free them from the excess surfactant. Furthermore, they are often not redispersible once precipitated and have to be stored in solution for further use.

## 4.3 Microgel-Stabilized Metal Nanoclusters

The generation of metal nanoclusters within microgels is performed in solution. Generally, the microgels are first loaded with a predetermined amount of a metal precursor upon interaction of the functional groups in the microgel with suitable metal ions or complexes through coordination or ion exchange. Straightforward, high-yielding reactions that provide quantitative incorporation of the metal precursor are best utilized in this step, since they allow one to reliably predict the loading of metal precursor into the microgel. This makes it necessary to have suitable functional groups available in the microgel network, which can be either already contained in the main employed monomer for microgel synthesis or can be purposely introduced by postfunctionalization of the microgel or by co-polymerization of suitable functional monomers. The latter approach (Fig. 4.2), pioneered by our group [13], offers the greatest flexibility in microgel design, since it allows one to tune the microgel to the specific metal precursor to be employed upon tailoring the nature and amount of the functional co-monomer.

In several instances, also weaker coordinative and/or electrostatic interactions of the metal precursors with the microgel chains, especially those containing polar groups such as poly(*N*-

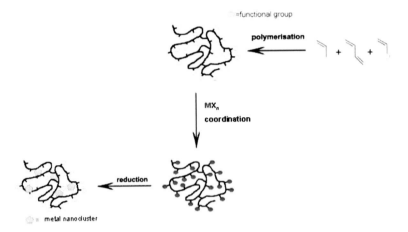

**Figure 4.2** Generation of metal nanoparticles inside microgel stabilizers.

isopropylacrylamide) [14] or poly(N-vinylcaprolactam) [15], or with a second component present within the microgel, for example, positively charged, oxidized poly(3,4-ethylenedioxythiophene) (PEDOT) nanorods [16] or negatively charged laponite nanoclay [17], have been utilized for this purpose. In these cases, though, the reaction leading to the incorporation of the metal precursor into the microgel is often only partial, and control of the loading of metal precursor into the microgel is consequently not possible. Furthermore, the excess metal precursor not anchored to the microgel has to be removed from the solution in order to have best results in the subsequent reduction process leading to metal nanoclusters. This could complicate the overall procedure.

In the case of gold, the employed precursors are tpically gold(III) compounds such as $AuCl_3$, $HAuCl_4$ (which provides anionic $AuCl_4^-$), or $Au(en)_2Cl_3$, (en = ethylenediamine), which provides cationic $Au(en)_2^{3+}$. Clearly, the choice of the actual precursor depends on the nature of the anchoring groups present in the microgel. The resulting metal content in the microgel may range widely from around 0.1% by weight up to several tens of percent.

Subsequent reduction of the metal precursors loaded into the microgel yields microgel-stabilized metal nanoclusters. Care must be taken that the anchored metal precursor remains stable until reduction is deliberately started, since slow spontaneous decomposition to metal can produce nucleation centers before the actual reduction. This may result in the growth of very polydisperse and large metal colloids [13a, 18]. This is particularly true for gold, since the Au(III) precursors that are usually employed for this scope are easily reduced to metal; for example, microgel-bound trialkylamino groups, which are frequently employed to anchor Au(III) precursors to microgels, also act as reducing agents toward the metal [19]. Nevertheless, deliberate reduction of gold(III) to gold(0) upon reaction with functional groups present in the microgel has been occasionally employed as a means for Au nanocluster formation without the addition of a reducing agent [20]. Usually, though, a suitable reducing agent, such as sodium borohydride, sodium citrate, hydrazine, etc., is added in excess in order to ensure fast and complete reduction to metal.

The morphology and allocation of the metal nanophase resulting from reduction is actually dependent on several factors such as the morphology of the microgel, the nature and strength of the microgel interactions with the metal in both oxidized and reduced forms, the reactivity of the reducing agent, and the amount of metal. This was already apparent in the first report describing the preparation of microgel-stabilized metal nanoclusters, which was published by Antoniettti in 1997 [21]. Antonietti utilized polystyrene-based microgels prepared by emulsion polymerization and fully sulfonated in a second step by treatment with concentrated sulfuric acid. The sulfonated microgels were soluble in water and could be loaded with gold precursors upon addition of $AuCl_3$ in a molar ratio slightly lower than 1:3 with respect to the available sulfonate groups (no study concerning the extent of gold loading and the speciation of gold anchored to the microgel was actually carried out). Subsequent reduction produced widely different results, depending on the nature of the reducing agent (Fig. 4.3).

With a very reactive reducing agent such as $NaBH_4$ under neutral conditions, microgel-stabilized, ca. 5 nm sized gold nanoclusters homogeneously dispersed within the microgel network were formed. Under basic conditions (0.1 M NaOH), in which $NaBH_4$ is less reactive, the gold nanoclusters did coalesce to form "nanonetworks" extending over more than one microgel. With an even less reactive

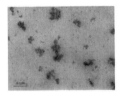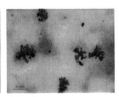

**Figure 4.3** Different morphologies of Au metal generated inside sulfonated microgels. Reproduced with permission from Ref. [21]. Copyright Wiley-VCH Verlag GmbH & Co. KGaA.

reducing agent such as hydrazine, 25 nm sized agglomerated gold "nanonuggets" were formed within the microgels. Furthermore, in the two latter cases a strongly uneven distribution of gold in the microgels was observed, the nanonetworks or nanonuggets accumulating within 15%–20% of the microgels, leaving the others completely empty. The reported results highlight that (1) highly reactive reducing agents are preferable in order to obtain small, monodisperse metal nanoparticles and (2) functional groups within the microgel should significantly interact with the metal precursors and perhaps also with the zerovalent metal centers initially generated upon reduction in order to prevent their migration from one microgel macromolecule to another, which results in an uneven metal distribution with formation of large metal nanoparticles or even of superstructures of coalesced nanoparticles.

The crosslinking degree of the microgel is another parameter to be taken into account as it influences the size of the metal nanoparticles generated upon reduction in a very remarkable way. In his 1997 report, Antonietti already noticed that empolying sulfonated polystyrene microgels of higher crosslinking degree "thinner" gold nanostructures could be produced [21]. Later, in systematic studies employing microgels prepared upon co-polymerization of defined amounts of functional monomers strongly interacting with metal precursors, it has been demonstrated that indeed the average size of metal nanoclusters generated inside microgels upon reduction considerably decreases with increasing crosslinking degree of the microgel [18, 22, 23]. Remarkably, the observed change in nanocluster size was found to depend solely on the nanomorphology of the employed microgel: the chemical nature of the microgel, the nature and loading of metal precursor, and the reaction conditions for nanocluster formation were kept the same, so the different "mesh sizes" of the polymer network building up the swollen microgel molecules remained the only factor that actually determined the average nanocluster size.

The reason of such a size control is to be traced to the elastic forces exerted by the microgel network on the growing metal nanocluster, which limit the growth process; the extent of this limiting effect clearly depends on the rigidity of the polymer chains as well as on the strength of the interactions between the surface of the growing nanoparticle and the polymer chains, which explains the different sensitivity of the final metal nanocluster size from the crosslinking degree that has been recorded with microgels of different nature.

The constraints imposed by the microgel network cannot, however, be too strong, otherwise diffusion of the reducing agent toward the microgel interior becomes the limiting factor for the overall nanocluster formation process involving metal reduction, nucleation, and growth. In this situation, formation of comparatively large metal nanoparticles at the periphery of the microgel particles is often observed. This feature has been demonstrated by Akamatsu, who prepared poly(2-vinylpyridine) microgels stabilized by a corona of PEG chains in order to keep the microgel in water solution at any pH. Loading with $HAuCl_4$ was performed at an acidic pH, at which the microgels were completely swollen because of the protonation of the pyridyl groups, and resulted in a high loading of Au precursor homogeneously distributed within the microgel. Subsequent reduction with dimethylamine-borane was instead carried out at neural pH, at which the microgels were in the collapsed state. The result was that the reducing agent was unable to diffuse toward the interior of the microgel, whereas the gold precursor slowly migrated from the interior of the microgel to the exterior, resulting at the end in the generation of well-defined but rather large (20–30 nm) gold nanoparticles on the external surface of the collapsed microgel (Fig. 4.4) [24].

Microgel-supported metal nanoclusters with a similar distribution of metal have been generated also upon simple absorption on/into the microgels of preformed metal nanoclusters dispersed

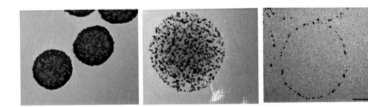

**Figure 4.4** TEM micrographs of microgel-supported Au nanoclusters prepared by Akamatsu et al. deposited at pH 8 (left) and pH 2 (middle); TEM cross-sectional view of the microgels (right). Reproduced with permission from Ref. [24]. Copyright (2010) American Chemical Society.

in solution [25]. In this case, though, the exotemplating effect of the microgel is obviously lost and the loading of metal nanoparticles that can be obtained turns out to be quite low.

Summarizing, the morphology and allocation of metal nanoparticle generated inside microgels upon reduction of metal precursors are the result of the combined influence of more than one parameter, the most important of which are (a) the kinetics of formation of the zerovalent metal precursors, (b) the kinetics of nucleation and growth of the metal nanoparticles, (c) the strength of the interaction of the polymer stabilizer with the metal precursors and with the growing nanoparticle surface, and finally (d) the steric effect ("cage" effect) exerted by the polymer chains surrounding the growing nanoparticles.

The resulting microgel-stabilized nanoclusters can be purified by ultrafiltration, ultracentrifugation, or precipitation. In the third case, they can be often obtained as powders and redispersed in suitable solvents for the microgel molecules, where they are stable for months without noticeable precipitation of metal. Care must be taken, however, in the case of microgels with functionalities that interact quite strongly with the metal nanoclusters, since such interactions may result in drastic variations in the solubility of the resulting nanocomposites after isolation. This is particularly true for microgels based on polymer chains featured by a low $T_g$ and high hygroscopicity, such as the widely employed polydimethylaminoethylmethacrylate (DEAEMA). Such microgels tend to precipitate as viscous oils or sticky solids from solution, which easily results in the occurrence of intermolecular physical crosslinking (because of the entanglements between polymer chains from different microgels) or even chemical crosslinking (because of the occurrence of interactions between functional groups of a microgel molecule and metal nanoparticles located in a second microgel molecule) between the microgels.

## 4.4 Advantages of Microgel-Stabilized Metal Nanoclusters

As outlined in the previous section, the use of microgel stabilizers in the preparation of metal nanoparticles offers the opportunity not only to stabilize the resulting metal nanoclusters against aggregation but also to control the size of the metal nanoclusters through the exotemplating effect of the microgel, as well as their distribution throughout the microgel particle. Aside from this, the peculiar properties of the employed microgel stabilizer may also have important implications for the applications of the resulting metal nanoparticles, in particular for catalytic applications. Indeed, through the choice of a proper microgel stabilizer it is possible, for example, to easily recover the catalytically active metal nanoparticles, switch the catalytic activity on and off, or use the microgel/metal nanocluster nanocomposites as suitable building blocks for the production of other kinds of catalytically active materials.

Many microgels are known to respond to external stimuli such as changes of temperature, pH, and ionic strength by typically changing their swelling degree, sometimes quite abruptly, thus behaving as responsive systems [26]. This opens up interesting opportunities for catalysis, in that as soon as the microgel stabilizer de-swells and collapses, the metal nanoclusters contained in it become poorly accessible to the reactants and are consequently prevented to further act as catalysts; thus, such an event shuts down or at least considerably decreases the catalytic performance of the complexes; examples of this will be provided in the next section.

Another interesting feature of microgels is related to their partition behavior between two immiscible solvents. Some years ago, we were investigating on the partition behavior of Pd or Au nanoclusters stabilized by a microgel containing pendant tertiary amino groups between water and dichloromethane at different pHs, with the goal to determine conditions for switching the partition behavior of the microgel and consequently enabling catalyst recovery through extraction and recycling [18]. Surprisingly, when in contact with the second liquid phase, the microgel invariably and quantitatively remained in the liquid phase in which it was dissolved first, no matter which one it was and how long the system was stirred. Transitions could be induced only by changing the pH of the water phase and were easily monitored by following the migration of the nanocluster color (Fig. 4.5). For example, transition from water to dichloromethane occurred quite abruptly and quantitatively when the pH was raised above 10, corresponding to neutralization of the microgel-bound ammonium groups. Conversely, transition from dichloromethane to water was slower but could still be

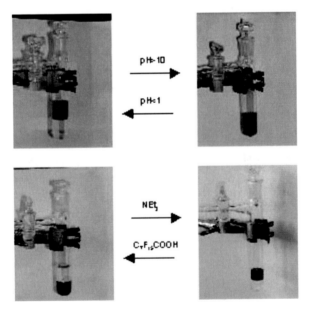

**Figure 4.5** Partition of microgel-stabilized Au nanoclusters (5 mg) in liquid–liquid biphasic systems under different conditions: water–dichloromethane (top, 2 mL of each solvent) and dichloromethane–perfluoro(methylcyclohexane) (bottom, 2 mL of each solvent).

quantitatively induced by lowering the pH below 1 upon addition of 1 M aqueous HCl. Thus, in the broad pH window between 1 and 10 the microgel showed the interesting property of remaining preferentially soluble in an organic or aqueous phase, depending on the phase in which it was dissolved first.

The observed behavior is the consequence of competitive solvation effects as well as of the acid–base equilibria in which the basic amino groups in the microgel are involved. When a microgels is dissolved in water, the pendant amino groups become partially protonated; consequently, the hydrophilicity of the microgel increases. The microgel becomes organophilic only when almost all ammonium groups in the polymer become neutral. On the other hand, when the microgel is dissolved first in dichloromethane, the amino groups need to become protonated in order for the microgel to migrate into the aqueous phase; however, the solvation of the microgel by dichloromethane efficiently prevents contact between the solvated protons present in the water phase and the amino groups contained in the microgel up to high proton concentrations.

Reversible microgel transfer has been induced also from dichloromethane to a fluorous phase (Fig. 4.5, bottom). Microgels are completely extracted into perfluoro(methylcyclohexane) by adding perfluorooctanoic acid (7 equivalents with respect to the microgel-bound amino groups). In this case, an acid–base reaction between the basic amino groups contained in the microgel and the perfluorooctanoic acid additive results in the formation of microgel-anchored trialkylammonium perfluorooctanoate moieties, which impart to the microgel a strong affinity for the fluorous phase [18]. The phase transfer can be reversed by simple addition of triethylamine to the organic–fluorous biphasic system, which cleaves the ammonium moieties liberating the microgel in neutral form.

Such a possibility to switch the partition behavior of microgel-stabilized metal nanoclusters between different phases (organic, aqueous, fluorous) may have obvious importance for their use as catalysts in view of their recovery and recycling.

Microgels can be also employed as building blocks for more extended structures still containing metal nanoclusters. For

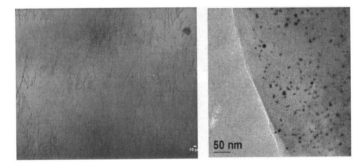

**Figure 4.6** Optical picture (left) and TEM micrograph (right) of a sample of microgel-stabilized Au nanoclusters after electrospinning.

example, many microgels exhibit excellent film-forming properties, which paves the way to their utilization in the preparation of composite films. Subsequent pyrolysis of the microgel should result in the deposition of size-controlled metal nanoclusters onto the film support. This possibility, though, has yet to be investigated in detail.

Interestingly, a gold-containing microgel solution can also be subjected to electrospinning (Fig. 4.6). The resulting fibers have a diameter of around 1 micron, and the metal nanoclusters are homogeneously dispersed throughout the fiber and maintain the average size that they possessed before electrospinning (Fig. 4.6) [27]. Also in this case, further studies are necessary to check possible applications of these composite fibers.

## 4.5 Microgel-Stabilized Gold Nanoclusters: Catalytic Applications

The number of reports concerning applications of microgel-stabilized metal nanoclusters, and particularly microgel-stabilized gold nanoclusters, is currently increasing at a steady pace and covers various technological fields. Though the majority of these reports actually concerns applications in medicinal chemistry (such as targeted imaging, drug delivery, photothermal therapy, etc.), which fall outside the scope of this chapter, several investigations have been carried out also on their potential as quasi-homogeneous catalysts, which are focused upon in this section

Our group was the first to investigate on the application of microgel-stabilized metal nanoclusters as catalysts [13]; initially, we successfully targeted the use of Pd nanoclusters stabilized by microgels as catalysts for C–C coupling reactions (Heck and Suzuki couplings). Building on this experience, in 2005 we reported also the first catalytic application of microgel-stabilized gold nanoclusters, in the frame of a systematic study devoted to the development of recoverable and recyclable microgel-stabilized metal nanocluster catalysts for the aerobic oxidation of alcohols in water [28]. In this work, gold nanoclusters were prepared by spontaneous reduction of $AuCl_3$ inside poly($N$, $N$-dimethylacrylamide)-co-($N$, $N$-dimethylaminoethylmethacrylate) microgels. Slow reduction of the gold(III) precursor by the tertiary amino groups present in the microgel caused in this case the formation of rather large and polydisperse Au nanoclusters (ca. 10 nm), which exhibited low catalytic activity in the aerobic oxidation of 1-phenylethanol at neutral pH (13% yield after 24 hours at 100°C with 1 mol% metal). The activity could be significantly enhanced by raising the pH of the solution, as it is known that a basic pH promotes the reaction in the case of gold metal catalysts [29]; however, above pH 10 the microgel precipitated as a result of the de-protonation of the microgel-bound trialkylamino groups. Remarkably, the role of the microgel support in this study was not only to stabilize the metal nanoclusters during the catalytic event but also to impart them a solubility preference for water, which allowed us to completely recover and recycle the catalyst upon simple extraction with diethylether of the reaction product from the aqueous reaction solution.

Two years later, we published a second contribution in which considerable amelioration of the performance of the microgel-stabilized Au nanocatalyst was carried out [30]. Indeed, substitution of the coordinating functional monomer $N$, $N$-dimethylamino-ethylmethacrylate with 4-vinylpyridine in the microgel as well as use of $HAuCl_4$ instead of $AuCl_3$ as a precursor of the gold nanoclusters resulted in microgel-gold(III) adducts that were perfectly stable in solution. Subsequent reduction with excess $NaBH_4$ resulted in Au nanoclusters that were much smaller and less polydisperse than in the previous report (ca. 2.5 nm in size) and could be employed as catalysts even at high pH without precipitation (provided that the temperature was kept below 80°C). The resulting catalytic activities in the oxidation of various alcohols were comparable or higher (in particular with hydrophobic alcohols) than those recorded under identical reaction conditions with a benchmark Au/C catalyst with a Au nanoparticle size very similar to those in the microgel; activities compared well also with the activity of the best catalysts known at the time for aerobic alcohol oxidation.

More recently, we returned to investigate this catalytic system, particularly the effect of an increase in the content of pyridyl functional groups in the microgel on the performance of the microgel-stabilized gold nanoclusters [31]. The aim was to gain more control on the hydrophilic/hydrophobic properties of the microgel support upon pH changes in order to maintain the possibility of efficient catalyst recovery, while at the same time ensuring a high affinity of the microgel-supported catalyst for the hydrophobic alcohol substrate. Unfortunately, although the generation of Au nanoclusters inside such novel microgels again worked well and produced 3–5 nm nanoparrticles, depending on the Au content, on the other hand such nanoclusters proved quite inactive in aerobic alcohol oxidations under a variety of conditions. The logical conclusion was that interaction of the excess pyridyl functional groups with the Au nanocluster surface efficiently suppressed its catalytic activity. This unexpected result represents a caveat for further studies in this direction.

Au nanoclusters stabilized by thermosensitive core–shell microgels were also successfully applied as catalysts for aerobic alcohol oxidation. In 2009, Yu et al. reported the preparation of metal nanoclusters, including gold ones, inside core–shell microgels featured by a polystyrene core and a thermosensitive poly(N-isopropylacrylamide) shell [32]. The microgels were prepared by emulsion polymerization seeded with the polystyrene cores and were subsequently loaded with metal nanoclusters by addition the metal precursor ($HAuCl_4$ in the case of gold) to the microgel dispersion, followed by addition of the reducing agent ($NaBH_4$). In spite of the fact that incorporation of the metal precursor inside the microgel was most probably only partial before reduction (no

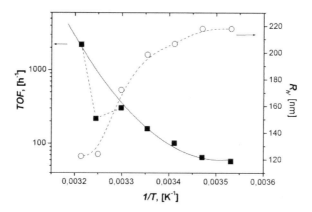

**Figure 4.7** TOF (squares) and hydrodynamic radius (circles) versus $1/T$ for the aerobic oxidation of benzyl alcohol in water in the presence of Au nanoclusters stabilized by thermosensitive core–shell microgels. Reproduced from Ref. [32] with permission of the Royal Society of Chemistry.

strongly coordinating or ion-exchanging groups were present in the microgel, apart from those deriving from the employed polymerization initiator), rather small Au nanoclusters could be obtained (4.5 ± 1.2 nm in size), probably thanks to the comparatively small loading of metal employed (0.4% for gold).

The catalytic activity of the resulting gold nanoclusters was tested in the aerobic oxidation of benzyl alcohol in water, using 1 equivalent $K_2CO_3$ with respect to the alcohol as basic promoter and 1 atm of air. The activity of the catalyst was highest with Au nanoclusters, reaching an average TON after 20 hours of 159 $h^{-1}$ at 25°C and 2203 $h^{-1}$ at 40°C. Such data also indicate a strong dependance of catalytic activity from the temperature, which was investigated in greater detail (Fig. 4.7). First of all, the increase in catalytic activity with temperature, measured through the turnover frequency (TOF) value, was found on the whole to be greater than exponential; on the other hand, the increase in catalytic activity with temperature was not monotonic but went through a local minimum corresponding to the temperature at which the volume phase transition of the thermosensitive microgel shell occurs.

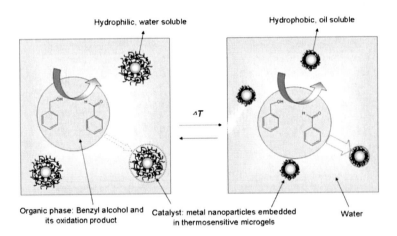

**Figure 4.8** Representative picture of the effect of a temperature increase on the catalytic behavior of thermoresponsive core–shell microgels in the aerobic oxidation of benzyl alcohol in water. Reproduced from Ref. [32] with permission of the Royal Society of Chemistry.

This behavior has been rationalized by taking into account the increase in hydrophobicity of the microgel shell upon de-swelling with increasing temperature, which favors absorption of the poorly water soluble substrate by the microgel and consequently its subsequent oxidation (Fig. 4.8). Such a polarity effect accounts for the more than exponential increase in catalytic activity with temperature, whereas the decreased accessibility of the Au nanoclusters following the volume-phase transition of the microgel shell explains the observed local minimum in catalytic activity.

Starting from 2006, another test reaction was intensively investigated with microgel-stabilized Au nanocluster catalysts, namely the reduction of *p*-nitrophenol with sodium borohydride. Rather than for its intrinsic synthetic value, this reaction is a very useful mean for rapidly assessing the reactivity of Au nanoclusters, since it takes place under mild conditions and it is easily followed by monitoring the absorption band of the *p*-nitrophenolate anion via ultraviolet-visible (UV-Vis) spectroscopy. In this way, differences in the catalytic efficiency of microgel-supported Au nanoclusters can be spotted and traced back to differences in parameters such as the size of the

Au nanoclusters or the chemical and morphological features of the microgel stabilizer.

The groups of Pich and Lu were the first to investigate this reaction with microgel-stabilized Au nanoclusters. They initially employed core–shell microgels with a poly-(acetoacetoxyethyl metacrylate) core and a loosely crosslinked poly-(*N*-vinylcaprolactam) shell [33]. The microgels were contacted with a HAuCl$_4$ solution and subsequent addition of a reducing agent (trisodium citrate) resulted in the production of Au nanoclusters located in the poly-(*N*-vinylcaprolactam) shell. The catalytic activity of such composite nanosystems was found to be higher than that of citrate-stabilized Au colloids prepared without the microgel stabilizer (though no data were available on the average size of the Au nanoparticles in the two cases). Most notably, the catalytic activity of the microgel-stabilized Au nanoclusters increased with increasing temperature much more than that of the Au colloids (Fig. 4.9).

The authors explained this observation with the affinity of the relatively hydrophobic microgel shell for the susbtrate, which

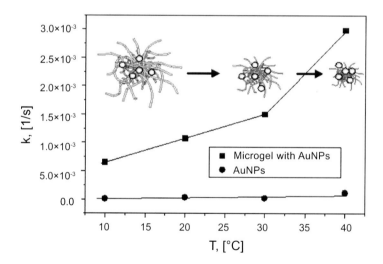

**Figure 4.9** Temperature dependence of the rate constants for *p*-nitrophenol reduction recorded with microgel-stabilized Au nanoclusters and with citrate-stabilized Au colloids. Reproduced from Ref. [33] by courtesy of Andrij Pich.

increases with temperature as the consequence of temperature-induced de-swelling.

Two years later the group of Pich re-investigated this reaction with a catalytic system made out of Au nanoclusters stabilized by similar core–shell microgels, which, however, also contained PEDOT nanorods as an additional component [18]. The resulting ternary nanocomposites were slightly more active (by a factor of 2) than the related microgel-stabilized Au nanoclusters without the PEDOT component. The temperature dependence of the catalytic efficiency of the ternary nanocomposites was also more pronounced. Moreover, an overall increase with increasing temperature has been shown but with a local minimum in correspondence to the volume-phase transition temperature of the poly(N-vinylcaprolactam) shell. Such a behavior is fully analogous to that reported in Fig. 4.7 in the case of poly(*N*-isopropylacrylamide)-based core–shell microgel catalysts for aerobic alcohol oxidations and indicates that de-swelling of the microgel stabilizer does not shut down completely the reactivity of the Au nanoclusters. This is probably the consequence of the fact that the Au nanoclusters in these core–shell systems are located in the microgel shell. Indeed, a much more pronounced effect of microgel de-swelling was recorded by Zhang and Shi, who investigated as a catalyst for the same reaction Au nanoclusters stabilized by microgels made out of a poly(*N*-isopropylacrylamide) shell and a poly(glycidylmethacrylate) core crosslinked with diethylenetriamine [14]. In this case, the presence of amino groups in the crosslinked core probably determines the location of the Au nanoclusters in the core itself. Consequently, de-swelling of the microgel shell shields the Au nanoclusters more effectively, and indeed in this case the activity of the catalyst in the reduction of *p*-nitrophenol drops considerably on going from room temperature (25 °C) to 75 °C.

More recently, the group of Pich introduced novel core–shell microgel stabilizers in which acrylic acid was employed as an additional functional co-monomer [34]. In this way, carboxylic acid groups were preferentially introduced in the microgel core, which acted both as anchoring groups for the gold precursors as well as reducing agents. The result is the preferential formation of gold nanoclusters in the microgel core. The nanocomposite catalysts were again active in the reduction of *p*-nitrophenol and were featured by a notable stability: they could be recovered and recycled several times with only a negligible variation of catalytic activity.

## 4.6 Conclusions

Microgel-stabilized gold nanoclusters represent nanocomposite colloidal materials with manifold applications in current technologies. Thanks to the flexibility that can be achieved in microgel synthesis, they can be easily prepared on a large scale, with microgel stabilizers displaying a wide range of properties. The exploitation of these nanocomposite colloids in the field of catalysis is still in its infancy, but it is likely to grow considerably in the next future, as smart catalytic systems featured by novel properties such as switchable catalytic activity, ease of recovery and recycling, etc., are increasingly needed for advanced applications.

## Acknowledgments

This work was partially supported by the University of Padova (PRAT 2010).

## References

1. Feldheim, D.L., and Foss, C.A. (2002). Metal Nanoparticles: Synthesis, Characterization and Applications (Marcel Dekker, New York).
2. Crooks, R.M., Zhao, M., Sun, L., Chechik, V., and Yeung, L.K. (2001). Dendrimer-encapsulated metal nanoparticles:? synthesis, characterization, and applications to catalysis, *Acc. Chem. Res.*, **32**, pp. 181–190; Myers, S.V., Weir, M.G., Carino, E.V., Yancey, D.F., Pande, S., and Crooks, R.M. (2011). Dendrimer-encapsulated nanoparticles: new synthetic and characterization methods and catalytic applications, *Chem. Sci.*, **2**, pp. 1632–1646
3. Antonietti, M., Förster, S., and Oestreich, M. (1997). Micellization of amphiphilic block copolymers and use of their micelles as nanosized reaction vessels, *Macromol. Symp.*, **121**, pp. 75–88.

4. Schrinner, M., Proch, S., Mei, Y., Kempe, R., Miyajima, N., and Ballauff, M. (2008). Stable bimetallic gold–platinum nanoparticles immobilized on spherical polyelectrolyte brushes: synthesis, characterization, and application for the oxidation of alcohols, *Adv. Mater.*, **20**, pp. 1928–1933
5. Fernandez-Nieves, A., Wyss, H., Mattsson, J., and Weitz, D.A. (2011). *Microgel Suspensions* (Wiley-VCH, Weinheim); Pich, A., and Richtering, W. (2010). Chemical design of responsive microgels, in *Advances in Polymer Science Series*, Vol. 234 (Springer Verlag, Heidelberg).
6. Funke, W., Okay, O., and Joos-Müller, B. (1998). Microgels-intramolecularly crosslinked macromolecules with a globular structure, *Adv. Polym. Sci.*, **136**, pp. 139–234; Saunders, B.R., and Vincent, B. (1999). Microgel particles as model colloids: theory, properties and applications, *Adv. Colloid Interface Sci.*, **80**, pp. 1–25; Thorne, J.B., Vine, G.J., and Snowden, M.J. (2011). Microgel applications and commercial considerations, *Colloid Polym. Sci.*, **289**, pp. 625–646.
7. For recent review articles specifically addressing nanocomposites of microgels with inorganic nanoparticles, including metals, see Karg, M. (2012). Multifunctional inorganic/organic hybrid microgels. An overview of recent developments in synthesis, characterization, and application, *Colloid Polym. Sci.*, **290**, pp. 673–688; Perez-Juste, J. Pastoriza-Santos, I., and Liz-Marzan, L.M. (2013). Multifunctionality in metal@microgel colloidal nanocomposites, *J. Mater. Chem. A* **1**, pp. 20–26.
8. Chern, C.S. (2008). *Principles and Applications of Emulsion Polymerization* (Wiley, Chichester).
9. Landfester, K., Bechthold, N., Tiarks, F., and Antonietti, M. (1999). Formulation and stability mechanisms of polymerizable miniemulsions, *Macromolecules*, **32**, pp. 5222–5228
10. Barrett, K.E.J., and Thomas, H.R. (eds.) (1975). *Dispersion Polymerization in Organic Media* (Wiley, London).
11. Brandrup, J., and Immergut, E.H. (1989). *Polymer Handbook*, 3rd ed. (Wiley, New York).
12. Frank, R.S., Downey, J.S., Yu, K., and Stöver H.D.H. (2002). Poly (divinylbenzene-alt-maleic anhydride) microgels: intermediates to microspheres and macrogels in cross-linking copolymerization, *Macromolecules*, **35**, pp. 2728–2735
13. (a) Biffis, A. (2001). Functionalised microgels: novel stabilisers for catalytically active metal colloids, *J. Mol. Catal. A: Chem.*, **165**, pp. 303–307; (b) Biffis, A., Orlandi, N., and Corain, B. (2003). Microgel-stabilized

metal nanoclusters: size control by microgel nanomorphology, *Adv. Mater.*, **15**, pp. 1551–1555

14. Jiang, X., Xiong, D., An, Y., Zheng, P., Zhang W., and Shi, L. (2010). Thermoresponsive hydrogel of poly(glycidylmethacrylate-co-N-isopropylacrylamide) as a nanoreactor of gold nanoparticles, *J. Polym. Sci. A: Polym. Chem.*, **45**, pp. 2812–2819.

15. Pich, A., Karak, A., Lu, Y., Ghosh, A.K., and Adler, H.-J.P. (2006). Hybrid microgels containing gold nanoparticles, *E-Polymers*, **018**, pp. 1–16.

16. Hain, J., Schrinner, M., Lu, Y., and Pich, A. (2008). Design of multicomponent microgels by selective deposition of nanomaterials, *Small*, **4**, pp. 2016–2024.

17. Contin, A., Biffis, A., Sterchele, S., Dörmbach, K., Schipmann, S., and Pich, A. (2014). Metal nanoparticles inside microgel/clay nanohybrids: synthesis, characterization and catalytic efficiency in cross-coupling reactions, *J. Colloid Interface Sci.*, **414**, pp. 41–45.

18. Minati, L., and Biffis, A. (2005). A polymer support with controllable solubility in mutually immiscible solvents, *Chem. Commun.*, pp. 1034–1036.

19. See, for example, McCrindle, R., Ferguson, G., Arsenault G.J., and

20. McAlees, A.J. (1983). Reaction of tertiary amines with bis(benzonitrile)dichloropalladium(II). Formation and crystal structure analysis of di-μ -chloro-dichlorobis[2-(N,N-di-isopropyliminio)ethyl-C]dipalladium(II), *J. Chem. Soc., Chem. Commun.*, pp. 571–572.

21. Oishi, M., Hayashi, H., Uno, T., Ishii, T., Iijima, M., and Nagasaki, Y. (2007). One-pot synthesis of pH-responsive PEGylated nanogels containing gold nanoparticles by autoreduction of chloroaurate ions within nanoreactors, *Macromol. Chem. Phys.*, **208**, pp. 1176–1182

22. Antonietti, M., Gröhn, F., Hartmann, J., and Bronstein, L. (1997). Nonclassical shapes of noble-metal colloids by synthesis in microgel nanoreactors, *Angew. Chem., Int. Ed. Engl.*, **36**, pp. 2080–2083.

23. Biffis A. (2008). Microgels as exotemplates in the synthesis of size-controlled metal nanoclusters, in *Metal nanoclusters in Catalysis and Materials Science: The Issue of Size Control*, eds. Corain, B., Schmid, G., and Toshima, N. (Elsevier, Amsterdam), pp. 341–346.

24. Lu, Y., Mei, Y., Ballauff, M., and Drechsler, M. (2006). Thermosensitive core–shell particles as carrier systems for metallic nanoparticles, *J. Phys. Chem. B*, **110**, pp. 3930–3937.

25. Akamatsu, K., Shimada, M., Tsuruoka, K., Nawafune, H., Fujii, S., and Nakamura, Y. (2010). Synthesis of pH-responsive nanocomposite

microgels with size-controlled gold nanoparticles from ion-doped, lightly cross-linked poly(vinylpyridine), *Langmuir*, **26**, pp. 1254–1259.

26. See, for example, Gawlitza, K., Turner, S.T., Polzer, F., Wellert, S., Karg, M., Mulvaney, P., and von Klitzing, R. (2013). Interaction of gold nanoparticles with thermoresponsive microgels: influence of the cross-linker density on optical properties, *Phys. Chem. Chem. Phys.*, **15**, pp. 15623–15631, and references cited therein.

27. (a) Pelton, R.H. (2000). Temperature-sensitive aqueous microgels, *Adv. Colloid Interface Sci.*, **85**(2000), pp. 1–33; (b) Lu, Y., and Ballauff, M. (2011). Thermosensitive core–shell microgels: from colloidal model systems to nanoreactors, *Prog. Polym. Sci.*, **36**, pp. 767–792.

28. Piperno, S., Gheber, L.A., Canton, P., Pich, A., Dvorakova, G., and Biffis, A. (2009). Microgel electrospinning: a novel tool for the fabrication of nanocomposite fibers, *Polymer*, **50**, pp. 6193–6197

29. Biffis, A., and Minati, L. (2005). Efficient aerobic oxidation of alcohols in water catalysed by microgel-stabilised metal nanoclusters, *J. Catal.* **236**, pp. 405–409.

30. See, for example, Biella, S., Castiglioni, G.L., Fumagalli, C., Prati, L., and Rossi, M. (2002). Application of gold catalysts to selective liquid phase oxidation, *Catal. Today*, **72**, pp. 43–49.

31. Biffis, A., Cunial, S., Spontoni, P., and Prati, L. (2007). Microgel-stabilised gold nanoclusters: powerful "quasi-homogeneous" catalysts for the aerobic oxidation of alcohols in water, *J. Catal.*, **251**, pp. 1–6.

32. Cera, G. (2013).Tesi di Laurea, University of Padova, Italy.

33. Lu, Y., Proch, S., Schrinner, M., Drechsler, M., Kempe, R., and Ballauff, M. (2009). Thermosensitive core–shell microgels as a "nanoreactor" for catalytically active metal nanoparticles, *J. Mater. Chem.*, **19**, pp. 3955–3961.

34. Pich, A., Karak, A., Lu, Y., Ghosh, A.K., and Adler, H.-J.P. (2006). Tuneable catalytic properties of hybrid microgels containing gold nanoparticles, *J. Nanosci. Nanotechnol.*, **6**(12), pp. 3763–3769.

35. Agrawal, G., Schurings, M.P., van Rijn, P., and Pich, A. (2013). Formation of catalytically active gold-polymer microgel hybrids via a controlled in situ reductive process, *J. Mater. Chem. A*, **1**, pp. 13244–13251.

# Chapter 5

# Miscellaneous

Carine E. Chan-Thaw, Laura Prati, and Alberto Villa

*Dipartimento di Chimica, Università degli Studi di Milano, Via Golgi, 19-20133 Milano, Italy*
alberto.villa@unimi.it

## 5.1 Introduction

Gold-based catalysts have attracted great interest for diverse reactions such as CO oxidation [1], propene oxidation [2], and alcohol [3] and polyol [4] liquid-phase oxidation. This constant fervor on gold-based catalysts started with the pioneer work of the Haruta [5] and Hutchings [6] groups, who demonstrated that gold is a catalytically active metal. To obtain highly active Au catalysts, it was compulsory to achieve proper metal dispersion (i.e., small gold nanoparticles [AuNPs] ranged between 1 and 20 nm). For gold, this represents high limiting factors, but up to now, several methods have been set up to prepare active heterogeneous gold catalysts.

In this chapter, we will have an overview of methods that are producing active gold catalysts, and enter into more detail for a few of them that were not specifically presented in a dedicated chapter.

*Gold Catalysis: Preparation, Characterization, and Applications*
Edited by Laura Prati and Alberto Villa
Copyright © 2016 Pan Stanford Publishing Pte. Ltd.
ISBN 978-981-4669-28-3 (Hardcover), 978-981-4669-29-0 (eBook)
www.panstanford.com

**Table 5.1** Summary of the most used gold catalysts' preparation

| Methods | Particle size range (nm) | Ref. |
| --- | --- | --- |
| Co-precipitation | 4 | [1, 7, 8] |
| Deposition-precipitation | 2–3 | Chapter 1 |
| Sol immobilization | 2–10 | Chapter 5 |
| Chemical vapor deposition (CVD) | 2–20 | [9–13] |
| Physical vapor deposition (PVD) | 10 | [2, 3, 14–23] |
| Solvated metal atoms | 2–40 | Chapter 3 |
| Polymer encapsulation | 2–10 | Chapter 4 |
| Solid grinding (SG) | 2–4 | [24–26] |

There are at least eight methods to obtain highly dispersed gold catalysts. Each of these methods has been successful in depositing gold, on diverse supports, as small particles with a mean diameter ranging from 3 to 10 nm. The methods are not cited in order of importance and are co-precipitation, deposition-precipitation, co-sputtering, chemical vapor deposition (CVD), physical vapor deposition (PVD), sol immobilization, solid grinding (SG), polymer encapsulation, etc. (Table 5.1).

In this chapter, a fast overview of co-precipitation, CVD, SG, and PVD will be presented.

## 5.2 Co-Precipitation

This method is one of the first that were successful in obtaining good dispersion of gold nanoparticles on metal oxides [1, 5].

The co-precipitation method consists of pouring an aqueous solution of HAuCl$_4$ and a water-soluble metal salt, generally nitrates in an alkaline solution under a fine control of the pH using, for example, NaOH. However, carbonates such as Na$_2$CO$_3$ and K$_2$CO$_3$ are also widely used to adjust the pH during the co-precipitation process. Indeed, contrary to NaOH, carbonates increase the stability of the pH. It is clear that the pH at which the precipitation is performed is important as well as the temperature, as described by Haruta [7]. After precipitation, the hydroxides or carbonates are filtered, washed, dried, and then calcined. This technique can be applied to salts of metals in the first row of the transition series in groups IV–XII and also to Al and Mg, which can be precipitated as hydroxides or hydrated oxides. During calcination in air, the formed hydroxides or carbonates of metal components of the supports are converted to metal oxides. According to the chosen calcination temperature, the size of the AuNPs can be tuned.

Haruta et al. [1] reported a calcination temperature in air of 673 K to obtain gold crystallites in the range of 3–5 nm in Au/$\alpha$-Fe$_2$O$_3$ (5 wt% Au loading).

Nguyet et al. [8] reported that calcination at 393 K or 473 K is not enough to obtain the formation of crystallites in the case of Au/$\alpha$-Fe$_2$O$_3$. With a temperature of 573 K, they obtained two phases of crystals. Most of the time, the co-precipitated precursors are calcined in air at 573 K or even higher temperatures [1] to produce the desired crystalline metal oxide supports. However, above 673 K, coagulation of Au nanoparticles accelerates.

This method of preparation has also been used to prepare Au on MnOx [27] and ZnO [28] with good dispersion.

To conclude with this brief description of co-precipitation, the size of AuNPs is correlated to not only the pH at which the process occurs but also the temperature. And last but not the least, the affinity between gold and the support is to be taken into consideration.

## 5.3 Chemical Vapor Deposition

The CVD technique is a process where one or more volatile precursors are transported via the vapor phase to the reaction chamber. Once there, they decompose by heating in the presence of a support. A surface chemical reaction is occurring, leading to the formation of a solid film and gas by-products. These gas by-products are desorbed and transported out from the reactor. This method is depicted in Fig. 5.1. In the case of metal oxides, the support is pretreated to remove adsorbed water and organic residues from the surface. Once the support is ready, vapors of the gold precursor adsorb on the support, which is then calcined. This leads to the decomposition of the precursor(s) into metallic gold particles. This

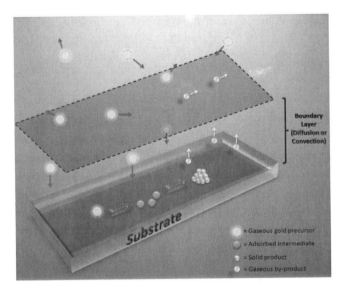

**Figure 5.1** Scheme representation of different steps of chemical vapor deposition (CVD).

technique is applicable to a wide range of metal oxides and is easy to use with high-surface-area materials.

Haruta et al. reported the use of dimethyl gold (III) $\beta$-diketone for preparing Au on $TiO_2$ [9]. When amorphous $TiO_2$ is calcined or reduced, very fine Au particles with diameters smaller than 2 nm were obtained. In the case of anatase $TiO_2$, $O_2$ pretreatment was preferable and a mean diameter of 7 to 9 nm was observed.

Okumura et al. [11] reported that the CVD technique was successful in preparing highly dispersed AuNPs of a mean diameter of 6.6 nm on $SiO_2$, which is active for CO oxidation at a temperature below 273 K.

Later, Haruta et al. [12] were able to deposit, by CVD, an acetylacetonate complex of gold, not only on $SiO_2$, but also on MCM-41, $SiO_2$-$Al_2O_3$, and activated carbon. On all these supports, gold nanoparticles with a mean diameter of above 5 nm (but less than 10 nm) were observed. They demonstrated that it is possible to deposit AuNPs with strong interaction, even on $SiO_2$. The strong metal support interaction (SMSI) induces high activity in CO oxidation. This SMSI was lacking with MOF-5, a **m**etal **o**rganic **f**ramework, a highly porous support. This was surprising as AuNPs were spread through the intact lattice of MOF-5. Hermes et al. [10] used CVD to deposit the gold precursor (CH$_3$)Au(PMe$_3$) on MOF-5. They observed polydisperse gold particles in the range of 5–20 nm. These gold atoms or clusters might be mobile and aggregate to form bigger particles inside the pores. Moreover, gold particles can diffuse and form large particles of about 20 nm outside the pores of MOF-5.

In comparison to the liquid-phase method, a less homogeneous dispersion of Au particles on the support is observed using the CVD method. The size distribution of gold particles is relatively broad as it is linked to the interaction between the gold precursor and the OH that the support contains [13]. The CVD method is anyhow able to produce Au nanoparticles smaller than 2 nm [9].

The final metal loading depends on the support and often how its pretreatment was.

## 5.4 Solid Grinding

The SG method consists of mixing directly a solid gold complex together with a support material using an agate mortar or a ball mill. Haruta et al. [24] developed the first SG method to deposit directly gold clusters on porous coordinated polymers (PCPs), including MOF-5. For this purpose, a volatile organogold complex without any organic solvent was mixed with a PCP at room temperature. Thermal treatment at 393 K for two hours in diluted H$_2$ is necessary to finally obtain Au/PCPs with small Au clusters of 2.2 nm mean diameter.

However, it should be mentioned that the size of Au strongly depends on the choice of the PCP. Indeed, the metallic species, porous structures, and pore size of the support material strongly influence the Au size [10]. Carbon [24, 25], oxides [24], and cellulose [26] have been also reported as support and gold NPs in the range of 1.5–4 nm are often obtained. Haruta et al. [25] reported that in the case of metal oxides, ball milling is preferred to agate mortar. Indeed, metal oxides are harder materials than organic polymers or carbon, which are softer and for which grinding in an agate mortar is enough.

It is also possible to perform SG of the gold complex with the support in the presence of organic solvents such as acetone [25] or ethanol. When a metal oxide is used as the support material, such as nonconductive $Al_2O_3$ and $ZrO_2$, the ball milling does not lead to uniform grinding. Indeed, large AuNPs together with small Au clusters are obtained. To overcome this situation, acetone is added to the gold complex and metal oxide mixture to obtain uniform grinding and to avoid Au nanoparticle aggregation. In such a way, while the grinding is proceeding, the gold complex vaporizes and may move over the surface of the supports and adsorb. Particles of 2.6 nm size were obtained.

However, it is worth mentioning that SG is not suitable for all oxides. This method is thus not appropriate to obtain small AuNPs on semiconductive supports such as $TiO_2$ and $CeO_2$ due to the uncontrolled reduction of gold during the grinding.

## 5.5 Physical Vapor Deposition

PVD consists of evaporating or sputtering the target molecule onto a substrate. Indeed, PVD methods involve the energetic removal of normally nonvolatile species from a source material, that is, the target, by heating or using energetic particle bombardment by ions, electrons, photons, or atoms [15]. Gas-phase gold atoms or clusters are thus produced. The reader should not confuse PVD techniques with CVD methods, as CVD relies on the decomposition of an organometallic precursor to form a film or clusters.

PVD methods can be divided into two main categories, thermal evaporation, which is the simplest one to obtain gas-phase gold atoms, and sputtering.

### 5.5.1 *Thermal Evaporation*

Evaporation is based on two basic processes, which are the evaporation of a source material (e.g., the gold precursor) and the subsequent condensation of the evaporated material on the substrate (i.e., the chosen support). The condensation is directly followed by nucleation and growth. Evaporation takes place in high vacuum in order to avoid nonuniform deposition. In high vacuum, vapors that are not from the source material are previously completely removed from the chamber.

Together with the vacuum pump, energy sources are needed. An electron or laser beam can heat the surface source with an energy up to 15 keV. A crucible that contains the material source can be heated by induction, that is, an electric filament. Alternatively, the same crucible can be heated by passing a large current through a resistive wire of foil containing the material to be deposited.

Valden and Goodman in 1998 [16] reported the deposition on Au nanoparticles in a range of 1–6 nm on $TiO_2$ by PVD.

### 5.5.2 Sputtering

Sputtering consists of ejecting material from a target material source onto a substrate, that is, the support. The atoms are physically removed from the target and then collected on the support. Different sputtering systems exist, that is, the conventional diode sputtering, the triode sputtering, and finally the magnetron sputtering. In the case of magnetron sputtering, argon ions are accelerated out of plasma toward a high-purity gold metal target. Gold is then sputtered from the target and is subsequently deposited onto a support material (Fig. 5.2). Like in the case of thermal evaporation, the process takes place in a close recipient, which is under vacuum pressure before any deposition procedure.

Using this method, Veith et al. demonstrated [14] that it is possible to deposit gold nanoparticles, with an average diameter of 2–3 nm, on high-surface-area $\gamma$-$Al_2O_3$. The resulting catalyst is as active as (e.g., CVD) the conventionally prepared one in CO oxidation catalysts. The magnetron sputtering method as reported by Veith et al. has the main advantages (a) to be economically and environmentally friendly as it is possible to recover the excess of gold from the chamber and no liquid waste is produced, (b) to not require heat to decompose the precursors, and (c) to avoid any contamination from solvent or precursors molecules on the surface.

Magnetron sputtering can also be applied to prepare supported metal nanoparticles on a wide range of materials, including $WO_3$ and carbon [13]. Magnetron sputtering preparation is indeed able

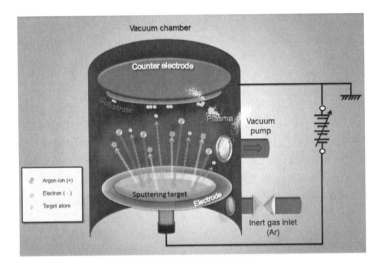

**Figure 5.2** Representation of sputtering.

to form a Au/WO$_3$ active catalyst that is not possible to obtain by the classical deposition-precipitation method because of its low isoelectric point. Concerning Au/C, very small (1 nm) AuNPs were obtained, showing high activity in the glycerol oxidation [17].

The particle size distribution is related to the substrate but also to the time of deposition. However, almost always a narrow distribution of nanoparticle sizes is obtained. By increasing the weight loading of the samples, particle size increases, but up to 11 wt%, has little effect on the particle size distribution [18]. The sputtering method was also used to prepare gold catalysts on other supports such as TiO$_2$ [19], SiO$_2$ [20], C$_3$N$_4$ [21], and steel fiber [22].

Using sputtering instead of thermal evaporation allows one to obtain better uniformity over a large size. Furthermore, it is possible to perform a presputter cleaning of the surface. However, certain sputtering systems, for example, the glow discharge plasma, require a medium-level vacuum that can promote contamination over evaporation.

Cathodic arc plasma deposition (CAPD), a kind of PVD, was applied by Fujitani et al. [23] to deposit gold nanoparticles on single-crystal TiO$_2$ (110) surfaces. They could accurately control the size distribution, which is 4.2 ± 0.3 nm, while maintaining the same gold loading. This method consists of using an electric arc to vaporize material from a cathode target. The so-obtained vaporized material is then condensed on the substrate in order to form a thin film.

## 5.6 Conclusions

Several preparation methods have been developed in order to obtain gold particles smaller than 5 nm with homogeneous distributions on supports. Scientists have therefore a lot of options for preparing active gold-supported catalysts. The choice depends strictly on the supports but also on the requirements of the reaction. Indeed, the catalytic activity of gold nanoparticles can be tuned by three main parameters, namely the choice of the support, the size of the AuNPs and their dispersion, and finally the interaction between gold and the support.

## Acknowledgments

Sebastiano Campisi is gratefully acknowledged for graphical support.

## References

1. Haruta, M., Yamada, N., Kobayashi, T., and Iijima, S. (1989). Gold catalysts prepared by coprecipitation for low-temperature oxidation of hydrogen and of carbon monoxide, *J. Catal.*, **115**, pp. 301–309.
2. Hayashi, T., Tanaka, K., and Haruta, M. (1998). Selective vapor-phase epoxidation of propylene over Au/TiO$_2$ catalysts in the presence of oxygen and hydrogen, *J. Catal.*, **178**, pp. 566–576.
3. Prati, L., Villa, A., Lupini, A. R., and Veith, G. M. (2012). Gold on carbon: one billion catalysts under a single label, *Phys. Chem. Chem. Phys.*, **14**, pp. 2969–2978.
4. Carrettin, S., McMorn, P., Johnston, P., Griffin, K., and Hutchings, G. J. (2002). Selective oxidation of glycerol to glyceric acid using a gold catalyst in aqueous sodium hydroxide, *Chem. Commun.*, **7**, pp. 696–697; Porta, F., and Prati, L. (2004). Selective oxidation of glycerol to

sodium glycerate with gold-on-carbon catalyst: an insight into reaction selectivity, *J. Catal.*, **224**, pp. 397–403.

5. Haruta, M., Kobayashi, T., Sano, H., and Yamada, N. (1987). Novel gold catalysts for the oxidation of carbon-monoxide at a temperature far below 0°C, *Chem. Lett.*, pp. 405–408.

6. Hutchings, G. J. (1985). Vapor phase hydrochlorination of acetylene: correlation of catalytic activity of supported metal chloride catalysts, *J. Catal.*, **96**, pp. 292–295.

7. Haruta, M. (2004). Nanoparticulate gold catalysts for low-temperature CO oxidation, *J. New Mater. Electrochem. Syst.*, **7**, pp. 163–172.

8. Nguyet, T. T. M., Trang, N. C., Huan, N. Q., and Xuan, N. (2008). Preparation of gold nanoparticles, Au/Fe$_2$O$_3$ by using a co-precipitation method and their catalytic activity, *J. Korean Phys. Soc.*, **52**, pp. 1345–1349.

9. Okumura, M., Tanaka, K., Ueda, A., and Haruta, M. (1997). The reactivities of dimethyl gold (III)$\beta$-diketone on the surface of TiO$_2$. A novel preparation method for gold catalysts, *Solid State Ionics*, **95**, pp. 143–149.

10. Hermes, S., Schröter, M.-K., Schmid, R., Khodeir, L., Muhler, M., Tissler, A., Fischer, R. W., and Fischer R. A. (2005). Metal@MOF: loading of highly porous coordination polymers host lattices by metal organic chemical vapor deposition, *Angew. Chem., Int. Ed.*, **44**, pp. 6237–6241.

11. Okumura, M., Nakamura, S., Tsubota, S., Nakamura, T., Azuma, M., and Haruta M. (1998). Chemical vapor deposition of gold (CH$_3$)$_2$Au(CH$_3$COCH$_2$COCH$_3$) on Al$_2$O$_3$, SiO$_2$, and TiO$_2$ for the oxidation of CO and of H$_2$, *Catal. Lett.*, **51**, pp. 53–58.

12. Okumura, M., Tsubota, S., and Haruta, M. (2003). Preparation of supported gold catalysts by gas-phase grafting of gold acethylacetonate for low-temperature oxidation of CO and of H$_2$, *J. Mol. Catal. A: Chem.*, **199**, pp. 73–84.

13. Serp, P., Kalck, P., and Feurer, R. (2002). Chemical vapor deposition methods for the controlled preparation of supported catalytic materials, *Chem. Rev.*, **102**, pp. 3085–3128.

14. Veith, G. M., Lupini, A. R., Pennycook, S. J., Ownby, G. W., and Dudney N. J. (*2005*). Nanoparticles of gold on $\gamma$-Al$_2$O$_3$ produced by dc magnetron sputtering, *J. Catal.*, **231**, pp. 151–158.

15. Rossnagel S. M. (2003) Thin film deposition with physical vapor deposition and related technologies, *J. Vac. Sci. Technol. A*, **21**, S74.

16. Valden, M., Lai, X., and Goodman, D. W. (1998). Onset of catalytic activity of gold clusters on titania with the appearance of nonmetallic properties, *Science*, **281**, pp. 1647–1650.

17. Veith, G. M., Lupini, A. R., Pennycook, S. J., Villa, A., Prati, L., and Dudney N. J. (2007). Magnetron sputtering of gold nanoparticles onto $WO_3$ and activated carbon, *Catal. Today*, **122**, pp. 248–253.
18. Veith, G. M., Lupini, A. R., and Dudney N. J. (2008) Magnetron sputtering to prepare supported metal catalysts, in *Metal Nanoclusters in Catalysis and Materials Scicence: The Issue of Size Control*, eds. Corain, B., Schmid, G., and Toshima, N. (Elsevier, Amsterdam), pp. 347–353.
19. Veith, G. M., Lupini, A. R., and Dudney N. J. (2009). Role of pH in the formation of structurally stable and catalytically active $TiO_2$-supported gold catalysts, *J. Phys. Chem. C*, **113**, pp. 269–280.
20. Veith, G. M., Lupini, A. R., Rashkeev, S., Pennycook, S. J., Mullins, D. R., Schwartz, V., Bridges C. A., and Dudney N. J. (2009). Thermal stability and catalytic activity of gold nanoparticles supported on silica, *J. Catal.*, **262**, pp. 92–101.
21. Singh, J. A., Overbury, S. H., Dudney, N. J., Li, M., and Veith, G. M. (2012). Gold nanoparticles supported on carbon nitride: influence of surface hydroxyls on low temperature carbon monoxide oxidation, *ACS Catal.*, **2**, pp. 1138–1146.
22. Guo, H., Kemell, M., Al-Hunaiti, A., Rautiainen, S., Leskelä, M., and Repo, T. (2011). Gold–palladium supported on porous steel fiber matrix: structured catalyst for benzyl alcohol oxidation and benzyl amine oxidation, *Catal. Commun.*, **12**, pp. 1260–1264.
23. Fujitani, T., Nakamura, I., Akita, T., Okumura, M., and Haruta, M. (2009). Hydrogen dissociation by gold clusters, *Angew. Chem., Int. Ed.*, **48**, pp. 9515–9518.
24. Ishida, T., Nagaoka, M., Akita, T., and Haruta, M. (2008). Deposition of gold clusters on porous coordination polymers by solid grinding and their catalytic activity in aerobic oxidation of alcohols, *Chem. Eur. J.*, **14**, pp. 8456–8460.
25. Ishida, T., Kinoshita, N., Okatsu, H., Akita, T., Takei, T., and Haruta M. (2008). Influence of the support and the size of gold clusters on catalytic activity for glucose oxidation, *Angew. Chem., Int. Ed.*, **47**, pp. 9265–9268.
26. Ishida, T., Watanabe, H., Bebeko, T., Akita, T., and Haruta M. (2010). Aerobic oxidation of glucose over gold nanoparticles deposited on cellulose, *Appl. Catal. A: Gen.*, **377**, pp. 42–46.
27. Lee, S.J., Gavriilidis, A., Pankhurst, Q.A., Kyek, A., Wagner, F.E., Wong, P.C.L., and Yeung, K.L. (2001). Effect of drying conditions of Au-Mn Co-precipitates for low-temperature CO oxidation, *J. Catal.*, **200**, pp. 298–308.

28. Wang, G.Y., Zhang, W.X., Lian, H.L., Jiang, D.Z., and Wu, T.H. (2003). Effect of calcination temperatures and precipitant on the catalytic performance of Au/ZnO catalysts for CO oxidation at ambient temperature and in humid circumstances, *Appl. Catal. A*, **239**, pp. 1–10; Bailie, J.E., Abdullah, H.A., Anderson, J.A., Rochester, C.H., Richardson, N.V., Hodge, N., Zhang, J.G., Burrows, A., Kiely, C.J., and Hutchings, G.J. (2001). Hydrogenation of but-2-enal over supported Au/ZnO catalysts, *Phys. Chem. Chem. Phys.*, **3**, pp. 4113–4121.

# Chapter 6

# Transmission Electron Microscopy on Au-Based Catalysts

**Di Wang**

*Institute of Nanotechnology, Karlsruhe Institute of Technology,
Hermann-von-Helmholtz Platz 1, D-76344 Eggenstein-Leopoldshafen, Germany*
di.wang@kit.edu

This chapter is intended to provide a brief introduction of transmission electron microscopy techniques with respect to their application in studying the structures of supported nanocatalysts. Mainly it is focused on basic principles and the latest development in this field to show the capabilities of a modern electron microscope. Examples of TEM characterization are selected from Au-based catalysts rather for some common interests relevant to structures than for specific catalytic reactions.

## 6.1 Introduction

With fast development of transmission electron microscopy (TEM) and more understanding of heterogeneous catalysts, TEM has become a routine and very important technique for the catalyst

---

*Gold Catalysis: Preparation, Characterization, and Applications*
Edited by Laura Prati and Alberto Villa
Copyright © 2016 Pan Stanford Publishing Pte. Ltd.
ISBN 978-981-4669-28-3 (Hardcover), 978-981-4669-29-0 (eBook)
www.panstanford.com

characterization, especially complementary to other in situ and ex situ spectroscopic techniques, aiming at the understanding of the structures of catalysts at different spatial scales from the atomic level to the macroreactor, as well as different time scales from elementary steps on the single active site to the life cycle in a real reactor [1]. The correlation between the structures down to the atomic level with the activity and selectivity of the catalyst is expected to ultimately lead to the capability of tailoring the highly efficient multifunctional and synergistic catalysts.

Since the first practical transmission electron microscope was constructed in 1933, TEM has undergone fast development in its resolution, from resolution similar to optic microscopy until nowadays, 0.5 Å [2, 3], which is much shorter than the interatomic distance in most of the materials. The ultrahigh resolution enables structure determination at the atomic level for the supported nanoparticles, especially for the configuration of atoms on the surface steps and kinks, at the interface and the perimeter between metal particle and support, disorders from equilibrium lattice positions, including vacancies, interstitial atoms, dislocation, and stacking faults. Moreover, the transmission electron microscope is far beyond an imaging instrument; it also offers powerful analytic tools, from routine energy-dispersive X-ray (EDX) and electron energy loss spectroscopy (EELS) spectrometers to more unconventional detectors, such as cathodoluminescence [4, 5] and secondary electron detectors [6]. The application of TEM on catalysis started already in the 1940s [7, 8] and developed quickly from simple imaging of catalyst morphology to more advanced techniques, including high-resolution transmission electron microscopy (HRTEM), high-angle annular dark field (HAADF) scanning transmission electron microscopy (STEM), analytic spectroscopy and environmental transmission electron microscopy (ETEM) [9–12]. These developments directly lead to progress in nanotechnology, including the exciting discovery of fullerene, carbon nanotubes (CNTs), and graphene, which attract tremendous interest to use them as supports or even as metal-free catalysts [13–16].

## 6.2 The Principle of Transmission Electron Microscopy

Most of the heterogeneous catalysts consist of support and metallic nanoparticles on it. To achieve high surface area and high dispersion, it is necessary to control the particle size in nanometer or even subnanometer range. With such small particle size, X-ray diffraction (XRD) only shows fairly broad peaks, which makes the measurement of size distribution difficult. Therefore HRTEM and STEM are more suitable methods since they offer direct measurement of the particle size from electron micrographs. Particularly with the development of probe-aberration-corrected STEM, single metal atoms, or small clusters consisting of few atoms can be resolved [17]. The contrast in TEM comes from mass-thickness contrast, diffraction contrast, and phase contrast [18], which makes the image interpretation not straightforward. Without the assistance of analytic methods, like spectrum imaging, distinguishing metal particles from support sometimes is difficult, especially when the aggregates are very thick. While HAADF STEM, due to its contrast mainly coming from incoherent thermal scattered electrons at high angle, the intensity can be directly correlated to the atomic number and sample thickness along the beam direction, forming the so-called Z-contrast image [19]. Heavy metal particles can be easily distinguished from the support, which usually consists of light atoms, like in the cases of C supports and metal oxide supports. Recently, probe-corrected STEM has improved the resolution greatly and HAADF STEM becomes more and more popular in characterizing catalysts not only for morphology but also for crystal structures. Energy-filtered transmission electron microscopy (EFTEM) and spectrum imaging provide additional element distribution at high spatial resolution.

### 6.2.1 *Electron Diffraction*

Electron diffraction is widely used to determine the crystallinity as well as the phases in catalysts. Comparing to XRD, it is more sensitive to nanometer-sized particles and local structures. In an electron diffraction pattern, a diffraction spot arises from the reflection by a series of crystallographic planes (hkl) and the geometry is described by Bragg's law, $2d\sin\theta = \lambda$, where $\lambda$ is the wavelength of the incident electrons, $d$ the lattice planar spacing of hkl crystallographic planes, and $\theta$ the scattering angle. The electron wavelength, $\lambda$, is determined by the accelerating voltage, and the diffraction angle, $\theta$, can be calculated from the diffraction pattern by measuring the distance from the incident beam spot to the diffraction spot when the camera length is known. The lattice planar spacing, $d_{hkl}$, therefore can be derived. Phase determination is achieved by comparing the $d$ values with the crystallographic data of possible phases. Due to the complexity a diffraction pattern often exhibits, chemistry information about the sample is often helpful.

For a structural study of small features in catalysts, for example, nanoparticles with a size of a few to several tens of nanometers, electron nanodiffraction is particularly useful [18, 20]. By choosing a small spot size, the electron beam can be converged into a probe, illuminating only the area of interest. With a convergence angle smaller than the Bragg diffraction angle, diffraction disks are resolved individually, that is, the so-called Kossel–Möllenstedt pattern [21] can be obtained. Indexing the zero-order Laue zone (ZOLZ) in such a pattern is similar to that of the selected-area electron diffraction (SAED) pattern. In addition, the dynamical contrast in diffraction disks can be used for symmetry determination [20].

### 6.2.2 HRTEM and HAADF STEM

HRTEM and HAADF STEM are very important techniques to study the structure of nanoparticles at the atomic level and have been extensively applied to determine the configuration of supported metal catalysts [12, 22–24]. The successful application of the spherical aberration (Cs) corrector [25, 26] has significantly improved the point resolution of the electron microscope down to the subangstrom. More importantly, it eliminates the delocalization [27, 28], which is mainly caused by spherical aberration of objective lens and represented by image details appearing beyond the actual boundary. When the features of interest are of nanometer or subnanometer scale, the interference from the de-localized contrasts makes the precise determination of the structure of nanoparticle complex as well as surface or defect characterization almost impossible. With aberration-corrected HRTEM and HAADF STEM [29], the atomic configurations of nanoparticles and clusters have been successfully resolved.

HRTEM directly reveals the local structures of catalysts, which are highly relevant to special properties and offer information about reaction mechanisms. Crystallinity, lattice spacings, and crystallographic orientations of the sample, as well as defect structures such as dislocation, planar defect, interface, and cluster configuration, can be readily resolved. Surface facets and features such as roughness and decoration by other species can also be resolved by profile imaging [30]. Care must be taken when interpreting HRTEM images. Only for a sample thin enough, satisfying weak-phase object approximation (WPOA) [31] or pseudo-WPOA [32], the contrast in an HRTEM image taken near the Scherzer focus condition [33] corresponds to the positions of atomic columns.

STEM images are formed by detecting selectron scattered to a certain range of solid angles after interacting with the specimen. By different detector configurations, bright-field (BF)-STEM, annular dark field (ADF)-STEM, annular bright field (ABF)-STEM and HAADF STEM can be obtained [34]. Among them, HAADF STEM is most widely applied to catalyst characterization due to several advantages in contrast to conventional the TEM imaging technique. Firstly, by using a high-angle annular detector, coherent Bragg diffraction can be excluded from contributing to STEM images. When a high collection angle is selected for an ADF detector, typically >100 mrad, incoherent thermal diffuse scattering originating from the vibration of lattice atoms dominates in the total scattering. Under this condition, the integrated intensity over the detector area at a very high scattering angle is proportional to $Z^2$ in approximation of unscreened Rutherford scattering, but the real exponent is less than 2, depending on many factors [35, 36]. The fact that the intensity is highly dependent on $Z$ and much less dependent on the orientation of crystals and focusing conditions offers an easier interpretation of HAADF STEM images and additional chemistry information about element distributions. Also due to this reason and its high contrast, HAADF STEM becomes a more preferred imaging mode for particle size measurement and electron tomography [37]. On the basis of it, several analytic methods have been developed to precisely measure the particle size, for example, by integrating the intensity from the area of the cluster to quantify the number of atoms in it or by the blurring propagation method [38, 39]. Secondly, in STEM mode, an electron beam is focused onto a very small probe with the size down to the subnanometer. An EDX spectrum can be collected from a very local site in the specimen. Due to the annular type of the STEM detector, electron energy loss spectra and/or nanodiffraction patterns can also be acquired at the same time as scanning. This offers versatile possibilities of multidimensional mapping at high spatial resolution, such as spectrum imaging, spectrum line profile, and diffraction mapping, from which phase maps and orientation maps can be derived. Lastly, for some beam-sensitive materials, it is convenient to control the total dose for imaging by controlling the dwell time on each pixel. Therefore the electron dose received before the final exposure is equal to the one received during the search at low magnification and therefore greatly reduced. With imaging processing techniques, useful structure information can be extracted even from very noisy images [40].

### 6.2.3 Analytic TEM

Coulomb interaction between a fast incident electron and an atomic electron may result in excitation of the atomic electron to an unoccupied electronic state at a high-energy level and creating an electron hole. An electron from an outer, higher-energy shell then fills the hole, and the difference in energy between the higher-energy shell and the lower-energy shell may be released in the form of X-rays. This amount of energy is characteristic for each element and can be detected by EDX spectroscopy. At the same time the incident electron loses the same characteristic amount of energy. The energy loss in the electron beam passing through thin TEM sample can be detected by EELS [41], reflecting also composition of the specimen. These two spectroscopic methods are available in most analytic TEM nowadays. To some extent, they are complementary to each other. EDX has a wide energy range covering typically 0–40 keV. Within this range, except for H, He, and Li, almost all other elements have excitations from one or two of the K, L, and M shells. However, the energy resolution is only about 130 eV and different excitation lines from different elements are often found overlapped with each other, which makes element identification and quantification difficult. EELS detects only an energy loss range of 0–2000 eV; therefore not all elements have distinguishable edges for composition analysis. Nevertheless the energy resolution in EELS is much higher than in EDX, which is normally better than 1 eV in a transmission electron microscope equipped with a field emission gun and could be reduced to below 0.2 eV for a transmission electron microscope equipped with a monochromator [42]. In addition, energy loss near edge structure (ELNES) reflects the unoccupied states of a higher-energy level and therefore can be used for probing the electronic configuration. For example, ELNES on K edges of oxygen and carbon and L edges on transition metals in the fourth period from Sc to Cu, etc., can serve as "fingerprints," reflecting changes in the oxidation state, chemical bonding, and coordination environment of the detected species. Furthermore, by filtering electrons at an energy loss corresponding to a characteristic inner-shell excitation and background energy windows before the edge, it is possible to obtain images representing the distribution of the interested element [43]. This technique is called EFTEM and such element mapping can reach a lateral resolution down to 1 nm.

Both EDX and EELS can be used for quantitative analysis of the local composition of catalysts. As mentioned in Section 6.2.2, in STEM mode, EDX and EELS spectrum imaging or line profiles can be obtained at the same time. Multivariate statistics analysis [44, 45] to these spectra series not only gives element distribution but also gives the constitution of components, which will help to determine the phases, especially for unknown structures. Recently, the application of probe correctors in combination has enabled high current in subangstrom probes. By detecting X-rays with the high efficient in-column silicon drift detector (SDD) or by EELS, atomically resolved element mapping has been achieved [46, 47]. It is an exciting technique that makes it possible to study the element distribution in a single nanometer-sized multimetallic particle, as well as the surface decoration by some species and enrichment of some groups at the interface.

## 6.2.4 Electron Tomography

The above-described TEM characterization techniques for thin specimens reveals the projected structure along the incident electron beam direction. However, nanocatalyst particles are usually deposited on supports with a high surface area and pores of sizes from micrometers down to nanometers, such as carbon materials, silica, alumina, and titania. The preferred anchoring position of metal particles on supports has attracted great interest due to the possible confinement effect and shape effect introduced by supports. The distribution of particles at special locations is expected to help in the understanding of the activity and stability of these particles. Electron tomography is therefore valuable for studying the 3D structure of supported catalysts at a resolution down to ~1 nm [48, 49].

Electron tomography provides an approach to explore 3D structures by acquiring a series of 2D images representing projected structures at different tilting angles, typically in a range of $\pm$ 80° with a step of 1–2 degrees. After the alignment, usually done by tracking fiducial markers or catalyst particles themselves, tilting angles, image shifts, and other image distortions can be corrected and an aligned tilt series of images is obtained. Several algorithms can be used to calculate the 3D object on the basis of the aligned projected images. The most widely used methods are weighted back projection (WBP) and the simultaneous iterative reconstruction technique (SIRT) [49, 50]. The last step is the visualization of the 3D structure. Sliced cross sections, rendered surfaces of the interested object parts, are normally used to represent both the interior and the external morphology of the objects. With segmentation, different materials of different densities, usually metal particles and supports, including voids as well, can be distinguished and the surface of each material and the interface between them can be visualized. Furthermore, quantitative analysis of particle size, shape, loading, percentage of particles situated at certain locations, and proportion of the pore volume can be applied [51, 52].

Basically, any imaging type available in TEM, including TEM, STEM, EFTEM, and STEM-EDX/EELS maps, can be used for tomography. Practically, TEM images show strong diffraction contrast, depending on the orientation of crystals, and therefore are not recommended for tomography reconstruction for crystallized catalysts. HAADF STEM, due to its high contrast related with density and much less dependence of crystallographic orientation, becomes the standard imaging mode for electron tomography. Acquiring a tilt series of element maps either by EFTEM or by spectrum imaging is more challenging because of the prolonged acquisition time and demanding stability of the catalysts exposed to long-time electron beam illumination. Nevertheless, tomography with element maps provides 3D compositional information and has shown interesting applications in resolving 3D structures consisting of different elements being similar in density [37, 53, 54].

As an example, HAADF STEM electron tomography was used to study the 3D structure of carbon nanofiber (CNF)-supported Au catalysts to show the selective deposition of the particles by different synthesis procedures, that is, incipient wetness impregnation and deposition from poly(vinyl alcohol) (PVA)-protected Au colloids. The cross-sectional slice intersecting a few particles on one representative CNF for each catalyst is shown in Fig. 6.1. For impregnated Au/CNFs, particles were found both inside and outside the CNFs, while for Au(PVA)/CNFs, particles were found exclusively on the outer surface of the CNFs. One possible explanation for the selective anchoring is that the PVA polymer around the preformed Au nanoparticles increases the total particle size to about 20–30 nm. Since the diameter of CNFs is about 40–80 nm, particles can hardly immigrate into the inner space of the fiber.

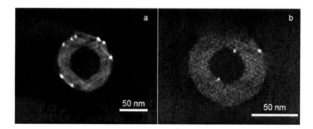

**Figure 6.1** Cross-sectional slices derived from electron tomography showing (a) preformed Au(PVA) particles exclusively situated on the outer surface of N-CNFs and (b) Au particles prepared by impregnation situated on both inner and outer surfaces of N-CNFs.

## 6.3 Structures of Au-Based Catalysts

### 6.3.1 *Crystal Structure of Au Nanoparticles*

Bulk gold has a face-centered cubic (fcc) structure. It can be envisaged as sequential stacking of close-packed atomic planes along the [111] direction. The change of sequence leads to the formation of a stacking fault or twin boundary in a grain. Since twinning does not change the interatomic distance between the nearest atoms at the twin boundary, the formation energy of a twin boundary is very small; therefore twinning and de-twinning take place easily in a Au crystal. The equilibrium shape of a Au particle in vacuum is determined by the minimum total surface free energy of the crystal with a given volume. The resultant shape is well known as the Wulff construction [55, 56] for a large single-crystal particle. When the metal particle size decreases, the structure becomes much more complicated than expected from equilibrium considerations [56]. Firstly, with a decrease of particle size, the energy of edge and corner atoms becomes more important since the facet area shrinks along with the particle size. Secondly, (111) faceting becomes more favorable at the expense of (100) and (110) facets, which leads to the formation of multiply twinned particles, for example, decahedra and icosahedra particles [57, 58], when the decrease in total free energy is enough to compensate the extra energy due to the introduced twin boundary and distortion. Furthermore, for supported nanoparticles, the influence from interface and perimeter atoms is of more importance. The adsorbents also changes the morphology of the metal particles.

With these complicating factors, Au nanoparticles exhibit various shapes. Generally cuboctahedron, decahedron, and icosahedron, as well as various combinations and modified types based on them, are mostly observed [57, 59]. A cuboctahedral particle is a single crystal and can be recognized by eight triangular {111} facets and six square {100} facets. A decahedral particle consists of five tetragonal parts of fcc Au, each having four {111} facets. This five tetragonal parts are joined together around a fivefold zone axis, which is along one edge from each tetragon in the <110> direction. Each two neighboring tetragons shares a common {111} plane as the twinning interface. Such a decahedron then has all exposed surfaces being {111} facets. As a request of an ideal fivefold symmetry, the angle between the neighboring twinning interfaces needs to be 72°, while the angle between two {111} planes is 70.53°. Obviously, a decahedral particle is strained and the strain increases with the radius [60]. Therefore a decahedral particle is not favored in energy when its size is above a certain diameter. Very often, truncation takes place close to the outermost twinning interface to reduce its area and therefore to lower the strain. Icosahedral particles can be built by 20 tetragons, with all of them sharing one apex in the center of the particle. It is also distorted compared to an fcc lattice but in a more complicated way. Depending on the projection direction, either an HRTEM image or nanodiffraction may reflect the twofold, threefold, and fivefold symmetry of the particle along different directions. Schematic diagrams of these three basic types of Au particle configuration are shown in Fig. 6.2a–c and the corresponding HRTEM images by aberration corrected TEM are shown in Fig. 6.2d–f. The structures of real Au catalyst particles are more complicated by truncating the tetrahedron to form new exposed edges and by formation of steps, kinks, and even isolated

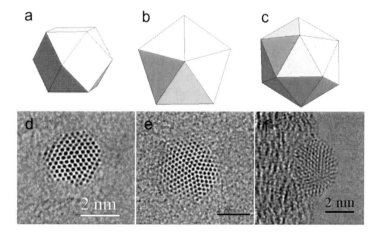

**Figure 6.2** Schematic diagrams (a–c) and corresponding experimental images (d–f) of three basic types, that is, cuboctahedron, decahedron, and icosahedron, of Au particle configuration.

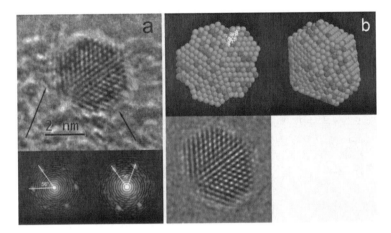

**Figure 6.3** (a) The representative HRTEM image of a multiply twinned particle in the form of a truncated decahedron configuration with insets showing FFTs from both sides of the visible twin boundary; (b) the modeled truncated decahedron particle viewed along the fivefold axis (top left) and the orientation that is analogous to the projection of the HRTEM image (top right); the simulated HRTEM image according to the model in (b) using microscope parameters of Philips CM200 FEG and the focus value of −130 nm, indicating good agreement with the experimental image.

atoms on the surface. Moreover, for small clusters and particles, the structure can be deformed from the highly symmetric shape to form disordered structural isomers.

Since the structure of most of Au nanoparticles is much more complicated than a single-crystal fcc structure, the interpretation of HRTEM images of randomly oriented particles is not straightforward. Very often the twinning parts are overlapped in the projection; therefore the measured lattice spacings and the angle between lattice planes could be an artifact that is originated from the interference of lattices in different twinning parts. For example, the HRTEM image of a Au-Pd particle, as shown in Fig. 6.3a, is frequently observed. A twin boundary is visible through the middle of the particle. The fast Fourier transforms (FFTs) of the parts at both sides of the twin boundary are also inserted in Fig. 6.3a. Two sets of (111) reflections are visible in both FFT diffractograms. The angle between the two sets of (111) planes in both FFT maps is ca. 56° instead of ca. 71°, as expected in an fcc single crystal. Modeling can prove that such lattice characteristics are produced by two overlapping fcc lattices forming a twin boundary in between. On the basis of the HRTEM images, a multiply twinned structure owning the shape of truncated decahedra is suggested. A model of the structure along the fivefold symmetric axis is shown in Fig. 6.3b (top left). In Fig. 6.3b (top right), the same particle is tilted to a specific orientation to give the projection analogous to the experimental HRTEM image. HRTEM images are simulated for a series of focus values in this projection. The simulation with a defocus value of −130 nm shown in Fig. 6.3b (bottom) is in good agreement with the contrast variations observed in the experimental image [24]. In the case of icosahedral particles, the lattice fringes in HRTEM images are even more complicated, especially when the particle is on random orientation other than the symmetric axes. HRTEM simulation is very useful to determine the structure of a multiply twinned particle. The simulation results of metals particles on a number of projections have been nicely published in [58, 61].

### 6.3.2 Interface and Surface Structures of Supported Au Nanoparticles

The support plays an essential role in tuning the fine structures of Au particles supported on it. One important fact is that the activity of Au depends strongly on the support and preparation methods [62–64]. Additionally, only the *size effect* is not sufficient to explain the catalytic behavior of supported Au particles. This implies that the activity is determined by specific sites either on the Au surface or on the interface between Au and the support. Recently, it has been reported that CO adsorption is the same for Au monolayer islands and extended Au surfaces. It is suggested that the difference in activity comes from uncoordinated Au sites [65]. The number of such sites will be strongly dependent on the shape and roughness of the particle. Hereon the support properties can have a big influence. There are also experiments and theoretical calculations showing that Au adatoms and clusters on ultrathin oxide films on metal substrates can be charged either positively or negatively, depending on the work function between the oxide and the metal substrate [66]. The charged Au atoms form a regular array due to electrostatic repulsion. This effect may deeply modify the structures and, therefore, properties of the supported species.

Gold nanoparticles can be synthesized on a crystalline oxide support such as $TiO_2$, $Al_2O_3$, $SiO_2$, $Fe_2O_3$, and $CeO_2$ and on a variety of carbon materials by using methods such as impregnation, sol immobilization, vacuum evaporation, etc. [67–72]. To understand the metal support interaction and its effects on catalyst activity, HTRTEM and HAADF STEM have been widely used to explore the atomic structure at the metal support interface by so-called profile imaging [30, 67, 68, 73]. This technique is to search for the outermost part of the supported catalysts, normally avoiding overlapping with other crystallites and carbon film from the TEM grid, and to image the interface being parallel to the electron beam as well as the surface configurations. Importantly, the application of an image corrector is able to eliminate the phenomenon known as *de-localization*, as mentioned in Section 6.2.2, to provide electron micrographs where interface configuration can be sharply resolved. Achieving subangstrom resolution, aberration-corrected HAADF STEM is also more and more used for profile imaging. One must be aware that the profile image shows the projected structure. It is not straightforward to distinguish whether the appeared surface of a particle is an edge of two intersected facets or is a real terminating plane that is parallel to the beam. Considering this issue, the HAADF STEM image has the advantage that the intensity of an atom column can be quantified to obtain the number of atoms in one column; therefore the shape of nanoparticles can be envisaged. With HRTEM, image simulation is often helpful to verify the constructed interface model [74, 75].

The equilibrium shape of a supported macroscopic crystal is defined by the surface energy of the facets and the interaction with the substrate as quantified by the adhesion energy [55]. When the particle size is reduced to the nanometer range, the equilibrium shape is modified due to the nonnegligible edge energies. In addition, the adsorption of foreign atoms or molecules, and the presence of strain at the interface due to a misfit between the lattices of the support and of the deposited crystal, further modifies the shape of the supported particles. The adsorbed molecules often reduce the anisotropy of the particle and lead to a more spherical

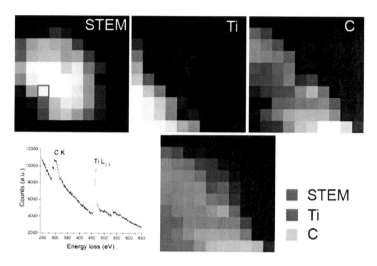

**Figure 6.4** STEM-EELS spectrum imaging of one Au particle supported on TiO$_2$. An EELS spectrum including a C K-edge and a Ti L$_{2,3}$-edge is acquired from each pixel. A Ti map and a C map can be derived, where the C species are enriched at the interface.

shape [76, 77]. Au particles synthesized in the existence of a protective layer clearly show the decoration of the residue on their surfaces, even after washing the catalysts under mild conditions. The protection layer was also observed to decorate the perimeters at the interface, which could assist in the adhesion of particles to the support. As an example, Au colloidal particles with PVA immobilized on TiO$_2$ are investigated by STEM-EELS spectrum imaging. A carbon species at Au/TiO$_2$ has been clearly revealed, as shown in Fig. 6.4. In contrast, Au particles deposited on polished TiO$_2$ surfaces by vacuum evaporation show a different interface structure dependent on the particle size. When the particle is smaller than a few nanometers, gold atoms preferentially attach to specific sites on the TiO$_2$ surface and form an epitaxial hetero-interface. As the gold size becomes larger, the gold–TiO$_2$ interface loses lattice coherency in order to accommodate the large lattice mismatch between the two dissimilar crystals [68].

Carbon-based materials, like activated carbon (AC), CNTs, and CNFs, have been widely used as catalyst supports as it is possible to manipulate them by specific chemical and thermal treatments aimed to tune the surface chemistry. Theoretical studies on the interaction between Au clusters and graphite show that the stability and structure of Au are highly dependent on the Au cluster size and the defects on graphitic layers [78–80]. The exact type of carbon support exerts a great influence on the catalytic properties, even when the Au particles are preformed by the sol immobilization method [81]. Therefore, the preformed nanoparticles offer an ideal initial structure to investigate the structural modification introduced by different carbon supports.

Detailed characterization by aberration-corrected HRTEM has been done on two catalysts consisting of 1 wt% Au on two different types of CNF supports. Au particles were synthesized by a two-step methodology (generation of nanoparticles, followed by their immobilization on the support). Two commercial CNFs were chosen as the support: The one denoted as PR24-PS is produced by pyrolytically stripping the as-produced fiber to remove polyaromatic hydrocarbons from the fiber surface, and the other one PR24-LHT is produced by heat-treating the fiber at 1500°C. The PR24-PS nanofiber has about a 2 nm deposited layer of carbon on the surface, which consists of highly defective and discontinuous graphitic layers. PR24-LHT exhibits a more ordered and graphitized surface. The layer of deposited carbon is only about 0.5 nm. It was found that the final shape of the Au particles is related to the surface structures of the support. In the case of Au/PR24-PS, the Au particles were attached to the PR24-PS nanofibers without a distinguishable orientation relation between Au low-index crystallographic planes and the CNF surface. Among the examined 167 particles, only 36 particles (21.6%) were found to have one set of {111} planes parallel to the CNF surface and the rest were randomly oriented (Fig. 6.5). Most of the particles exhibited a polyhedral shape, for example, cuboctahedron, decahedron, icosahedron, and truncated alternatives. Figures 6.6a and 6.6b show two Au particles of a size of about 3 nm with a cuboctahedron and an icosahedron shape, respectively. The Au particle with the cuboctahedron shape has clean surfaces. However, looking carefully at the upper-right corner where the two {111} surfaces and two {100} surfaces intersect at one common point, there is one missing atom (arrow). Furthermore, this particle is situated on the CNF surface, not with its (100)

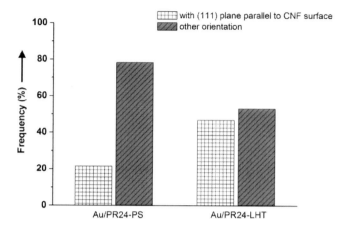

**Figure 6.5** 167 particles from Au/PR24-PS and 218 particles from Au/PR24-LHT were imaged by HRTEM and the percentage of the particles with (111) planes parallel to the CNF surface and the particles with other orientation was evaluated statistically for both catalysts. The Au/PR24-LHT shows a much higher percentage of particles, with (111) planes parallel to the CNF surface than the Au/PR24-PS.

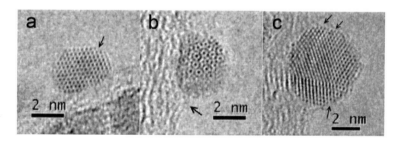

**Figure 6.6** Aberration-corrected HRTEM images of Au particles supported on PR24-PS with (a) modified cuboctahedral configuration, (b) icosahedral configuration showing interaction between particle and CNF surface, and (c) surface carbon decoration having uncovered windows (indicated by arrows).

plane or (111) plane parallel to the CNF surface, but it is inclined. Consequently, the atoms around the intersection ridge of the (100) and (111) surfaces in contact with the CNF are missing so that the interface between the particle and the CNF is smoothed without any sharp edge. Such a configuration reduces the contacting area.

In Fig. 6.6b, the icosahedron nanoparticle was imaged on neither its sixfold nor its fivefold axes, therefore exhibiting a complicated projected contrast. Here we see again the modification of the interface between the particle and the CNF support. The contacting plane on the particle is more flattened than the side exposed to the vacuum and some carbon decoration is observed at the perimeter of the interface (indicated by the arrow in Fig. 6.6b), due to either strong metal–carbon interaction or the residual protection agent, PVA. These two examples offer direct evidence that both metal particles and the CNF support undergo considerable reconstruction, which modifies the exposed surfaces and interfaces.

It is also frequently observed that the particle is well covered by a layer of carbonaceous species but usually not completely. Figure 6.6c is such an example showing a particle decorated with the layer. In the places where indicated by arrows, windows without a covering can be distinguished. This is due to the residual PVA applied when the metallic Au sol is formed. It is therefore apparent that the contact between the metal Au and the support surface is mediated by the protector (PVA). This is in agreement with the observation reported in Ref. [82], where the extent of wettability of Au particles was differentiated between carbon and $TiO_2$ supports. From Fig. 6.6c, we can see that the surface decoration is directly bound to the CNF support at the perimeter of the interface. The holes in the capping are essential to the accessibility of the reactant to the particle. Moreover, it should be noticed that such carbonaceous surface decoration also makes the particle surface rougher so that it has less flat facets and more uncoordinated sites.

The most noticeable difference between Au/PR24-LHT and Au/PR24-PS is that statistically the {111} planes of Au particles on PR24-LHT are more frequently observed to be parallel to the CNF surface than in the case of PR24-PS. For the 218 particles imaged by HRTEM, 102 particles (46.8%) were found to have one set of {111} planes parallel to the CNF surface (Fig. 6.5). This percentage is much higher than in the case of Au/PR24-PS (21.6%). Such correlation is extraordinarily prominent for Au particles with a size smaller than 3 nm. Within this range, the percentage of the oriented particles reaches 72.1%, in contrast to 44.8% for particles with a size between 3 nm and 4 nm and only 21.2% for particles bigger than

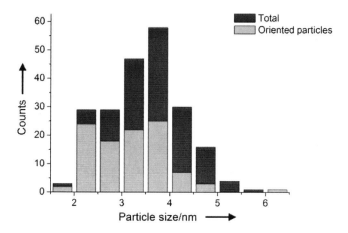

**Figure 6.7** The histograms of size distribution for the epitaxially oriented particles and for all the particles in the case of Au/PR24-LHT.

4 nm. The histograms of the size distribution for oriented particles and for all the particles are shown in Fig. 6.7, respectively. Figure 6.8a demonstrates a representative particle with one set of {111} lattice planes directly sitting on the CNF surface. The particle has a configuration similar to the cuboctahedron shape, with {111} and {100} surfaces exposed. A structure model has been constructed and shown in Fig. 6.8b. The particle has a relatively large (111) base plane in contact with the supporting CNF. However, the contact angle is larger than 90° and there is no distinguishable carbonaceous adlayer binding the Au metal particle to the CNF support. Moreover, a stacking fault is observed at the second stacking plane on the CNF. These observations strongly suggest that the more graphitized CNF surface stabilizes the first few layers of Au (111) stacks and the whole particle keeps more or less the regular shape and less surface-decorated carbon is found on such particles. This tendency is more prominent for smaller particles. Similar to the case of Au/PR24-PS, the lattice spacing in Au particles stays constant from top to bottom within the accuracy of the measurement, which means that no carbon is dissolved in Au. Statistically particles larger than 4 nm are more inclined to form spherical shapes as the one shown in Fig. 6.8c. Among the 218 imaged particles, 52 particles (25%) are larger than 4 nm, and 39 of them (75%) were found in spherical shape. Due

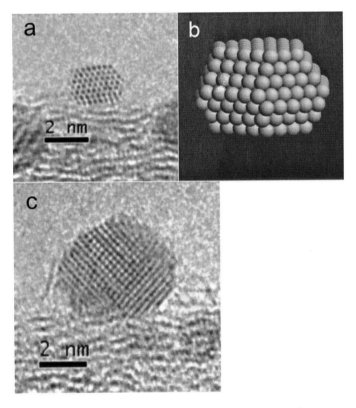

**Figure 6.8** Aberration-corrected HRTEM images of Au particles supported on PR24-LHT. (a) A representative 2–4 nm particle with the (111) surface epitaxially parallel to the graphitic layer of the CNF. (b) Structure model derived from the image in (a). (c) A larger particle with spherical shape shows carbon binding to the CNF surface.

to the relatively low fraction of the interface area, carbon binding is essential for the stabilization of these particles on the support surface against leaching and coarsening. Previous experimental data showed the increasing of particle dispersion with increasing acidic groups, thus suggesting that such particles sit over stable oxygen functional groups due to the *oxophilicity* of gold [83].

With the two catalysts possessing a similar size but different shapes, testing in catalytic oxidation of glycerol in the liquid phase has shown similar activity but different selectivity. Au/PR24-PS presents selectivity toward glyceric acid, similar to that observed in the case of similarly sized AuNPs supported on AC (58%). Formiate and glycolate, probing the C–C bond cleavage, are formed in this case with a selectivity of 19% and 22%, respectively. Au/PR24-LHT produced these latter compounds with considerably higher selectivity (40% and 35%, respectively). The observation on the PR24-LHT-supported Au particles exposing low-index surfaces suggested that the active sites on these surfaces could promote the C1 and C2 products, that is, the C–C bond cleavage.

Because of the vacuum environment in most of the electron microscopes, it is challenging to study the morphology of nanoparticles in different atmospheres at an elevated temperature. Environmental TEM therefore has been developed to bridge this gap and to explore the real structure of catalysts under reaction conditions. In general there are two types of setups to perform environmental TEM experiments. One is based on a closed-cell transmission electron microscope holder [84] and the other is based on a differential pumping scheme [85]. For the closed cell, a specially designed transmission electron microscope holder allows gas to be injected onto the catalyst specimen enclosed between two thin electron transparent windows (e.g., C or SiN), which are separated by a spacer of appropriate size. This setup allows a gas pressure up to 1 bar, as well as the observation of the catalyst in liquid environments. The other setup is to construct a dedicated environmental transmission electron microscope by modifying the column. In this case, the gas pressure is usually controlled below 3000 Pa in the vicinity of the sample [86].

Au nanoparticles on different supports exposed to various atmospheres have been studied with ETEM. For CO oxidation on Au/CeO$_2$, with 1 vol % CO in 1 mbar CO/air, Au particles showed well-faceted surfaces. And in pure O$_2$, the particles tended to be round. This trend was also observed that the particles became less faceted with decreasing CO partial pressure. In contrast, the morphology of Au/TiC, which is not active in CO oxidation, seemed not affected by the gas environments [87]. A catalyst of Au/TiO$_2$ exhibited dynamically changed morphology during CO oxidation in a windowed environmental cell [88]. The application of aberration-corrected ETEM has achieved atomic resolution for surface reconstruction due to the presence of CO molecules [89]. More and more application of ETEM techniques will undoubtedly provide unique structural information under reaction conditions and therefore will help unravel the origin of gold's special catalytic properties.

### 6.3.3 Structure of Au-Based Bimetallic Catalysts

Alloying with noble metals, Au-based catalysts such as Au-Pd and Au-Pt, show extra high activity and high resistance to deactivation due to synergistic effects in contrast to the monometallic counterparts [82, 90–92]. The synergistic effect could be attributed to either geometric or electronic modifications on some specific sites. Although the reaction mechanisms are not clarified, the combination of two or more metals has increased largely the possibilities to tailor the structures therefore the reactivity of the catalysts. In addition to particle size and shape, the composition, distribution, and surface ordering should be taken into account for affecting the activity, selectivity, and stability of the catalysts.

Efforts have been dedicated to synthesize different types of bimetallic nanoparticles and to characterize them with advanced TEM techniques [93–98]. HAADF STEM is able to distinguish heavier atoms, which are brighter than the lighter ones, offering an opportunity to study the distribution of different metals. STEM-EDX offers element analysis within one single nanoparticle. Probe corrector has improved the spatial resolution down to the atomic level. HRTEM imaging on a single particle also helps to distinguish the different phases by measuring the characteristic lattice spacings. By these methods, it has been unraveled that the structure of bimetallic catalysts is dependent on support materials, synthesis procedures, and the posttreatment to the as-prepared catalysts. The inhomogeneity in composition and distribution is usually related to the ratio of alloyed metals and the specific particle size. For example, for the catalysts prepared by co-impregnation of the supports using incipient wetness with aqueous solutions of $PdCl_2$ and $HAuCl_4$ and following calcination at 400°C, the Au-Pd particles on $TiO_2$ and $Al_2O_3$ as supports were found to exhibit a core–shell structure, Pd being concentrated on the surface. In contrast, the Au-Pd/carbon catalyst exhibited Au-Pd nanoparticles, which were homogeneous alloys. The structure of bimetallic catalysts synthesized by the sol immobilization method can be controlled by the sequence of reduction of metal precursors, namely $PdCl_2$ and $HAuCl_4$. A random alloy, a Au core–Pd shell and a Pd core–Au shell structure, can be formed by reduction of a mixed precursor solution, reduction of $PdCl_2$ in the presence of Au(0) and reduction of $HAuCl_4$ in the presence of Pd(0), respectively [96]. A similar successive reduction method has been used to produce three-layer core–shell particles, which consist of an alloyed inner core, an Au-rich intermediate layer, and a Pd-rich outer shell, as revealed by HAADF STEM and EDX analysis [99]. The order of metal addition and reduction during initial sol formation affects not only the activity but also the selectivity.

To systematically study the ratio of metal components on the nanoparticle structure, a series of catalysts with Au:Pd ratios varying from 9.5:0.5 to 2:8 were synthesized following a two-step procedure: immobilization of Au sol onto AC followed by immobilization of Pd(0). These catalysts were characterized by TEM, HRTEM, EDX, and X-ray mapping techniques to obtain morphological information, particle size distributions, crystalline structure, and distribution of the two metals [24]. For catalysts with Au:Pd ratios of 9:1, 8:2, and 6:4, the HRTEM images reveal uniform lattice spacings between the Pd (111) plane (2.25 Å) and the Au (111) plane (2.35 Å), which implies the alloying state, in good agreement with Vegard's law [100, 101]. EDX from a number of different-sized particles suggested more or less a uniform composition and no segregation of Au or Pd. The catalyst with Au:Pd = 9.5:0.5 consists of mainly small particles as well as some big irregularly shaped particles (>10 nm). The small particles are Au rich and the lattice spacings are also close to that of Au (111) planes. EDX spectra showed varying Au:Pd ratios on some big particles. Pd-dominating particles were identified by lattices close to Pd (111) planes and by EDX in this catalyst. For catalysts with Au:Pd = 2:8, various lattice spacings were observed for different particles, indicating fluctuation in composition. Pd-rich particles are more often observed than in other catalysts. The results from the series catalysts have clearly confirmed the uniform alloy formation for certain Au:Pd ratios. These alloyed catalysts also

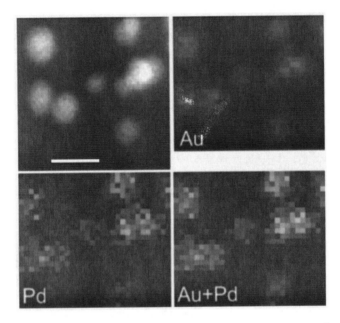

**Figure 6.9** STEM images, Au and Pd maps, and two superimposed maps from one piece of original Au/AC in the physically mixed catalyst after 1-hour reaction of benzyl alcohol oxidation.

show high activity in oxidation of glycerol, while those with the composition Au:Pd = 9.5:0.5 and Au:Pd = 2:8 deactivated quickly. The fact that the uniform Au-rich bimetallic catalysts were more active suggested that the surface Pd monomer in contact with Au has a prominent promoting effect on activity and stability, particularly reducing the leaching of Pd [102].

Another interesting experiment using physically mixed Au/AC and Pd /AC as a starting catalyst has shown that Au-Pd bimetallic active sites can be formed during the liquid-phase oxidation of alcohols [103]. TEM observations confirmed the monometallic nature of the fresh, physically mixed catalyst. The EDX spectrum imaging on a few particles from one piece of original Au/AC in the physically mixed catalyst after 0.5 hours in oxidation of benzyl alcohol has already confirmed reprecipitation of Pd on Au particles. The quantitative analysis of the integrated EDX spectrum from the mapped particles suggests a Au:Pd atomic ratio equal to 96.6:3.4.

Structures of Au-Based Catalysts | 159

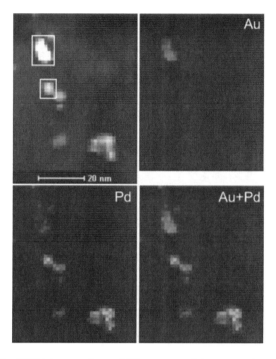

**Figure 6.10** STEM image, Au and Pd maps, and two superimposed maps from one piece of original Pd/AC in the physically mixed catalyst after 1-hour reaction of benzyl alcohol oxidation.

After a one-hour reaction, similar Au-Pd alloyed particles were observed on the original Au/AC. Figure 6.9 shows the STEM image, Au and Pd maps, and two superimposed maps of a few alloy particles formed after a one-hour reaction. The quantitative analysis of the integrated spectrum from these particles suggests a Au:Pd atomic ratio equal to 91.9:8.1, clearly indicating the increase of Pd concentration with reaction progress. The STEM image, Au and Pd maps, and superimposed map from a piece of original Pd/AC after one hour of reaction are shown in Fig. 6.10. A big particle was found to consist of both Au and Pd but Au is the majority. This particle detached the original Au/AC piece. Other particles were Pd dominating and contained no detectable Au. The evolution of the activity is also in good consistency with the structure evolution. The physically mixed Au+Pd catalyst showed a prominent increase in conversion compared to the monometallic catalysts. Especially, from 0.5 to 1 hour, the activity of the physically mixed catalyst increases drastically, which is largely different from the performance of the alloyed bimetallic catalyst. The results again reinforced the synergistic effect alloyed Au-Pd, which is the present case formed by soluble atomic Pd species (like $Pd^{2+}$) reprecipitated on supported Au nanoparticles forming alloyed species.

## 6.4 Outlook

With the improvement of resolution, both for TEM and for STEM, the transmission electron microscope has become a powerful yet routine instrument for catalyst characterization. The determination of the configuration in individual nanoparticles, including surface structure, interface structure between metal particle and support, distribution of different metal atoms, defects, and local strain, has already been realized by a variety of techniques available in TEM. On the one hand, our knowledge about supported catalysts has been greatly extended; on the other hand, we are more aware of the complexity of the system. For example, only with atomic resolution, we are able to observe the single atom and small clusters consisting of very few atoms. Whether these isolated atoms play a big role in reactions is hardly learned. Therefore simplified systems and theoretical simulations are necessary to interpret the experimental results.

TEM usually reveals various nanoparticle types in a catalyst. The quantitative analysis of the data represents an important direction of development, in addition to the instrument itself. Population statistics of a certain type of particles, with respect to size, shape, exposed facets, anchoring locations, etc., pictures a more complete image of the catalyst structure. With a tailored structure modifying selected parameters, one is able to correlate the structure with catalytic properties more decisively.

In situ TEM, not only ETEM, but also other electrochemistry experiments on MEMS chips, is becoming more and more important as it provides the structural information under reaction conditions. However, there are still considerable pressure and material gaps.

Moreover TEM only explore very thin specimens for an area of only a few tens of microns and the signals come from the whole depth rather than only from the surface. Therefore other complementary techniques, such as XRD, X-ray photoelectron spectroscopy (XPS), extended X-ray absorption fine structure (EXAFS) and X-ray absorption near edge structure (XANES), Raman, IR and X-ray ptychography, can be beneficial for investigating catalyst structures and understanding the reaction mechanisms. Many of these characterization techniques can be used for in situ experiments in different types of reactors, where reaction dynamics plays a more important role.

## Acknowledgments

The author appreciates Prof. Laura Prati and Dr. Alberto Villa from the University of Milan for their long-term and productive collaborations. The author also appreciates greatly the support from Prof. Robert Schlögl from Fritz Haber Institute of the Max Plank Society, Prof. Dangsheng Su from the Institute of Metal Research, and the Chinese Academy of Sciences for frequent, constructive discussions.

## References

1. Schlögl, R., and Abd Hamid, S.B. (2004). Nanocatalysis: mature science revisited or something really new?, *Angew. Chem., Int. Ed.*, **43**, pp. 1628–1637.
2. Tanaka, N. (2008). Present status and future prospects of spherical aberration corrected TEM/STEM for study of nanomaterials, *Sci. Technol. Adv. Mater.*, **9**, p. 014111.
3. Smith, D.J. (2008). Development of aberration-corrected electron microscopy, *Microsc. Microanal.*, **14**, pp. 2–15.
4. Mahfoud, Z., Dijksman A.T., Javaux, C., Bassoul, P., Baudrion, A.-L., Plain, J., Dubertret, B., and Kociak, M. (2013). Cathodoluminescence in a scanning transmission electron microscope: a nanometer-scale counterpart of photoluminescence for the study of II–VI quantum dots, *J. Phys. Chem. Lett.*, **4**, pp. 4090–4094.

5. Tizei, L.H.G., and Kociak, M. (2013). Spatially resolved quantum nano-optics of single photons using an electron microscope, *Phys. Rev. Lett.*, **110**, p. 153604.
6. Zhu, Y., Inada H., Nakamura, K., and Wall, J. (2009). Imaging single atoms using secondary electrons with an aberration-corrected electron microscope, *Nat. Mater.*, **8**, pp. 808–812.
7. V. Ardenne, M., and Beischer, D. (1940). Untersuchungen von Katalysatoren mit dem Universal-Elektronenmikroskop, *Angew. Chem.*, **53**, pp. 103–107.
8. Turkevich, J. (1945). Electron microscopy of catalysts, *J. Chem. Phys.*, **13**, p. 235.
9. Liu, J. (2005). Scanning transmission electron microscopy and its application to the study of nanoparticles and nanoparticle systems, *J. Electron Microsc.*, **54**, pp. 251–278.
10. Yang, J.C., Small, M.W., Grieshaber, R.V., and Nuzzo, R.G. (2012). Recent developments and applications of electron microscopy to heterogeneous catalysis, *Chem. Soc. Rev.*, **41**, pp. 8179–8194.
11. Thomas, J.M., and Midgley, P.A. (2004). High-resolution transmission electron microscopy: the ultimate nanoanalytical technique, *Chem. Commun.*, pp. 1253–1267.
12. Datye, A.K., and Smith, D.J. (1992). The study of heterogeneous catalysts by high-resolution transmission electron microscopy, *Catal. Rev.*, **34**, pp. 129–178.
13. Rodríguez-reinoso, F. (1998). The role of carbon materials in heterogeneous catalysis, *Carbon*, **36**, pp. 159–175.
14. Su, D.S., Zhang, J., Frank, B., Thomas, A., Wang, X., Paraknowitsch, J., and Schlögl, R. (2010). Metal-free heterogeneous catalysis for sustainable chemistry, *ChemSusChem*, **3**, pp. 169–180.
15. Tessonnier, J.-P., Rosenthal, D., Hansen, T.W., Hess, C., Schuster, M.E., Blume, R., Girgsdies, F., Pfänder, Timpe, N.,O., Su, D.S., and Schlögl, R. (2009). Analysis of the structure and chemical properties of some commercial carbon nanostructures, *Carbon*, **47**, pp. 1779–1798.
16. Serp, P., Corrias, M., and Kalck, P. (2003). Carbon nanotubes and nanofibers in catalysis, *Appl. Catal. A*, **253**, pp. 337–358.
17. Krivanek, O.L., Chisholm, M.F., Nicolosi, V., Pennycook, T.J., Corbin, G.J., Dellby, N., Murfitt, M.F., Own, C.S., Szilagyi, Z.S., Oxley, M.P., Pantelides, S.T., and Pennycook, S.J. (2010). Atom-by-atom structural and chemical analysis by annular dark-field electron microscopy, *Nature*, **464**, pp. 571–574.

18. Williams, D.B., and Carter, C.B. (2009). *Transmission Electron Microscopy* (Springer, Boston, MA).
19. Pennycook, S.J., and Jesson, D.E. (1991). High-resolution Z-contrast imaging of crystals, *Ultramicroscopy*, **37**, pp. 14–38.
20. Spence, J.C.H., and Zuo, J.M. (1992). *Electron Microdiffraction* (Springer, Boston, MA).
21. Kossel, W., and Möllenstedt, G. (1939). Elektroneninterferenzen im konvergenten Bündel, *Ann. Phys.*, **428**, pp. 113–140.
22. Mejía-Rosales, S.J., Fernández-Navarro, C., Pérez-Tijerina, E., Blom, D.A., Allard, L.F., and José-Yacamán, M. (2007). On the structure of Au/Pd bimetallic nanoparticles, *J. Phys. Chem. C*, **111**, pp. 1256–1260.
23. Thomas, J.M., and Gai, P.L. (2004). Electron microscopy and the materials chemistry of solid catalysts, *Adv. Catal.*, **48**, pp. 171–227.
24. Wang, D., Villa, A., Porta, F., Prati, L., and Su, D.S. (2008). Bimetallic gold/palladium catalysts: correlation between nanostructure and synergistic effects, *J. Phys. Chem. C*, **112**, pp. 8617–8622.
25. Freitag, B., Kujawa, S., Mul, P.M., Ringnalda, J., and Tiemeijer, P.C. (2005). Breaking the spherical and chromatic aberration barrier in transmission electron microscopy, *Ultramicroscopy*, **102**, pp. 209–214.
26. Haider, M., Rose, H., Uhlemann, S., Kabius, B., and Urban, K. (1998). Towards 0.1 nm resolution with the first spherically corrected transmission electron microscope, *J. Electron Microsc.*, **47**, pp. 395–405.
27. Marks, L.D. (1984). Dispersive equations for high resolution imaging and lattice fringe artifacts, *Ultramicroscopy*, **12**, pp. 237–242.
28. Zandbergen, H.W., Tang, D., and Van Dyck, D. (1996). Non-linear interference in relation to strong delocalisation, *Ultramicroscopy*, **64**, pp. 185–198.
29. Krivanek, O.L., Corbin, G.J., Dellby, N., Elston, B.F., Keyse, R.J., Murfitt, M.F., Own, C.S., Szilagyi, Z.S., and Woodruff, J.W. (2008). An electron microscope for the aberration-corrected era, *Ultramicroscopy*, **108**, pp. 179–195.
30. Zhou, W., and Thomas, J.M. (2001). HRTEM surface profile imaging of solids, *Curr. Opin. Solid State Mater. Sci.*, **5**, pp. 75–83.
31. Cowley, J.M. (1995). *Diffraction Physics, Third Edition (North-Holland Personal Library)* (North Holland).
32. Li, F.H., and Tang, D. (1985). Pseudo-weak-phase-object approximation in high-resolution electron microscopy. I. Theory, *Acta Crystallogr. Sect. A Found. Crystallogr.*, **41**, pp. 376–382.

33. Scherzer, O. (1949). The theoretical resolution limit of the electron microscope, *J. Appl. Phys.*, **20**, p. 20.
34. Liu, J., and Cowley, J.M. (1993). High-resolution scanning transmission electron microscopy, *Ultramicroscopy*, **52**, pp. 335–346.
35. Kirkland, E.J., Loane, R.F., and Silcox, J. (1987). Simulation of annular dark field stem images using a modified multislice method, *Ultramicroscopy*, **23**, pp. 77–96.
36. Wang, Z.L., and Cowley, J.M. (1989). Simulating high-angle annular dark-field stem images including inelastic thermal diffuse scattering, *Ultramicroscopy*, **31**, pp. 437–453.
37. Midgley, P.A., and Weyland, M. (2003). 3D electron microscopy in the physical sciences: the development of Z-contrast and EFTEM tomography, *Ultramicroscopy*, **96**, pp. 413–431.
38. Reed, B.W., Morgan, D.G., Okamoto, N.L., Kulkarni, A., Gates, B.C., and Browning, N.D. (2009). Validation and generalization of a method for precise size measurements of metal nanoclusters on supports, *Ultramicroscopy*, **110**, pp. 48–60.
39. Carlsson, A., Puig-Molina, A., and Janssens, T.V.W. (2006). New method for analysis of nanoparticle geometry in supported fcc metal catalysts with scanning transmission electron microscopy, *J. Phys. Chem. B*, **110**, pp. 5286–5293.
40. Buban, J.P., Ramasse, Q., Gipson, B., Browning, N.D., and Stahlberg, H. (2010). High-resolution low-dose scanning transmission electron microscopy, *J. Electron Microsc.*, **59**, pp. 103–112.
41. Egerton, R.F. (2011). *Electron Energy-Loss Spectroscopy in the Electron Microscope* (Springer, Boston, MA).
42. Tiemeijer, P.C. (1999). Measurement of Coulomb interactions in an electron beam monochromator, *Ultramicroscopy*, **78**, pp. 53–62.
43. Shuman, H., Chang, C.-F., and Somlyo, A.P. (1986). Elemental imaging and resolution in energy-filtered conventional electron microscopy, *Ultramicroscopy*, **19**, pp. 121–133.
44. Grogger, W., Hofer, F., and Kothleitner, G. (1998). Quantitative chemical phase analysis of EFTEM elemental maps using scatter diagrams, *Micron*, **29**, pp. 43–51.
45. Bonnet, N., Simova, E., Lebonvallet, S., and Kaplan, H. (1992). New applications of multivariate statistical analysis in spectroscopy and microscopy, *Ultramicroscopy*, **40**, pp. 1–11.
46. Allen, L.J., D'Alfonso, A.J., Freitag, B., and Klenov, D.O. (2012). Chemical mapping at atomic resolution using energy-dispersive x-ray spectroscopy, *MRS Bull.*, **37**, pp. 47–52.

47. Muller, D.A. (2009). Structure and bonding at the atomic scale by scanning transmission electron microscopy, *Nat. Mater.*, **8**, pp. 263–270.
48. Friedrich, H., de Jongh, P.E., Verkleij, A.J., and de Jong, K.P. (2009). Electron tomography for heterogeneous catalysts and related nanostructured materials, *Chem. Rev.*, **109**, pp. 1613–1629.
49. Zečević, J., de Jong, K.P., and de Jongh, P.E. (2013). Progress in electron tomography to assess the 3D nanostructure of catalysts, *Curr. Opin. Solid State Mater. Sci.*, **17**, pp. 115–125.
50. Gilbert, P. (1972). Iterative methods for the three-dimensional reconstruction of an object from projections, *J. Theor. Biol.*, **36**, pp. 105–117.
51. Friedrich, H., Sietsma, J.R.A., de Jongh, P.E., Verkleij, A.J., and de Jong, K.P. (2007). Measuring location, size, distribution, and loading of NiO crystallites in individual SBA-15 pores by electron tomography, *J. Am. Chem. Soc.*, **129**, pp. 10249–10254.
52. Biermans, E., Molina, L., Batenburg, K.J., Bals, S., and Van Tendeloo, G. (2010). Measuring porosity at the nanoscale by quantitative electron tomography, *Nano Lett.*, **10**, pp. 5014–5019.
53. Genc, A., Kovarik, L., Gu, M., Cheng, H., Plachinda, P., Pullan, L., Freitag, B., and Wang, C. (2013). XEDS STEM tomography for 3D chemical characterization of nanoscale particles, *Ultramicroscopy*, **131**, pp. 24–32.
54. Jarausch, K., Thomas, P., Leonard, D.N., Twesten, R., and Booth, C.R. (2009). Four-dimensional STEM-EELS: enabling nano-scale chemical tomography, *Ultramicroscopy*, **109**, pp. 326–337.
55. Henry, C.R. (2005). Morphology of supported nanoparticles, *Prog. Surf. Sci.*, **80**, pp. 92–116.
56. Marks, L.D. (1994). Experimental studies of small particle structures, *Reports Prog. Phys.*, **57**, pp. 603–649.
57. Ajayan, P.M., and Marks, L.D. (1990). Phase instabilities in small particles, *Phase Trans.*, **24–26**, pp. 229–258.
58. Urban, J., Sack-Kongehl, H., and Weiss, K. (1993). Computer simulations of HREM images of metal clusters, *Z. Phys. D Atoms, Mol. Clust.*, **28**, pp. 247–255.
59. Yacamań, M.J., Ascencio, J.A., Liu, H.B., and Gardea-Torresdey, J. (2001). Structure shape and stability of nanometric sized particles, *J. Vac. Sci. Technol. B Microelectron. Nanom. Struct.*, **19**, p. 1091.
60. Johnson, C.L., Snoeck, E., Ezcurdia, M., Rodríguez-González, B., Pastoriza-Santos, I., Liz-Marzán, L.M., and Hÿtch, M.J. (2008). Effects

of elastic anisotropy on strain distributions in decahedral gold nanoparticles, *Nat. Mater.*, **7**, pp. 120–124.

61. Koga, K., and Sugawara, K. (2003). Population statistics of gold nanoparticle morphologies: direct determination by HREM observations, *Surf. Sci.*, **529**, pp. 23–35.

62. Haruta, M. (2004). Gold as a novel catalyst in the 21st century: preparation, working mechanism and applications, *Gold Bull.*, **37**, pp. 27–36.

63. Hutchings, G.J. (2004). New directions in gold catalysis, *Gold Bull.*, **37**, pp. 3–11.

64. Bond, G.C., and Thompson, D.T. (1999). Catalysis by gold, *Catal. Rev.*, **41**, pp. 319–388.

65. Lemire, C., Meyer, R., Shaikhutdinov, S., and Freund, H.-J. (2004). Do quantum size effects control CO adsorption on gold nanoparticles?, *Angew. Chem., Int. Ed.*, **43**, pp. 118–121.

66. Freund, H.-J., and Pacchioni, G. (2008). Oxide ultra-thin films on metals: new materials for the design of supported metal catalysts, *Chem. Soc. Rev.*, **37**, pp. 2224–2242.

67. Akita, T., Tanaka, K., Kohyama, M., and Haruta, M. (2008). HAADF-STEM observation of Au nanoparticles on $TiO_2$, *Surf. Interface Anal.*, **40**, pp. 1760–1763.

68. Shibata, N., Goto, A., Matsunaga, K., Mizoguchi, T., Findlay, S., Yamamoto, T., and Ikuhara, Y. (2009). Interface structures of gold nanoparticles on TiO2 (110), *Phys. Rev. Lett.*, **102**(1-4), pp. 136015.

69. Zhou, W., Wachs, I.E., and Kiely, C.J. (2012). Nanostructural and chemical characterization of supported metal oxide catalysts by aberration corrected analytical electron microscopy, *Curr. Opin. Solid State Mater. Sci.*, **16**, pp. 10–22.

70. Miedziak, P.J., Tang, Z., Davies, T.E., Enache, D.I., Bartley, J.K., Carley, A.F., Herzing, A.A., Kiely, C.J., Taylor, S.H., and Hutchings, G.J. (2009). Ceria prepared using supercritical antisolvent precipitation: a green support for gold–palladium nanoparticles for the selective catalytic oxidation of alcohols, *J. Mater. Chem.*, **19**, p. 8619.

71. Wang, D., Villa, A., Porta, F., Su, D.S., and Prati, L. (2006). Single-phase bimetallic system for the selective oxidation of glycerol to glycerate, *Chem. Commun.* pp. 1956–1958.

72. Gajan, D., Guillois, K., Delichère, P., Basset, J.-M., Candy, J.-P., Caps, V., Copéret, C., Lesage, A., and Emsley, L. (2009). Gold nanoparticles supported on passivated silica: access to an efficient aerobic epoxidation

catalyst and the intrinsic oxidation activity of gold, *J. Am. Chem. Soc.*, **131**, pp. 14667–14669.

73. López-Haro, M., Cíes, J.M., Trasobares, S., Pérez-Omil, J.A., Delgado, J.J., Bernal S., Bayle-Guillemaud, P., Stéphan, O., Yoshida, K., Boyes, E.D., Gai, P.L., and Calvino, J.J. (2012). Imaging nanostructural modifications induced by electronic metal-support interaction effects at Au||cerium-based oxide nanointerfaces, *ACS Nano*, **6**, pp. 6812–6820.

74. Bernal, S., Botana, F.J., Calvino, J.J., López-Cartes, C., Pérez-Omil, J.A., and Rodríguez-Izquierdo, J.M. (1998). The interpretation of HREM images of supported metal catalysts using image simulation: profile view images, *Ultramicroscopy*, **72**, pp. 135–164.

75. Bernal, S., Baker, R.T., Burrows, A., Calvino, J.J., Kiely, C.J., López-Cartes, C., Pérez-Omil, J.A., and Rodríguez-Izquierdo, J.M. (2000). Structure of highly dispersed metals and oxides: exploring the capabilities of high-resolution electron microscopy, *Surf. Interface Anal.*, **29**, pp. 411–421.

76. Wang, T., Lee, C., and Schmidt, L.D. (1985). Shape and orientation of supported Pt particles, *Surf. Sci.*, **163**, pp. 181–197.

77. Wang, D., Villa, A., Su, D.S., Prati, L., and Schlögl, R. (2013). Carbon-supported gold nanocatalysts: shape effect in the selective glycerol oxidation, *ChemCatChem*, pp. 2717–2723.

78. Jiang, K., Eitan, A., Schadler, L.S., Ajayan, P.M., Siegel, R.W., Grobert, N., Mayne, M., Reyes-Reyes, M., Terrones, H., and Terrones, M. (2003). Selective attachment of gold nanoparticles to nitrogen-doped carbon nanotubes, *Nano Lett.*, **3**, pp. 275–277.

79. Akola, J., and Häkkinen, H. (2006). Density functional study of gold atoms and clusters on a graphite (0001) surface with defects, *Phys. Rev. B*, **74**, p. 165404.

80. McKenna, K.P., and Shluger, A.L. (2007). Shaping the morphology of gold nanoparticles by CO adsorption, *J. Phys. Chem. C*, **111**, pp. 18848–18852.

81. Porta, F., Prati, L., Rossi, M., and Scari, G. (2002). New Au(0) sols as precursors for heterogeneous liquid-phase oxidation catalysts, *J. Catal.*, **211**, pp. 464–469.

82. Enache, D.I., Edwards, J.K., Landon, P., Solsona-Espriu, B., Carley, A.F., Herzing, A.A., Watanabe, M., Kiely, C.J., Knight, D.W., and Hutchings, G.J. (2006). Solvent-free oxidation of primary alcohols to aldehydes using Au-Pd/TiO2 catalysts, *Science*, **311**, pp. 362–365.

83. Prati, L., Villa, A., Chan-Thaw, C.E., Arrigo, R., Wang, D., and Su, D.S. (2011). Gold catalyzed liquid phase oxidation of alcohol: the issue of selectivity, *Faraday Discuss.*, **152**, p. 353.
84. Creemer, J.F., Helveg, S., Hoveling, G.H., Ullmann, S., Molenbroek, A.M., Sarro, P.M., and Zandbergen, H.W. (2008). Atomic-scale electron microscopy at ambient pressure, *Ultramicroscopy*, **108**, pp. 993–998.
85. Boyes, E.D., and Gai, P.L. (1997). Environmental high resolution electron microscopy and applications to chemical science, *Ultramicroscopy*, **67**, pp. 219–232.
86. Wagner, J.B., Cavalca, F., Damsgaard, C.D., Duchstein, L.D.L., and Hansen, T.W. (2012). Exploring the environmental transmission electron microscope, *Micron*, **43**, pp. 1169–1175.
87. Uchiyama, T., Yoshida, H., Kuwauchi, Y., Ichikawa, S., Shimada, S., Haruta, M., and Takeda, S. (2011). Systematic morphology changes of gold nanoparticles supported on $CeO_2$ during CO oxidation, *Angew. Chem., Int. Ed.*, **50**, pp. 10157–10160.
88. Kawasaki, T., Ueda, K., Ichihashi, M., and Tanji, T. (2009). Improvement of windowed type environmental-cell transmission electron microscope for in situ observation of gas-solid interactions, *Rev. Sci. Instrum.*, **80**, p. 113701.
89. Yoshida, H., Kuwauchi, Y., Jinschek, J.R., Sun, K., Tanaka, S., Kohyama, M., Shimada, S., Haruta, M., and Takeda, S. (2012). Visualizing gas molecules interacting with supported nanoparticulate catalysts at reaction conditions, *Science*, **335**, pp. 317–319.
90. Bianchi, C.L., Canton, P., Dimitratos, N., Porta, F., and Prati, L. (2005). Selective oxidation of glycerol with oxygen using mono and bimetallic catalysts based on Au, Pd and Pt metals, *Catal. Today*, **102**, pp. 203–212.
91. Guczi, L. (2003). AuPd bimetallic nanoparticles on $TiO_2$: XRD, TEM, in situ EXAFS studies and catalytic activity in CO oxidation, *J. Mol. Catal. A Chem.*, **204-205**, pp. 545–552.
92. Marx, S., and Baiker, A. (2009). Beneficial interaction of gold and palladium in bimetallic catalysts for the selective oxidation of benzyl alcohol, *J. Phys. Chem. C*, **113**, pp. 6191–6201.
93. Tsuji, M., Ikedo, K., Matsunaga, M., and Uto, K. (2012). Epitaxial growth of Au@Pd core–shell nanocrystals prepared using a PVP-assisted polyol reduction method, *Cryst. Eng. Commun.*, **14**, p. 3411.
94. Herzing, A.A., Watanabe, M., Edwards, J.K., Conte, M., Tang, Z.-R., Hutchings, G.J., and Kiely, C.J. (2008). Energy dispersive X-ray

spectroscopy of bimetallic nanoparticles in an aberration corrected scanning transmission electron microscope, *Faraday Discuss.*, **138**, pp. 337–351.

95. Wang, D., Villa, A., Porta, F., Prati, L., and Su, D.S. (2008). Bimetallic gold/palladium catalysts: correlation between nanostructure and synergistic effects, *J. Phys. Chem. C*, **112**, pp. 8617–8622.

96. Tiruvalam, R.C., Pritchard, J.C., Dimitratos, N., Lopez-Sanchez, J.A., Edwards, J.K., Carley, A.F., Hutchings, G.J., and Kiely, C.J. (2011). Aberration corrected analytical electron microscopy studies of sol-immobilized Au + Pd, Au{Pd} and Pd{Au} catalysts used for benzyl alcohol oxidation and hydrogen peroxide production, *Faraday Discuss.*, **152**, p. 63.

97. Dash, P., Bond, T., Fowler, C., Hou, W., Coombs, N., and Scott, R.W.J. (2009). Rational design of supported PdAu nanoparticle catalysts from structured nanoparticle precursors, *J. Phys. Chem. C*, **113**, pp. 12719–12730.

98. Heggen, M., Oezaslan, M., Houben, L., and Strasser P. (2012). Formation and analysis of core–shell fine structures in Pt bimetallic nanoparticle fuel cell electrocatalysts, *J. Phys. Chem. C*, **116**, pp. 19073–19083.

99. Ferrer, D., Torres-Castro, A., Gao, X., Sepúlveda-Guzmán, S., Ortiz-Méndez, U., and José-Yacamán, M. (2007). Three-layer core/shell structure in Au-Pd bimetallic nanoparticles, *Nano Lett.*, **7**, pp. 1701–1705.

100. Tsen, S.-C.Y., Crozier, P.A., and Liu, J. (2003). Lattice measurement and alloy compositions in metal and bimetallic nanoparticles, *Ultramicroscopy*, **98**, pp. 63–72.

101. Denton, A., and Ashcroft, N. (1991). Vegard's law, *Phys. Rev. A*, **43**, pp. 3161–3164.

102. Villa, A., Wang, D., Dimitratos, N., Su, D.S., Trevisan V., and Prati, L. (2010). Pd on carbon nanotubes for liquid phase alcohol oxidation, *Catal. Today*, **150**, pp. 8–15.

103. Wang, D., Villa, A., Spontoni, P., Su, D.S., and Prati, L. (2010). In situ formation of Au-Pd bimetallic active sites promoting the physically mixed monometallic catalysts in the liquid-phase oxidation of alcohols, *Chem. Eur. J.*, **16**, pp. 10007–10013.

# Chapter 7

# X-Ray Photoelectron Spectroscopy Characterization of Gold Catalysts

**Gabriel M. Veith**

*Materials Science and Technology Division, Oak Ridge National Laboratory, Oak Ridge, TN 37831, USA*
veithgm@ornl.gov

## 7.1 Introduction

A continuing focus of gold catalyst work is attempting to correlate catalytic activity and selectivity with a nanoparticle's physical properties. These properties include particle size, particle shape, gold oxidation state, support oxidation state, particle stability, and catalyst electronic structure. Electron microscopy studies are useful for estimating catalyst particle sizes and how they evolve with time/temperatures. However, they suffer from a severe limitation: they only sample a small fraction of the gold particles. Indeed, if one were to sample 1000 2.5 nm hemispherical particles using transmission electron microscopy (TEM) this would only account for around 10%–16% of the total particles in a 1 g gold catalyst sample with 1 wt% gold loading. X-ray absorption

---

*Gold Catalysis: Preparation, Characterization, and Applications*
Edited by Laura Prati and Alberto Villa
Copyright © 2016 Pan Stanford Publishing Pte. Ltd.
ISBN 978-981-4669-28-3 (Hardcover), 978-981-4669-29-0 (eBook)
www.panstanford.com

spectroscopy (XAS) will provide information about the oxidation state and particle size (based on extended X-ray absorption fine structure measurements) but requires a synchrotron radiation source [1, 2]. These measurements have the advantage of being able to be performed in situ through specially designed cells [3]. $^{197}$Au-Mössbauer spectroscopy is a nuclear resonance technique suitable for identifying oxidation states and estimating chemical environments around the isotope of interest [4–6]. In this method $^{197}$Pt is irradiated with neutrons to generate a monochromatic gamma source. The monochromaticity derives from the nuclear transition from gamma decay. Fitting Mössbauer data results in two variables for gold, the isomer shift (IS) and the quadrapole splitting (QS). The IS is a measure of electron density in the 6$s$ orbital of gold. The QS provides an estimate of the asymmetric distribution of electrons around the gold atom. Larger QS is due to lower symmetry around the gold atom due to ligands. Together the IS and QS results provide an unequivocal description of the metal oxidation state and its local symmetry. The difficulties with Au-Mössbauer spectroscopy are the need to perform the measurements at low temperatures (~4.2 K), increase sensitivity, and produce the Pt source on a neutron reactor and the short half-life of the Pt source (18 hours).

X-ray photoelectron spectroscopy (XPS), the topic of this chapter, does not suffer from these limitations. Indeed, XPS methods average over the entire sample or sampling volume, providing a more statistically relevant snapshot of the gold catalyst. Furthermore, XPS can be applied in situ to gain insights into the origin of catalytic activity as a function of reaction conditions. For example, XPS can estimate the average oxidation state of the catalyst and support as well as provide evidence of catalyst stability. Naturally, given the insights available from this variety of applications of XPS, the literature concerning the use of these techniques is large and diverse.

To make this chapter usable to the average gold catalysis scientist it will focus on spectroscopic characterization of bulk gold catalysts prepared in air, for example, 1 wt% Au on $TiO_2$ powder. There are two reasons for this arbitrary selection. First, spectroscopic characterizations of model ultrahigh-vacuum (UHV) gold catalyst systems have been reviewed previously by Mayer et al. [7], and second, the chemistry of gold catalysts prepared in a UHV environment differs significantly from those prepared/stored at atmospheric conditions. These differences limit the applicability of UHV results to bulk catalysts. The initial section will begin with a description of the XPS technique, how the measurement is performed, and what information can typically be gained. This introductory portion is directed toward nonexperts in the field. References are provided for those interested in more details. Subsequent sections will highlight some types of studies that can be performed using examples gleaned from the gold catalyst literature. Given the large number of studies using XPS within the literature, only specific examples highlighting the utility of the method with respect to gold catalysts will be presented.

## 7.2 X-Ray Photoelectron Spectroscopy

XPS, also known as electron spectroscopy for chemical analysis (ESCA), involves irradiating a sample with an X-ray source [8]. If the energy of the X-ray source is large enough a photoelectron will be emitted from atomic orbitals. Most of the emitted electrons are recaptured by the sample. However, electrons close to the surface can be excited from the sample and into vacuum. The kinetic energy (speed) of the photoemitted electrons is measured by the spectrometer. The kinetic energy of the emitted electron is equal to the source energy less the binding energy and instrument work function constant ($\phi$) (Eq. 7.1):

$$E_{\text{kinetic energy}} = E_{\text{photon}} - E_{\text{binding}} - \phi_{\text{work function (constant)}} \quad (7.1)$$

The source energy $E_{\text{photon}}$ is a constant and depends on the X-ray source; typical sources are Al (1486.7 eV) and Mg (1253.4 eV), with more specialized sources usually used by XPS labs. The instrument work function ($\phi$) is a small correction factor that accounts for electron absorption on the detector walls. Those new to the field should be aware that XPS spectra may be presented as a function of either kinetic energy or binding energy. In this chapter we will use the more widely reported binding energies.

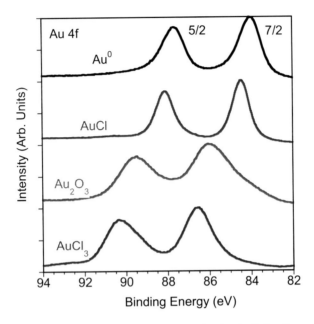

**Figure 7.1** Representative XPS data collected for Au, AuCl, Au$_2$O$_3$, and AuCl$_3$ reference materials using a pass energy of 5.85 eV and an Al X-ray source. The shifts to higher binding energy are due to changes in the gold oxidation state and ligand-withdrawing strength. Note the wide peaks evident from Au$_2$O$_3$ are due to the reduction of Au$^{3+}$ to Au$^0$.

The reported binding energies are unique to specific orbitals of individual elements. In the case of gold the most common binding energy reported originates from the filled Au $4f$ orbitals because this has the largest signal in the energy regime typically sampled by XPS (Fig. 7.1). The Au $4f$ spectrum is comprised of a doublet described as Au $4f_{7/2}$ and Au $4f_{5/2}$. This doublet is due to the energy differences originating from spin–orbit coupling, which is best described using the standard notation from quantum mechanics. The sampled electron originates from the $f$-orbital (orbital momentum, $l = 3$) with the spin ($s$) of the sampled electron represented as $+1/2$ or $-1/2$. The total momentum ($j$) of these electrons is equivalent to $l + s$ and results in spin orbit values of 3.5 (7/2) or 2.5 (5/2). One will notice the differences in relative intensity of these orbitals (4:3), which is due to the degeneracy of these two levels. The degeneracy is defined as $2j+1$ so that for $l=7/2$, the degeneracy is 8 and for $l=5/2$ the degeneracy is 6. This results in an intensity ratio for the spin–orbit doublet of 8:6 or 4:3. Similar derivations can be made for all other photoelectron transitions.

The binding energy position is a function of the oxidation state. Typically, more oxidized species have a shift to positive binding energies, while reduced species are shifted toward smaller binding energies. These shifts are due to the loss or gain of screening electrons from the valence band, making it harder or easier to remove a core electron. The magnitude of the shifts is related to the electron-withdrawing strength of the ligands. Representative data from several gold samples, measured in the author's laboratory, are presented in Fig. 7.1. The magnitude of these shifts is related to the oxidation state of the gold species as well as the ligands surrounding the gold atom. In the case of gold compounds the binding energies shift from 84.0 eV for $Au^0$ to 86.7 for $Au^{3+}$. Measuring these binding energies is a good indication of the average oxidation state of the gold catalyst. The inelastic mean free path of the Au $4f$ electron is about 1.4 nm with a standard laboratory source. This means that virtually all of the emitted electrons measured with the spectrometer originate from the top 4.2–5.0 nm of the sample. Consequently gold catalysts buried within pores or under other material are not sampled.

## 7.3 XPS of Gold Catalysts: Study of Gold Oxidation States

XPS has been used extensively as a characterization tool for gold catalysts grown on metal oxides and carbon. In general, most of the reports involve studying the evolution of the gold oxidation state during synthesis or identification of the gold oxidation state before use in the catalytic reactor. Highlighted below are, in the author's opinion, some of the more interesting uses of XPS in the study of gold catalysts in order to help the reader envision the types of experiments that can be done. Unfortunately, the desire for brevity precludes highlighting all of the many excellent studies in the literature. Several examples are taken from the authors own work, where access to the raw data was considered to be of pedagogical value.

### 7.3.1 Evolution of Gold Oxidation State with Synthesis

Most gold catalysts are grown on reducible oxides (i.e., $Fe_2O_3$, $CeO_2$, etc.) because they usually, but not always, result in the most active gold catalysts [9]. There are a large number of synthetic methods used to prepare these catalysts. One of the most popular methods is the deposition-precipitation (DP) method [10]. The DP method entails adjusting the pH of an aqueous suspension of support material and $HAuCl_4$ to precipitate a gold hydroxide species onto the support surface [10]. The gold hydroxide species is then reduced to form the gold catalyst. This transition in the gold oxidation state is easily followed by XPS. For example, Zwijnenburg et al. deposited $Au(OH)_3$ on $TiO_2$ and followed the Au $4f$ signal as a function of thermal treatment [11]. They report the Au $4f$ binding energy shifting from 84.2 eV to 83.3 eV upon annealing from 150°C to 400°C, indicating a reduction in the gold oxidation state. Similar studies were performed by Park and Lee looking at the formation of catalysts on $Fe_2O_3$, $TiO_2$, and $Al_2O_3$ [12]. Similarly, Yuan et al. [13] and Choudhary et al. [14] studied similar the decomposition of Au–phosphine complexes. Again the as-deposited Au species had a binding energy of around 84.7 eV but after thermal treatment at 450°C there was a reduction in the gold binding energy (83.6 eV). In addition to simply studying the Au oxidation state Choudhary et al. also followed the evolution of the phosphorous $2p$ signal with thermal treatment. They found a shift to higher binding energies of the P signal (131.6 to 134.0 eV) due to the oxidation of phosphorous.

We highlight this example from Choudhary et al. [14] because most do not report XPS spectra for other precursor components such as Na, Cl, and P. Given the importance of residual ions to catalyst stability and activity, the surface sensitivity of XPS should be a vital tool in identifying impurity effects that bulk methods such as energy-dispersive X-rays (EDX) would miss. Indeed, the author has found that a number of commercially prepared gold catalysts from DP methods have a high concentration of residual sodium on the catalyst surface (Fig. 7.2). The role of this sodium is not clear; however, it may act as a co-catalyst or even poison a catalytic site. Understanding if there is an effect on activity from this residual species could be an excellent way to advance the field of gold catalysis.

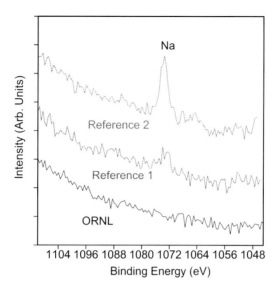

**Figure 7.2** XPS data collected for two gold reference catalysts showing a large concentration of Na$^+$ on the surface. A comparison to specially synthesized low-Na catalyst is made.

### 7.3.2 Correlation of Oxidation State to Catalytic Activity

As discussed above XPS can be a valuable tool to identify oxidation states, but the reader will note that we have very carefully avoided assigning oxidation states to gold catalysts. The reason for this will become clear in the next section. Instead of assigning specific oxidation states, we have described samples as oxidized or reduced. The oxidation states of gold catalysts have been a source of constant discussion throughout the literature. Some researchers maintain that oxidized gold is the catalytic species, while others claim that reduced gold is the active species. The examples below highlight studies that use XPS to correlate activity to oxidation state.

Hutchings et al. reported a very interesting study where Au/Fe$_2$O$_3$ catalysts were prepared by co-precipitation and then subjected to annealing at 120°C and 400°C [15]. As discussed above, higher temperatures result in a more reduced type of gold species. Using XPS the authors measured the Au 4$d$ 5/2 binding energies of the gold species. They correlated the binding energies with the

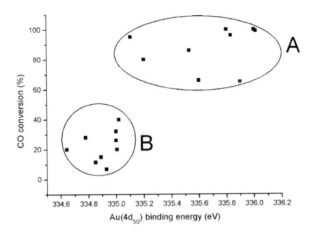

**Figure 7.3** Correlation between activity for CO conversion and Au (4d 5/2) binding energy. The data for a series of catalysts (5 wt% Au supported on Fe$_2$O$_3$) are shown; the cluster of points labeled A corresponds to catalysts dried at 393 K, and the cluster B to those calcined at 673 K. Reprinted from Ref. [15], Copyright (2006), with permission from Elsevier.

catalytic activity of these materials for the oxidation of CO (Fig. 7.3). These data show the samples annealed at 120°C have higher gold binding energies (more oxidized) and concomitantly higher catalytic activity toward CO. Those samples annealed at 400°C have little to no oxidized gold and little to no CO activity. Those skilled in the art would recognize that annealing at higher temperatures may also lead to sintering of the gold particles. Through careful scanning transmission electron microscopy (STEM) studies the authors were able to show similar-size gold particles (7.0 and 5.4 nm), indicating that particle size is not the major factor in the resulting activity. Instead their XPS studies, coupled with XAS studies pointed to cationic gold as the active species for the catalysts. In this particular report the authors did not use XPS to characterize the surface chemistry of the iron oxide. This is unfortunate since support chemistry plays a major role in catalytic properties. Indeed other authors looking at FeOOH-based samples clearly showed an evolution of the O 1s species with temperature from a change in surface OH [16]. Future researchers investigating Au on Fe$_2$O$_3$ should examine the Fe 3s line (~93 eV) [17]. The Fe 3s has a doublet structure, which has its origin in the exchange coupling between the 3s vacancy and the 3d electrons [18]. The value of the splitting is sensitive to the number of unpaired 3d electrons of the metal atom, so it can provide information about the formal oxidation state of the transition metal [18, 19]. Generally this splitting is about 6.2 eV for $Fe^{3+}$ and about 5.5 eV for $Fe^{2+}$ in simple iron oxides [20, 21]. This peak would also be useful for following the evolution of Fe-OH- to Fe-O-type chemistry which may occur at higher temperatures.

Casaletto et al. used XPS studies to identify the effect of reduced and oxidized Au on a variety of supports [22]. They found a different distribution of oxidized and reduced gold species, depending on the support chemistry. Furthermore, samples with the most $Au^+$ were the most active at low temperatures for CO oxidation. Catalytic studies of gold and gold oxide clusters grown on $TiO_2$ showed similar trends [23, 24]. In these studies the $Au^{3+}$ catalysts were over 180 times more active than the $Au^0$ samples. Finally, Fu, Saltsburg, and Flytzani-Stephanopoulos used XPS to show that $Au^{\delta+}$ were the catalytically active species for water–gas shift (WGS) reactions after an elegant leaching experiment with cyanide [25]. The NaCN selectively removed all the $Au^0$ from a catalyst bed, leaving only the oxidized gold as measured by XPS.

### 7.3.3 Identification of Supported Gold Nanoparticle Oxidation States

In the studies of the gold catalysts described above, and many others, the reported binding energies of the reduced gold species were less than 84.0 eV. Table 7.1 lists a sample of some of the Au binding energies and the reported particle sizes measured by STEM/TEM. As these data show, there is a large variation in reported binding energies for the gold catalysts. To a first-order approximation this lower binding energy would indicate that the gold particles are significantly more reduced than $Au^0$, that is, $Au^{\delta-}$. The assignment of $Au^{\delta-}$ is not supported by Au-XAS or $^{197}$Au-Mössabuer studies [11].

The origin of these binding energy shifts has been the subject of much debate and discussion in the literature. Some investigators attribute the shift to electron transfer from the support to the metallic gold due to the larger electronegativity of gold relative

**Table 7.1** Reported Au binding energies for gold catalysts grown on various supports

| Support | Preparation method | Au 4f 7/2 Binding energy (eV) | TEM (nm) | Reference |
| --- | --- | --- | --- | --- |
| $TiO_2$ | DP | 83.3 | 3–6 | [11] |
| $TiO_2$ | DP | 83.7 | 3.8 | [9] |
| $TiO_2$ | PLD | 83.4 | 2.9 | [26] |
| $TiO_2$ | DP | 83.9 | 2–3 | [27] |
| $TiO_2$ | CVD | 83.3 | 2.7 | [28] |
| $TiO_2$ | Sol–gel | 83.3 | 1.1 | [29] |
| $TiO_2$ | I | 83.0 | 2.0 | [30] |
| $TiO_2$ | MS | 83.2 | 2.4 | [24] |
| $TiO_2$ | MS | 82.9 | 22 | [24] |
| $TiO_2$ | MS | 83.0 | 1.2 | [24] |
| $Al_2O_3$ | MS | 83.7 | 2.3 | [31] |
| $Al_2O_3$ | Au colloid | 83.3 | 4.0 | [32] |
| $Al_2O_3$ | PLD | 83.1 | 2.6 | [26] |
| $SiO_2$ | CVD | 83.2 | 1.4 | [28] |
| $SiO_2$ | SMAD | 84.2 | 2–5 | [22] |
| $CeO_2$ | DP | 84.5 | NR | [33] |
| $MnO_2$ | DP | 83.9 | 4.4 | [34] |
| NiO | DP | 83.9 | 3.5 | [35] |
| $ZrO_2$ | PLD | 83.9 | 2.9 | [26] |
| $Fe_2O_3$ | CP | 83.8 | 10 | [16] |
| Au | Foil | 84.0 | – | [8] |

DP = deposition–precipitation; MS = magnetron sputtering; SMAD = solvated metal atom dispersion; CVD = chemical vapor deposition; I = impregnation; PLD = pulsed laser deposition

to metals such as $Al^{3+}$ [26, 36, 37]. Such an electron transfer would modify the electronic properties of the supported gold catalyst possibly activating it for a catalytic reaction [38–40]. Other researchers attribute the changes in electron density to Coulombic forces [41]. In these models increasing support alkalinity, such as through the introduction of Na discussed above, results in an increase in electron density on the oxygen atoms of the support. This extra electron density partially transfers to the metal, where it is sampled in the XPS and appears as a lower-binding-energy peak. Another, more general way to view this effect would be through the polarization of a metal cluster by the Madelung potential of the support [42]. In this model changes in the local chemistry from doping or impurities results in a change in electric potential for all the atoms in the lattice, including those of the catalyst cluster [43].

Others attribute these shifts to initial or final state effects [7, 37]. Initial state effects would lead to negative binding energy shifts and are due to quantum size effects. Smaller particles have a higher fraction of undercoordinated surface atoms, resulting in the Fermi level shifting to lower energy. Shifting the particles Fermi level would reduce the energy needed to emit surface electrons from the metal, resulting in a lower binding energy [7]. Work by Radnik et al. indicates that particle shapes could be a critical factor in determining the negative binding energy shifts of gold-supported nanoparticles [28]. Radnik et al. claim that the more spherical the nanoparticles, the lower the binding energy (down to 83.0 eV for Au on $TiO_2$) [28]. This shift is due to initial state effects where spherical particles have a larger fraction of undercoordinated surface atoms, which reduces their binding energies relative to nanoparticles with large faces [28, 44].

Alternative models of initial state effects attribute the shifts to lower binding energy to lattice strain [45]. As the gold cluster relaxes on the metal oxide the gold–gold bonds change to conform to the oxide surface. The shorter metal oxide bonds cause a reduction in the Au–Au bond length. This reduction in bond length increases electron density, resulting in easier removal by the incident X-ray.

Final state effects lead to a positive binding energy shift and are due to the difficulty in neutralizing the electron hole produced on the gold particle with an electron from the support [7]. Given that many of the catalyst supports are electrically insulating oxides, and the fact that these binding energy shifts are observed for all catalysts despite dramatic differences in activity, the electronic effects may play a secondary role in the observed XPS data. Alternatively, the smaller a gold particle becomes, the lower the screening efficiency the remaining gold atoms provide to neutralize the hole created by the emitted electron. This process would also result in a positive binding energy shift [46].

All these different models of the origins of binding energy shifts have their supporters and detractors. The problem is that no model describes all of the experimental data appropriately. Take for example Au on $TiO_2$ prepared by magnetron sputtering [24].

The initial binding energy reported for Au was 83.2 eV for 2.4 nm particles, consistent with initial state effects. Over time the gold clusters coarsened to form 22 nm particles but the binding energy was reduced further to 82.9 eV [24]. These particles are too large to exhibit initial state effects or final state effects due to a lack of screening electrons. The best way to determine which model is correct is to measure the Auger parameter of the catalyst particle [47–49]. The Auger parameter ($a$) was introduced by Wagner in 1971 and relies on the production of an Auger electron from the same orbital as the photoelectron excitation [50]. The Auger electron is produced when an electron from a core level, like a $1s$ orbital, is excited. An electron from a $2s$ or $2p$ orbital falls into the $1s$ hole. The energy of this transition is transferred to another $2p$ electron, which is emitted from its orbital and measured by the spectrometer. This type of Auger transition is described as KLL because by convention they are described using X-ray notation. In the case of the Au $4f$ electron the corresponding Auger transition is $M_5N_{67}N_{67}$, where an electron excited from the $M_5$ subshell produces a $d$-hole. Electrons from the $N_{67}$ subshell ($4f$) fill the M hole and are emitted and measured by the spectrometer. The $M_5N_{67}N_{67}$ Auger transition has a kinetic energy of 2015 eV.

The Auger parameter ($\alpha$) is determined using Eq. 7.2:

$$\alpha = BE(Au4f7/2) + KE(AuM_5N_{67}N_{67}) \qquad (7.2)$$

There is a fixed energy difference between the photoelectron excitation and the Auger transition from the same orbital. Since the difference is fixed any shifts in binding energy due to initial or final state effects, Coulombic interactions, or lattice strain will be quantifiable. A negative shift will be due to more electron density on the gold atom. A benefit of this analysis is charge corrections are unnecessary; thus it can be used to analyze insulators or metals on insulators to elucidate final state effects.

The challenge of performing these measurements comes from the high energies of the Au $M_5N_{67}N_{67}$ transition (2015 eV). Conventional Al and Mg X-ray sources only operate at 1486.7 and 1253.4 eV, respectively. As a result higher-energy sources, such as Zr (L$\alpha$ X-rays, 2042 eV) or Bremsstrahlung radiation are required. To our knowledge only one XPS study investigating the Auger

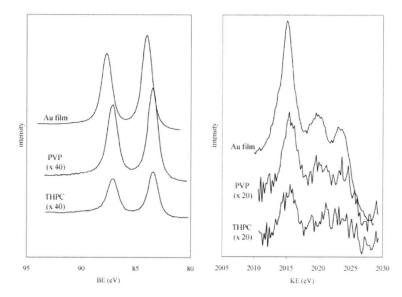

**Figure 7.4** XPS spectra of Au $4f$ (left) and Auger $M_5N_{67}N_{67}$ (right) for 10 wt% Au on $TiO_2$ catalysts prepared via THPC or PVP route compared to Au reference (Au film). BE = binding energy, KE = kinetic energy. Reprinted with permission from Ref. [11]. Copyright (2002) American Chemical Society.

parameter for supported gold catalysts has been attempted [11]. Figure 7.4 shows experimental data from this work. These data show the challenge of measuring the relatively low-sensitivity $M_5N_{67}N_{67}$ transition compared to the photoelectron spectra. The observed $\alpha$ parameters are within range of what is observed for pure gold ($\alpha$ = −0.07 to −0.27 eV), indicating that the surface gold was metallic $Au^0$ [11]. More studies like this one are required to start identifying how trends in catalyst properties correlate with spectroscopy data.

From the data above there are no clear trends correlating Au binding energy and catalytic activity or catalyst support. There are often changes in binding energy that may indicate changes in a particle's shape or size or modification of the interactions with the support. Identifying these trends may introduce new methods to optimize catalyst production or properties. As noted above most experimental studies have focused on the catalyst metal chemistry alone and neglected other aspects of the surface chemistry, such as salts and metal oxide supports. Those entering the field or in the field should be encouraged to explore these variables with as much vigor as possible.

## 7.4 Postmortem Analysis

Most of the studies discussed above and in the literature were performed on as-prepared catalysts or catalysts as they were prepared. Often these reports focused only on the gold spectra, neglecting other components of the surface chemistry. In the author's opinion the real power and benefit of XPS studies come from evaluating catalysts after use and comparing these results to the as-prepared samples. Unfortunately, this is often not done entirely or with enough specificity. The following sections highlight the type of studies where experimental insights into the catalyst properties could be gained by careful postmortem analysis.

### 7.4.1 Evolution of Gold Oxidation State with Reaction

One of the most common uses of XPS is evaluating the change in the oxidation state after a catalytic reaction. Figure 7.5 shows Au $4f$ data collected for Au–O catalysts before and after a catalytic reaction [24]. In this case the initial Au $4f$ species had a binding energy of 85.1 eV consistent with oxidized gold. After a catalytic reaction for 24 hours the binding energy shifted to 83.3 eV, consistent with the reduction of the $Au^+$ to more metallic Au. Concomitant to these shifts there was a loss in catalytic activity due to the reduction in the gold catalyst. In a similar study, Herzing et al. attributed the activity of gold catalysts supported on $FeOOH/Fe_2O_3$ to the formation of ~0.5 nm bilayer clusters with about 10 gold atoms [51]. The initial gold clusters reported by Herzing et al. were oxidized (as evidenced by a Au$4f_{7/2}$ binding energy of 85.1 eV) [51]. Upon annealing the gold clusters were reduced, as evidenced by a Au $4f$ binding energy to around 83.8 eV, and there was a loss of activity that was attributed to the loss of the bilayer clusters [51] and presumably the reduction of the gold.

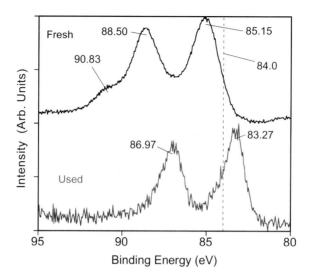

**Figure 7.5** Normalized XPS data of fresh AuO$_x$ on pH 10 treated TiO$_2$ (top) and AuO$_x$ on pH 10 treated TiO$_2$ after it was used for the oxidation of CO. Adapted with permission from Ref. [24]. Copyright (2009) American Chemical Society.

## 7.4.2 Changes in Support Oxygen Chemistry with Synthesis

One of the assumptions often made in gold catalyst studies is that the support is a passive element and does not change with synthesis. XPS studies enable us to test these theories and find examples where these assumptions are wrong. For example, Fig. 7.6 shows the O 1$s$ data recorded for gold nanoparticles grown on native Degussa P25 TiO$_2$ and P25 treated at pH 4 and 10 to modify the support basicity [24]. XPS data reported for three P25 support samples show no change in the O 1$s$ or Ti 2$p$ spectra [24]. However, with the addition of gold there is a statistically significant shift to higher binding energies observed for both the O 1$s$ and Ti 2$p$ spectra. The Ti peak shifts by 0.22 eV for Au-P25, 0.18 eV for Au-pH 4, and 0.08 eV for Au-pH 10 compared to 458.50 eV for any of the TiO$_2$ supports without Au. Similar shifts were observed for the O 1$s$ photoelectron line originating from the support: 0.25 eV for Au-P25, 0.19 eV for Au-pH 4, and 0.09 eV for Au-pH 10 compared to 529.75 eV for the TiO$_2$ supports without Au. It is tempting to relate these small changes

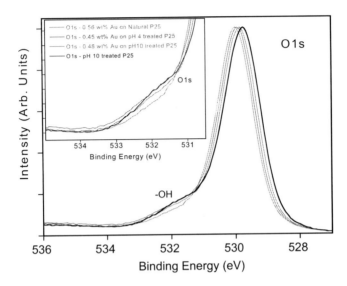

**Figure 7.6** O 1s XPS data for treated and natural P25 support gold catalysts. For comparison the O 1s spectra of pH 10-treated TiO$_2$ is included. Inset is an expanded view of the –OH portion of the O 1s spectra. Reprinted with permission from [24]. Copyright (2009) American Chemical Society.

to errors associated with the charging correction methodology; however, the binding energy of the hydroxyl component in the O XPS data remains constant with the addition of gold, 531.43 [24], shown in Fig. 7.6. If the charge correction was wrong the OH peaks would shift as well. The impact of these shifts is not clear. Obviously, shifts in the O 1s binding energy to higher energies would be correlated with electron density removal from the O atoms. A shift in O electron density to the gold could partially explain the low binding energies observed for Au (83.2–83.0 eV). As noted earlier, shifts in electron density for more basic supports could be due to Coulombic forces [41]. However, the smaller shift observed for the more basic pH 10-treated TiO$_2$ compared to acidic P25 may indicate this model does not accurately capture the complexity of the gold–titania interactions.

Numerous other groups have also studied the O 1s spectra as a function of catalyst synthesis, particularly for Fe-based supports where FeOOH transforms to Fe$_2$O$_3$ [51]. For example, Wang et al.

studied the O 1s species of their MnO$_2$ support material [34]. Their data showed a dramatic increase in the O 1s species attributed to water or surface hydroxyls after the addition/formation of gold nanoparticles using the DP method [34]. They found certain MnO$_2$ sources have an ability to absorb more OH or water groups than others. Interestingly, they found the MnO$_2$-supported gold particles with the most OH/H$_2$O exhibited the highest activity for the oxidation of benzyl alcohol. A different group studying Au/SnO$_2$ catalysts found a similar change in the O 1s spectra associated with hydroxyls [52]. For these catalysts the residual OH decreased with increasing annealing temperature. The authors correlated the maximum catalytic activity of their 5–10 nm gold particles with supports containing a maximum concentration of OH. One final example of changes in the support O 1s spectra involves the growth of gold particles on NiO and Ni(OH)$_2$ support materials by gold sols [35]. XPS studies on these materials showed the growth of a O 1s and Ni 2$p_{3/2}$ species attributed to the formation of OH on NiO (Fig. 7.7). As a result what was a NiO starting material evolved to be more Ni(OH)$_2$-like. After the material was used as a catalyst for the oxidation of benzyl alcohol the Ni 2$p_{3/2}$ and O 1s spectra changed further, resulting in evidence for the formation of NiOOH due to the oxidation of NiO/Ni(OH)$_2$. In addition to the oxidation of the Ni support XPS data collected for the Au nanoparticles revealed their oxidation under reaction conditions (Au 4$f$ binding energy $= 86.1$ eV). It was hypothesized that the NiOOH could act as a co-catalyst for the reaction.

In all these studies referenced above residual OH/H$_2$O was observed in the XPS. To a first approximation this is surprising because XPS is a UHV method and one might expect all the surface water to be removed. However, surfaces are not terminated with oxygen atoms. Under atmospheric conditions the surfaces of typical metal–oxygen support materials are terminated by hydrogen and coated with physisorbed coordinating water molecules [53–58]. The surface hydration/protonation is often strong enough such that the surface OH species are observed on bulk metal oxides like TiO$_2$ under UHV conditions. Surface scientists have to work very hard to remove these OH species from the surface of an oxide. Figure 7.8 shows representative O 1s spectra collected for P25-type TiO$_2$ and

**188** | *X-Ray Photoelectron Spectroscopy Characterization of Gold Catalysts*

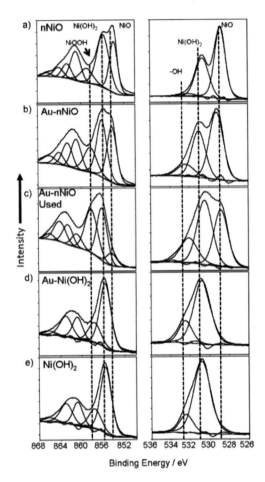

**Figure 7.7** Ni 2p3/2 and O 1s XPS data for (a) nNiO, (b) Au/nNiO, (c) Au/nNiO used, (d) Au/Ni(OH)$_2$, and (e) Ni(OH)$_2$. Dashed lines are added to aid position identification. Adapted with permission from Ref. [35]. Copyright Wiley-VCH Verlag GmbH & Co. KGaA.

a model TiO$_2$ (110) surface prepared under UHV conditions. It is evident in these data that the surface OH groups are removed in the UHV-prepared material. It is this loss of OH that results in a dramatically different surface chemistry and is the main reason UHV model surface science studies have limited applicability to bulk powder-based catalysts. Indeed numerous works by several groups

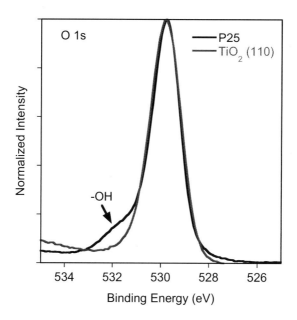

**Figure 7.8** O 1s data collected for Degussa P25 and TiO$_2$ (110) single crystal.

have attributed the variability in catalytic activity of gold catalysts to the hydroxyl species of the oxide support [59–63]. In addition several papers have reported a similar increase in nanoparticle stability through the introduction of strongly bound OH groups [64–66].

### 7.4.3 Changes in Support Metal Oxide Chemistry after Catalytic Reaction

In addition to evidence for changes in the O 1s data there are several studies that investigated the metal of the metal oxide support before and after reaction. One of the most commonly investigated is the conversion of FeOOH to Fe$_2$O$_3$. Horváth et al. studied FeOOH/Fe$_2$O$_3$ before and after a catalytic reaction and found the formation of FeO/Fe$_3$O$_4$ [16]. They hypothesized that the FeO promotes electron transfer to the Au, resulting in the most active gold catalyst. Arrii et al. followed the Ti, Al, and Zr binding energies after reactions

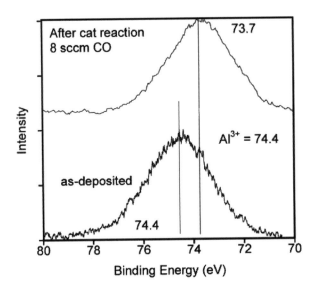

**Figure 7.9** Al $2p$ XPS data for Au-$\gamma$-Al$_2$O$_3$ as-deposited and after reaction. Reproduced from Ref. [31], Copyright (2005), with permission from Elsevier.

at elevated temperatures [26]. They reported a significantly higher energy shift in Ti $2p$ binding energy (458.8 to 459.2 eV) after the reaction with a concomitant reduction in the Au $4f$ binding energy (83.4 to 82.9 eV). The origin of this shift is not clear but would indicate a modification of the Au–TiO$_2$ interface chemistry.

Gold catalysts supported on $\gamma$-Al$_2$O$_3$ have been reported to exhibit peculiar surface chemistry [31]. Indeed, as shown in Fig. 7.9, after a 2.3 nm Au-$\gamma$-Al$_2$O$_3$ was used in the catalytic oxidation of CO there was a 0.7 eV decrease in the Al $2p$ binding energy (74.4 to 73.7 eV). This decrease mirrored a similar 0.7 eV shift observed for the gold clusters (83.7 to 83.0 eV), which could indicate a charge correction error. However, the O $1s$ and C $1s$ spectra from adventitious hydrocarbons remained unchanged. The origin of this shift might be attributed to increased interactions between Au and Al$_2$O$_3$ [67]; however, the shift could also be due to subtle structural changes of the Al$_2$O$_3$ at the surface, such as the $\gamma \rightarrow \delta \rightarrow \theta$ Al$_2$O$_3$ phase transition resulting in a slightly modified Al$^{3+}$ binding energy [68, 69]. The large amount of energy released during the CO oxidation reaction ($\Delta G = -257$ kJ/mol $CO_2$) could be enough to promote local surface changes after several hours on stream [68]. Another possibility could be the formation of a Au–Al alloy [70]. However this seems unlikely since STEM results show no indication of alloy formation. In addition, the majority of the Al XPS signal would still be from the $Al_2O_3$ and it would be expected that a shoulder or a broader peak would occur for the Al.

### 7.4.4 Quantifying Catalyst Coarsening

One of the biggest problems confronting the use of gold catalysts is the stability of the catalysts as a function of temperature and time [10, 71–73]. Coarsening studies on gold catalysts rely on three main methodologies: electron microscopy, extended X-ray absorption fine structure (EXAFS), and X-ray diffraction. However, XPS should be considered to evaluate gold coarsening due to the ease in interpreting the data and sample handling. Furthermore, XPS samples a much larger area than microscopy studies, improving statistics, and the measurement can be performed much more quickly. There are three mechanisms by which supported nanoparticles coarsen: sintering, Ostwald ripening, and migration/coalescence, which have all been described in numerous publications [74, 75]. These processes occur when there is enough thermal energy to promote the migration of and reaction between metal particles. The temperature of the coarsening is often related to the Tammann temperature, where bulk atoms begin to diffuse (roughly 1/2 of the bulk melting point of the metal) but will also depend on the wetting energy of the metal and the support material. On the basis of simple temperature arguments gold catalysts ($T_{\text{Tammann}} = 395\,°C$) should be quite stable from room temperature to $200\,°C$. Figure 7.10 shows results from a coarsening study performed at room temperature without heating [24]. In these plots the Ti $2p$ to Au $4f$ photoelectron intensity ratio as a function of time is plotted for a light-sensitive gold catalyst grown on native $TiO_2$ (P25) and pH 10–treated P25. When the sample is 1 hour old the Ti:Au ratio is 10 for the catalyst on native P25. After 24-hour storage in air at atmospheric pressure the Ti:Au ratio increases to 29 for the sample stored in the light and 19 for the sample stored in the dark. This dramatic change in the Ti:Au ratio is due to the loss of

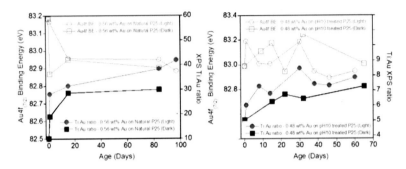

**Figure 7.10** (Left-Right) XPS Ti $2p$:Au $4f$ intensity ratios for 0.56 wt% Au on P25 as a function of time and light exposure. (Left-Left) Plot of Au $4f$ binding energies for 0.56 wt% Au on P25 as a function of time and storage conditions. (Right-Right) XPS Ti $2p$ and Au $4f$ intensity ratios for 0.48 wt% Au on pH 10-treated P25 as a function of time and light exposure. (Right-Left) Plot of Au $4f$ binding energies for 0.48 wt% Au on pH 10-treated P25 as a function of time and storage conditions. Reprinted with permission from Ref. [24]. Copyright (2009) American Chemical Society.

small gold particles, which results in more of the Ti being exposed on the surface and a much lower percentage of gold atoms on the surface of a gold cluster [76]. This trend continues to evolve with time; however, the rate slows substantially. Electron microscopy studies showed the evolution of 2.4 nm gold clusters to large 22 nm clusters. This growth in particles requires the consumption of approximately 400 2.4 nm gold clusters to form the 22 nm species. In contrast the Ti:Au ratio measured for pH 10-treated supports show a slow and steady increase in the Ti:Au ratio, confirming that the gold particles are still agglomerating as confirmed with electron microscopy. However, the smaller change in the XPS Ti:Au intensity ratio with time for Au/pH 10 $TiO_2$ (5.0 → 7.5) compared to Au native $TiO_2$ (10 → 40) is consistent with the reduced rate of particle coarsening observed in the microscopy data. The smaller initial Ti:Au ratio for the catalysts on treated supports is due to smaller gold (1.2 nm) particles. In addition, treating the $TiO_2$ surface has virtually eliminated the enhancing effect of light on the coarsening of the gold particles observed for the catalysts supported on natural P25.

Similar comparisons of gold support ratios were performed for gold catalysts grown on $SiO_2$ and a reference gold on $TiO_2$

[77]. These studies reported the initial Si $2p$:Au $4f$ ratio to be around 0.90. After two weeks at 500°C, well above the Tammann temperature, the ratio was 0.97 consistent with the slight evolution in particle sizes, 2.5 nm to 3.2 nm. Further annealing at 500°C for an additional two weeks resulted in virtually the same ratio, 0.98, indicating that the catalyst was stable over long periods of time. In contrast, the Au/TiO$_2$ reference material had a Ti $2p$:Au $4f$ ratio of 19.6, which increased to 44.7 after four weeks at 500°C. Microscopy data indicated the gold cluster supported on TiO$_2$ increased from 3.7 nm to 13.9 nm.

### 7.4.5 Chemical Deactivation/Blocking of Gold Catalysts

Another method to deactivate a gold catalyst is through the irreversible absorption of a reactant. Using XPS Kim et al. studied the surface chemistry of a gold catalyst supported on CeO$_x$ before and after a WGS reaction [78]. Analyzing the Au $4f$ XPS data showed the gold species remained fully oxidized (Au $4f$ binding energy = ~89 eV). Further characterization of the C $1s$ data showed the increase in –CO$_3$/–HCOO species after reaction. This increase in the C–O species is evidence of the formation of carbonate and formate molecules during the reaction. The residual species block the active sites, causing catalyst deactivation. Similar results were reported for other Au–CeO$_2$-based catalysts used for WGS [79]. In liquid-phase catalysts there is often a wide variation in catalyst activities despite the same-size gold clusters. For example, when the samesize Au catalyst was on the same carbon using *N*-dodecyl-*N*, *N*-dimethyl-3-ammonio-1-propane sulfonate (SB) or poly(vinyl alcohol) (PVA) as the protecting group, the two catalysts showed significantly different activities in the oxidation of ethylene glycol [80] XPS showed a higher concentration of exposed gold on the SBgrown catalyst leading to more active sites, resulting in a more active catalyst.

### 7.4.6 XPS to Understand the Nucleation and Growth of Gold Nanoparticles

One constant problem for catalyst researchers is identifying the binding sites of catalyst clusters on a metal oxide surface. This information is important to develop accurate density functional theory models as well as understand the electronic effects discussed above. XPS is a useful tool for addressing this question. In this example three different MgAl$_2$O$_4$ spinel-type support materials with different Mg surface enrichments were investigated to understand how surface basicity influences catalytic reactions [81]. By XPS all the supports were slightly rich in Al, with Al:Mg ratios of 3.5, 4.0, and 6.9, depending on how the supports were fabricated. Catalysts were grown on these materials by DP, as sol-protected particles, and by sputtering. After nanoparticle fabrication the Al:Mg ratios were remeasured with XPS. A surprising trend was observed: the gold catalyst grown from solution had much higher Al:Mg ratios than those prepared by vapor deposition. These higher Al:Mg ratios indicate that there is less Mg on the surface to be sampled by XPS. This loss of Mg intensity is due to the preferential bonding of Au on the more basic Mg sites. This Au blocks the emitted electrons from the Mg, lowering its apparent intensity in the XPS data. Interestingly, the vapor-deposited catalysts, which were produced in the absence of water, had the exact same Al:Mg ratios as the starting materials. This indicates that all the gold is homogenously dispersed on the surface of the support. Previous investigations of gold catalysts on monometallic metal oxides or carbon supports showed that catalytic activity and selectivity for the oxidation of aqueous glycerol varied as a function of metal nanoparticle sizes [80]. In general smaller catalysts (<2 nm) promoted the overoxidation of glycerol to C2- and C1-containing products. However, these results demonstrated that the synthesis-dependent enrichment of the MgAl$_2$O$_4$ surface with Al species results in a catalyst that promotes the cleavage of C–C bonds with large (>9 nm) gold clusters. This activity is not due to the formation of H$_2$O$_2$, which would promote the oxidative cleavage of C–C bonds, but instead is due to the increase in Lewis acidity with the increase in Al$^{3+}$ content, which likely increases the residence time of glycerol absorbed on the catalyst surface.

## 7.5 Frontiers in XPS Instrumentation

The above examples were all collected using laboratory-based XPS instruments. There are several next-generation XPS instruments coming online at national user facilities that offer the opportunity to perform XPS experiments at close to ambient conditions [82, 83]. The potential experiments include exploring the hydroxylation of support materials [84] and understanding changes in electronic structure with exposure to reactant gases. Jiang et al. have explored the activation of $O_2$ on gold. They have found that $O_2$ does not interact strongly with the gold surface [85]. Willneff et al. studied gold catalysts supported on P25 as a function of temperature and $CO + O_2$ pressure [86]. They found the electronic structure of the gold catalysts changed as a function of temperature. Indeed there appears to be two gold species at room temperature but only one gold species at 350 K with a slightly higher binding energy than $Au^0$. With regard to synthesizing gold nanoparticles, XPS microjet studies offer the potential to study ion distribution in the electrochemical double layer of a metal oxide support [87]. This insight could provide a new understanding into the stabilization of gold precursors during synthesis, leading to more stable and active gold catalysts.

## 7.6 Conclusions and Perspective

With this chapter the author tried to convey the utility and insights that can be gained from applying XPS to practical catalyst problems. Indeed, insights from XPS can exceed simple identification of metal oxidation states and lead to a greater understanding of gold catalyst deactivation, gold catalyst surface chemistry, and the growth of gold catalysts. Most research institutions have their own XPS instruments, and using these tools is far easier and more cost effective than traveling to synchrotron facilities or operating state-of-the-art microscopy facilities. With care and focus on detail tremendous insights can be made to advance the science of gold.

## Acknowledgments

My work, which was included as examples in this chapter, was supported by the Materials Science and Engineering Division of the United States Department of Energy's Office of Basic Energy Sciences. The author would like to thank Dr. Nancy J. Dudney (ORNL) for her support during these studies, as well as Drs. Harry Meyer III, Rose E. Ruther, and Ethan J. Crumlin for their comments and suggestions, which improved the work significantly.

## References

1. Calvin, S. (2013). *EXAFS for Everyone* (CRC Press, Boca Raton).
2. Weiher, N., Bus, E., Delannoy, L., Louis, C., Ramaker, D.E., Miller, J.T., and van Bokhoven, J.A. (2006). Structure and oxidation state of gold on different supports under various CO oxidation conditions, *J. Catal.*, **240**, pp. 100–107.
3. Ketchie, W.C., Maris, E.P., and Davis, R.J. (2007). In-situ X-ray absorption spectroscopy of supported Ru catalysts in the aqueous phase, *Chem. Mater.*, **19**, pp. 3406–3411.
4. Parish, R.V. (1982). Gold and Mössbauer spectroscopy, *Gold Bull.*, **15**, pp. 51–63.
5. Faltens, M.O., and Shirley, D.A. (1970). Mössbauer spectroscopy of gold compounds, *J. Chem. Phys.*, **53**, pp. 4249–4264.
6. Daniells, S.T., Overweg, A.R., Makkee, M., and Moulijn, J.A. (2005). The mechanism of low-temperature CO oxidation with Au/Fe$_2$O$_3$ catalysts: a combined Mossbauer, FT-IR, and TAP reactor study, *J. Catal.*, **230**, pp. 52–65.
7. Meyer, R., Lemire, C., Shaikhutdinov, S.K., and Freund, H.-J. (2004). Surface chemistry of catalysis by gold, *Gold Bull.*, **37**, pp. 72.
8. Moulder, J.F., Stickle, W.F., Sobol, P.E., and Bomben, K.D. (1992). *Handbook of X-Ray Photoelectron Spectroscopy* (Perkin-Elmer).
9. Comotti, M., Li, W.-C., Spliethoff, B., and Schüth, F. (2006). Support effect in high activity gold catalysts for CO oxidation, *J. Am. Chem. Soc.*, **128**, pp. 917–924.
10. Zanella, R., Giorgio, S., Henry, C.R., and Louis, C. (2002). Alternative methods for the preparation of gold nanoparticles supported on TiO$_2$, *J. Phys. Chem. B*, **106**, pp. 7634–7642.
11. Zwijnenburg, A., Goossens, A., Sloof, W.G., Crajé, M.W.J., van der Kraan, A.M., de Jongh, L.J., Makkee, M., and Moulijn, J.A. (2002). XPS and Mössbauer characterization of Au-TiO$_2$ propene epoxidation catalysts, *J. Phys. Chem. B*, **106**, pp. 9853–9862.
12. Park, E.D., and Lee, J.S. (1999). Effects of pretreatment conditions on CO oxidation over supported Au catalysts, *J. Catal.*, **186**, pp. 1–11.

13. Yuan, Y., Asakura, K., Kozlova, A.P., Wan, H., Tsai, K., and Iwasawa, Y. (1998). Supported gold catalysis derived from the interaction of a Au–phosphine complex with as-precipitated titanium hydroxide and titanium oxide, *Catal. Today*, **44**, pp. 333–342.
14. Choudhary, T.V., Sivadinarayana, C., Chusuei, C.C., Datye, A.K., Fackler, Jr., J.P., and Goodman, D.W. (2002). CO oxidation on supported nano-Au catalysts synthesized from a [Au$_6$(PPh$_3$)$_6$](BF$_4$)$_2$ complex, *J. Catal.*, **207**, pp. 247–255.
15. Hutchings, G.J., Hall, M.S., Carley, A.F., Landon, P., Solsona, B.E., Kiely, C.J., Herzing, A., Makkee, M., Moulijn, J.A., and Overweg, A. (2006). Role of gold cations in the oxidation of carbon monoxide catalyzed by iron oxide-supported gold, *J. Catal.*, **242**, pp. 71–81.
16. Horváth, D., Toth, L., and Guczi, L. (2000). Gold nanoparticles: effect of treatment on structure and catalytic activity of Au/Fe$_2$O$_3$ catalyst prepared by co-precipitation, *Catal. Lett.*, **67**, pp. 117–128.
17. Zhou, H., Nanda, J., Martha, S.K., Adcock, J., Idrobo, J.C., Baggetto, L., Veith, G.M., Dai, S., Pannala, S., and Dudney, N.J. (2013). Formation of iron oxyfluoride phase on the surface of nano-Fe$_3$O$_4$ conversion compound for electrochemical energy storage, *J. Phys. Chem. Lett.*, **4**, pp. 3798–3805.
18. Van Vleck, J.H. (1934). The dirac vector model in complex spectra, *Phys. Rev.*, **45**, pp. 405–419.
19. Kozakov, A.T., Kochur, A.G., Googlev, K.A., Nikolsky, A.V., Raevski, I.P., Smotrakov, V.G., and Yeremkin, V.V. (2011). X-ray photoelectron study of the valence state of iron in iron-containing single-crystal (BiFeO$_3$, PbFe$_{1/2}$Nb$_{1/2}$O$_3$), and ceramic (BaFe$_{1/2}$Nb$_{1/2}$O$_3$) multiferroics, *J. Electron Spectrosc. Relat. Phenom.*, **184**, pp. 16–23.
20. Chiuzbaian, S.G., Neumann, M., Waldmann, O., Schneider, B., Bernt, I., and Saalfrank, R.W. (2001). X-ray photoelectron spectroscopy study of a cyclic hexanuclear cluster, *Surf. Sci.*, **482–485**(Part 2), pp. 1272–1276.
21. McIntyre, N.S., and Zetaruk, D.G. (1977). X-ray photoelectron spectroscopic studies of iron oxides, *Anal. Chem.*, **49**, pp. 1521–1529.
22. Casaletto, M.P., Longo, A., Martorana, A., Prestianni, A., and Venezia, A.M. (2006). XPS study of supported gold catalysts: the rold of Au$^o$ and Au$^{+1}$ species as active sites, *Surf. Interface Anal.*, **38**, pp. 215–218.
23. Veith, G.M., Lupini, A.R., Pennycook, S.J., and Dudney, N.J. (2010). Influence of support hydroxides on the catalytic activity of oxidized gold clusters, *ChemCatChem*, **2**, pp. 281–286.

24. Veith, G.M., Lupini, A.R., and Dudney, N.J. (2009). Role of pH in the formation of structurally stable and catalytically active TiO$_2$-supported gold catalysts, *J. Phys. Chem. C*, **113**, pp. 269–281.
25. Fu, Q., Saltsburg, H., and Flytzani-Stephanopoulos, M. (2003). Active nonmetallic Au and Pt species on ceria-based water-gas shift catalysts, *Science*, **301**, pp. 935–938.
26. Arrii, S., Morfin, F., Renouprez, A.J., and Rousset, j.L. (2004). Oxidation of CO on gold supported catalysis prepared by pulsed laser deposition, *J. Am. Chem. Soc.*, **126**, pp. 1199–1205.
27. Schumacher, B., Plzak, V., Cai, J., and Behm, R.J. (2005). Reproducibility of highly active Au/TiO$_2$ catalyst preparation and conditioning, *Catal. Lett.*, **101**, pp. 215–224.
28. Radnik, J., Mohr, C., and Claus, P. (2003). On origin of binding enrgy shifts of core levels of supported gold nanoparticels and dependence of pretreatments, *Phys. Chem. Chem. Phys.*, **5**, pp. 172–177.
29. Schimpf, S., Lucas, M., Mohr, C., Rodemerck, U., Brücker, A., Radnik, J., Hofmeister, H., and Claus, P. (2002). Supported gold nanoparticles: in-depth catalysts characterization and application in hydrogen and oxidation reactions, *Catal. Today*, **72**, pp. 63–78.
30. Claus, P., Brückner, A., Mohr, C., and Hofmeister, H. (2000). Supported gold nanoparticles from quantum dot to mesoscopic size scale: effect of electronic and structural properties on catalytic hydrogenation of conjugated functional groups, *J. Am. Chem. Soc.*, **122**, pp. 11430–11439.
31. Veith, G.M., Lupini, A.R., Pennycook, S.J., Ownby, G.W., and Dudney, N.J. (2005). Nanoparticles of gold on $\gamma$-Al$_2$O$_3$ produced by DC magnetron sputtering, *J. Catal.*, **231**, pp. 151–158.
32. Han, Y.-F., Zhong, Z., Ramesh, K., Chen, F., and Chen, L. (2007). Effects of different types of $\gamma$-Al$_2$O$_3$ on the activity of gold nanoparticles for CO oxidation at low-temperatures, *J. Phys. Chem. C*, **111**, pp. 3163–3170.
33. Casaletto, M.P., Longo, A., Venezia, A.M., Martorana, A., and Prestianni, A. (2006). Metal-support and preparation influence on the structural and electronic properties of gold catalysts, *Appl. Catal. A: Gen.*, **302**, pp. 309–316.
34. Wang, L.-C., Liu, Y.-M., Chen, M., Cao, Y., He, and Fan, K.-N. (2008). MnO$_2$ nanorod supported gold nanoparticles with enhanced activity for solvent-free aerobic alcohol oxidation, *J. Phys. Chem. C*, **112**, pp. 6981–6987.
35. Villa, A., Chan-Thaw, C.E., Veith, G.M., More, K.L., Ferri, D., and Prati, L. (2011). Au on nanosized NiO: a cooperative effect between Au and

nanosized NiO in the base-free alcohol oxidation, *ChemCatChem*, **3**, pp. 1612–1618.
36. Shukla, S., and Seal, S. (1999). Cluster size effect observed for gold annoparticles for XPS, *Nano Struct. Mater.*, **11**, pp. 1181.
37. Mason, M.G. (1983). Electronic structure of supported small metal clusters, *Phys. Rev. B*, **27**, p. 748.
38. Guczi, L., Horváth, D., Pászti, Z., and Petö, G. (2002). Effect of treatments on gold nanoparticles: Relation between morphology, electron structure and catalytic activity in CO oxidation, *Catal. Today*, **72**, pp. 101–105.
39. Weiher, N., Beesley, A.M., Tsapatsaris, N., Delannoy, L., Louis, C., vanBokhoven, J.A., and Schroeder, S.L.M. (2007). Activation of oxygen by metallic gold in Au/TiO$_2$ catalysts, *J. Am. Chem. Soc.*, **129**, pp. 2240–2241.
40. Chen, M.S., Cai, Y., Yan, Z., and Goodman, D.W. (2006). On the origin of the unique properties of supported Au nanoparticles, *J. Am. Chem. Soc.*, **128**, pp. 6341–6346.
41. Mojet, B.L., Miller, J.T., Ramaker, D.E., and Koningsberger, D.C. (1999). A new model describing the metal-support interaction in noble metal catalysts, *J. Catal.*, **186**, pp. 373–386.
42. Ramaker, D.E., Teliska, M., Zhang, Y., Stakheev, A.Y., and Koningsberger, D.C. (2003). Understanding the influence of support alkalinity on the hydrogen and oxygen chemisorption properties of Pt particles Comparison of X-ray absorption near edge data from gas phase and electrochemical systems, *Phys. Chem. Chem. Phys.*, **5**, pp. 4492–4501.
43. Koningsberger, D.C., de Graaf, J., Mojet, B.L., Ramaker, D.E., and Miller, J.T. (2000). The metal–support interaction in Pt/Y zeolite: evidence for a shift in energy of metal d-valence orbitals by Pt–H shape resonance and atomic XAFS spectroscopy, *Appl. Catal. A: Gen.*, **191**, pp. 205–220.
44. Henry, C.R. (1998). Surface science of catalysts, *Surf. Sci. Rep.*, **31**, pp. 231.
45. Richter, B., Kuhlenbeck, H., Freund, H.-J., and Bagus, P.S. (2004). Cluster core-level binding-energy shifts: the role of lattice strain, *Phys. Rev. Lett.*, **93**, p. 026805.
46. Lim, D.C., Lopez-Salido, I., Dietsche, R., Bubek, M., and Kim, Y.D. (2006). Electronic and chemical properties of supported Au nanoparticles, *Chem. Phys.*, **330**, pp. 441–448.
47. Zafeiratos, S., and Kennou, S. (1999). A study of gold ultrathin film gorwth on yttria-stabilized ZrO$_2$ (100), *Surf. Sci.*, **443**, p. 238.

48. Moretti, G. (1998). Auger parameter and Wagner plot in the characterization of chemical states by X-ray photoelectron spectroscopy: a review, *J. Electron Spectrosc. Relat. Phenom.*, **95**, pp. 95–144.
49. Wertheim, G.K. (1987). Auger shifts in metal clusters, *Phys. Rev. B*, **36**, p. 9559.
50. Wagner, C.D. (1975). Chemical shifts of Auger lines, and the Auger parameter, *Faraday Discuss.*, **60**, p. 291–300.
51. Herzing, A.A., Kiely, C.J., Carley, A.F., Landon, P., and Hutchings, G.J. (2008). Identification of active gold nanoclusters on iron oxide supports for CO oxidation, *Science*, **321**, pp. 1331–1335.
52. Wang, S., Huang, J., Geng, L., Zhu, B., Wang, X., Wu, S., Zhang, S., and Huang, W. (2006). Tin dioxide supported nanometric gold: synthesis, characterization, and low temperature catalytic oxidation of CO, *Catal. Lett.*, **108**, pp. 97–102.
53. Machesky, M.L., Predota, M., Wesolowski, D.J., et al. (2008). Surface protonation at the rutile (110) interface: explicit incorporation of solvation structure within the refined MUSIC model framework, *Langmuir*, **24**, pp. 12331–12339.
54. Mamontov, E., Vlcek, L., Wesolowski, D.J., Cummings, P.T., Rosenqvist, J., Wang, W., Cole, D.R., Anovitz, L.M., and Gasparovic, G. (2009). Suppression of the dynamic transition in surface water at low hydration levels: a study of water on rutile, *Phys. Rev. E*, **79**, p. 051504.
55. Mamontov, E., Vlcek, L., Wesolowski, D.J., Cummings, P.T., Wang, W., Anovitz, L.M., Rosenqvist, J., Brown, C.M., and Sakai, V.G. (2007). Dynamics and structure of hydration water on rutile and cassiterite nanopowders studied by quasielastic neutron scattering and molecular dynamics simulations, *J. Phys. Chem. C*, **111**, pp. 4328–4341.
56. Mamontov, E., Wesolowski, D.J., Vlcek, L., Cummings, P.T., Rosenqvist, J., Wang, W., and Cole, D.R. (2008). Dynamics of hydration water on rutile studied by backscattering neutron spectroscopy and molecular dynamics simulation, *J. Phys. Chem. C*, **112**, pp. 12334–12341.
57. Zhang, Z., Fenter, P., Sturchio, N.C., Bedzyk, M.J., Machesky, M.L., and Wesolowski, D.J. (2007). Structure of rutile TiO2 (110) in water and 1 molal Rb+ at pH 12: inter-relationship among surface charge, interfacial hydration structure, and substrate structural displacements, *Surf. Sci.*, **601**, pp. 1129–1143.
58. Henderson, M.A. (2002). The interaction of water with solid surfaces: fundamental aspects revisited, *Surf. Sci. Rep.*, **46**, pp. 1–308.

59. Edwards, J.K., Ntainjua N, E., Carley, A.F., Herzing, A.A., Kiely, C.J., and Hutchings, G.J. (2009). Direct synthesis of $H_2O_2$ from $H_2$ and $O_2$ over gold, palladium, and gold-palladium catalysts supported on acid-pretreated $TiO_2$, *Angew. Chem., Int. Ed.*, **48**, pp. 8512–8515.
60. Qian, K., Zhang, W., Sun, H., Fang, J., He, B., Ma, Y., Jiang, Z., Wei, S., Yang, J., and Huang, W. (2011). Hydroxyls-induced oxygen activation on "inert" Au nanoparticles for low-temperature CO oxidation, *J. Catal.*, **277**, pp. 95–103.
61. Singh, J.A., Overbury, S.H., Dudney, N.J., Li, M., and Veith, G.M. (2012). Gold nanoparticles supported on carbon nitride: influence of surface hydroxyls on low temperature carbon monoxide oxidation, *ACS Catal.*, **2**, pp. 1138–1146.
62. Karwacki, C.J., Ganesh, P., Kent, P.R.C., Gordon, W.O., Peterson, G.W., Niu, J.J., and Gogotsi, Y. (2013). Structure-activity relationship of $Au/ZrO_2$ catalyst on formation of hydroxyl groups and its influence on CO oxidation, *J. Mater. Chem. A*, **1**, pp. 6051–6062.
63. Moreau, F., Bond, G.C., van der Linden, B., Silberova, B.A.A., and Makkee, M. (2008). Gold supported on mixed oxides for the oxidation of carbon monoxide, *Appl. Catal. A*, **347**, pp. 208–215.
64. Jiang, D.-e., Overbury, S.H., and Dai, S. (2011). Interaction of gold clusters with a hydroxylated surface, *J. Phys. Chem. Lett.*, **2**, pp. 1211–1215.
65. Zhao, K., Qiao, B., Wang, J., Zhang, Y., and Zhang, T. (2011). A highly active and sintering-resistant Au/FeOx-hydroxyapatite catalyst for CO oxidation, *Chem. Commun.*, **47**, pp. 1779–1781.
66. Brown, M.A., Carrasco, E., Sterrer, M., and Freund, H.J. (2010). Enhanced stability of gold clusters supported on hydroxylated MgO(001) surfaces, *J. Am. Chem. Soc.*, **132**, p. 4064.
67. Ishizaka, T., Muto, S., and Kurokawa, Y. (2001). NLO and XPS prop of Au and Ag nanometer particles on Al2O3, *Opt. Commun.*, **190**, pp. 385.
68. Wilson, S.J., and Mc Connell, J.D.C. (1980). Kinetic study of the system AlOOH - Al2O3, *J. Solid State Chem.*, **34**, p. 315.
69. Böse, O., Kemnitz, E., Lippitz, A., and Unger, W.E.S. (1997). C1s and Au4f referenced XPS BE data, *Fresenius J. Anal. Chem.*, **358**, p. 175.
70. Piao, H., and McIntyre, N.S. (2001). Oxidation studies of Au-Al alloys using XPS and XANES, *Surf. Interface Anal.*, **31**, p. 874.
71. Daté, M., Ichihashi, Y., Yamashita, T., Chiorino, A., Boccuzzi, F., and Haruta, M. (2002). Performance of Au/TiO2 catalyst under ambient conditions, *Catal. Today*, **72**, pp. 89–94.

72. Moreau, F., and Bond, G.C. (2006). Gold on titania catalysts, influence of some physicochemical parameters on the activity and stability for the oxidation of carbon monoxide, *Appl. Catal. A*, **302**, pp. 110–117.
73. Zanella, R., and Louis, C. (2005). Influence of the conditions of thermal treatments and of storage on the size of the gold particles in Au/TiO$_2$ samples, *Catal. Today*, **107–108**, pp. 768–777.
74. Bartholomew, C.H. (2001). Mechanisms of catalyst deactivation, *Appl. Catal. A: Gen.*, **212**, pp. 17–60.
75. Cao, A., Lu, R., and Veser, G. (2010). Stabilizing metal nanoparticles for heterogeneous catalysis, *Phys. Chem. Chem. Phys.*, **12**, pp. 13499–13510.
76. Frydman, A., Castner, D.G., Schmal, M., and Campbell, C.T. (1995). A method for accurate quantitative XPS analysis of multimetallic or multiphase catalysts on support particles, *J. Catal.*, **157**, pp. 133–144.
77. Veith, G.M., Lupini, A.R., Rashkeev, S.N., Pennycook, S.J., Mullins, D.R., Schwartz, V., Bridges, C.A., and Dudney, N.J. (2009). Thermal stability and catalytic activity of gold nanoparticles supported on silica, *J. Catal.*, **262**, pp. 92–101.
78. Kim, C.H., and Thompson, L.T. (2005). Deactivation of Au/CeO$_x$ water gas shift catalysts, *J. Catal.*, **230**, pp. 66–74.
79. Wang, H., Zhu, H., Qin, Z., Liang, F., Wang, G., and Wang, J. (2009). Deactivation of a Au/CeO$_2$–Co$_3$O$_4$ catalyst during CO preferential oxidation in H$_2$-rich stream, *J. Catal.*, **264**, pp. 154–162.
80. Prati, L., Villa, A., Lupini, A.R., and Veith, G.M. (2012). Gold on carbon: one billion catalysts under a single label, *Phys. Chem. Chem. Phys.*, **14**, pp. 2969–2978.
81. Villa, A., Gaiassi, A., Rossetti, I., Bianchi, C.L., van Benthem, K., Veith, G.M., and Prati, L. (2010). Au on MgAl$_2$O$_4$ spinels: the effect of support surface properties in glycerol oxidation, *J. Catal.*, **275**, pp. 108–116.
82. Teschner, D., Borsodi, J., Wootsch, A., Révay, Z., Hävecker, M., Knop-Gericke, A., Jackson, S.D., and Schlögl, R. (2008). The roles of subsurface carbon and hydrogen in palladium-catalyzed alkyne hydrogenation, *Science*, **320**, pp. 86–89.
83. Starr, D.E., Liu, Z., Havecker, M., Knop-Gericke, A., and Bluhm, H. (2013). Investigation of solid/vapor interfaces using ambient pressure X-ray photoelectron spectroscopy, *Chem. Soc. Rev.*, **42**, pp. 5833–5857.
84. Newberg, J.T., Starr, D.E., Yamamoto, S., et al. (2011). Formation of hydroxyl and water layers on MgO films studied with ambient pressure XPS, *Surf. Sci.*, **605**, pp. 89–94.

85. Jiang, P., Porsgaard, S., Borondics, F., Köber, M., Caballero, A., Bluhm, H., Besenbacher, F., and Salmeron, M. (2010). Room-temperature reaction of oxygen with gold: an in situ ambient-pressure X-ray photoelectron spectroscopy investigation, *J. Am. Chem. Soc.*, **132**, pp. 2858–2859.
86. Willneff, E.A., Braun, S., Rosenthal, D., Bluhm, H., Havecker, M., Kleimenov, E., Knop-Gericke, A., Schlogl, R., and Schroeder, S.L.M. (2006). Dynamic electronic structure of a $Au/TiO_2$ catalyst under reaction conditions, *J. Am. Chem. Soc.*, **128**, pp. 12052–12053.
87. Brown, M.A., Beloqui Redondo, A., Sterrer, M., Winter, B., Pacchioni, G., Abbas, Z., and van Bokhoven, J.A. (2013). Measure of surface potential at the aqueous–oxide nanoparticle interface by XPS from a liquid microjet, *Nano Lett.*, **13**, pp. 5403–5407.

## Chapter 8

# FTIR Techniques for the Characterization of Au(-Ceria)-Based Catalysts

**Maela Manzoli and Floriana Vindigni**

*Chemistry Department, NIS Centre of Excellence, University of Torino,*
*Via Pietro Giuria 7, Torino 10125, Italy*
maela.manzoli@unito.it

FTIR spectroscopy represents a powerful technique to investigate the surface sites at an atomic level. In particular, the analysis of FTIR spectra of adsorbed probe molecules allows us to understand the nature and abundance of exposed active sites as well as to have detailed information on their structure and chemical environment. Therefore, FTIR spectroscopy can usefully assist in the comprehension of the parameters ruling the unique catalytic properties of gold catalysts, as well as in implementing the knowledge in the design of new systems. Keeping in mind these purposes, the present chapter tries to provide some insights into the spectroscopic characterization of supported gold nanoparticles, and we choose an intriguing and quite investigated system such as Au/CeO$_2$ as a *case history* to show what can be learned from the analysis of FTIR spectra collected on gold catalysts in different experimental conditions.

---

*Gold Catalysis: Preparation, Characterization, and Applications*
Edited by Laura Prati and Alberto Villa
Copyright © 2016 Pan Stanford Publishing Pte. Ltd.
ISBN 978-981-4669-28-3 (Hardcover), 978-981-4669-29-0 (eBook)
www.panstanford.com

The results concerning Au catalysts supported on pure ceria and on differently modified ceria will be discussed in order to evaluate the effect of the support on the nature of the gold species and on the reactivity in different reactions. More in detail, the data have been organized and will be illustrated on the basis of the following kinds of experimental setup:

(i) Ex situ measurements at low temperature and at room temperature (r.t.)
(ii) In situ measurements at increasing temperature
(iii) Operando measurements

## 8.1 An Overview of Gold/Ceria-Based Catalysts

Catalysis by gold represents a fascinating research field, driven not only by the economic attractions of the material, which among precious metals, is one of the more abundant and cheaper ones, but also by its highly desirable propensity to catalyze reactions at relatively low temperature. Different explanations have been given to justify the dramatic catalytic activity displayed by gold nanoparticles, observed for the first time by Haruta et al. [1]. Besides the extreme dependence of the catalytic activity on the size of the Au particles, which is a generally accepted phenomenon, other interpretations are based on structural effects, that is, the presence of Au reactive steps and corner sites [2]; on electronic effects [3, 4], due to the presence of gold nonmetallic and ionic sites; and on support effects [5], where the support is able to stabilize unusual structural/electronic properties of the metal or to favor reactions occurring at the particle/support interface [6]. The most likely portrait may be that all such phenomena are important to different extents, depending on the kind of reaction we are considering. For all these reasons, the selection of the support is a critical factor. A suitable support should be able to avoid coalescence and agglomeration of the gold nanoparticles, as well as to participate actively in the reaction mechanism by redox cycling of the support metal ions [7].

The fortune of ceria in several catalytic applications is mainly due to its ability to shift easily between oxidized and reduced states ($Ce^{4+} \leftrightarrow Ce^{3+}$) [8], given its unique oxygen storage capacity (OSC). This OCS, strictly connected to high oxygen defectivity, makes $CeO_2$ an appropriate support for precious metals, promoting noble metal activity and dispersion [9].

A variety of experimental studies has provided evidence to suggest that gold nanoparticles supported on ceria can be of great interest from the catalytic viewpoint [10–16]. It has been reported that oxygen vacancies, which are the most relevant surface defects, play a crucial role in binding catalytically active species [17, 18]: When Au nanoparticles are supported on ceria, the system exhibits a high activity for the water–gas shift (WGS) reaction over a wide temperature range [19]. The concentration of oxygen vacancies can be increased by modifying ceria either by doping or by simply adding an opportune amount of another oxide. The dopant cations with ionic radius and electronegativity close to those of the cerium cation are thought to be the most appropriate modifiers of structural and chemical properties of ceria. This is related to the ability of the heterocations to cause structural distortions inside ceria, producing strain in the oxide lattice and favoring oxygen vacancy formation. Promising performance in the preferential CO oxidation (PROX) reaction of gold catalysts supported on ceria doped by some transition or rare-earth cations has been recently reported [16, 20, 21]. Laguna et al. [22, 23] have shown that the oxygen release capability is enhanced when iron is added as a dopant to ceria, and $Au/FeO_x/CeO_2$ catalysts demonstrated high activity for total CO oxidation and PROX. In fact, the presence of iron in the ceria structure was found to increase Au dispersion by creating sites with an increased electronic density, which act in a similar way to oxygen vacancies. Recent studies have shown that high CO oxidation activity during the PROX reaction could be achieved after deposition of gold nanoparticles on $\alpha$-$Mn_2O_3$ [24] or $MnO_2$–$CeO_2$ [25]. Wang et al. [26] have reported that $Au/CeO_2$–$Co_3O_4$ with an appropriate Ce/Co atomic ratio exhibited much higher activity than $Au/Co_3O_4$ and $Au/CeO_2$. In contrast, Liotta et al. [27] have observed the highest CO oxidation activity of $Au/CeO_2$ in a comparative study of the support effect on the catalytic performance of gold-based catalysts supported on $Co_3O_4$, on $CeO_2$, and on mixed $Co_3O_4$–$CeO_2$. In addition, it has to be considered that pure cerium dioxide is poorly thermostable as it undergoes sintering at high temperature, thereby losing its crucial oxygen storage and release ability [28, 29]. Various additives, such as samaria, lanthana, and zirconia, have been incorporated in a ceria lattice in order to improve its reducibility, oxygen mobility, and thermal stability [30, 31].

All known WGS catalysts based on ceria show similar problems related to the deactivation with the time-on-stream and/or shutdown restart operation. For example, Kim and Thompson [32] reported fast deactivation of their Au–ceria catalyst, which was attributed to the blockage of the active sites by carbonates and/or formates formed during the WGS reaction. They reported that oxygen deficiency is an important factor with regard to the formation of carbonate and formate species on the oxide surface. It was also found that the initial activities for the $Au/CeO_2$ catalysts could be fully recovered by calcination of the deactivated catalysts in flowing air at elevated temperatures. They also found that the deposition of carbonates and/or formates is facilitated by oxygen-deficient sites on the catalyst surface. As a consequence, the deactivation can be, in principle, influenced by the addition of other constituents in order to control oxygen deficiency and decrease the formation of carbonates species. It was previously reported that the incorporation of $ZrO_2$ into a solid solution with $CeO_2$ has a significant effect on both OSC and stability of this defective oxide [19, 33]. In particular, the Zr presence, unlike on pure ceria, allows the storage and release of oxygen also from the bulk lattice structure and not just from surface layers. However, $ZrO_2$ addition gives rise to a modification of the support surface and, as a consequence, could also affect the gold dispersion. Tibiletti et al. [34] in a density functional theory (DFT) and in situ extended X-ray absorption fine structure (EXAFS) study on a $Au/CeZrO_4$ catalyst evidenced that the presence of zirconium in the oxide support stabilizes the oxygen defects surrounding the gold particles. These defects are critical in the reaction mechanism and promote WGS activity. Later, some of the same authors reported that the catalyst deactivation was due to the detachment of the gold particles from the $CeZrO_4$ support in the presence of water [35]. This is due to the hydrolysis of the interface between the gold and the oxide that decreases the metal support interaction by a breakage of the Au–support link at the metal oxide interface.

Therefore, the present chapter encompasses a detailed characterization study on several gold catalysts supported on bare ceria and on ceria modified by addition of various dopants (Zn, Sm, La) or of different amounts of oxides such as $ZrO_2$, $Fe_2O_3$, $TiO_2$, and $Co_3O_4$, trying to find a spectroscopic explanation for which the substituted systems displayed better catalytic activity than the simple $Au/CeO_2$ catalyst. Such an approach is timely as it facilitates structure–property relationship comprehension of paramount importance to optimize the catalytic performances of novel heterogeneous catalysts, suitably designing their chemical synthesis.

### 8.1.1 Preparation of the Samples

The gold catalysts were provided by the researchers of the Institute of Catalysis, Bulgarian Academy of Sciences, Sofia, Bulgaria. The syntheses of the bare supports as well as of the catalysts were carried out in a Contalab laboratory reactor, enabling complete control of the reaction parameters (pH, temperature, stirrer speed, reactant feed flow, etc.) and high reproducibility.

The catalysts were prepared by deposition-precipitation [36] of gold on the different supports suspended in water, via interaction of $HAuCl_4 \cdot 3H_2O$ and $K_2CO_3$ at a constant pH of 7.0 and at a temperature of 333 K. After aging for 1 h, the precipitates were washed, dried in vacuo at 353 K, and calcined under air at 673 K for 2 h.

As shown in Fig. 8.1, generally the color of the samples, which are in the form of nanocrystalline powders, turns from pale yellow (bare ceria) to dark violet (Au/ceria) after the addition of gold. The dark violet is due to the presence of gold nanoparticles, which are able to give a plasmonic absorption band around 18,000 $cm^{-1}$ in the ultraviolet-visible (UV-Vis) spectroscopic range, as demonstrated by the diffuse reflectance UV-Vis spectrum reported in the right panel of the figure.

The samples investigated in the present chapter (Table 8.1) all have a 3 wt% gold loading, with the exception of those in which

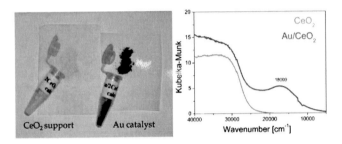

**Figure 8.1** (Left) Images of the powders. (Right) DRUV-Vis spectra of pure ceria support and of Au/ceria.

**Table 8.1** Examined Au catalysts

| Sample | Name | Ceria: Oxide ratio | Au size (nm) | Ref. |
|---|---|---|---|---|
| Au/CeO$_2$ | AuCe | – | >10 and <1 | [37] |
| Au/Zn-CeO$_2$ | AuZn-Ce | a | b | [16, 38] |
| Au/Sm-CeO$_2$ | AuSm-Ce | a | 3.5–4 | [16, 38] |
| Au/La-CeO$_2$ | AuLa-Ce | a | around 5 | [16, 38] |
| Au/CeO$_2$-Fe$_2$O$_3$ | AuCe75Fe25 | 75:25 | around 10 | [39, 40] |
| Au/CeO$_2$-Fe$_2$O$_3$ | AuCe50Fe50 | 50:50 | 1–1.8 | [39, 40] |
| Au/CeO$_2$-Fe$_2$O$_3$ | AuCe25Fe75 | 25:75 | 10–15 | [39, 40] |
| Au/CeO$_2$-TiO$_2$ | AuCe50Ti50 | 50:50 | ≤ 2.0 | [41] |
| Au/CeO$_2$-ZrO$_2$ | AuCe80Zr20 | 80:20 | 1.55 ± 0.3 | [42] |
| Au/CeO$_2$-ZrO$_2$ | AuCe50Zr50 | 50:50 | 10–15 | [42] |

[a] Atomic ratio M/(M + Ce) = 0.05 (M: Sm, Zn, or La).
[b] Very high, escaping from HRTEM detection.

ceria has been modified by the addition of iron oxide, where the Au loading is 3.5 wt%.

The results on gold catalysts supported on mixed CeO$_2$–Me$_x$O$_y$ (Me = Fe, Mn, and Co) that were prepared by the mechanochemical mixing procedure [43] will be also summarized in Section 8.3.4.

The Fourier transform infrared spectroscopy (FTIR) spectra were taken on a Perkin–Elmer 1760 and on a Perkin–Elmer 2000 spectrometer (equipped with mercury cadmium telluride (MCT) detectors) with the samples in self-supporting pellets introduced in cells, allowing thermal treatments in controlled atmospheres and spectrum scanning at r.t. or at controlled temperatures (from 100 to 773 K).

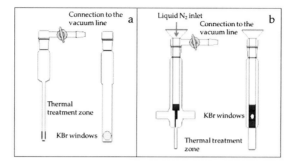

**Figure 8.2** FTIR cells. (a) Cell for room-temperature analysis. (b) Cell for low-temperature analysis.

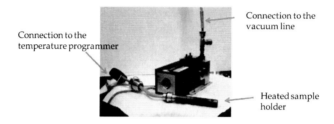

**Figure 8.3** AABSPEC heated stainless steel cell.

The quartz cells, used either for the thermal treatment or the spectroscopic measurements at r.t. and at low temperature are depicted in Figs. 8.2a and 8.2b, respectively.

In addition, a stainless steel AABSPEC 2000 cell allowing us to run the spectra in situ in controlled atmospheres and temperatures is shown in Fig. 8.3. This kind of cell has been also employed for operando experiments.

Band integration was carried out by Curvefit, in Spectra Calc (Galactic Industries Co.), by means of Lorentzian curves. From each spectrum, the spectrum of the sample before the inlet of the reactants was subtracted. The spectra were normalized with respect to the gold content of each pellet, unless otherwise specified. All samples were previously submitted to an oxidative treatment, eventually followed by a reductive one, or simply to an outgassing at r.t. (as-received samples). The thermal treatment with oxygen is necessary to clean the surface of the catalysts, which is covered with

**Table 8.2** Oxidative treatment

| Step | Description |
|---|---|
| 1 | r.t. → 473 K under outgassing |
| 2 | 473 K → 673 K in 40 mbar of $O_2$ |
| 3 | 4 inlets of $O_2$ at 673 K (10 min each ones) |
| 4 | 673 K → r.t. in 40 mbar of $O_2$ |

**Table 8.3** Reductive treatment

| Step | Description |
|---|---|
| 1 | r.t. → 423/473 K in 20 mbar of $H_2$ |
| 2 | 2 inlets of $H_2$ at 423/473 K (10 min each one) |
| 3 | 423/473 K → r.t. under outgassing |

water and carbonate-like species because of air exposure. Therefore, to make the sites available to the reactant, cleaning treatment is needed. In Tables 8.2 and 8.3 the steps of oxidative and reductive thermal treatments are summarized.

## 8.2 Ex situ CO Adsorption at 100 K

### 8.2.1 Some Insights into CO Adsorption

Infrared (IR) spectroscopy of probe molecules, CO especially, is one of the most useful techniques to deeply characterize nanostructured catalysts. CO adsorption can be nonreactive (carbonyls species are formed) and reactive (CO undergoes chemical transformations on the catalyst surface). The use of CO as a probe is based on nonreactive adsorption. CO is an ideal probe molecule in the case of gold because the position of the CO bands ($\nu_{CO}$) can provide specific information on both oxidation and coordination states and on the electrophilic properties of the accessible gold sites [44]. Additionally, CO is also a reactant, for example, in CO oxidation and in the WGS reaction.

The vibrational spectrum of CO is simple. The $\nu_{CO}$ stretching vibration is sensitive to the strength of the bond formed with

**Figure 8.4** Metal–carbonyl bond: formation of a $\sigma$ metal $\leftarrow$ carbon bond by using a lone pair on the C atom. The other CO molecular orbitals are omitted for the sake of clarity; the red orbitals are full, and the yellow one is empty.

the surface, and most of the supports are transparent in the C–O stretching region. In surface and catalytic chemistry, a change in the position of the band related to adsorbed CO with respect to the molecule in the gaseous phase (2143 cm$^{-1}$) can be interpreted by the *Bhyholder model*, which considers the adsorbing site (a metal site)–CO bond as a result of two main contributions [45]:

(i) A *σ bonding interaction*, which originates from the overlapping between the $5\sigma$ full orbital of the carbon atom (with weak antibonding nature) and the empty $d$-orbital of the metal site with an opportune symmetry ($d_z^2$), as shown in Fig. 8.4, which implies an electron density transfer from the CO molecule (lone pair on the C atom) to the metal center with an increase in bond strength. As a consequence, the position of the CO band undergoes a blue shift if compared to the free molecule.

(ii) A *π back-donation* ($d_\pi \rightarrow p_\pi$) with bonding character, due to the overlapping between two full $d$-orbitals of the metal and the antibonding $2\pi^*$ degenerate molecular orbitals of CO (Fig. 8.5). This interaction introduces electron density into the antibonding CO orbital, and as a consequence, the (C–O) bond strength is decreased and the $\nu_{CO}$ is red-shifted with respect to the free molecule.

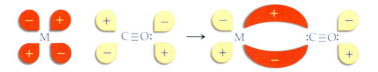

**Figure 8.5** Metal–carbonyl bond: formation of a $\pi$ metal $\rightarrow$ carbon bond by using a lone pair on the C atom. The other CO molecular orbitals are omitted for the sake of clarity; the red orbitals are full, and the yellow one is empty.

Depending on the nature of the adsorbing site, one of these two factors, that is, $\sigma$ donation or $\pi$ back-donation, can prevail on the other: metal ions with a high oxidation state give mainly donation, whereas metal atoms, neutral or partially negatively charged, bestow back-donation. Generally, the CO extinction coefficient is high. However, it strongly depends on the nature of the bond with the surface. In addition, the CO molecule is small and its adsorption is not hindered by steric hindrance. Therefore, CO is one of the most frequently used probes in IR spectroscopy. Linear-bonded CO molecules are associated with wavenumbers in the approximate range of 2000–2170 cm$^{-1}$, doubly bridged ones in the 1880–2000 cm$^{-1}$ region, and finally multiply bridged CO species below 1880 cm$^{-1}$. Due to the nature of our samples, in this chapter we will manage with linear-bonded carbonyls.

It is well known that CO adsorption at low temperature (the working temperature reached by our cell is around 100 K) is able to put in evidence sites on which the probe molecule is weakly adsorbed. Hadjiivanov and Vayssilov [46] reported previously that CO is able to reduce cationic gold, that is, Au$^{3+}$ and/or Au$^{+}$ sites [47]; therefore it is advantageous to perform experiments at low temperature to minimize reactive adsorption of CO. Moreover, to get further information on the stability of all surface sites, spectroscopic analysis was carried out also during the heating from 100 K to r.t In such a case, sites that are weakly bonded to CO are progressively depleted as a consequence of the temperature increase, allowing the complete assignment of the bands in the carbonylic stretching region.

### 8.2.2 As-Received Au/CeO$_2$ Catalyst

Here we report the results obtained for the as-received AuCe catalyst, that is, for at the catalyst surface simply exposed to air, without any thermal treatment. The spectrum collected after the adsorption of 0.5 mbar CO at 100 K on the as-received sample is shown in Fig. 8.6 (orange line). An intense and asymmetric band, related to CO interacting with the support cations [48], is observed at 2158 cm$^{-1}$. In addition, a broad absorption at 2110–2120 cm$^{-1}$ is

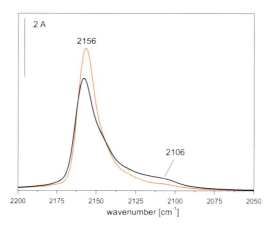

**Figure 8.6** FTIR spectra after the adsorption of 0.5 mbar of CO at 100 K on AuCe as-received (orange line) and reduced in CO at r.t. (black line).

also produced. On the basis of its position, this band can be related to the presence of partially oxidized gold [2].

Thereafter, the sample was heated up to r.t. in a CO atmosphere and cooled down to 100 K again (Fig. 8.6, black line). The heating in CO has a reductive effect, even if less effective than that carried out by $H_2$, and it puts in evidence the presence of $Au^0$ sites, as demonstrated by the increase in intensity of the component centered at 2106 cm$^{-1}$. Such intensification is an indirect indication that, on the as-received sample, most of the gold particles are covered by adsorbed oxygen as a consequence of their small size. This phenomenon occurs when the Au particles have a size below 2 nm [49].

It can be hypothesized that the oxygen initially covering the small gold particles of the $Au/CeO_2$ catalyst is removed by the reductive interaction with CO, as further confirmed by simultaneous $CO_2$ formation (data not shown). Hence, the gold reactive sites become available to the adsorption by mild interaction with the CO molecule.

### 8.2.3 Effect of Pre-Oxidation at 473 K on the Exposed Sites

In Fig. 8.7 the spectrum produced after CO adsorption at 100 K on a previously oxidized AuCe catalyst (bold line) is shown, as well as

## 216 | FTIR Techniques for the Characterization of Au(-Ceria)-Based Catalysts

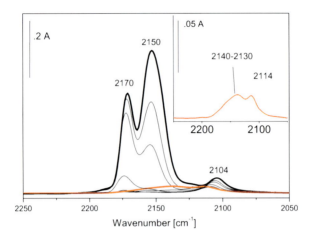

**Figure 8.7** FTIR spectra of AuCe oxidized at 673 K after the adsorption of 0.5 mbar of CO at 100 K (bold line) and at increasing temperature (fine lines) up to r.t. (orange line). In the zoom, the spectrum of adsorbed CO (0.5 mbar) after heating up to r.t. is reported.

the spectra collected at increasing temperature (fine lines) up to r.t. (orange line).

The 2170 cm$^{-1}$ and 2150 cm$^{-1}$ bands are associated to CO adsorbed on Ce$^{4+}$ cations with different unsaturated coordination. The feature at 2104 cm$^{-1}$ is due to CO on Au sites exposed at the surface of metallic nanoparticles. Interestingly, a broad band centered at 2130–2140 cm$^{-1}$ (see the zoom in the figure) appears after the heating up to r.t., in addition to an increase of the carbonate/hydrogen carbonate bands in the 1900–1000 cm$^{-1}$ range (not shown for the sake of brevity). In addition, this band shows also a quite high stability toward the outgassing at the same temperature. It is well known that the most stable gold carbonyls are those of the Au$^+$ cations. This is due to the bonding of CO to cationic gold sites by $\sigma$- and $\pi$-back bonds. The increase of the effective charge of the gold sites reflects in a strengthening of the $\sigma$-bond and in an increase of the CO stretching frequency. The $\pi$-back-donation should lead to some decrease of the CO stretching frequency and also to an increase of the overall bond strength. However, the increase of the effective charge of the cation alone should reflect in the decrease of the $\pi$-bond order. A detailed analysis of these effects has been provided for the $Cu^+$–CO and $Ag^+$–CO systems. It has been concluded that the $\pi$-bond for the $Cu^+$–CO system is only slightly enhanced with the increase of the effective cationic charge. In the case of $Au^+$, where the $\pi$-back-donation is less important than for the $Cu^+$ cations, the change of the $\pi$-bond order should be even weaker. Hence, the stability and frequency of the $Au^+$–CO species can well be interpreted by the covalent $\sigma$-bond alone. It may be relevant to observe that also cationic copper binds CO irreversibly at r.t., differently from metallic copper. Looking at the paper of Wu et al. [50], where DFT calculations are applied to carbon monoxide adsorption on small cationic, neutral, and anionic clusters, an assignment of that absorption can be made. The authors reported that the adsorption energies of CO on the cationic clusters are greater than those on the neutral and anionic complexes and also the calculated CO vibrational frequencies are larger than those on neutral clusters. Very recently, Fielicke et al. [51] estimated experimentally the CO vibrational frequencies of $Au_n(CO)^{m+}$ complexes in the gas phase, where $3 \leq n \leq 10$ and $3 \leq m \leq 8$. In particular, they observed that the range of frequencies is between 2180 cm$^{-1}$ and 2120 cm$^{-1}$ and also that $\nu_{CO}$ decreases when the nuclearity of the gold clusters increases. Therefore, the band at 2130–2140 cm$^{-1}$ can be assigned to $Au_n(CO)^{m+}$ species. We can conclude that heating from 100 K to r.t. has the effect to allow the gold atoms to reach the surface from subsurface positions, giving rise to the above band.

### 8.2.4 Au/CeO$_2$ Catalyst Reduced in H$_2$

In Fig. 8.8 the spectra collected after CO adsorption at 100 K on the reduced AuCe catalyst, and during the heating in CO, are reported.

The CO adsorption gives rise to two bands at 2160 cm$^{-1}$ and 2140 cm$^{-1}$. The former is due to CO on $Ce^{4+}$ sites, while the latter is ascribed to CO on $Ce^{3+}$ sites, formed during the reduction [48]. These bands display low stability, since they disappear when r.t. is reached. Moreover, the reductive treatment gives rise to a strong modification of the carbonylic species on Au sites. In particular, if compared to the band obtained in the case of the oxidized sample (see Fig. 8.7), the band is changed as for its position and shape and

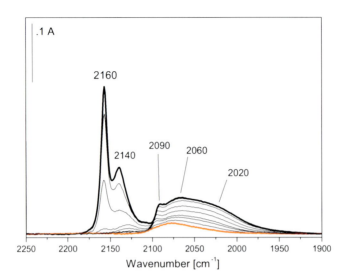

**Figure 8.8** FTIR spectra of AuCe reduced at 473 K after the adsorption of 0.5 mbar of CO at 100 K (black line) and at increasing temperature (fine lines) up to r.t. (orange line).

it seems to be made by at least three components. It is possible to associate the component at 2090 cm$^{-1}$ to Au nanoparticles affected by the presence of Ce$^{3+}$, as remarked before [2], while the component centered at 2060 cm$^{-1}$ can be due to Au$_n$(CO)$^{m-}$ species [51], which are produced by the conversion of Au$_n$(CO)$^{m+}$ into negatively charged species by the reductive treatment.

In addition, the presence of a component centered at 2020 cm$^{-1}$, which is a quite low frequency for gold carbonyl species, seems to be an indication that ceria has been seriously reduced, possibly up to the formation of metallic cerium and a Ce–Au alloy. It is worth noting that evidence of this alloy is reported in the work of Zhao et al. [52], where the reaction of water with the Ce–Au(111) and CeO$_x$(111) surface has been studied by photoemission and scanning tunneling microscopy (STM) techniques. This hypothesis can be confirmed looking at the spectra collected on AuCe after reductive treatments at increasing temperature and reported in Fig. 8.9.

Going from 373 K up to 473 K, an intensity decrease of the 2090 cm$^{-1}$ band with a parallel increase of the component at lower

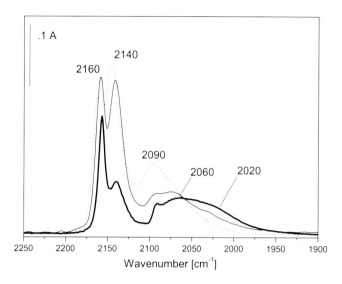

**Figure 8.9** FTIR spectra of CO adsorbed at 100 K on AuCe reduced at 373 K (dotted line), at 423 K (fine line), and at 473 K (bold line).

frequency is evident, confirming a progressive charge transfer from the support to the gold. Moreover, it is also interesting to observe the effect of the reduction temperature on the band at 2140 cm$^{-1}$, associated to CO on $Ce^{3+}$ cations. The intensity of this band is strongly decreased after the treatment at 473 K, possibly because $Ce^{3+}$ ions are almost completely reduced.

On the basis of all the above findings and of previous results on other gold catalysts supported on ceria [53], we can propose that negatively charged clusters or flattened, thin gold particles are present at the surface of defective ceria. It has been reported that nanosized ceria shows higher activity than bulk phases because of its easier reducibility. The consequent presence of $Ce^{3+}$ and oxygen vacancy defects stabilizes the highly dispersed gold species differently from other samples. It can be proposed that after reduction, Au spreads on the reduced ceria and the 3D gold particles, possibly present before the reductive treatment, may become very thin films. Akita et al. [54] studied by high-resolution transmission electron microscopy (HRTEM) the gold nanoparticles supported on a model $CeO_2$ exposing low-index flat faces, typically in the {111} and {100} planes, to overcome the problem of the poor difference in contrast between the two phases and of the consequent difficulty in detecting gold particles on ceria. The authors observed that the smallest and thin gold nanoparticles disappeared under the beam and in vacuo, shrinking layer by layer down to a monoatomic layer. The flattening of Au nanoparticles on reduced ceria was also observed by us by HRTEM analysis on a Au catalyst supported on a mixed ceria–titania oxide [41].

## 8.3 CO–$O_2$ Interaction at Low Temperature up to Room Temperature

The admission of $^{18}O_2$ at 100 K on pre-adsorbed CO at the same temperature produced immediately the erosion of the broad adsorption at 2080–2020 cm$^{-1}$, and a new, more intense band at 2100 cm$^{-1}$ is produced (see black and orange lines of Fig. 8.10 [55]), evidencing that the Au sites are no longer negatively charged.

This feature is evidence of the reoxidation of the support by the oxygen molecules, dissociated at the oxygen vacancies. This finding is further confirmed by the depletion of the 2140 cm$^{-1}$ band and the slight intensification of the 2159 cm$^{-1}$ component, related to CO–

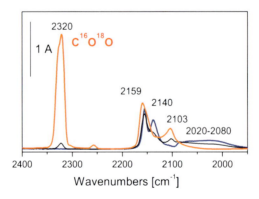

**Figure 8.10** Evolution of FTIR absorbance spectra at 100 K collected on AuCe reduced at 373 K after the inlet of $^{18}O_2$ on pre-adsorbed CO (blue line) and after 20 min (orange line). Reprinted from Ref. [55], Copyright (2009), with permission from Elsevier.

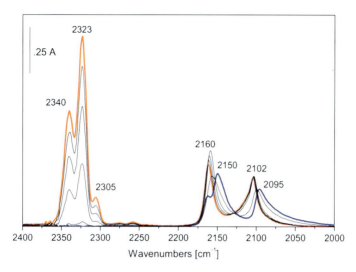

**Figure 8.11** Evolution of FTIR absorbance spectra at 100 K collected on AuZn–Ce reduced at 373 K after the inlet of $^{18}O_2$ on pre-adsorbed CO (blue line) and after 20 min (orange line). Reprinted with permission from Ref. [56]. Copyright (2010) American Chemical Society.

$Ce^{4+}$ interaction. Moreover, the presence of the 2320 cm$^{-1}$ peak, due to $C^{16}O^{18}O$ on $CeO_2$ sites, indicates that the catalyst is able to produce $CO_2$ already at low temperature and that no exchange reaction with the oxygen of the support occurs, differently from what was previously described on a $Au/TiO_2$ catalyst [2].

### 8.3.1 *Effect of Doping: Au Supported on Zn-Modified Ceria*

Spectroscopic evidence of the effect of the support composition on oxygen mobility in the CO oxidation reaction is reported in Fig. 8.11 [56], where the evolution of the bands during CO and $^{18}O_2$ interaction on the AuZn–Ce sample reduced at 373 K within 20 min is shown.

The analysis of the spectra showed that a rapid exchange between the oxygen of ceria and the $^{18}O_2$ molecules coming from the gas phase occurs already at 100 K. The inlet of $^{18}O_2$ at increasing diffusion times (black lines up to the orange line) over the sample previously saturated by CO at 100 K (blue line) caused strong erosion from the low-frequency side of the band at 2095 cm$^{-1}$. After the $^{18}O_2$ interaction at 100 K, the band shifted from 2095 to 2102 cm$^{-1}$, and it appeared to be narrow and with a maximum in the usual position of CO adsorbed on the Au$^0$ sites. The band at 2150 cm$^{-1}$ disappeared, and a new band at 2160 cm$^{-1}$ related to CO adsorbed on Ce$^{4+}$ sites was produced.

The aim of this experiment was to compare the bands at higher frequencies, namely those related to $CO_2$ formation, during the CO oxidation on AuZn–Ce and AuCe. Three isotopomers of $CO_2$ were registered in the spectra of AuZn–Ce (red curve): a growing band at 2323 cm$^{-1}$, assigned to the $C^{16}O^{18}O$ solid-like phase, accompanied by weaker bands at 2340 and 2305 cm$^{-1}$, assigned to the $C^{16}O_2$ and $C^{18}O_2$ solid-like phase, respectively [57]. The high intensity of the band ascribed to $C^{16}O^{18}O$ indicated that oxygen participating in the reaction at 100 K mainly comes from the gas phase. At the same time, the appearance of the band at 2340 cm$^{-1}$ evidenced the fact that the modification with Zn led to increased oxygen mobility because of improved exchange properties of ceria support. This result demonstrated the enhanced ability of a doped ceria support to supply active lattice oxygen that is beneficial for the reaction. Here we recall that these experiments were performed on catalysts reduced at 373 K, that is, the temperature of optimum performance for the PROX reaction [16]. The presence of a multiplet of bands in the $CO_2$ stretching region after CO interaction with $^{18}O_2$ over AuZn–Ce, unlike over AuCe, clearly reveals that the doping of ceria by Zn facilitates the exchange reaction with the oxygen atoms of the support. The surface state of this catalyst after the PROX reaction at 400 K was also studied by FTIR spectroscopy. In these experiments, AuZn–Ce was contacted with the PROX mixture, CO–O$_2$–H$_2$ (1:1:7), in the presence or in the absence of water and heated at 400 K for 20 min. Once the catalyst was cooled to r.t., the spectra were collected (not shown). The most interesting results are those obtained by CO adsorption at 100 K after reaction at 400 K and outgassing.

Examining the spectra in the 2200–1900 cm$^{-1}$ range (see Fig. 8.12) suggests that no agglomeration and/or deactivation by carbonates of highly dispersed gold species occurred either after the PROX reaction at 400 K or after the same reaction in the presence of

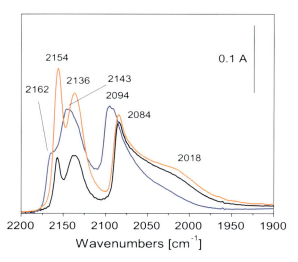

**Figure 8.12** FTIR absorbance spectra of AuZn–Ce after CO (0.5 mbar) adsorption at 100 K on freshly reduced catalyst (blue line), after PROX at 400 K and outgassing (black line), and after PROX in the presence of water and outgassing (orange line). Reprinted from Ref. [16], Copyright (2008), with permission from Elsevier.

water. In particular, the position, intensity, and shape of the band of CO adsorbed on gold species were practically the same after PROX (black line) as in the freshly reduced sample (blue line). As for the carbonyls on the support, their intensity decreased after PROX, due to the presence of carbonates partially covering the support surface sites.

When the PROX reaction was performed at 400 K in the presence of water, the CO adsorbed after the reaction produced a blue-shifted band at 2096 cm$^{-1}$ with a shoulder toward lower frequencies (orange line). Moreover, the component at 2018 cm$^{-1}$, related to electron-rich Au sites (possibly Au–Ce alloy clusters), was totally depleted. At the same time, a new intense band at 2143 cm$^{-1}$ and a weak component at 2162 cm$^{-1}$ were produced. Specifically, this behavior is due to the presence of water, which reacts with the reduced support and the Au–Ce alloy, leading to O–H bond breaking and formation of Ce(OH)$_x$ species [52].

Finally, the comparison of the integrated areas of the bands produced after CO adsorption at 100 K on Au sites on (i) freshly reduced AuZn–Ce (blue line), (ii) the sample outgassed after the PROX reaction (black line), and (iii) the sample outgassed after the PROX reaction in the presence of water (orange line) showed very similar values. This indicates that almost all of the active gold sites detected before the PROX reaction were still available. These results correlate with the catalytic activity/selectivity tests demonstrating the ability of Au/doped ceria catalysts to tolerate significant amounts of $CO_2$ and $H_2O$ in the feed [16]. Gold catalysts on modified ceria showed better activity and selectivity in the presence of both $CO_2$ and $H_2O$ compared to undoped AuCe.

### 8.3.2 Effect of Doping by Other Elements (Sm, La)

In previous studies [16, 38] some of us investigated the modification of ceria by the addition of various cations ($Sm^{3+}$, $La^{3+}$, and $Zn^{2+}$) in order to obtain a defective fluorite structure with increased oxygen mobility that could result in activity enhancement and improved resistance toward deactivation caused by the presence of $CO_2$ and $H_2O$ in the PROX feed. A detailed characterization study has been undertaken to investigate the effect of modification of ceria on structural properties of the catalysts and to find a correlation with catalytic performance in the PROX reaction.

The dopants addition to ceria influenced to a different extent the catalytic activity in the PROX reaction of the $Au/CeO_2$ catalysts. More specifically, they caused either an increase (Sm- and Zn-doped catalysts) or a decrease (La-doped catalysts) of activity. Moreover, doping of ceria had a beneficial effect on $CO_2$ and $H_2O$ tolerance of the doped gold catalysts. The activity order observed in the PROX reaction was AuZn–Ce >AuSm–Ce >AuCe >Au/La–Ce.

The origin of the observed differences can be related to a modification of the defective structure of ceria. The Raman data collected along the characterization of the bare supports allowed us to evidence a change of the ceria environment in the presence of dopants due to the formation of solid solutions as well as an increase in the number of oxygen vacancies formed in the lattice [58, 59]. FTIR spectroscopic findings indicated that the concentration of $Ce^{3+}$-defective sites is higher on the surface of AuCe than on Au/Sm- and Zn-doped ceria (data not shown for the sake of brevity). The

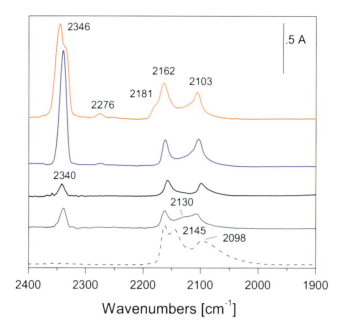

**Figure 8.13** FTIR absorption spectra after 20 min of $O_2$ admission at 100 K on 0.5 mbar of pre-adsorbed CO on the reduced catalysts (spectrum of AuCe reported as a reference, dashed line): AuCe (fine black line), AuLa–Ce (bold black line), AuZn–Ce (blue line), and AuSm–Ce (orange). All spectra are normalized to the weight of the pellet. Reprinted from Ref. [16], Copyright (2008), with permission from Elsevier.

trend was opposite than that concerning the catalytic results for CO oxidation rate, where an enhancement by Sm- and Zn-doped ceria catalysts was observed. Indeed, FTIR spectra showed the largest intensity and broadness of the absorption related to CO on gold for reduced AuCe, indicating the presence of Au clusters on the surface of this catalyst (see the dashed line in Fig. 8.13). This result may imply that gold clusters and $Ce^{3+}$-defective sites are not the active sites in the PROX reaction, while the step sites of gold particles are the active sites for both CO and oxygen activation and play a decisive role. Temperature-programmed reduction (TPR) measurements on the doped and undoped catalysts evidenced a different reducibility of the materials, also indirectly confirmed by the different relative intensities of the FTIR CO bands related to $Ce^{4+}$ and $Ce^{3+}$ on the catalysts after reduction and depending possibly from differences in the gold dispersion, as shown by HRTEM analysis. One of the most often debated question concerns the identification of the active sites in CO oxidation, that is, whether ionic or metallic gold is the active species. FTIR spectra, collected during CO–$O_2$ interaction at 100 K, can put in evidence the important role of metallic gold particles.

An IR experiment was performed to explore the role of dopants on the activity of the modified Au/$CeO_2$ catalysts. $O_2$ was admitted at 100 K on CO pre-adsorbed at the same temperature on the catalysts reduced previously at 373 K. Significant changes in the spectra were observed after 20 min of CO–$O_2$ interaction, as reported in Fig. 8.13.

Starting from the spectrum related to CO adsorbed on reduced AuCe as a reference (dashed line), a weak absorption at 2340 $cm^{-1}$, related to molecular $CO_2$, was noted after 20 min of CO–$O_2$ interaction (fine black line), and the broad absorption at 2098 $cm^{-1}$ related to CO adsorbed on small clusters was depleted, as was the band at 2145 $cm^{-1}$ assigned to CO adsorbed on $Ce^{3+}$ sites (see also dashed line). Moreover, two components at 2103 $cm^{-1}$ and at 2130 $cm^{-1}$, related to CO on mildly oxidized gold, were observed. As for the doped samples (bold black, blue, and orange lines), greater amounts of $CO_2$ were produced in the presence of a high concentration of metallic gold particles on the surface of some of modified Au catalysts (blue and orange lines), as indicated by the higher intensity of the band at 2103 $cm^{-1}$, confirming the important role of metallic gold in the CO oxidation reaction. It can be seen that this band is more intense in the spectra of AuZn–Ce and AuSm–Ce. This observation is of particular importance because it correlates well with the registered higher CO oxidation rate of the above-mentioned catalysts and reveals the decisive role of metallic gold in CO oxidation reactions. At the same time, the lowest intensity of the band, related to CO adsorption on metallic gold particles on the surface of AuLa–Ce, agrees with the lowest catalytic performance of this sample. In the low-frequency range, bands at 1134 and 840 $cm^{-1}$, assigned to superoxo and peroxo species adsorbed on the reduced support sites [60], increased simultaneously (spectra not shown). After 20 min of reaction at 100 K these bands show the highest intensity on the surface of AuSm–Ce, whereas the lowest was found on AuLa–Ce. These species are related to the reduced sites at the interface with gold particles. Bands related to carbonate-like species also appeared on the surface of all catalysts in the 1800–800 cm$^{-1}$ range.

The spectroscopic characterization, corroborated by HRTEM analysis, demonstrated that the dispersion of gold species formed on ceria doped with Sm$^{3+}$, La$^{3+}$, and Zn$^{2+}$ cations was influenced by the dopant. Both large gold particles (>10 nm) and highly dispersed gold clusters (of about 1 nm) were found on the AuCe catalyst, whereas the gold particle size was around 5 nm for AuLa-Ce and 3.5–4 nm for AuSm-Ce. The highest dispersion of gold was found on the AuZn-Ce catalyst.

### 8.3.3 On the Modification of the Support by Iron Oxide

Some of our previous results [39] revealed that the variation of the composition of the support may produce significant differences in the gold particle size and, as a consequence, in the catalytic activity of Au/CeO$_2$–Fe$_2$O$_3$ catalysts under PROX (in the presence of H$_2$ excess) conditions. The following activity order was observed: AuCe $\cong$ AuCe50Fe50 > AuCe75Fe25 > AuCe25Fe75 > Au/Fe$_2$O$_3$.

The variation in the catalytic behavior of the gold catalysts supported on Ce-Fe mixed oxides originates from the difference in the gold particle size and the varying ability of the supports to activate oxygen. In particular, AuCe50Fe50 showed high PROX activity, similar to that of AuCe, and its catalytic behavior differed from that of the other gold catalysts examined. In particular, this catalyst did not follow the trend of decreased catalytic activity by increasing the Fe$_2$O$_3$ amount in the support composition. The spectra reported in Fig. 8.14 and collected after 20 min CO-O$_2$ interaction at 100 K give information about the higher reactivity of AuCe50Fe50 (orange line) in comparison to AuCe75Fe25 (black line) and AuCe25Fe75 (blue line). The production of CO$_2$ is evidenced by the strong band at 2340 cm$^{-1}$, which is most intense in the spectrum of AuCe50Fe50. The intensity of the band at 2340 cm$^{-1}$ decreased in the following order: AuCe50Fe50 > AuCe75Fe25 > AuCe25Fe75.

The divergences in the CO oxidation activity could be explained by considering the different ability of the supports to supply reactive

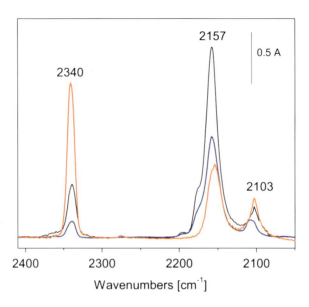

**Figure 8.14** FTIR absorption spectra collected 20 min after admission of $O_2$ at 100 K on 3.5 mbar pre-adsorbed CO on AuCe50Fe50 (orange line), AuCe75Fe25 (black line), and AuCe25Fe75 (blue line). Reprinted from Ref. [39], Copyright (2010), with permission from Elsevier.

oxygen species [61–63]. To confirm this hypothesis, FTIR studies on CO–$^{18}O_2$ interaction were also performed. In this way, we can try to find spectroscopic evidence for the effect of the support composition on the catalytic activity in the CO oxidation reaction.

The evolution of the spectra of AuCe50Fe50 during CO–$^{18}O_2$ interaction at 100 K and up to 20 min (black lines up to the orange line) is shown in Fig. 8.15. A decrease in intensity and a simultaneous shift of the band at 2101 cm$^{-1}$ up to 2104 cm$^{-1}$ were immediately observed after the inlet of $^{18}O_2$ at 100 K over the sample saturated previously by CO at the same temperature (blue line). Additionally, a broad shoulder at about 2123 cm$^{-1}$ indicated the formation of Au species positively charged. This new feature is in agreement with the HRTEM observations as for the presence of small gold particles on the surface of this catalyst that are able to interact immediately with oxygen [49].

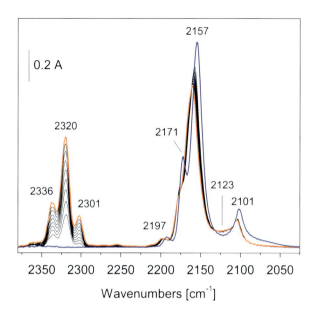

**Figure 8.15** FTIR spectra of AuCe50Fe50 after inlet of 3.5 mbar CO at 90 K (blue line) and evolution after $^{18}O_2$ adsorption on pre-adsorbed CO (black lines) up to 20 min at the same temperature (orange line). Reprinted from Ref. [39], Copyright (2010), with permission from Elsevier.

However, our goal is to observe the bands at higher frequencies, namely those related to $CO_2$ formation during CO oxidation. The analysis of the spectra show a rapid exchange between the oxygen of ceria and that coming from the $^{18}O_2$ molecules of the gas phase already at 100 K. Three isotopomers of solid-like $CO_2$ (orange line) were registered: a growing band at 2320 cm$^{-1}$, assigned to the $C^{16}O^{18}O$ solid-like phase, accompanied by weaker bands at 2338 cm$^{-1}$ and 2302 cm$^{-1}$, assigned to the $C^{16}O_2$ and $C^{18}O_2$ solid-like phase, respectively.

The comparison between the spectra collected after CO–$^{18}O_2$ interaction at 100 K on AuCe50Fe50 (blue line) and AuCe75Fe25 (orange line) is shown in Fig. 8.16. The bands at 2338, 2320, and 2302 cm$^{-1}$ related to $CO_2$ isotopomers are significantly more intense on AuCe50Fe50 than those observed for AuCe75Fe25. At the same time, a general lower intensity of the bands in the carbonate region is observed in the spectrum of AuCe50Fe50.

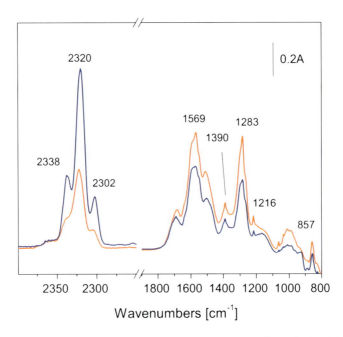

**Figure 8.16** Comparison of FTIR absorption spectra collected after 20 min CO–$^{18}O_2$ interaction at 100 K on AuCe50Fe50 (blue line) and AuCe75Fe25 (orange line). Reprinted from Ref. [39], Copyright (2010), with permission from Elsevier.

The structural features of mixed oxide supports, particularly the presence of $Ce^{3+}$ defects and oxygen vacancies, controlled the performance of $Au/CeO_2$–$Fe_2O_3$ catalysts. FTIR findings demonstrated an enhanced reactivity of AuCe50Fe50. In particular, the support with composition 50 wt% $CeO_2$–50 wt% $Fe_2O_3$ appeared beneficial not only for nucleation and growth of highly dispersed gold particles (1–1.8 nm) [39] but also for oxygen activation and mobility. Mössbauer spectra collected at r.t. and 100 K showed that different amounts of cubic $CeO_2$-like solid solution and hematite co-existed in $CeO_2$–$Fe_2O_3$ supports [40]. The analysis of the characterization data suggested that $Fe^{3+}$ were distributed at $Ce^{4+}$ and interstitial sites by a dopant compensation mechanism. This mechanism of solid solution formation accounted for the lower oxygen vacancy concentration and $Ce^{3+}$ amount on all gold catalysts supported on mixed $CeO_2$–$Fe_2O_3$ materials, which affected the catalytic performance. Moreover, the presence of $Fe_2O_3$ improved the resistance toward deactivation by $CO_2$. As discussed in previous reports [18, 19], the negative effect of $CO_2$ is due to the competitive adsorption of CO (and $H_2$) and $CO_2$ on the catalyst surface [64, 65]. The nature of the support is thought to affect the catalyst stability in the presence of $CO_2$ in the feed; acidic supports were found to be more resistant to deactivation than basic ones [66]. Due to its basic character ceria strongly binds carbonates. Hence, the lower amount of ceria in the mixed oxides is favorable for the stability in the presence of $CO_2$.

To find additional evidence for the different ability of AuCe50Fe50 and AuCe75Fe25 to tolerate $CO_2$ in the reaction mixture, the cell was evacuated after the PROX reaction and 3.5 mbar CO were admitted. The comparison of the spectra produced after this CO admission on AuCe50Fe50 and on AuCe75Fe25 is shown in Fig. 8.17.

A quite symmetric band at 2098 $cm^{-1}$, due to CO adsorbed on metallic gold corner and edge sites, has been observed on AuCe50Fe50 (blue line). On AuCe75Fe25 (black line), the same band was centered at 2105 $cm^{-1}$ and had a lower intensity if compared to that related to AuCe50Fe50, indicating that on AuCe75Fe25 the number of active gold sites available after use is noticeably lower than that on AuCe50Fe50. The comparison of the spectra produced after admission of CO at 100 K on AuCe50Fe50 after oxidative treatment (orange dashed line) and after PROX and evacuation (orange solid line) is reported in the inset. The similar intensity and shape of the bands related to CO adsorbed on gold sites is a strong indication that all gold edge and corner sites able to activate CO are still available on AuCe50Fe50 after the reaction. On the contrary, the same comparison on AuCe75Fe25 put in evidence that on this catalyst a large fraction of gold sites is no more available after PROX, possibly due to the blockage of these sites by carbonates.

### 8.3.4 Modification of Ceria by Other Oxides

Very recently, we studied catalytic performance in the total oxidation of CO and methanol over gold catalysts supported on ceria doped by different metal oxides (Me = Fe, Mn, and Co) [43]. A strong influence

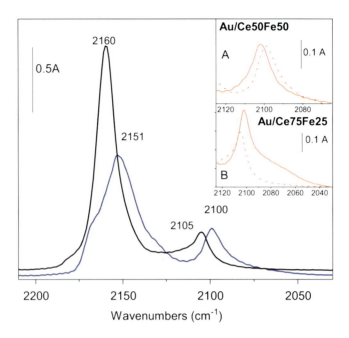

**Figure 8.17** FTIR spectra produced after admission of 3.5 mbar CO at 100 K on AuCe50Fe50 (blue line) and AuCe75Fe25 (black line) after PROX and evacuation. Inset: FTIR spectra produced after admission of 3.5 mbar CO at 100 K on AuCe50Fe50 (A) and AuCe75Fe25 (B) after oxidative treatment (orange dashed lines) and after PROX and evacuation (orange solid line). Reprinted from Ref. [39], Copyright (2010), with permission from Elsevier.

of the nature of the dopant was observed. The Au90Ce10Co (i.e., the Au catalyst supported on 90 wt% CeO$_2$, 10 wt% Co$_3$O$_4$) catalyst exhibited superior low-temperature CO oxidation activity (a 100% conversion degree was obtained at 298 K) and almost 100% total oxidation of CH$_3$OH at about 313 K. The effect of modification with Co$_3$O$_4$ of Au/CeO$_2$ catalysts on their CO oxidation activity was further studied by varying of the dopant content (5, 10, and 15 wt% Co$_3$O$_4$). The high defectivity of ceria-exposed faces, the presence of gold particles with the smallest size, and a highly reducible CoO$_x$ phase resulted in a larger enhancement of mobility/reactivity of surface oxygen. The supply of active oxygen species could be performed by the participation of the oxygen vacancies on modified ceria as well as by the surface of a separate, highly reducible CoO$_x$

phase. The synergy between gold and a Co-doped ceria support causes significant enhancement of reducibility and capability for oxygen activation, which resulted in improved oxidation activity.

The support with composition 90 wt% $CeO_2$–10 wt% $Co_3O_4$ appeared beneficial for nucleation and growth of highly dispersed gold particles, as put in evidence by X-ray diffraction (XRD) and HRTEM analyses. Moreover, higher hydrogen consumption was estimated by means of TPR over this catalyst. The evolution of the FTIR spectra run at liquid nitrogen temperature after admission of $O_2$ on pre-adsorbed CO on the most active catalyst demonstrated its ability to oxidize CO at very low temperature. Further FTIR measurements of CO–$^{18}O_2$ interaction (from 100 K up to r.t.) revealed that on all doped supports different amounts of $C^{16}O^{18}O$ and $C^{16}O_2$ are produced, indicating the participation of the O atoms of the supports. Bicarbonate species and bidentate carbonate species are simultaneously produced in different amounts, depending on the $Co_3O_4$ amount in the support, indicating that cobalt oxide lowers the basicity of ceria on Au90Ce10Co and less carbonate is formed, despite the highest $CO_2$ amount produced.

Gold catalysts supported on ceria modified by addition of different amounts of $ZrO_2$ (i.e., 50% and 20%) were previously synthesized and tested in the WGS reaction at low temperature [42]. These systems displayed better catalytic activity than gold supported on pure ceria. In particular, the AuCe50Zr50 sample showed the best catalytic performance, both for the activity and for the stability. Morphologic, textural, and spectroscopic analysis, undertaken to examine the impact of zirconia addition to $CeO_2$, revealed different gold dispersion, depending on the zirconia amount. In particular, Au agglomerates (10–20 nm) and nanoparticles (about 2.5 nm) have been detected on the sample with 20% of zirconia, while nanoparticles and Au clusters have been found on the catalyst with 50% of zirconia. Moreover, FTIR measurements also revealed that the addition of zirconia modifies the acid/base properties of the mixed oxides, influencing the stability of carbonate-like species adsorbed on the catalyst's surface. All the experimental results have allowed us to state that the lowest stability of the carbonate-like species on the surface of the catalyst with 50% of zirconia, along with the presence of high gold dispersion, makes this system the best catalyst in terms of activity and stability.

## 8.4 Ex situ CO Adsorption at Room Temperature

Figure 8.18a shows the FTIR spectra produced by the adsorption of 15 mbar CO at r.t. (after 1 hour) and at decreasing coverages at the same temperature on AuCe (Fig. 8.18a) and on AuCeTi (Fig. 8.18b), both being samples previously oxidized at 673 K.

It has been reported previously [37] that on AuCe a few large gold particles with size >10 nm, as well as very highly dispersed gold clusters ($d_m \approx 1$ nm) are present. Because of their size, the big particles expose only Au terrace sites that are not able to adsorb CO. Therefore, the spectroscopic features observed on this sample are exclusively due to the CO adsorption on the gold clusters, which expose mainly corner sites. On the contrary, on AuCe50Ti50 only few particles, with a size around 2 nm, have been observed, suggesting that most of the gold escapes from HRTEM detection [41]. This feature can explain the lower intensity of the IR bands on AuCe in comparison to those on AuCe50Ti50. Besides the usual bands related to CO on the support and on metallic gold that will be discussed later, a broad and complex absorption band centered at 2166 cm$^{-1}$ is observed on both catalysts [41]. This band, that is likely due to contributions of a different nature, is to a large extent irreversible to the outgassing at r.t. Whereas the small reversible portion is evidence of CO adsorbed on Ce$^{4+}$ sites [48], the resistance to the outgassing indicates a strong bond between CO and the involved adsorption sites, and the band at 2166 cm$^{-1}$ can be assigned to CO on Au$_n^{m+}$ clusters, where $3 \leq n \leq 10$ and $3 \leq m \leq 8$ [50, 51]. Moreover, the band at 2166 cm$^{-1}$ displays a quite different intensity on the two samples, indicating that on AuCeTi the gold clusters are more abundant than on AuCe, and we found that their abundance appears related to the amount of ceria in the samples [41]. Possibly, the gold clusters are stabilized by interaction with nanodispersed ceria. The lack of a band at 2166 cm$^{-1}$ upon reduction supports the assignment to the cationic gold clusters, present only at the surface of the oxidized sample.

*Ex situ CO Adsorption at Room Temperature* | 235

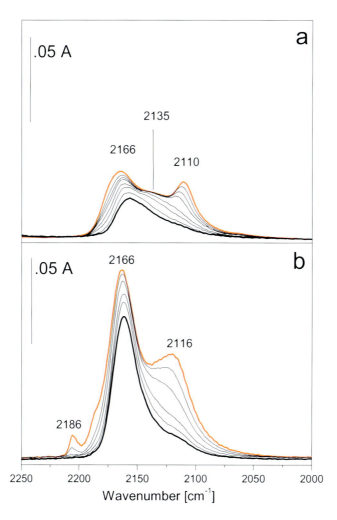

**Figure 8.18** FTIR spectra of AuCe (a) and of AuCeTi (b) oxidized at 673 K after the adsorption of 15 mbar of CO at r.t. (orange line), at decreasing pressures (fine lines), and under outgassing at r.t. (black line). Reprinted with permission from Ref. [41]. Copyright (2006) American Chemical Society.

Going back to the other features evidenced in Fig. 8.18, the bands at 2110–2116 cm$^{-1}$ are due to CO molecules adsorbed on Au sites; their frequency can be taken as an indication that the sites are partially oxidized. It has been observed by Bondzie et al. [49] that Au particles with a size below 2 nm are able to bind oxygen. As confirmed by HRTEM, most of the gold particles are very small and, as a consequence, likely able to bind oxygen.

Other interesting observations on the irreversible component at 2166 cm$^{-1}$ can be made looking at its behavior at increasing CO contact times. We choose to show only the experiment performed on AuCe50Ti50, because of the presence of the largest amount of Au clusters on this catalyst. The spectra collected immediately after the CO inlet (dashed blue line) after 10 min (blue line) and after 1 h (orange line) are shown in Fig. 8.19.

Briefly, an intensity increase of the band at 2166 cm$^{-1}$ with the contact time was observed, providing evidence that the gold atoms, giving rise to clusters responsible for this band by interaction with CO, are not initially on the surface. Because of the high basicity of ceria and its high oxygen storage capacity (OSC), CO molecules are able to reduce the ceria surface, leading up to the formation of oxygen vacancies on the support. As a consequence, the subsurface gold can slowly migrate toward the surface, giving rise to the feature at 2166 cm$^{-1}$.

Even more interesting is the marked change of the carbonyl bands that appears, further increasing the time of contact, as shown in Fig. 8.19b, where the evolution of the bands after one night in CO at r.t. is reported. After such a long contact time, the bands due to CO adsorbed on support cations disappear, probably because the huge amount of carbonate/hydrogen carbonate species, produced by the CO interaction with the surface, has covered the support sites. Moreover, we observed the disappearance of the band at 2166 cm$^{-1}$ and the appearance of a broad band at 2120 cm$^{-1}$, irreversible to outgassing. As demonstrated before [51], the frequency of Au$_n$(CO)$^{m+}$ species is related to the $n$ index; particularly, $\nu_{CO}$ decreases when $n$ increases. Therefore, the 2120 cm$^{-1}$ feature can be assigned to cationic gold clusters with higher nuclearity in respect to those associated to the band at 2166 cm$^{-1}$. The cluster nuclearity can increase if more gold atoms reach the surface from

*Ex situ CO Adsorption at Room Temperature* | 237

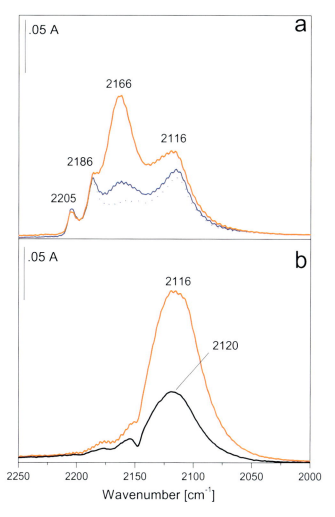

**Figure 8.19** FTIR spectra of AuCe50Ti50 oxidized at 673 K (a) immediately after the adsorption of 15 mbar of CO at r.t. (dashed blue line), after 10 min (blue line), and after 1 h (orange line) and (b) after one night in CO at r.t. (orange line) and after outgassing at r.t. (black line).

subsurface positions: this is very likely what happens with the longer contact time with CO.

## 8.5  In situ FTIR Measurements at Increasing Temperature

What is the stability of these gold clusters? Are they present in hypothetic reaction conditions? In this section we would like to show the ability of in situ FTIR measurements to thoroughly investigate the behavior of active gold sites in reaction conditions. Therefore, to obtain further information about cationic gold clusters, particularly on their thermal stability, spectroscopic analysis, using a commercial heated stainless steel cell (AABSPEC), was undertaken. We recall that this equipment allows in situ thermal treatments in vacuum or in controlled atmospheres up to 873 K and, at the same time, spectra collection. The AuCe50Ti50 catalyst was submitted to an oxidizing treatment, then it was submitted to 1 h of CO interaction at r.t., and then it was outgassed for a long time in order to be sure that only CO molecules adsorbed on Au clusters remain on the surface. After such pretreatment, the sample underwent two procedures:

(i)  Heating from r.t. up to 423 K under outgassing
(ii) Heating from r.t. up to 423 K in an oxygen atmosphere (3 mbar)

In Fig. 8.20 the spectra of AuCe50Ti50, collected under outgassing (panel a) or in an oxygen atmosphere (panel b) at r.t. (black line), at 373 K (orange line), and at 423 K (dotted line) are reported.

At 373 K, the intensity of the band at 2166 cm$^{-1}$ detected in an oxygen atmosphere (Fig. 8.20b, orange line) is at least two times lower than that observed under outgassing (see Fig. 8.20a, orange line). This is due to the fact that in the presence of oxygen CO oxidation to $CO_2$ occursdespite the fact that the band related to adsorbed $CO_2$ cannot be detected as a consequence of the high temperature at which the measurement has been performed. The decrease in intensity observed in an oxygen atmosphere is also a strong indication that at 373 K, the gold clusters are reactant species and not simply "spectators." In both cases, the total depletion of

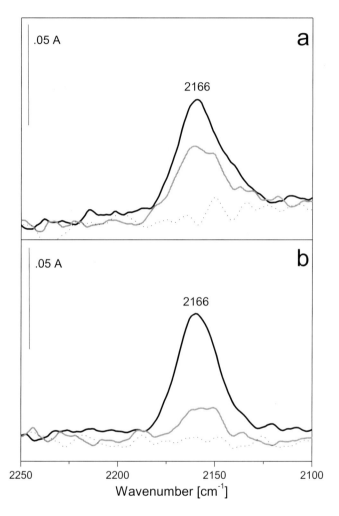

**Figure 8.20** FTIR spectra of AuCe50Ti50 under outgassing (a) or in oxygen (b) at r.t. (black line), at 373 K (orange line), and at 423 K (dotted line).

the band occurs at 423 K, indicating that at this temperature the different experimental conditions do not have influence anymore. Moreover, these results are evidence of the high stability of cationic gold clusters.

## 8.6 Operando Measurements of $CO_2$ Uptake

At the beginning of this chapter we reported that WGS catalysts based on ceria, due to blocking of active sites by carbonates and/or formates species produced during the reaction, undergo deactivation with the time-on-stream and/or shut-down restart operation [32]. Ceria modification by the addition of constituents, as other metal oxides, that is, $ZrO_2$, $Fe_2O_3$, or $TiO_2$, has been proposed as a possible solution to overcome this problem. It has been reported that carbonate-like species are considered reaction intermediates [67, 68]; therefore their formation can be connected to the catalytic activity. As a consequence, improved stability of ceria-modified catalysts can be reasonably linked to a decrease in the stability of the carbonate-like species.

Therefore, more detailed studies on the formation/adsorption and stability of these carbonate-like species can be useful to improve the knowledge to design and synthesize new gold catalysts. The $CO_2$ molecule may act as a Lewis acid toward both $O^{2-}$ surface ions, forming carbonate species, and residual basic OH surface species, producing hydrogen carbonate species. Binet et al. [48] reported a detailed attribution of these adsorbed species, formed by $CO_2$ adsorption on unreduced and reduced ceria at r.t. Therefore, we performed operando FTIR measurements of $CO_2$ uptake, followed by a temperature-programmed desorption (TPD) experiment, in order to investigate the carbonate-like species stability. The measurements have been undertaken on AuCe and on two ceria catalysts, modified by the addiction of $ZrO_2$ and $Fe_2O_3$. All the samples have been previously submitted to an oxidation treatment. For the sake of clarity, the steps of the oxidative thermal pretreatment, of $CO_2$ uptake, and of the TPD experiment are reported in Fig. 8.21. After the oxidative thermal pretreatment, the FTIR spectra, normalized to the pellet weight, of AuCe, AuCe50Zr50, and AuCe50Fe50 in the carbonate-like region show a residual broad and complex adsorption. However, comparison with the spectra collected before the oxidative treatment puts in evidence the good effectiveness of our cleaning procedure (data not shown for the sake of brevity). Generally, it is quite arduous to precisely assign different

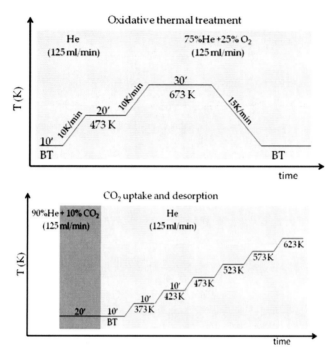

**Figure 8.21** Steps of oxidative thermal treatment, of CO$_2$ uptake, and of the TPD experiment (BT = beam temperature). The increase of the temperature in the TPD was undertaken at 15 K/min.

peaks to determinate carbonate or bicarbonate species: However, we found that the relative intensity of the bands seem to be different on each sample and that the highest intensity of the bands was observed in the case of AuCe50Zr50.

The CO$_2$ uptake was undertaken by flowing a mixture of He and CO$_2$ (90:10) at the IR beam temperature (i.e., 303 K) for 20 min, and then the flow was switched to pure He and the spectra were collected after 10 min. The results are shown in Fig. 8.22. Several kinds of carbonate-like species were formed on all the samples, and it is clearly evident that their relative abundance is different on the three catalysts. By focusing on AuCe (black line), due to the low transmittance (<5%) in the 1550–1430 cm$^{-1}$ region, it is not possible either to recognize the absorption bands or to make

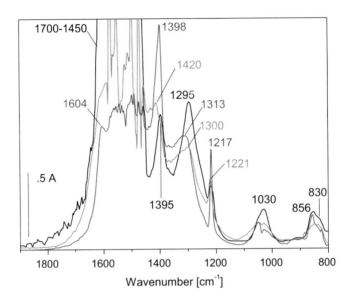

**Figure 8.22** FTIR spectra collected upon $CO_2$ uptake on AuCe (black line), AuCe50Zr50 (orange line), and AuCe50Fe50 (blue line) after 10 min of evacuation in He flow. All spectra are normalized to the weight of the pellet.

consideration of the total amount of carbonate-like species upon $CO_2$ uptake.

However, we can tentatively assign the peaks at 856 cm$^{-1}$ ($\pi CO_3$) and 1295 cm$^{-1}$ ($\upsilon CO_3$) to bidentate carbonates adsorbed on the support surface.

The stretching mode of these species consists of three different components [48]. In the 1650–1450 cm$^{-1}$ range, accompanied by the peak at 1295 cm$^{-1}$, one band could be reasonably present. However, it is not detectable because of the out-of-scale intensity in this region. The second component, that Binet et al. have found, in their experimental conditions, at 1014 cm$^{-1}$ could be located as a component of the broad absorption centered at 1030 cm$^{-1}$. Behind these features, the peaks at 830 cm$^{-1}$ ($\pi CO_3$), 1395 cm$^{-1}$ ($\upsilon CO_3$), and 1216 cm$^{-1}$ ($\delta OH$), related to hydrogen carbonate species formed upon $CO_2$ adsorption, are evident. The other two stretching mode bands are present at 1030 cm$^{-1}$ and in the 1650–1450 cm$^{-1}$ range. The same authors above mentioned have reported that hydrogen carbonates are present as two different species, named I and II; the former species disappears upon evacuation at r.t, while heating at 373 K is needed to eliminate the latter one. In our case, whether for the frequency or for the behavior toward heating, which will be discussed later, species II are adsorbed on support surface. These results put in evidence that bidentate carbonates seem to be the most abundant species formed on the pure ceria surface after $CO_2$ uptake.

The addition of another oxide to ceria has a remarkable influence on the FTIR spectra collected after $CO_2$ interaction. On the AuCe50Fe50 sample (Fig. 8.22, blue line) the high intensity of the bands at 1398 cm$^{-1}$ and at 1217 cm$^{-1}$ clearly demonstrates that $CO_2$ mainly turns into hydrogen carbonate species on the surface. This feature is also confirmed by the low intensity of the broad absorption in the 1650–1450 cm$^{-1}$. In this case, the third stretching mode of hydrogen carbonates is easily distinguishable at 1604 cm$^{-1}$. On the basis of the observed frequency of these peaks, along with the results published by Busca and Lorenzelli [69] on $Fe_2O_3$, it can be confirmed that these species are exclusively adsorbed on the ceria phase. The slight shift to 1313 cm$^{-1}$ gives evidence of an effect due to the addition of $Fe_2O_3$ consisting of the presence of different amounts of cubic $CeO_2$-like solid solution and hematite, as discussed in Section 8.3.3.

On the AuCe50Zr50 catalyst (orange line), a lower amount of bidentate carbonate species is adsorbed on the support surface if compared to AuCe (black line). The influence of the presence of zirconia is confirmed by the shift to 1300 cm$^{-1}$ also in this case. On the contrary, the bands related to hydrogen carbonate species are shifted to higher wavenumbers, that is, 1420 cm$^{-1}$ ($\upsilon CO_3$) and 1220 cm$^{-1}$ ($\delta OH$), than those observed on AuCe. Interestingly, these values are comparable to those reported by Bachiller-Baeza et al. [70] in a study on $CO_2$ adsorption with a zirconia polymorph. These findings suggest that hydrogen carbonates are formed and adsorbed on the zirconia phase. The stability of these adsorbed species, formed by $CO_2$ uptake, was tested by a TPD experiment by heating in He flow the samples from the IR beam temperature up to 623 K. In Fig. 8.23 the spectra collected at increasing temperature are reported in the 1900–1100 cm$^{-1}$ range.

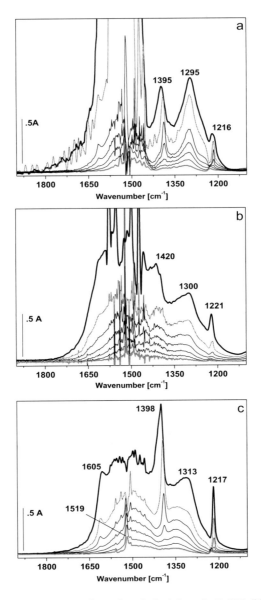

**Figure 8.23** FTIR spectra collected on AuCe (a), on AuCe50Zr50 (b), and on AuCe50Fe50 (c) at beam temperature (black bold line), at 373 K (blue line), upon heating (black fine lines), and at 623 K (orange line) in He flow. All spectra are normalized to the weight of the pellet.

At 373 K (blue lines), the bands related to bidentate carbonates (i.e., 1650–1450 cm$^{-1}$ and around 1300 cm$^{-1}$) are strongly decreased in intensity on AuCeZr and AuCe50Fe50 (Figs. 8.23b and 8.23c, respectively) if compared to AuCe (Fig. 8.23a), indicating lower stability of these species on the mixed oxides surface. The behavior upon the heating of the bands related to hydrogen carbonates, that is, ~1395 cm$^{-1}$ and ~1216 cm$^{-1}$, adsorbed on AuCe (Fig. 8.23a) and on AuCe50Fe50 (Fig. 8.23b) confirm that they are species II [48]. It is interesting to note that these species are very stable up to 473 K. On the contrary, these bands strongly decrease at 373 K and disappear at 473 K in the case of AuCe50Zr50, further confirming that hydrogen carbonates are adsorbed on zirconia rather than on ceria.

The operando FTIR measurements demonstrated that the addition of another oxide to ceria has a marked effect in terms of decreasing the amount and stability of carbonate/bicarbonate species adsorbed on the catalyst surface. As a consequence, such kinds of modification can strongly influence both stability and reactivity of gold supported on ceria systems in the WGS reaction.

## 8.7 Final Remarks

The results on gold dispersed on pure, doped, and modified $CeO_2$ demonstrate that FTIR spectroscopy can be effectively used to investigate surface sites at an atomic level. Keeping in mind that the spectroscopic approach needs to be corroborated by additional experimental and theoretical findings to point out a relationship between the nanostructure and the physical and chemical properties of supported gold, FTIR spectroscopy of adsorbed probe molecules has provided information on the involved active sites exposed at the surface of catalysts.

A detailed analysis of the FTIR spectra collected at different temperatures (starting from 100 K up to 473 K) and of the modification induced on the CO bands by co-interaction with oxygen, as well as $^{18}O_2$, as been presented. By varying the thermal pretreatment atmosphere it is possible, for example, to acquire qualitative knowledge on the oxidation state and on the coordinative unsaturation of the sites. Isotopic experiments elegantly allowed us to find a relationship between the nature of the dopant or the added oxide and catalytic performances.

The high stability of cationic gold clusters has been proved by in situ measurements at increasing temperature. Moreover, it has been fully stated that at 373 K, Au clusters act as reactant species and not as simply "spectators."

Finally, operando measurements are fundamental studies to study the abundance and stability of carbonate/bicarbonate species adsorbed on the catalyst surface in reaction conditions.

## Acknowledgments

The authors are grateful to Prof. Tatyana Tabakova (Institute of Catalysis, Bulgarian Academy of Sciences, Sofia, Bulgaria) for kindly preparing and providing the samples. The authors would like to thank Prof. Flora Boccuzzi (Dipartimento di Chimica, Università degli Studi di Torino, Italia) and Prof. Anna Chiorino (Dipartimento di Chimica, Università degli Studi di Torino, Italia) for fruitful discussions. Dr. Francesco Paolo Armigliato (Università degli Studi di Torino, Italia) is gratefully acknowledged for operando measurements.

## References

1. Haruta, M., Kobayashi, T., Sano, H., and Yamada, N. (1987). Novel gold catalysts for the oxidation of carbon-monoxide at a temperature far below 0-degrees-C, *Chem. Lett.*, **16**, pp. 405–408.
2. Boccuzzi, F., Chiorino, A., Manzoli, M., Lu, P., Akita, T., Ichikawa, S., and Haruta, M. (2001). Au/TiO2 nanosized samples: a catalytic, TEM, and FTIR study of the effect of calcination temperature on the CO oxidation, *J. Catal.*, **202**, pp. 256–267.
3. Chen, M.S., and Goodman, D.W. (2006). Structure-activity relationships in supported Au catalysts, *Catal. Today*, **111**, pp. 22–33.
4. Fu, Q., Saltsburg, H., and Flytzani-Stephanopoulos, M. (2003). Active nonmetallic Au and Pt species on ceria-based water-gas shift catalysts, *Science*, **301**, pp. 935–938.

5. Rodriguez, J.A., Ma, S., Liu, P., Hrbek, J., Evans, J., and Perez, M. (2007). Activity of CeO$_x$ and TiOx nanoparticles grown on Au(111) in the water-gas shift reaction, *Science*, **318**, pp. 1757–1760.

6. Green, I.X., Tang, W.J., Neurock, M., and Yates, J.T. (2011). Spectroscopic observation of dual catalytic sites during oxidation of CO on a Au/TiO$_2$ catalyst, *Science*, **333**, pp. 736–739.

7. Corma, A., and Garcia, H. (2008). Supported gold nanoparticles as catalysts for organic reactions, *Chem. Soc. Rev.*, **37**, pp. 2096–2126.

8. Trovarelli, A., deLeitenburg, C., and Dolcetti, G. (1997). Design better cerium-based oxidation catalysts, *Chemtech*, **27**, pp. 32–37.

9. Trovarelli, A. (2002). Structural properties and nonstoichiometric behavior of CeO$_2$, in *Catalysis by Ceria and Related Materials*, Vol. 2, pp. 15-50, ed. Trovarelli, A. (Imperial College Press, London).

10. Sakurai, H., Akita, T., Tsubota, S., Kiuchi, M., and Haruta, M. (2005). Low-temperature activity of Au/CeO$_2$ for water gas shift reaction, and characterization by ADF-STEM, temperature-programmed reaction, and pulse reaction, *Appl. Catal. A: Gen.*, **291**, pp. 179–187.

11. Miller, J.T., Kropf, A.J., Zha, Y., Regalbuto, J.R., Delannoy, L., Louis, C., Bus, E., and van Bokhoven, J.A. (2006). The effect of gold particle size on Au-Au bond length and reactivity toward oxygen in supported catalysts, *J. Catal.*, **240**, pp. 222–234.

12. Burch, R. (2006). Gold catalysts for pure hydrogen production in the water-gas shift reaction: activity, structure and reaction mechanism, *Phys. Chem. Chem. Phys.*, **8**, pp. 5483–5500.

13. Park, J.B., Graciani, J., Evans, J., Stacchiola, D., Senanayake, S.D., Barrio, L., Liu, P., Sanz, J.F., Hrbek, J., and Rodriguez, J.A. (2010). Gold, copper, and platinum nanoparticles dispersed on CeO$_x$/TiO$_2$(110) surfaces: high water-gas shift activity and the nature of the mixed-metal oxide at the nanometer level, *J. Am. Chem. Soc.*, **132**, pp. 356–363.

14. Zhang, C.J., Michaelides, A., and Jenkins, S.J. (2011). Theory of gold on ceria, *Phys. Chem. Chem. Phys.*, **13**, pp. 22–33.

15. Manzoli, M., Boccuzzi, F., Chiorino, A., Vindigni, F., Deng, W., and Flytzani-Stephanopoulos, M. (2007). Spectroscopic features and reactivity of CO adsorbed on different Au/CeO$_2$ catalysts, *J. Catal.*, **245**, pp. 308–315.

16. Avgouropoulos, G., Manzoli, M., Boccuzzi, F., Tabakova, T., Papavasiliou, J., Ioannides, T., and Idakiev, V. (2008). Catalytic performance and characterization of Au/doped-ceria catalysts for the preferential CO oxidation reaction, *J. Catal.*, **256**, pp. 237–247.

17. Vindigni, F., Manzoli, M., Damin, A., Tabakova, T., and Zecchina, A. (2011). Surface and inner defects in Au/CeO$_2$ WGS catalysts: relation between raman properties, reactivity and morphology, *Chem. Eur. J.*, **17**, pp. 4356–4361.
18. Rodriguez, J.A., Wang, X., Liu, P., Wen, W., Hanson, J.C., Hrbek, J., Perez, M., and Evans, J. (2007). Gold nanoparticles on ceria: importance of O vacancies in the activation of gold, *Top. Catal.*, **44**, pp. 73–81.
19. Kaspar, J., and Fornasiero, P. (2002). Structural properties and thermal stability of ceria-zirconia and related materials, in *Catalysis by Ceria and Related Materials*, Vol. 2, pp. 217–242, ed. Trovarelli, A. (Imperial College Press, London).
20. Ilieva, L., Pantaleo, G., Ivanov, I., Zanella, R., Venezia, A.M., and Andreeva, D. (2009). A comparative study of differently prepared rare earths-modified ceria-supported gold catalysts for preferential oxidation of CO, *Int. J. Hydrogen Energy*, **34**, pp. 6505–6515.
21. Ilieva, L., Pantaleo, G., Ivanov, I., Maximova, A., Zanella, R., Kaszkur, Z., Venezia, A.M., and Andreeva, D. (2010). Preferential oxidation of CO in H$_2$ rich stream (PROX) over gold catalysts supported on doped ceria: effect of preparation method and nature of dopant, *Catal. Today*, **158**, pp. 44–55.
22. Laguna, O.H., Sarria, F.R., Centeno, M.A., and Odriozola, J.A. (2010). Gold supported on metal-doped ceria catalysts (M = Zr, Zn and Fe) for the preferential oxidation of CO (PROX), *J. Catal.*, **276**, pp. 360–370.
23. Laguna, O.H., Centeno, M.A., Arzamendi, G., Gandia, L.M., Romero-Sarria, F., and Odriozola, J.A. (2010). Iron-modified ceria and Au/ceria catalysts for total and preferential oxidation of CO (TOX and PROX), *Catal. Today*, **157**, pp. 155–159.
24. Wang, L.C., Huang, X.S., Liu, Q., Liu, Y.M., Cao, Y., He, H.Y., Fan, K.N., and Zhuang, J.H. (2008). Gold nanoparticles deposited on manganese(III) oxide as novel efficient catalyst for low temperature CO oxidation, *J. Catal.*, **259**, pp. 66–74.
25. Tu, Y.B., Luo, J.Y., Meng, M., Wang, G., and He, J.J. (2009). Ultrasonic-assisted synthesis of highly active catalyst Au/MnO$_{x-}$CeO$_2$ used for the preferential oxidation of CO in H$_2$-rich stream, *Int. J. Hydrogen Energy*, **34**, pp. 3743–3754.
26. Wang, H., Zhu, H.Q., Qin, Z.F., Wang, G.F., Liang, F.X., and Wang, J.G. (2008). Preferential oxidation of CO in H$_2$ rich stream over Au/CeO$_2$-Co$_3$O$_4$ catalysts, *Catal. Commun.*, **9**, pp. 1487–1492.

27. Liotta, L.F., Di Carlo, G., Longo, A., Pantaleo, G., and Venezia, A.M. (2008). Support effect on the catalytic performance of Au/Co$_3$O$_4$-CeO$_2$ catalysts for CO and CH$_4$ oxidation, *Catal. Today*, **139**, pp. 174–179.
28. Schmieg, S.J., and Belton, D.N. (1995). Effect of hydrothermal aging on oxygen storage release and activity in a commercial automotive catalyst, *Appl. Catal. B: Environ.*, **6**, pp. 127–144.
29. Di Monte, R., Fornasiero, P., Kaspar, J., Graziani, M., Gatica, J.M., Bernal, S., and Gomez-Herrero, A. (2000). Stabilisation of nanostructured Ce$_{0.2}$Zr$_{0.8}$O$_2$ solid solution by impregnation on Al$_2$O$_3$: a suitable method for the production of thermally stable oxygen storage/release promoters for three-way catalysts, *Chem. Commun.*, pp. 2167–2168.
30. Trovarelli, A. (1996). Catalytic properties of ceria and CeO$_2$-containing materials, *Catal. Rev.: Sci. Eng.*, **38**, pp. 439–520.
31. Mogensen, M., Sammes, N.M., and Tompsett, G.A. (2000). Physical, chemical and electrochemical properties of pure and doped ceria, *Sol. St. Ion.*, **129**, pp. 63–94.
32. Kim, C.H., and Thompson, L.T. (2005). Deactivation of Au/CeO$_x$ water gas shift catalysts, *J. Catal.*, **230**, pp. 66–74.
33. Duprez, D., and Descrome, C. (2002). Oxygen storage/redox capacity and related phenomena on ceria-based catalysts, in *Catalysis by Ceria and Related Materials*, Vol. 2, pp. 243–280, ed. Trovarelli, A. (Imperial College Press, London).
34. Tibiletti, D., Amieiro-Fonseca, A., Burch, R., Chen, Y., Fisher, J.M., Goguet, A., Hardacre, C., Hu, P., and Thompsett, A. (2005). DFT and in situ EXAFS investigation of gold/ceria-zirconia low-temperature water gas shift catalysts: identification of the nature of the active form of gold, *J. Phys. Chem. B*, **109**, pp. 22553–22559.
35. Goguet, A., Burch, R., Chen, Y., Hardacre, C., Hu, P., Joyner, R.W., Meunier, F.C., Mun, B.S., Thompsett, A., and Tibiletti, D. (2007). Deactivation mechanism of a Au/CeZrO$_4$ catalyst during a low-temperature water gas shift reaction, *J. Phys. Chem. C*, **111**, pp. 16927–16933.
36. Tabakova, T., Boccuzzi, F., Manzoli, M., Chiorino, A., and Andreeva, D. (2005). Characterization of nanosized gold, silver and copper catalysts supported on ceria, *Oxide Based Mater.*, **155**, pp. 493–500.
37. Tabakova, T., Boccuzzi, F., Manzoli, M., Sobczak, J.W., Idakiev, V., and Andreeva, D. (2004). Effect of synthesis procedure on the low-temperature WGS activity of Au/ceria catalysts, *Appl. Catal. B: Environ.*, **49**, pp. 73–81.

38. Manzoli, M., Avgouropoulos, G., Tabakova, T., Papavasiliou, J., Ioannides, T., and Boccuzzi, F. (2008). Preferential CO oxidation in $H_2$-rich gas mixtures over Au/doped ceria catalysts, *Catal. Today*, **138**, pp. 239–243.
39. Tabakova, T., Avgouropoulos, G., Papavasiliou, J., Manzoli, M., Boccuzzi, F., Tenchev, K., Vindigni, F., and Ioannides, T. (2011). CO-free hydrogen production over Au/CeO$_2$-Fe$_2$O$_3$ catalysts: part 1. Impact of the support composition on the performance for the preferential CO oxidation reaction, *Appl. Catal. B: Environ.*, **101**, pp. 256–265.
40. Tabakova, T., Manzoli, M., Paneva, D., Boccuzzi, F., Idakiev, V., and Mitov, I. 2011. CO-free hydrogen production over Au/CeO$_2$-Fe$_2$O$_3$ catalysts: part 2. Impact of the support composition on the performance in the water-gas shift reaction, *Appl. Catal. B: Environ.*, **101**, pp. 266–274.
41. Vindigni, F., Manzoli, M., Chiorino, A., Tabakova, T., and Boccuzzi, F. (2006). CO adsorption on gold clusters stabilized on ceria-titania mixed oxides: comparison with reference catalysts, *J. Phys. Chem. B*, **110**, pp. 23329–23336.
42. Vindigni, F., Manzoli, M., Tabakova, T., Idakiev, V., Boccuzzi, F., and Chiorino, A. (2012). Gold catalysts for low temperature water-gas shift reaction: effect of ZrO$_2$ addition to CeO$_2$ support, *Appl. Catal. B: Environ.*, **125**, pp. 507–515.
43. Tabakova, T., Dimitrov, D., Manzoli, M., Vindigni, F., Petrova, P., Ilieva, L., Zanella, R., and Ivanov, K. (2013). Impact of metal doping on the activity of Au/CeO$_2$ catalysts for catalytic abatement of VOCs and CO in waste gases, *Catal. Commun.*, **35**, pp. 51–58.
44. Mihaylov, M., Knozinger, H., Hadjiivanov, K., and Gates, B.C. (2007). Characterization of the oxidation states of supported gold species by IR spectroscopy of adsorbed CO, *Chem. Ing. Tech.*, **79**, pp. 795–806.
45. Hollins, P. (1992). The influence of surface defects on the infrared-spectra of adsorbed species, *Surf. Sci. Rep.*, **16**, pp. 51–94.
46. Hadjiivanov, K.I., and Vayssilov, G.N. (2002). Characterization of oxide surfaces and zeolites by carbon monoxide as an IR probe molecule, *Adv. Catal.*, **47**, pp. 307–511.
47. Concepcion, P., Carrettin, S., and Corma, A. (2006). Stabilization of cationic gold species on Au/CeO$_2$ catalysts under working conditions, *Appl. Catal. A: Gen.*, **307**, pp. 42–45.
48. Binet, C., Daturi, M., and Lavalley, J.C. (1999). IR study of polycrystalline ceria properties in oxidised and reduced states, *Catal. Today*, **50**, pp. 207–225.

49. Bondzie, V.A., Parker, S.C., Campbell, C.T. (1999). The kinetics of CO oxidation by adsorbed oxygen on well-defined gold particles on TiO$_2$(110), *Catal. Lett.*, **63**, pp. 143–151.
50. Wu, X., Senapati, L., Nayak, S.K., Selloni, A., and Hajaligol, M. (2002). A density functional study of carbon monoxide adsorption on small cationic, neutral, and anionic gold clusters, *J. Chem. Phys.*, **117**, pp. 4010–4015.
51. Fielicke, A., von Helden, G., Meijer, G., Pedersen, D.B., Simard, B., and Rayner, D.M. (2005). Gold cluster carbonyls: saturated adsorption of CO on gold cluster cations, vibrational spectroscopy, and implications for their structures, *J. Am. Chem. Soc.*, **127**, pp. 8416–8423.
52. Zhao, X.E., Ma, S.G., Hrbek, J., and Rodriguez, J.A. (2007). Reaction of water with Ce-Au(111) and CeO$_x$/Au(111) surfaces: photoemission and STM studies, *Surf. Sci.*, **601**, pp. 2445–2452.
53. Tabakova, T., Boccuzzi, F.B., Manzoli, M., and Andreeva, D. (2003). FTIR study of low-temperature water-gas shift reaction on gold/ceria catalyst, *Appl. Catal. A: Gen.*, **252**, pp. 385–397.
54. Akita, T., Okumura, M., Tanaka, K., Kohyama, M., and Haruta, M. (2006). Analytical TEM observation of Au nano-particles on cerium oxide, *Catal. Today*, **117**, pp. 62–68.
55. Chiorino, A., Manzoli, M., Menegazzo, F., Signoretto, M., Vindigni, F., Pinna, F., and Boccuzzi, F. (2009). New insight on the nature of catalytically active gold sites: quantitative CO chemisorption data and analysis of FTIR spectra of adsorbed CO and of isotopic mixtures, *J. Catal.*, **262**, pp. 169–176.
56. Tabakova, T., Manzoli, M., Vindigni, F., Idakiev, V., and Boccuzzi, F. (2010). CO-free hydrogen production for fuel cell applications over Au/CeO$_2$ catalysts: FTIR insight into the role of dopant, *J. Phys. Chem. A*, **114**, pp. 3909–3915.
57. Boccuzzi, F., Chiorino, A., Manzoli, M., Andreeva, D., and Tabakova, T. (1999). FTIR study of the low-temperature water-gas shift reaction on Au/Fe$_2$O$_3$ and Au/TiO$_2$ catalysts, *J. Catal.*, **188**, pp. 176–185.
58. Spanier, J.E., Robinson, R.D., Zheng, F., Chan, S.W., and Herman, I.P. (2001). Size-dependent properties of CeO$_{2-y}$ nanoparticles as studied by Raman scattering, *Phys. Rev. B*, **64**, p. 8.
59. Andreeva, D., Petrova, P., Sobczak, J.W., Ilieva, L., and Abrashev, M. (2006). Gold supported on ceria and ceria-alumina promoted by molybdena for complete benzene oxidation, *Appl. Catal. B: Environ.*, **67**, pp. 237–245.

60. Pushkarev, V.V., Kovalchuk, V.I., and d'Itri, J.L. (2004). Probing defect sites on the $CeO_2$ surface with dioxygen, *J. Phys. Chem. B*, **108**, pp. 5341–5348.
61. Haruta, M., Tsubota, S., Kobayashi, T., Kageyama, H., Genet, M.J., and Delmon, B. (1993). Low temperature oxidation of CO over gold supported on $TiO_2$, alpha-$Fe_2O_3$ and $Co_3O_4$, *J. Catal.*, **144**, pp. 175–192.
62. Schubert, M.M., Hackenberg, S., van Veen, A.C., Muhler, M., Plzak, V., and Behm, R.J. (2001). CO oxidation over supported gold catalysts: "inert" and "active" support materials and their role for the oxygen supply during reaction, *J. Catal.*, **197**, pp. 113–122.
63. Dekkers, M.A.P., Lippits, M.J., and Nieuwenhuys, B.E. (1999). Supported gold/$MO_x$ catalysts for $NO/H_2$ and $CO/O_2$ reactions, *Catal. Today*, **54**, pp. 381–390.
64. Panzera, G., Modafferi, V., Candamano, S., Donato, A., Frusteri, F., and Antonucci, P.L. (2004). CO selective oxidation on ceria-supported Au catalysts for fuel cell application, *J. Power Sources*, **135**, pp. 177–183.
65. Luengnaruemitchai, A., Osuwan, S., and Gulari, E. (2004). Selective catalytic oxidation of CO in the presence of $H_2$ over gold catalyst, *Int. J. Hydrogen Energy*, **29**, pp. 429–435.
66. Bond, G.C., and Thompson, D.T. (1999). Catalysis by gold, *Catal. Rev. Sci. Eng.*, **41**, pp. 319–388.
67. Goguet, A., Meunier, F.C., Tibiletti, D., Breen, J.P., and Burch, R. (2004). Spectrokinetic investigation of reverse water-gas-shift reaction intermediates over a $Pt/CeO_2$ catalyst, *J. Phys. Chem. B*, **108**, pp. 20240–20246.
68. Meunier, F.C., Tibiletti, D., Goguet, A., Reid, D., and Burch, R. (2005). On the reactivity of carbonate species on a $Pt/CeO_2$ catalyst under various reaction atmospheres: application of the isotopic exchange technique, *Appl. Catal. A: Gen.*, **289**, pp. 104–112.
69. Busca, G., and Lorenzelli, V. (1982). Infrared spectroscopic identification of species arising from reactive adsorption of carbon oxides on metal oxide surfaces, *Mater. Chem.*, **7**, pp. 89–126.
70. Bachiller-Baeza, B., Rodriguez-Ramos, I., and Guerrero-Ruiz, A. (1998). Interaction of carbon dioxide with the surface of zirconia polymorphs, *Langmuir*, **14**, pp. 3556–3564.

# Chapter 9

# Determination of Dispersion of Gold-Based Catalysts by Selective Chemisorption

**Michela Signoretto, Federica Menegazzo, Valentina Trevisan, and Francesco Pinna**

*Department of Molecular Sciences and Nanosystems, Ca' Foscari University Venice and Consortium INSTM, RU-Venice, Calle Larga S. Marta 2137, Venice 30123, Italy*
miky@unive.it

## 9.1 On Gold Dispersion

The activity of heterogeneous catalysts is known to be mainly a surface property. It is therefore often possible to reduce the amount of active components without affecting the overall activity by increasing its surface-to-volume ratio, commonly referred to as its dispersion [1]. In fact metal dispersion ($D$), namely the ratio between surface metal atoms ($N_S$) and total metal atoms ($N_T$), is a critical factor for several catalytic reactions and generally it has to be as high as possible:

$$D = \frac{N_S}{N_T} \quad (9.1)$$

---

*Gold Catalysis: Preparation, Characterization, and Applications*
Edited by Laura Prati and Alberto Villa
Copyright © 2016 Pan Stanford Publishing Pte. Ltd.
ISBN 978-981-4669-28-3 (Hardcover), 978-981-4669-29-0 (eBook)
www.panstanford.com

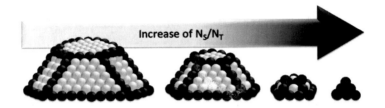

**Figure 9.1** Relationship between dispersion and particle size.

Dispersion is usually between 0 and 1 (or 0% and 100%) [2] and it is usually related to the size of metallic particles, as schematized in Fig. 9.1.

Optimum dispersion is of particular importance in the case of supported catalysts containing expensive active substances such as gold. Gold nanoparticles are typically present in supported catalysts, where metal particles having a size usually in the range of 1–10 nm are deposited on the external surface and/or in the porous texture of inert materials. However, deposition of noble metals, in particular of gold, with high dispersion on an active carbon or an oxidic carrier is not a trivial task, due to the tendency of metal nanoparticles toward agglomeration (Fig. 9.2). Metal dispersion measurements will allow the catalyst manufacturer to control the quality of the products and the customer to choose the best catalyst at the lower cost. In fact, gold particle nanosize is a critical factor for several catalytic reactions and it must be known with an accuracy as high as possible. Dispersion measurements are also of primary importance for scientists studying reaction mechanisms: the proper interpretation of heterogeneously catalyzed reaction kinetics requires a precise knowledge of the number of active atoms exposed on the surface [1]. The repertoire of reactions that gold supported on oxides or carbon can catalyze is really wide [3, 4]. Gold catalysts already have done it out of the lab and into commercial use, and a very large number of papers was published regarding catalysis by gold. Bulk gold itself is an inert material and the reasons for small gold particle activity are still a matter of debate. It has been shown that the catalytic activity of gold critically depends on the preparation method, on the nature and structure of the support, and on the pretreatment procedure. However, the most widely diffused explanation for the variability

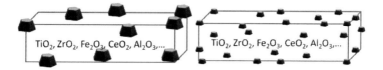

**Figure 9.2** Different types of gold dispersion.

of the catalytic properties of gold catalysts is focused on the size of the metal particles and on the amount of low-coordinated gold sites. Since gold particle size appears to be of crucial importance for the desired catalytic activity or selectivity, gold nanosize and surface area properties are of main importance in this field. Therefore, extensive quality controls before catalyst loading in the industrial plant are mandatory. Well-defined and widely accessible procedures of characterization are needed in order to understand the origin of the differences in the activity observed for different samples and to discriminate between conflicting explanations.

## 9.2 How to Measure Gold Dispersion

Each technique for particle size measurement has advantages and drawbacks; ideally, a combination of two or more methods would be necessary to obtain an unambiguous evaluation of particle size. The synergism produced by information coming from all the techniques is often crucial for identifying the structural features that really govern catalyst behavior.

### 9.2.1 *Electron Microscopy*

The size of gold metallic particles is, in most of the works, determined by transmission electron microscopy (TEM). Electron microscopy is the most general, reliable, and accurate technique for particle size measurements, since TEM images give the size, shape, and location of particles and analytical microscopy gives their composition [2]. However, the main drawback is that sample preparation, observation, and image analysis are not automated and therefore are time consuming. Besides, the number of measurements must be large enough to be sufficiently representative of the total gold population and they must be properly classified by histograms. Moreover, TEM, when performed accurately, provides a particle size distribution from which a mean size can be derived, but the very small metallic particles (<1 nm) are hardly detectable, so the method is applicable with difficulty for samples with highly dispersed gold. However, there is no doubt that TEM analyses provide valuable qualitative estimates of the dispersion to be compared with other characterization data.

## 9.2.2 XRD

X-ray diffraction (XRD) line broadening analysis is perhaps the oldest way to estimate the size of small crystallites and it has been used for many years for characterizing heterogeneous catalysts in conjunction with chemisorption measurements. XRD is a bulk analysis that is sensitive to all crystallites present in the sample, irrespective of their position either on the surface of the carrier or embedded in it, and therefore not accessible to the gas phase [1]. Therefore the particle size estimated by XRD might, in some cases, be smaller than the size derived from chemisorption measurements [1]. The applicability of XRD to catalysts is often restricted for a number of reasons [2]. For example, as the amount of the metallic phase decreases, the wings of the line profile become too weak to be measured and the smaller particles are not taken into account [2]. Moreover the XRD method hardly determines crystallite sizes smaller than about 1.5–2 nm [2]. Consequently, sometimes measurements of the crystallite sizes, performed only on the visible XRD peaks, overestimate the real average crystallite sizes. Furthermore, the XRD technique can give correct results, provided that the particles are monodomains, that is, single crystallites (Fig. 9.3).

This is a necessary condition for the correct use of the technique, but it does not always occur. In the case of polydomain particles XRD analysis, which measures the domain size, is unsuitable for obtaining reasonable gold particle sizes. In the latter case only small-angle X-ray scattering (SAXS), in the area of XRD techniques, can give accurate surface-weighted average particle sizes.

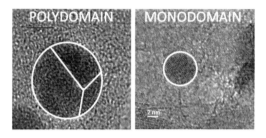

**Figure 9.3** Examples of poly- or monodomain metal particles.

## 9.2.3 Selective Chemisorption

The method traditionally used by the scientific community for measuring the metal particle size is certainly the selective chemisorption of a suitable gas on the surface of the active phase. Selective gas chemisorption remains the preferred technique for particle size and dispersion measurements in catalysis laboratories because it probes directly the catalyst active surface and because instruments are affordable, easily housed, and operated devices. Chemisorption is based on the formation of a strong chemical bond between the adsorbate molecules with the exposed metal surface. The selected gas molecule is chemisorbed under conditions that allow the formation of a monolayer coverage of gas on the metal surface without significant uptake by the nonmetallic part (support) through physical adsorption. This requires a careful selection of a suitable gas molecule and of the experimental conditions and procedures. In practice what is experimentally measured is the number of probe molecules that disappear from the gaseous phase in contact with the catalyst under the chosen experimental conditions of temperature and pressure [5].

Isothermal chemisorption analyses are obtained by two chemisorption techniques:

- Static (volumetric) chemisorption
- Dynamic (flowing gas) chemisorption

Both techniques can be used to measure the quantity of gas required to form a monolayer of chemisorbate on an active surface. In the first one a classical high-vacuum Brunauer–Emmett–Teller (BET) appa-

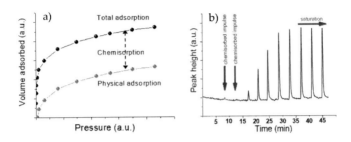

**Figure 9.4** Plot for static (a) and dynamic (b) chemisorption analysis.

ratus is used and usually two isotherms are measured (Fig. 9.4a): The first measures both chemisorbed and the physisorbed gas, and the second (run after pumping in proper condition) measures only the physically adsorbed gas. From the difference of the two isotherms the volume of the chemisorbed gas can be derived. The volumetric technique is convenient for obtaining a high-resolution measurement of the chemisorption isotherm from very low pressure to atmospheric pressure at essentially any temperature from near ambient to 1300 K. Obtaining a high-resolution isotherm requires many, precise dosing steps in pursuit of the equilibrium point, and many pressure steps that, without automation, would be a time-consuming and error-prone procedure; commercial embodiments of this technique are therefore almost exclusively automated.

The flowing gas (dynamic) technique is typically performed at ambient pressure. After the sample has been cleaned, small injections of accurately known quantities of adsorptive are administered in pulses until the sample is saturated, thus the name "pulse chemisorption." Injections may be made by syringe or by a loop injection valve. A calibrated thermal conductivity detector (TCD) is used to determine the quantity of adsorptive molecules taken up by active sites upon each injection. Initial injections may be adsorbed totally; ultimately none of the injections will be adsorbed, indicating saturation (Fig. 9.4b). The number of molecules of gas adsorbed is directly related to the exposed surface area of active material. The volume of gas adsorbed per gram of sample combined with the stoichiometry of the reaction and the quantity of active metal allows the percent metal dispersion to be calculated.

Before chemisorption a proper pretreatment procedure must be applied: the sample must be reduced at a proper temperature and then purged in inert flow at the same temperature in order to clean the surface. The estimation of active particle size is a geometrical calculation based on the assumption that the crystallite shape is of regular geometry, a sphere typically being the geometry of choice. From the assumed regular geometry, the diameter can be expressed in terms of the area and volume. Since the volume of active metal is not known, it can be expressed in terms of density ($\rho_m$). In particular, it is possible to determine the mean metal particle nanosize ($\Phi_{av}$) according to the equation

$$\phi_{av} = \frac{K V_m C_m}{V_g S_{av} N_A \rho_m} \tag{9.2}$$

- $V_g$ is the volume of gas chemisorbed per gram of metal;
- $S_{av}$ is the average chemisorption stoichiometry;
- $N_A$ is Avogardo's number;
- $\rho_m$ is the metal density;
- K is a constant
- $V_m$ is the molar volume;
- $C_m$ is the surface density of metal atoms.

The chemisorption technique requires a careful selection of a suitable gas molecule. The most used gas for chemisorption measurements is hydrogen, but carbon monoxide and oxygen are also used especially when hydrogen chemisorption give rise to problems. Other substances such as nitrous oxide, ethylene, carbon disulphide, and tiophene have been also used. The average chemisorption stoichiometry is the number of metal atoms that interact with one probe molecule. The chemisorption stoichiometry for hydrogen is 2, since the hydrogen molecule adsorbs dissociatively on metals and each hydrogen atoms is adsorbed on one metal atom. Unfortunately, the adsorption stoichiometry for gases such as CO and $O_2$ is not well defined. With oxygen, values from 2 to 1 or even lower (0.75) can be found according to the metal. Adsorption of oxygen on several metals causes formation of several layers of metal oxide. The stoichiometry of CO adsorption on metal is highly variable and the kind of CO–metal species formed depends on the metal, its dispersion, and its surface structure (Fig. 9.5).

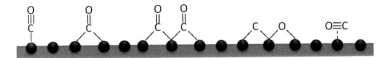

**Figure 9.5** Some different CO adsorption stoichiometries on metal particles.

Once the gas has been chosen, experimental variables also must be defined depending on the catalytic system. In particular, the optimum temperature and pressure for carrying out the chemisorption will depend on the sample and are usually established experimentally.

### 9.2.4 Other Techniques

Other techniques can be discarded for routine particle size measurements since they require both special equipment and skills that are usually not readily available [2]. Thus, although SAXS is certainly an accurate technique for measuring particle size and metallic surface areas, its use is restricted by the scarcity of equipment and specialists [2]. Other techniques as extended X-ray absorption fine structure (EXAFS) and X-ray absorption near-edge structure (XANES) can be employed, from which information on the coordination number of gold can be deduced. However, these measurements are not so widely diffuse and cannot be proposed as routine tests.

Assessment of routine dispersion measurements by complementary techniques is always rewarding. A valid comparison of dispersion values obtained from different investigation techniques is possible, provided some hypotheses are made at each step of the interpretation, as schematized in Table 9.1. Chemisorption, TEM, and XRD do not necessarily result in the same calculated crystal size for a given catalyst. For example, chemisorption is biased toward a smaller average crystallite size and XRD toward a larger size. In fact, XRD and chemisorption are not directly comparable, since the former is a volume-averaged measurement and the latter a surface-area-averaged measurement [6]. Sometimes a gold sample is monodispersed, namely its active fraction consists of particles

**Table 9.1** Comparison of dispersion values obtained from different investigation techniques

|  | TEM | Chemisorption | XRD |
|---|---|---|---|
| Representative of all the sample | × | ✓ | ✓ |
| Mono- or polydispersed Sample | ✓ | × | × |
| Very low metal size (<1nm) | × | ✓ | × |
| Very low metal amount | × | ✓ | × |
| Sensitive to all crystallites of the sample | ✓ | × | ✓ |
| Geometry (corners) | ✓ | × | × |
| Mono- or polydomains | ✓ | ✓ | × |
| Time consuming | × | ✓ | ✓ |

exhibiting a very high degree of homogeneity with respect to particle size. However, most gold catalysts are polydispersed, that is, they show a variety of particle sizes. The characterization of a population of particles exhibiting different sizes may follow different approaches according to the experimental technique available. TEM allows the determination of the size and of the shape of a discrete number of gold particles. On the contrary, chemisorption and XRD provide directly an average size of the gold nanoparticles. However, this gold average size will differ according to the technique as the sensitivity of each technique to the size of the particles differs according to the physical principle upon which it is based [1]. Methods such as chemisorption are primarily sensitive to the surface of the particles, providing the surface-average size. A method such as XRD, being sensitive to the bulk of the particles, provides the so-called volume-average size [1]. The wider the particle size distribution, the larger these differences.

Selective gas chemisorption remains the preferred technique for particle size and dispersion measurements in catalysis laboratories because it probes directly the catalyst active surface and because instruments are affordable, easily housed, and operated devices. Therefore chemisorption results are often the first data to be sought for in-metal catalysts; other methods will be usually involved afterward when the complexity of the problem requires further study.

## 9.3 Selective Chemisorption on Gold Catalysts

In spite of the great attention that gold-based catalysts have received in the past years, characterization of supported gold by chemisorption methods has not been widely investigated if compared to common applications of $H_2$, $O_2$, and CO chemisorption to characterize other metals. In fact, it is well known that gold does not chemisorb many molecules easily. Apparently, the application of the common probe molecules for chemisorption on gold is prevented by its nobility and relative inertness. Only few papers [7–12] have investigated chemisorption methods for the characterization of supported gold. Mostly, only physical methods such as TEM and XRD were used to characterize the structural properties of the gold particles besides their catalytic behavior.

### 9.3.1 $O_2$ Chemisorption

In 1953 Trapnell [13] stated that Au does not chemisorb oxygen up to 273 K. Several other authors obtained similar results and concluded that oxygen chemisorption on gold needs thermal activation [14, 15]. It has been found that a portion of oxygen can be stable-bound on Au up to 1073 K, as has been shown by low-energy electron diffraction (LEED) studies on a Au(111) crystal surface [16]. By temperature-programmed decomposition (TPD) studies on silica-supported gold [17] it has been shown that the transition of oxygen from a weakly to a strongly bound state is reversible, that is, the transfer from the surface to the subsurface also takes place conversely. Thus, there are some uncertainties concerning the formation of a defined oxygen monolayer on a gold surface. Moreover, there are also some uncertainties concerning the generation of an oxygen-free gold surface at the start of the oxygen adsorption measurement.

In principle, total oxygen adsorption can be determined by static measurements of oxygen consumption, as shown in a note of Fukushima et al. already in 1979 [8]. The authors observed by means of static adsorption measurements (3–6 Torr $O_2$) that the amount of adsorbed oxygen on Au supported on silica, magnesia, and alumina increases with increasing temperature. As the temperature is increased from 473 to 673 K the adsorption becomes reversible, while the isobar shows a maximum at about 610 K. The authors supposed a dissociative oxygen adsorption forming a monolayer with a stoichiometry $Au_s:O = 2$ at 473 K but with a stoichiometry $Au_s:O = 1$ at 573 K [8]. Whereas they reduced $Au/SiO_2$ samples by $H_2$ at 673 K, they evacuated Au/MgO samples at 623 K only, assuming that all oxygen was thermally removable and a reduction was not necessary.

More recently, Shastri et al. performed $O_2$ chemisorption at 473 K using a stoichiometry of four gold sites per oxygen molecule [11]. They also did not reduce $Au/TiO_2$ samples to avoid a partial reduction of titania that would lead to an incorrect determination of the oxygen adsorption on Au. However, some doubts have been expressed [7] concerning the small temperature difference applied to achieve complete oxygen removal from the Au surface (evacuation at 423 K) and complete oxygen coverage of the Au surface (adsorption at 473 K).

The pulse chemisorption technique has been applied to study oxygen adsorption on gold powder [18]. More recently, studies for determining the oxygen coverage on Au surfaces by a pulse flow method were carried out on $Au/Al_2O_3$ catalysts [7]. However, pre-investigations of $O_2$ pulse adsorption showed that oxygen adsorption occurs very slowly even at enhanced temperatures (473 K). Consequently an evaluation of the oxygen consumption would be very inexact. Therefore, static oxygen adsorption and hydrogen pulse titration of chemisorbed oxygen were used as tools for characterization of the surface area and average particle size of $Au/Al_2O_3$ catalysts [7]. In particular, the *dual-isotherm* method was applied to determine weakly and strongly adsorbed oxygen [7]. Using $Au_s:O = 2$ a good agreement was obtained between the average gold particle size calculated from oxygen chemisorption results and the particle size observed by TEM [7]. An alternative method to measure oxygen coverage on supported gold might be a *back-titration* of the oxygen monolayer on gold particles with $H_2$ pulses [7, 19]. In this way, pretreatments such as reductive or thermal removal of oxygen at high temperatures would be avoided [7]. The results of this method were comparable with the ones obtained by static oxygen adsorption; however, slightly higher oxygen coverages were observed by hydrogen titration of samples with lower gold dispersity [7].

### 9.3.2 H₂ Chemisorption

Chemisorption of hydrogen by supported gold has been occasionally mentioned, but some or no chemisorbed hydrogen has been reported [3, 20]. A H/Au ratio (14%) was measured by Jia et al. [21] for 3.8 nm Au particles on Al₂O₃. Bus et al. [22] demonstrated that hydrogen is dissociatively adsorbed on the gold particles in Au/Al₂O₃ catalysts, by a combination of in situ X-ray absorption spectroscopy, chemisorption, and H/D exchange experiments. Hydrogen chemisorption analyses were performed under static volumetric conditions after reduction and evacuation. H/Au values of 10%–73% were determined for the Au/Al₂O₃ catalysts. The H/Au values are up to five times lower than the EXAFS and scanning transmission electron microscopy (STEM) dispersions. It seems that the hydrogen atoms only adsorb on Au atoms at corner and edge positions. Therefore, the average number of adsorbed hydrogen atoms per surface gold atom increases with decreasing particle size. Later the same authors extended their study to Au/SiO₂ catalysts [23]. The ability of small gold clusters to dissociate and chemisorb hydrogen originates from the width and energy position of their 5$d$ band. In general, the strength of the adsorbate bonding is the strongest when the $d$ band is the closest to the Fermi level. The 5$d$ band of gold nanoclusters lies closer to the Fermi level and is narrower compared to bulk gold. Consequently, gold clusters can form bonds to reactants such as hydrogen and oxygen, whereas bulk gold cannot. Not all surface atoms of small gold clusters interact with hydrogen and the authors suggested that hydrogen chemisorbs only on the edges and corners [23]. These sites have a lower coordination number and therefore a more reactive $d$ band and, thus, can interact with adsorbates easier than the face atoms.

### 9.3.3 Methyl Mercaptane Chemisorption

Adsorption of methyl mercaptane (MM; CH₃SH) on Au/TiO₂ samples with an Au content of 0.8–9.9 wt% has been investigated, too [12].

Measurements were performed by thermogravimetry by injecting pulses of MM until no further mass increase could be observed. A prerequisite for determination of the surface area of the deposited gold nanoparticles is the proper discrimination of species adsorbing on the Au nanoparticles and the titania support. It has been shown that the adsorption of MM on the titania support strongly depends on the pretreatment temperature (303–673 K), whereas the adsorption on Au nanoparticles is virtually unaffected by this parameter [12]. Therefore if the sample is pretreated at 303 K, which is not high enough to remove water and carbon dioxide blocking the active support sites, the amount of MM adsorbed on the mere titania is negligible [12]. The observed mass gain during MM pulsing results from the adsorption of the thiol on the gold surface only. The MM adsorbed on Au forms an Au–S bond, as observed by Raman spectroscopy. The number of Au surface atoms required for the determination of the stoichiometric factor (MM/Au$_{SURF}$) was estimated assuming cuboctahedral particle shapes with an identical number for height and length of the sides of the hexagon [12]. The amount of MM adsorbed increases with the amount of Au in the system. Critical, however, is the fact that the MM/Au$_{SURF}$ ratio changes with particle size, ranging from 0.55 for the smallest to 0.39 for the largest average particle size found in this study [12]. Care should be taken when calculating the dispersion from MM uptakes, especially in the case of highly dispersed systems [12]. Since chemisorption of $H_2$ or CO demonstrated that these probe molecules chemisorb only on low-coordinated Au atoms, systems containing larger particles are difficult to characterize due to their relatively low amount of low-coordinated Au atoms [12]. Chemisorption of MM does not suffer from this drawback, making it more versatile in its application. It remains to be proven whether the method is also applicable for other supported Au systems [12].

### 9.3.4 CO Chemisorption

Carbon monoxide adsorption at 323 K on gold–titania samples was attempted by Shastri et al. [11]. The authors performed chemisorption experiments in a high-vacuum static volumetric system using pressures in the range of 0.5 to 10 Torr. Mere

TiO$_2$ showed some uptake of CO at room temperature. With increasing temperature, the uptake of CO decreased and became negligible at 323 K. Adsorption of CO was found to be weak and completely reversible in 5 min. The authors tentatively assumed a stoichiometry Au/CO = 1/1 and calculated the mean gold particle size. Their results are in qualitatively agreement with their oxygen chemisorption results.

Adsorption of CO on gold supported on TiO$_2$ was investigated by a constant pressure static system [9]. It was found that adsorption of CO on pre-oxidized Au/TiO$_2$ at 253–303 K takes place mainly as reversible Langmuir-type chemisorption. A large portion of CO is adsorbed on the TiO$_2$ support, which contributes to the reversible part of chemisorption. The irreversible part of CO adsorption, which is about 10% of the total amount of adsorbed CO, is related to CO$_2$ formation on the surface of gold particles and the accumulation of carbonate-like species on the surface of TiO$_2$.

More recently, Margitfalvi et al. [10] investigated CO chemisorption on different Au/MgO catalysts using volumetric equipment. The amount of chemisorbed CO was determined as a difference of total and physical adsorption by measuring the adsorption isotherms up to 1000 Torr CO pressure. The low value of the sticking coefficient for CO on gold made it necessary to use an unusually high pressure for the chemisorption measurements [10]. Results of CO chemisorption indicate that both Au/MgO catalysts and the MgO support chemisorb CO. However, the latter has much less chemisorption capacity than the gold-containing samples [10]. Under their experimental conditions, the amounts of CO chemisorbed on different samples roughly correspond to CO/Au (at/at) ratio = 0.10–0.17 [10].

## 9.4 Pulse Flow CO Chemisorption at Low Temperatures

We have proposed CO chemisorption by a pulse flow technique for determining the concentration of gold active sites and comparing different samples. The pulse flow chemisorption is very simple, and if performed under proper conditions, it is accurate and

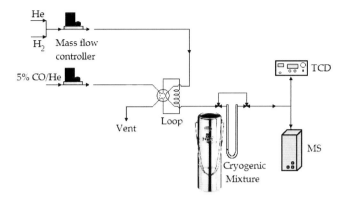

**Figure 9.6** Apparatus for pulse chemisorption.

reliable. Besides it is a widespread technique both in academic and in industrial laboratories. Chemisorption can be used for the measurements of metal particle nanosize, but it is often suitable also for the preliminary routine evaluation of the catalytic activity. Moreover, it is flexible, as proven by different examples of our recent experience [24–28]. We have investigated Au-based catalysts on various carriers ($TiO_2$, $ZrO_2$, $CeO_2$, $Fe_2O_3$). Samples had various metal contents and different gold nanosizes in order to check the validity of the test for a large variety of samples.

### 9.4.1 Apparatus for Pulse Flow CO Chemisorption

CO pulse chemisorption measurements were performed in an inexpensive lab-made equipment, as reported in Fig. 9.6. In principle the system is made of a U-shaped Pyrex reactor equipped with an oven controlled by a temperature programmer, mass flowmeters, a sampling valve, a Gow-Mac TCD detector, and/or a quadrupole mass detector.

Before the analysis, the following pretreatment was applied: The sample was reduced in a $H_2$ flow at 423 K, cooled in $H_2$ to room temperature, purged in He flow, and finally hydrated at room temperature. The hydration treatment was performed by contacting the sample with a He flow saturated with a proper amount of water. The sample was then cooled in He flow to the temperature chosen for CO chemisorption [25]. The chosen temperature can be easily attained and maintained by simply adding liquid nitrogen to the suitable solvent in a Dewar flask, and therefore no cryostatic equipment is required. So, the temperatures can be attained by a cryogenic mixture, respectively, of the following:

| | | |
|---|---|---|
| iso-pentane | & liquid nitrogen | $T = 110$ K |
| n-pentane | & liquid nitrogen | $T = 142$ K |
| ethanol | & liquid nitrogen | $T = 157$ K |
| acetone | & liquid nitrogen | $T = 179$ K |
| ethyl acetate | & liquid nitrogen | $T = 190$ K |
| ice | & $CaCl_2$ $6H_2O$ | $T = 223$ K |
| ice | | $T = 273$ K |

In practice what is experimentally measured is the number of CO molecules that disappear from the gaseous phase in contact with the gold catalyst at atmospheric pressure and proper temperature. It is possible to determine the mean metal particle nanosize ($\Phi_{av}$) according to the equation [29]

$$\phi_{av} = \frac{K V_m C_m}{V_g S_{av} N_A \rho_m}$$

where:

- $V_g$ is the volume of gas chemisorbed per gram of gold. This is the only experimental data.
- $S_{av}$ is the average chemisorption stoichiometry, that is, the ratio between total surface gold atoms and total chemisorbed CO molecules.
- $N_A$ is Avogadro's number ($6,022 \times 10^{23}$ atoms/mol).
- $\rho_m$ is the gold density ($19.32 \times 10^3$ Kg/m$^3$ for Au) [30].
- $K$ is a factor depending on the metal particle shape and the extent of surface contact with the support (Recommended value: $K = 5$).
- $V_m$ is the molar volume (22.414 L/mol).
- $C_m$ is the surface density of metal atoms. For Au, according to Anderson [31], $C_m$ is equal to $1.15 \times 10^{19}$ atoms/m$^2$.

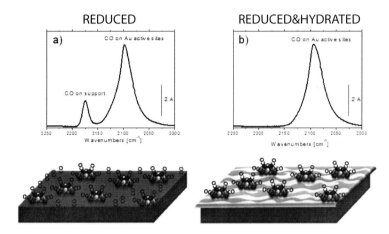

**Figure 9.7** FTIR spectra of CO adsorbed at 157 K on a typical gold-based sample (not calcined 2AuZ sample) and schematization of CO chemisorption for reduced (a) and reduced and hydrated (b) catalysts.

### 9.4.2 Pulse Flow CO Chemisorption Method

Au/TiO$_2$ and Au/Fe$_2$O$_3$ reference catalysts provided by the World Gold Council (WGC) were examined, together with different Au/TiO$_2$, Au/CeO$_2$, and Au/ZrO$_2$ catalysts prepared by deposition-precipitation (DP) in our lab. Volumetric measurements of CO chemisorption were performed on mildly reduced catalysts and after saturation of the surface with water in order to avoid CO chemisorption on uncoordinated support ions. In fact in the presence of water, as shown by Fourier transform infrared spectroscopy (FTIR) spectroscopic measurements [25, 32], the cations of the supports are covered by the water or by –OH groups and they are not more able to coordinate CO with a comparable strength than gold step sites. So, CO chemisorption occurs only on gold active sites, as schematized in Fig. 9.7.

First of all we have studied the effect of the chemisorption temperature. The CO chemisorption was preliminary investigated in a wide temperature range (77–273 K). As it can be seen in Fig. 9.8, the volume of chemisorbed CO remains constant in the range of 140–180 K, while at higher temperature (over 190 K) a significant decrease of CO chemisorbed occurs. On the other hand, at 110 K

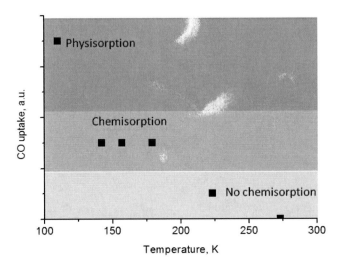

**Figure 9.8** Experimental data of CO uptake on a typical gold-based sample.

the large value obtained can be related to a physical adsorption. Measurements carried on at higher temperature (180 K) are, under the experimental error, identical to the ones obtained at 157 K. However, since the latter is a temperature that is easier to attain, we suggest using it for routine analyses. Preliminary runs carried out on the bare supports, in the same experimental conditions (meaning at 157 K after a reducing + hydrating pretreatment), have shown the absence of CO chemisorption on uncoordinated support ions (Fig. 9.9). On the contrary, measurements on the gold catalysts without performing the hydrating pretreatment have shown a very large amount of chemisorbed CO. Part of this is due to CO chemisorbed on the uncoordinated support ions, as demonstrated by FTIR analyses [25, 32] (Fig. 9.7).

### 9.4.2.1 Au/TiO$_2$ samples

TiO$_2$ is widely used for a variety of applications because of its high photocatalytic activity, nontoxicity, good availability, low cost, and stability. Its main characteristics strongly depend on its physicochemical properties, such as surface area, crystal structure (anatase, rutile, brookite), crystallite size, and surface hydroxyl

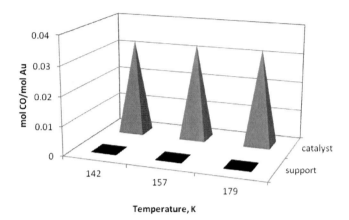

**Figure 9.9** Chemisorption results for Au/TiO$_2$ reference catalyst provided by the World Gold Council and the TiO$_2$ support.

groups. Au/TiO$_2$ catalysts have been used in many catalytic reactions such as oxidation of carbon monoxide [33], low-temperature water–gas shift (WGS) reaction [34], and direct epoxidation of propene [35]. Moreover, Au/TiO$_2$ by the WGC is commonly used by the scientific community as a reference catalyst. We have first of all investigated pulse flow CO chemisorption on such a commercial 1.51 wt% Au/TiO$_2$ reference catalyst [36]. As reported in the catalyst data sheets the sample exposes gold particles with mean diameters and standard deviations, determined by TEM, of 3.8 nm ± 1.5 nm, respectively. Volumetric measurements of CO chemisorption were performed on a mildly reduced catalyst and after saturation of the surface with water in order to avoid CO chemisorption on uncoordinated support ions. The results obtained are reported in Fig. 9.9. The volumes of CO adsorbed on the sample reduced in hydrogen and saturated by water do not change significantly with temperature in the 140–180 K range. Preliminary runs carried out on the titania support in the same experimental conditions have shown the absence of CO chemisorption.

FTIR spectra of CO adsorbed on the Au/TiO$_2$ WGC catalyst shown in Fig. 9.10 confirm that in the 140–180 K range the chemisorption data are related only to chemisorption on gold. CO adsorption on the Au/TiO$_2$ catalyst WGC at a lower temperature (120 K) produces

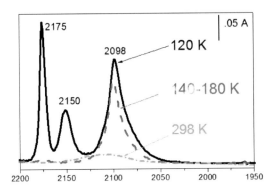

**Figure 9.10** FTIR spectra of CO adsorbed at different temperatures on Au/TiO$_2$ reference catalyst provided by the World Gold Council. Reprinted from Ref. [32], Copyright (2006), with permission from Elsevier.

in the FTIR spectra (bold curve) a band at 2098 cm$^{-1}$ related to CO on Au$^0$ sites exposed at the surface of the metallic particles [32] and also a band at about 2150 cm$^{-1}$, due to CO interacting with the –OH groups of the support. Moreover, at this temperature a residual band at 2175 cm$^{-1}$ is observed. This band, which is completely absent at the temperature of the quantitative chemisorption measurements (see dashed curves) and is very weak in comparison to that reported in the nonhydrated sample, has been assigned to CO on the cations of the supports [32]. On the contrary, a significant decrease of CO chemisorbed occurs at higher temperature (298 K).

Another Au/TiO$_2$ catalyst was synthesized by Ti(OH)$_4$ precipitation and calcination and successive addition of gold by DP [37]. After calcination of Au/TiO$_2$ at 573 K, high-resolution transmission electron microscopy (HRTEM) showed large gold nanoparticles with an average diameter of 4.2 ± 1.5 nm and a very broad particle size distribution (ranging between 2 and 11 nm). By chemisorption measurement, Au/TiO$_2$ presented a very low mol$_{CO}$/mol$_{Au}$ value (0.004), confirming that almost only large Au particles were present on this catalyst. Such particles are those detected by HRTEM. At the same time, the very low value of CO chemisorption indicates that no clusters are present on this sample. Therefore CO chemisorption measurements are reliable also for gold nanoparticles of 2–11 nm.

### 9.4.2.2 Au/ZrO$_2$ samples

We chose zirconia because it is of particular interest in a large field of applications, since it possesses desirable properties such as tunable surface acidity/basicity (controlled by the addition of different dopants), redox properties, and tunable porosity and surface area. It has been found that zirconia is a very efficient support for gold-based catalysts for the low-temperature WGS reaction [38]. Moreover, Au/ZrO$_2$ materials are of significant interest as catalysts for CO oxidation, butadiene hydrogenation, epoxidation of styrene, and liquid-phase oxidations [39–41]. Here we present a detailed investigation of samples with different metal content and various particle sizes. Catalysts have been deeply characterized by a number of techniques in order to improve, through a systematic study, the quantitative determination of gold sites able to activate CO and to further test the validity of the method.

In particular, Au/ZrO$_2$ samples with different gold amounts in the 0.5–2 wt% range were synthesized by DP [42] and finally calcined at a proper temperature in the 273–923 K range in order to modulate gold nanosizes [27]. HRTEM measurements were performed on all prepared catalysts. As regard to not calcined samples, no gold particles were detected, for example, on 0.5AuZ, indicating a very high metal dispersion on this catalyst (Fig. 9.11a). At the same time, by XRD patterns no peak related to the presence of gold crystallites is observable, the 2$\theta$ position corresponding to pure Au

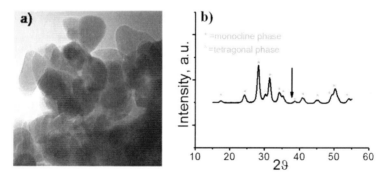

**Figure 9.11** HRTEM image and XRD diffraction pattern for not calcined 0.5AuZ catalyst.

**Table 9.2** Pulse flow CO chemisorption results for Au/ZrO$_2$ catalysts

| Sample | Au found (wt%) | Calcination temperature (K) | Au nanosizes by HRTEM (nm) | Chemisorption data (mol$_{CO}$/mol$_{Au}$) |
|---|---|---|---|---|
| 0.5AuZ | 0.5 | Not calcined | Not detected | 0.37 |
| 0.8AuZ | 0.8 | Not calcined | Not detected | 0.36 |
| 1 AuZ | 1.0 | Not calcined | Not detected | 0.35 |
| 1.5AuZ | 1.5 | Not calcined | Not detected | 0.33 |
| 2AuZ | 1.9 | Not calcined | 1.6 ± 0.6 | 0.30 |
| AuZ | 1.5 | 423 | 1.8 ± 0.2 | 0.24 |
| AuZ | 1.5 | 573 | 1.9 ± 0.4 | 0.23 |
| AuZ | 1.5 | 773 | 2.7 ± 0.7 | 0.03 |
| AuZ | 1.5 | 923 | 4.9 ± 1.3 | 0.01 |

being indicated by the arrow. This suggests again a high dispersion of gold nanoparticles on the support surface (Fig. 9.11b).

The results obtained by pulse flow CO chemisorption on these Au/ZrO$_2$ catalysts are reported in Table 9.2. First of all we would like to stress that chemisorption, if performed under proper conditions, is not only an easy, fast, and economic technique, but it is also very reproducible and so reliable. In fact we have estimated an experimental error in the mol$_{CO}$/mol$_{Au}$ ratio of ± 0.01 and ascertained that the main error of the methodology is the quantitative determination of the real gold content. Looking at chemisorption data on the different samples, reported in Table 9.2, we can make the following observation: Measurements performed on samples that HRTEM and XRD have indicated as highly dispersed give a very large chemisorption value, while in samples in which HRTEM analysis has revealed the presence of bigger gold particles, a significant decrease of chemisorbed CO occurs, indicating that the total adsorbing sites are less abundant. FTIR spectra have shown that the high amounts of chemisorbed CO on not calcined samples are due to chemisorption as carbonylic species on isolated Au clusters [26]. Therefore, CO chemisorption performed by a pulse flow system at 157 K on prehydrated samples can be taken as a method for the quantitative determination of the gold active sites also on Au/ZrO$_2$ catalysts where gold particles cannot be detected by TEM or XRD.

As regard to Au/ZrO$_2$ catalysts calcined at different temperatures, a detailed characterization has been carried out as well [27]. HRTEM analysis revealed only very small roundish Au particles and clusters with an average size of 1.8 ± 0.2 nm observed in the sample calcined at 423 K. The particle size distribution related to this sample is quite narrow and symmetric, indicating that the gold nanoparticles are very small, having a homogeneous size mainly between 1 and 2 nm. Nevertheless, very small gold species with a size below 1 nm (0.7 nm), that is, clusters, have been detected, too. Moreover, the presence of even smaller gold species cannot be excluded, since their size is below the detection limit of the microscope. Both gold size and morphology are unchanged after calcination treatment at 573 K, the average diameter $d_m$ being 1.9 ± 0.4 nm. The shape of the particle size distribution is similar to that related to the sample calcined at 423 K but slightly shifted toward higher sizes. Nevertheless, data evidence that gold dispersion is still very high also on this sample. On the contrary, the gold nanoparticles appear bigger when the calcination temperature is raised up to 773 K ($d_m$ = 2.7 ± 0.7 nm). The particle size distribution well describes the changes occurred on the sample during the calcination at 773 K. In fact, the shape of the distribution results markedly enlarged if compared to that related to samples calcined at lower temperatures, and gold particles with a size ranging from 1.5 to 4.5 nm have been observed. However, a large fraction of particles still remains very small, having a size below 2.5 nm, indicating that the thermal treatment at high temperature did not compromise totally the gold dispersion and that very small particles and, possibly, clusters coexist together with larger particles. The calcination at 923 K has a dramatic effect on the size of gold: big particles with average diameter $d_m$ = 4.9 ± 1.3 nm have been observed, while the gold particle size distribution extends in the size range between 3 and 9 nm, with the disappearance of small particles and clusters. The CO chemisorption is reported as a mol$_{CO}$/mol$_{Au}$ ratio, being an indication of the number of less coordinated gold sites able to absorb CO. As a consequence, CO chemisorption gives also a measure of gold dispersion. In fact, a high mol$_{CO}$/mol$_{Au}$ ratio can be related to the presence of clusters, exposing a high number of uncoordinated sites. Calcination at 923 K leads to a low mol$_{CO}$/mol$_{Au}$ value, meaning that larger gold particles are present. Samples calcined at 423 and 573 K have high chemisorption capability that indicates the presence of Au clusters, while the mol$_{CO}$/mol$_{Au}$ ratio is lower in the case of a catalyst calcined at 773 K and it can be related to the presence of both small Au nanoparticles and clusters. These findings well agree with the results of the HRTEM analyses.

### 9.4.2.3 Au/CeO$_2$ samples

CO chemisorption at low temperature can be used also for gold samples supported on ceria. CeO$_2$ is characterized by a high oxygen storage capacity and reducibility [43], and therefore it is useful in many oxidation reactions. For example, Au/CeO$_2$ samples find applications in CO oxidations and low-temperature WGS reactions.

A sample with a nominal value of 3 wt% Au was prepared by DP (3Au/CeO$_2$) [32]. HRTEM analysis and energy-dispersive X-ray spectroscopy (EDS) measurements evidenced the presence of both very small and highly dispersed clusters of about 1 nm and particles with a size of about 10 nm and 25 nm [32]. So, the particle size distribution is bimodal. On this Au/CeO$_2$ sample, the CO chemisorbed volume per gram of gold is almost three times higher than in the case of the Au/TiO$_2$ reference catalyst previously discussed. Therefore, even if a sample possesses a bimodal distribution of the particle size, chemisorption results provide directly an average size of the gold nanoparticles and the real amount of gold active sites.

If we consider a sample with a lower gold amount and a unimodal particle size distribution, chemisorption tests perform as well [37]. For example, for the 1.5Au/CeO$_2$ catalyst HRTEM analysis evidenced the presence of Au particles with an average size of 2.0 ± 1.4 nm. The particle size distribution was quite narrow and symmetric. However, we were not able to detect gold particles of even smaller size by means of HRTEM. Chemisorption data (mol$_{CO}$/mol$_{Au}$ = 0.06) indicate that on this sample not only the gold nanoparticles detected by HRTEM are present but also some gold clusters [37].

### 9.4.2.4 Au/Fe$_2$O$_3$ sample

Chemisorption measurements are effective also for catalysts with a very high gold content. As example, a 4.48 wt% Au/Fe$_2$O$_3$ reference catalyst provided by the WGC was analyzed [44]. As reported in the catalyst data sheets the sample was prepared by co-precipitation and exposes gold particles with mean diameters and standard deviations, determined by TEM, of 3.7 nm ± 0.9 nm, respectively. The volumes of CO adsorbed on the sample after reduction in hydrogen and saturation by water do not change significantly with the temperature in the 140–180 K range and are very similar to the value of the Au/TiO$_2$ WGC catalyst.

FTIR spectra of CO adsorbed on Au/Fe$_2$O$_3$ have confirmed that in the 140–180 K range the chemisorption data are related only to chemisorption on gold. On the contrary, in analogy with Au/TiO$_2$ samples, CO adsorption at lower temperature (120 K) produces in the FTIR spectra also a band due to CO interacting with the –OH groups of the support. Moreover, at this temperature a residual band at 2133 cm$^{-1}$ with a weak component at 2064 cm$^{-1}$ is observed. These bands have been assigned to CO on the cations of the supports [32].

In conclusion, by the use of a pulse flow system and after proper pretreatment, CO chemisorption at 140–180 K can be taken as a method for the quantitative determination of gold active sites. From our experience this pretreatment is the most suitable one for a comparison of different gold catalysts on different supports, since the mild reduction eliminates the oxygen bonded to the surface of very small gold particles, while the hydration saturates the support cation sites, preventing CO chemisorption. The technique can be used for a preliminary evaluation of catalytic properties. Moreover, CO chemisorption on prehydrated samples can be taken as a method for the quantitative determination of gold active sites also of gold clusters that cannot be detected by TEM.

### 9.4.3 Consideration on the Chemisorption Stoichiometry

It is known that when gold nanoparticles are supported on reducible oxides, they have a flattened shape. According to a Wulff-like model,

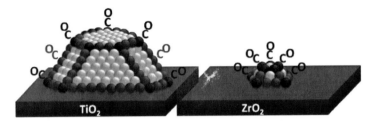

**Figure 9.12** CO chemisorption on bigger gold nanoparticles and small gold clusters. Reprinted from Ref. [45], Copyright (2009), with permission from Elsevier.

three-layer thick metal particles with a mean diameter of 3.8 nm and 1 nm high are assumed in our case on the basis of TEM data [32]. Therefore, as for the two reference samples Au/TiO$_2$ and Au/Fe$_2$O$_3$ by the WGC, some considerations regarding the stoichiometry of chemisorption can be made. The ratio between step-edge Au atoms and total Au atoms can be estimated to be 9.1%, neglecting those in direct contact with the support. Since only the poorly coordinated Au atoms (step-edge atoms) are able to chemisorb CO, a ratio (mol$_{CO}$/mol$_{Au}$) = 0.28–0.31 per step-edge Au atoms in the 140–180 K range can be calculated [32]. So the chemisorption stoichiometry in such a temperature range results to be approximately three Au step-edge atoms every CO, as schematized in Fig. 9.12 [45]. This value could appear rather high in comparison with the stoichiometries determined for other supported metals but is in agreement with the recent literature of CO on gold.

As regard to the 3Au/CeO$_2$ sample, we cannot refer to the mean particle size in order to discuss the chemisorption stoichiometry, taking into account that on this sample both very small clusters and quite big particles have been evidenced [32]. The larger chemisorbed volume must be mainly related to the presence of the very small Au clusters, where almost all the atoms are exposed at the surface, while the big particles do not contribute to the chemisorption at all. On such small clusters all the surface gold atoms are probably able to chemisorbed CO. Looking at the peculiar spectroscopic features of the Au/CeO$_2$ sample, we can assume a Au/CO ratio of approximately 1:1 for this catalyst [32]. At the same time, on not calcined Au/ZrO$_2$ samples, presenting very high chemisorption values, Au/CO = 1 can be reasonably assumed.

## 9.5 Final Remarks

Each technique of particle size measurement has advantages and drawbacks. Ideally, a combination of two or more methods would be necessary to obtain an unambiguous evaluation of particle size. A valid comparison of dispersion values obtained from different investigation techniques is possible, provided that some hypotheses are made at each step of the interpretation. Assessment of routine dispersion measurements by complementary techniques is always rewarding. In fact it is recommended to compare the results with those given by a different physical technique in order to avoid gross misinterpretation of data. However, selective gas chemisorption will remain the preferred technique for particle size and dispersion measurements in catalysis laboratories because it probes directly the catalyst active surface and because instruments are affordable, easily housed, and operated devices.

## Acknowledgments

We thank Prof. Flora Boccuzzi and Dr. Maela Manzoli for FTIR and HRTEM analyses.

## References

1. Lemaitre, J.L., Menon, P.G., and Delannay, F. (1984). The meaurement of catalyst dispersion, in *Characterization of Heterogeneous Catalysts*, Vol. 15, pp. 299–365, ed. Delannay (Marcel Dekker, New York).
2. Bergeret, G., and Gallezot, P. (1997). Particle size and dispersion measurements, in *Handbook of Heterogeneous Catalysis*, Vol. 2, pp. 439–464, eds. Ertl, G., Knozinger, H., and Weitkamp (Wiley VCH, Germany).
3. Bond, G.C., and Thompson, D.T. (1999). Catalysis by gold, *Catal. Rev. Sci. Eng.*, **41**, pp. 319–388.

4. Bond, G.C., Louis, C., and Thompson, D. (2006). *Catalysis by Gold* (Imperial College Press, London).
5. Pernicone, N. (2003). Catalysis at the nanoscale level, *Cattech*, **7**, pp. 196–204.
6. Bartholomew, C.H., and Farrauto, R.J. (2006). Catalyst characterization and selection, in*Fundamentals of Industrial Catalytic Processes,* 2nd ed., pp. 118–196, (Wiley Interscience, New Jersey).
7. Berndt, H., Pitsch, I., Evert, S., Struve, K., Pohl, M.-M., Radnik, J., and Martin, A. (2003). Oxygen adsorption on Au/Al$_2$O$_3$ catalysts and relation to the catalytic oxidation of ethylene glycol to glycolic acid, *Appl. Catal. A*, **244**, pp. 169–179.
8. Fukushima, T., Galvagno, S., and Parravano, G. (1979). Oxygen chemisorption on supported gold, *J. Catal.*, **57**, pp. 177–182.
9. Iizuka, Y., Fujiki, H., Yamauchi, N., Chijiiwa, T., Arai, S., Tsubota S., and Haruta, M. (1997). Adsorption of CO on gold supported on TiO$_2$, *Catal. Today*, **36**, pp. 115–123.
10. Margitfalvi, J.L., Fasi, A., Hegedus, M., Lonyi, F., Gobolos, S., and Bogdanchikova, N. (2002). Au/MgO catalysts modified with ascorbic acid for low temperature CO oxidation, *Catal. Today*, **72**, pp. 157–169.
11. Shastri, A.G., Datye, A.K., and Schwank, J. (1984). Gold-titania interactions: temperature dependence of surface area and crystallinity of TiO$_2$ and gold dispersion, *J. Catal.*, **87**, pp. 265–275.
12. van Vegten, N., Haider, P., Maciejewski, M., Krumeich, F., and Baiker, A. (2009). Chemisorption of methyl mercaptane on titania-supported Au nanoparticles: viability of Au surface area determination, *J. Colloid Interface Sci.*, **339**, pp. 310–316.
13. Trapnell, B.M.W. (1953). The activities of evaporated metal films in gas chemisorption, *Proc. R. Soc. London A*, **218**, pp. 566–577.
14. Eley, D., and Knights, C.F. (1966). The decomposition of nitrous oxide catalysed by palladium-gold alloy wires, *Proc. R. Soc. London A*, **294**, pp. 1–19.
15. Gonzales, O.D., and Parravano, G. (1956). Heats of adsorption of oxygen on nickel, platinum and silver, *J. Am. Chem. Soc.*, **78**, p. 4533–4537.
16. Chester, M.A., and Somorjai, G.A. (1975). The chemisorption of oxygen, water and selected hydrocarbons on the (111) and stepped gold surfaces, *Surf. Sci.*, **52**, pp. 21–28.
17. Choi, K.H., Coh, B.-Y., and Lee, H.-I. (1998). Properties of adsorbed oxygen on Au/SiO$_2$, *Catal. Today*, **44**, pp. 205–213.

18. MacDonald, W.R., and Hayes, K. (1970). A comparative study of the rapid adsorption of oxygen by silver and gold, *J. Catal.*, **18**, pp. 115–118.
19. Pansare, S.S., Sirijaruphan, A., and Goodwin, J.G. (2005). Au-catalyzed selective oxidation of CO: a steady-state isotopic transient kinetic study, *J. Catal.*, **234**, pp. 151–160.
20. Lin, S., and Vannice, M.A. (1991). Gold dispersed on $TiO_2$ and $SiO_2$: adsorption properties and catalytic behavior in hydrogenation reactions, *Catal. Lett.*, **10**, pp. 47–62.
21. Jia, J., Haraki, K., Kondo, J.N., Domen, K., and Tamaru, K. (2000). Selective hydrogenation of acetylene over $Au/Al_2O_3$ catalyst, *J. Phys. Chem. B*, **104**, pp. 11153–11156.
22. Bus, E., Miller, J.T., and van Bokhoven, J.A. (2005). Hydrogen chemisorption on $Al_2O_3$ supported gold catalysts, *J. Phys. Chem.*, **109**, pp. 14581–14587.
23. Bus, E., and van Bokhoven, J.A. (2007). Hydrogen chemisorption on supported platinum, gold and platinum gold alloy catalysts, *Phys. Chem. Chem. Phys.*, **9**, pp. 2894–2902.
24. Menegazzo, F., Pinna, F., Signoretto, M., Trevisan, V., Boccuzzi, F., Chiorino, A., and Manzoli, M. (2008). Highly dispersed gold on zirconia: characterization and activity in LT-WGS tests, *ChemSusChem*, **1**, pp. 320–326.
25. Menegazzo, F., Pinna, F., Signoretto, M., Trevisan, V., Boccuzzi, F., Chiorino, A., and Manzoli, M. (2009). Quantitative determination of sites able to chemisorb CO on $Au/ZrO_2$ catalysts, *Appl. Catal. A*, **356**, pp. 31–35.
26. Pinna, F., Olivo, A., Trevisan, V., Menegazzo, F., Signoretto, M., Manzoli, M., and Boccuzzi, F. (2013). The effects of gold nanosize for the exploitation of furfural by selective oxidation, *Catal. Today*, **203**, pp. 196–201.
27. Signoretto, M., Menegazzo, F., Contessotto, L., Pinna, F., Manzoli, M., and Boccuzzi, F. (2013). $Au/ZrO_2$: an efficient and reusable catalyst for the oxidative esterification of renewable furfural, *Appl. Catal. B*, **129**, pp. 287–293.
28. Zane, F., Trevisan, V., Pinna, F., Signoretto, M., and Menegazzo, F. (2009). Investigation on gold dispersion of $Au/ZrO_2$ catalysts and activity in the low-temperature WGS reaction, *Appl. Catal. B*, **89**, pp. 303–308.
29. Fagherazzi, G., Canton, P., Riello, P., Pernicone, N., Pinna, F., and Battagliarin, M. (2000). Nanostructural features of Pd/C catalysts investigated by physical methods: a reference for chemisorption analysis, *Langmuir*, **16**, pp. 4539–4546.

30. Anderson, J.B. (1975). *Structure of Metallic Catalysts* (Academic Press, London) p. 446.
31. Anderson, J.B. (1975). *Structure of Metallic Catalysts* (Academic Press, London) p. 296.
32. Menegazzo, F., Manzoli, M., Chiorino, A., Boccuzzi, F., Tabakova, T., Signoretto, M., Pinna, F., and Pernicone, N. (2006). Quantitative determination of gold active sites by chemisorption and by infrared measurements of adsorbed CO, *J. Catal.*, **237**, pp. 431–434.
33. Parida, K.M., Mohapatra, P., Moma, John, Jordaan, W.A., and Scurrell, Mike S. (2008). Effects of preparation methods on gold/titania catalysts for CO oxidation, *J. Mol. Catal. A: Chem.*, **288**, pp. 125–130.
34. Idakiev, V., Tabakova, T., Yuan, Z.-Y., and Su, B.-L. (2004). Gold catalysts supported on mesoporous titania for low-temperature water–gas shift reaction, *Appl. Catal. A*, **270**, pp. 135–141.
35. Nijhuis, T.A., and Weckhuysen B.M. (2006). The direct epoxidation of propene over gold–titania catalysts: a study into the kinetic mechanism and deactivation, *Catal. Today*, **117**, pp. 84–89
36. Sample number 17C, supplied by World Gold Council, http://www.gold.org.
37. Menegazzo, F., Signoretto, M., Pinna, F., Manzoli, M., Aina, V., Cerrato, G., and Boccuzzi, F. (2014). Oxidative esterification of renewable furfural on gold based catalysts: which is the best support?, *J. Catal.*, **309**, pp. 241-247.
38. Idakiev, V., Tabakova, T., Naydenov, A., Yuan, Z.-Y., and Su, B.-L. (2006). Gold catalysts supported on mesoporous zirconia for low-temperature water–gas shift reaction, *Appl. Catal. B*, **63**, pp. 178–186.
39. Moreau, F., and Bond, G.C. (2007). Influence of the surface area of the support on the activity of gold catalysts for CO oxidation, *Catal. Today*, **122**, pp. 215–221.
40. Patil, N.S., Uphade, B.S., McCulloh, D.G., Bhargava, S.K., and Choudary, V.R. (2004). Styrene epoxidation over gold supported on different transition metal oxides prepared by homogeneous deposition–precipitation, *Catal. Commun.*, **5**, pp. 681–685.
41. Zhang, X., Shi, H., and Xu, B. (2007). Comparative study of Au/ZrO$_2$ catalysts in CO oxidation and 1,3-butadiene hydrogenation, *Catal. Today*, **122**, pp. 330–337.
42. Manzoli, M., Boccuzzi, F., Trevisan, V., Menegazzo, F., Signoretto, M., and Pinna, F. (2010). Au/ZrO$_2$ catalysts for LT-WGSR: active role of sulfates during gold deposition, *Appl. Catal. B.*, **96**, pp. 28–33.

43. Trovarelli, A. (1996). Catalytic properties of ceria and $CeO_2$-containing materials, *Catal. Rev. Sci. Eng.*, **38**, pp. 439–520.
44. Sample type C, 5wt%Au/$Fe_2O_3$, supplied by World Gold Council, http://www.gold.org.
45. Chiorino, A., Manzoli, M., Menegazzo, F., Signoretto, M., Vindigni, F., Pinna, F., and Boccuzzi, F. (2009). New insight on the nature of catalytically active gold sites: quantitative CO chemisorption data and analysis of FTIR spectra of adsorbed CO and of isotopic mixtures, *J Catal.*, **262**(1), pp. 169–176.

# Chapter 10

# New Findings in CO Oxidation

**Yoshiro Shimojo and Masatake Haruta**

*Department of Applied Chemistry, Graduate School of Urban Environmental Sciences, Tokyo Metropolitan University, 1-1 Minami-osawa, Hachioji, Tokyo 192-0397, Japan*
haruta-masatake@tmu.ac.jp

## 10.1 Introduction

Gold has been attracting growing interest owing to the unique physicochemical properties generated by nanostructures and to its potential applications in wide areas of science and technology, for example, surface plasmonic effects leading to better electronic and photonic devices, catalysts for air and water purification and for green chemistry, and biomedicines for imaging, diagnosis, and disease treatment [1]. In particular, heterogeneous catalysis by gold nanoparticles (NPs) and clusters has already been launched to the stage of commercial applications. Accordingly, durability of gold NP catalysts has become one of the most important concerns. At the same time, the catalytic mechanism of gold NPs should be clarified under a variety of conditions, which leads to the improvement of catalytic performance by gold.

---

*Gold Catalysis: Preparation, Characterization, and Applications*
Edited by Laura Prati and Alberto Villa
Copyright © 2016 Pan Stanford Publishing Pte. Ltd.
ISBN 978-981-4669-28-3 (Hardcover), 978-981-4669-29-0 (eBook)
www.panstanford.com

CO oxidation can be regarded as the most useful test reaction for gold catalysts because it is the simplest reaction to discuss the catalytic mechanism and it can be catalyzed by gold catalysts at very low temperatures, at and below room temperature. The catalytic activity for this reaction usually reflects the dispersion of gold NPs when metal oxide supports are defined. The commercial applications of CO oxidation by gold catalysts have just started for safety gas masks, $CO_2$ lasers, and air purifiers. In this chapter some new findings concerning CO oxidation are described, focusing on atomic-scale observation of reaction pathways and useful hints for the preparation of long-life catalysts. It should be pointed out that our literature survey might not be perfect and many important findings are missing.

## 10.2 An Overview of Catalytic CO Oxidation

CO oxidation can be catalyzed at around room temperature by base metal oxides alone, in particular p-type semiconductive metal oxides such as $MnO_2$, $Co_3O_4$, and NiO. However, these metal oxides are severely deactivated by moisture in the reactant gas, even at a concentration of 10 ppm, which is the concentration of moisture in the standard gas cylinder with a guaranteed dew point of 213 K. In contrast, Pd and Pt catalysts are usually not active below 423 K but are activated by moisture to exhibit catalytic activity at around room temperature. Gold NPs supported on a variety of base metal oxides exhibit catalytic activity at and below room temperature with a metal loading of 1.0 wt% and are promoted by moisture.

On base metal oxides CO oxidation takes place at the oxygen defects between CO and oxygen molecules and atoms. Metal oxides having p-type semiconductivity can adsorb more oxygen than n-type semiconductive oxides and insulators, thus leading to higher catalytic activity. Since moisture preferentially adsorbs on the oxygen defects and disturbs the adsorption of CO, moisture deactivates base metal oxides, as shown in Fig. 10.1 for NiO [2].

There are three conditions which enable p-type semiconductive base metal oxides to be active at a temperature as low as 200 K. The first is to remove moisture below 1 ppm [2, 3]. This fact

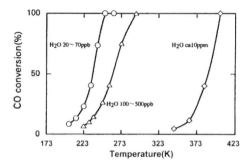

**Figure 10.1** Conversion of CO as a function of catalyst temperature in CO oxidation over NiO under different water content. Reaction gas, CO 0.9 vol% + O$_2$ 20 vol% + N$_2$ balance; SV, 2 × 10$^4$h$^{-1}$mL/g-catal. Reproduced from Ref. [2].

means that the surfaces of MnO$_2$, Co$_3$O$_4$, and NiO are intrinsically active for CO oxidation even at 200 K. The second is to control the morphology of p-type semiconductive oxides. A typical example is rod-like Co$_3$O$_4$, which showed high activity even at 193 K in the co-presence of moisture at a concentration of 10 ppm [4] (Fig. 10.2). The nanorods expose (110) planes with a fraction of 42% of total surfaces. Because this plane holds plenty of active sites (Co$^{3+}$), nanorod Co$_3$O$_4$ can tolerate moisture. The third is to pretreat p-type metal oxides at moderate temperature (423 K) under inert gas or air containing reducing gas (H$_2$ or CO) [5]. Figure 10.3 shows $T_{1/2}$ for base metal oxides after pretreatment in wet and dry gases. Dry gas pretreatment yields higher catalytic activity. Under these conditions, as shown in Fig. 10.4, oxygen defects are produced and maintained at temperatures below room temperature.

On unsupported single-crystal surfaces of Pd and Pt and on their NPs supported on Al$_2$O$_3$, SiO$_2$, or carbon, CO is preferentially adsorbed on the metal surfaces inhibiting O$_2$ dissociative adsorption. Accordingly, relatively high temperature is necessary for CO oxidation without moisture. Moisture enhances desorption of CO, leading to oxygen activation, and thus CO oxidation. When Pd or Pt is deposited or confined on base metal oxides (p-type and n-type semiconductors) or vice versa, the eighth group noble metals exhibit room-temperature activity [6–8]. In principle, the catalytic mechanism appears to be similar between Pd, Pt, and gold.

**288** | *New Findings in CO Oxidation*

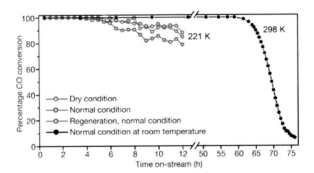

**Figure 10.2** Effects of moisture content, regeneration, and temperature on the oxidation of CO over $Co_3O_4$ nanorods. CO oxidation with a feed gas of 1.0 vol%, $CO/2.5$ vol%, $O_2/He$ under normal (moisture 3–10 ppm, blue symbols) and dry (moisture 1 ppm, red symbols) conditions at 196 K. The used sample was regenerated with a 20 vol% $O_2/He$ mixture at 723 K for 30 min and then tested for CO oxidation under normal conditions (green symbols) at 196 K. CO oxidation at 298 K (black symbols) was tested with the normal feed gas. Reprinted by permission from Macmillan Publishers Ltd: *Nature* (Ref. [4]), copyright (2009).

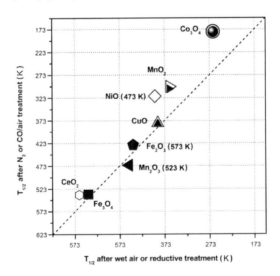

**Figure 10.3** Temperature for 50% CO conversion $(T_{1/2})$ after pretreatment in $N_2$ (solid mark) or 1.0 vol% CO/air (open mark) versus after pretreatment in wet air (or reductive atmosphere). The pretreatments were performed at 423 K unless otherwise noted in parentheses. Reprinted from Ref. [5], Copyright (2009), with permission from Elsevier.

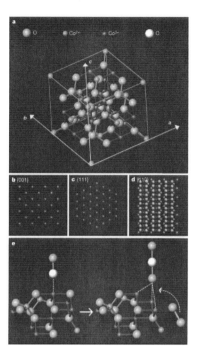

**Figure 10.4** Possible reaction pathway for CO oxidation on $Co_3O_4$ nanorod. (a) Spinel structure of a $Co_3O_4$ crystal. (b–d) The surface atomic configurations in the {001} (b), {111} (c), and {110} (d) planes. (e) A ball-and-stick model for CO adsorption and oxidation on the active $Co^{3+}$ site. Reprinted by permission from Macmillan Publishers Ltd: *Nature* (Ref. [4]), copyright (2009).

On gold NPs deposited on a single crystal of rutile $TiO_2$ (110), as indicated by Fig. 10.5, the reaction mechanism changes at a critical temperature of 333 K. Above this temperature CO oxidation takes place on the surfaces of gold and is not enhanced by moisture [9], while below 333 K the reaction takes place at the perimeter interfaces between CO adsorbed on the gold surfaces and oxygen molecules activated at the periphery and is enhanced by moisture. This mechanism has been proposed on the basis of a substantial set of experimental data and is supported by the fact that the reaction rate is in proportion to the surface-exposed gold atoms at 400 K and the number of atoms at the perimeter interfaces at 300 K [9, 10] (Fig. 10.6). It is interesting to note that at temperature below 200 K

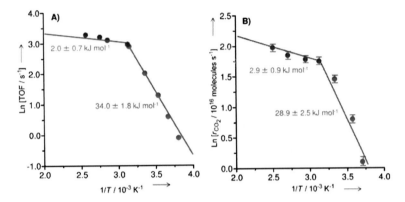

**Figure 10.5** Arrhenius plots for the formation of $CO_2$ over (A) a powdered Au/TiO$_2$ catalyst and (B) one MLE of Au/TiO$_2$ (110). The oxidation of CO was performed in batch mode under 25 Torr of CO, 625 Torr of $O_2$, and 0.1 Torr of $H_2O$. Reprinted with permission from Ref. [9]. Copyright (c) 2011, John Wiley and Sons.

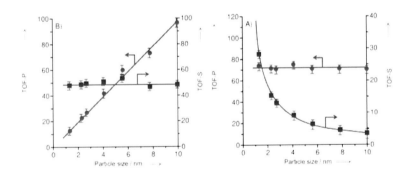

**Figure 10.6** Turnover frequencies (TOFs) for the formation of $CO_2$ over one MLE of Au/TiO$_2$-(110) as a function of the mean diameter of the gold particles at (A) 300 (top) and (B) 400 K (down). Red circles: By normalizing the number of $CO_2$ molecules formed per second to the total number of gold atoms at the perimeter interfaces (TOF-P). Blue squares: By normalizing the number of $CO_2$ molecules formed per second to the total number of exposed Au atoms at the gold particles (TOF-S). Reprinted with permission from Ref. [9]. Copyright (c) 2011, John Wiley and Sons.

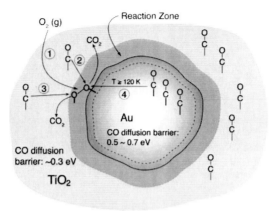

**Figure 10.7** Schematic of the mechanism of low-temperature CO oxidation over a Au/TiO$_2$ catalyst at a perimeter zone of reactivity. Experiments directly observing CO/TiO$_2$ and CO/Au surface species show that processes 2 and 3 are fast compared with process 4. From Ref. [11]. Reprinted with permission from AAAS.

the reaction proceeds on the surfaces of the metal oxide support through the spill-over of CO species [11] (Fig. 10.7).

## 10.3 Environmental TEM Observation under CO Oxidation

Atomic-scale structures of gold NPs deposited on CeO$_2$ have been observed by using an aberration-corrected environmental transmission electron microscope (TEM). The catalytic activity of the real catalyst sample was measured to be 0.24 s$^{-1}$ at 298 K. Under vacuum and under exposure to N$_2$, gold NPs held a polyhedral shape enclosed by the major {111} and {100} facets, while under O$_2$ they became rounded or fluctuating multifaceted [12]. On gold NPs supported on a nonoxide support such as TiC, switching the gas atmosphere did not induce any morphology change. It can be assumed that the O$_2$ molecule is activated or dissociated at the perimeter interfaces between gold NPs and the CeO$_2$ support and/or the surfaces of the CeO$_2$ support.

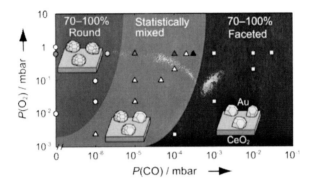

**Figure 10.8** Morphology of GNPs supported on $CeO_2$ as a function of the partial pressures of CO, $P_{CO}$, and $O_2$, $P_{O2}$, in CO/air gaseous mixtures. Reprinted with permission from Ref. [12]. Copyright (c) 2011, John Wiley and Sons.

Gold NPs remained faceted during CO oxidation under 1 vol% in air and became rounded with a decrease in the partial pressure of CO [13]. Figure 10.8 shows the morphology change of gold NPs as a function of the partial pressure of CO and $O_2$. This indicates that the oxygen molecule was activated at the perimeter interfaces and that CO is relatively strongly adsorbed on the surfaces of gold NPs.

The adsorption structure of CO during CO oxidation at room temperature has been observed for the first time [14].

Figure 10.9 shows that the adsorption of the CO molecule took place and caused the {100} facets of a gold NP to reconstruct. CO was adsorbed at the on-top sites of gold atoms, which was estimated to be energetically favorable by ab initio calculations and image simulations.

## 10.4 Stability of Nanoparticulate Gold Catalysts

Since CO oxidation is the most extensively and intensively studied reaction, plenty of gold catalysts have been prepared for this reaction until now. Recently, some efforts have been directed to the improvement of catalyst life for practical applications. Table 10.1 lists some recent work dealing with thermal stability and pretreatment of gold NP catalysts.

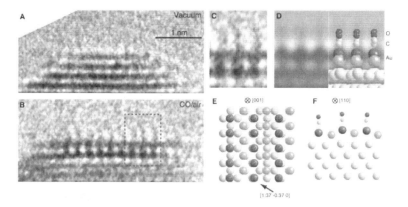

**Figure 10.9** Adsorbed CO molecules on a Au{100}-hex reconstructed surface under catalytic conditions. Aberration-corrected ETEM images in (A) a vacuum and (B) a reaction environment (1 vol% CO in air at 100 Pa at room temperature) taken using 80 keV electrons using an underdefocus condition. (C) The observed image in the rectangular region in (B) at higher magnification. (D) A simulated image based on an energetically favorable model. The model is superimposed on the simulated image. The model in (E) plan view along the [001] direction of crystalline gold and (F) cross-sectional view along the [110] direction of crystalline gold to show the undulating topmost Au layer. From Ref. [14]. Reprinted with permission from AAAS.

The deactivation of gold NP catalysts is mainly caused by the accumulation of carbonate and carboxylic species at the perimeter interfaces [15]. This deactivation takes place more rapidly at lower temperatures. The second reason is a decrease in moisture content in supported gold NPs during a reaction under dry conditions. The coagulation of gold NPs is usually not a serious problem as far as the catalysts are prepared by calcination at 573 K and above. The coprecipitated Au/Fe$_2$O$_3$ powder calcined at 673 K, which Haruta prepared 27 years ago and has been kept in a glass (not polymer) bottle, shows the same conversion versus temperature curve after pretreatment in a stream of dry air at 523 K.

### 10.4.1 Al$_2$O$_3$ Support

Aluminum oxide is not the best support for gold NPs to obtain catalytic activity at temperatures below 273 K; however, it can present

**Table 10.1** Examples of stable CO oxidation catalysts based on gold

| Catalyst composition | Au loading (wt%) | Au diameter (nm) | Preparation method | Calcination temperature (K) | Catalytic activity conv. or $T_{1/2}$ | Stability | Notes | Ref. |
|---|---|---|---|---|---|---|---|---|
| Au/$\gamma$-Al$_2$O$_3$ | 1.6–2.8 | 2.5 | CI | 463 | 257 | | | [16] |
| Au/$\delta$-Al$_2$O$_3$,Au/$\Delta$-Al$_2$O$_3$ | 1–2 | 3–5 | DP | 673 | 94% at 313 K | | Cl free | [17] |
| Au/fumed SiO$_2$ | 0.6–4.3 | 3.1–6.0 | DP at pH 10 | 773 | 265–267 | Stable up to 773 K | | [18] |
| Au/CuO, Co$_3$O$_4$/ mesoporous silica | 1.1–2 | | CI | 573 | 55% at 273 K | Stable up to 773 K | | [20] |
| Au/TiO$_2$ (AUROlite$^{TM}$) | 1 | | Commercial | | 1.4 s$^{-1}$ (MTY) | | Pretreatment in H$_2$ at 298 K | [15] |
| | 1 | 2–4 | Commercial | | 3.0 s$^{-1}$ (MTY) | | Pretreatment in air at 523 K | |
| Au/TiO$_2$ whisker | 1 | 6.5–11.5 | DP | 873 | 308 | | | [25] |
| Au/Na$_2$Ti$_3$O$_7$ nanotube | 2.53 | 1.5 | IE | 383 | 218 | ≤ 6 nm at 673 K | | [22] |
| Au/TiO$_2$/SiO$_2$ aerogel | 2 | 10 | CI | 673 | 210 | Stable up to 773 K | | [24] |
| Au/$\alpha$-Mn$_2$O$_3$ | 2.9 | 1–3 | DP | 537 | 261–283 | From higher activity | Pretreatment in O$_2$, He, none, H$_2$ | [28] |
| Au/$\alpha$-Mn$_2$O$_3$ | 5 | 2.2 | DP | 573 | 210 | | | [27] |
| Au/Ce-K-OMS-2 | 0.5 | 3 | WI | 383 | 338 | Active even below 273 K | | [29] |
| Au/Fe(OH)$_3$ | 2.5 | <1 (uncalcined) | CP | 298 | 243 | | Dropwise CP | [33] |

| | | | | | | |
|---|---|---|---|---|---|---|
| Au/FeO$_x$ | 2.1 | 1 | CP | 333 | 100% at 273 K (as-synthesized) | [34] |
| | 2.1 | 2.5 | CP | 673 | 12% at 273 K | |
| Au/$\alpha$-Fe$_2$O$_3$ | 3 | 0.7 | DP | 473 | 100% at 298 K | 100% at 298 K for 3,000 h | [32] |
| Au/$\alpha$-Fe$_2$O$_3$ (La$^{3+}$) | 1 | 4–8 | DP | 773 | 291 | Higher than none doping; High thermal stability | [35] |
| Au/MgFe$_2$O$_4$ nanocrystal | 0.8 | 1.5–5 | CI | Vacuum dried (r.t.) | 243 | From 100% to 90% (200 min); Pretreatment in O$_2$ | [37] |
| Au/FeO$_x$/hydroxyapatite | 5 | 2–3 | DP | 673 | 90% at 297 K (300 h) | Sintering resistance | [40] |
| Fe$_2$O$_3$/Au/Fe-La/Al$_2$O$_3$ | 1.36 | 1.6 | DP pH 9 | 573 | <248 | Post addition of Fe$_2$O$_3$ | [38] |
| Au/CeO$_2$ | 0.06–0.09 | 4.8 | DP | 873 | 70% at r.t. (70 min) | | [41] |
| Au/FeO$_x$/CeO$_2$ | 1 | <1 | DP | 573 | 320 | | [36] |

longer catalyst life at room temperature than semiconductive metal oxides. Deposition-precipitation (DP) and colloid immobilization (CI) methods are useful for preparing highly dispersed gold NPs on $Al_2O_3$. A catalyst having 3.5 wt% gold loading on $\gamma$-$Al_2O_3$, which was prepared by the CI method and by calcination at 463 K, could maintain nearly 100% CO conversion at 293 K for 1500 h [16].

A 1.4 wt%Au/$\delta$-$Al_2O_3$ catalyst prepared by DP method and by calcination at 673 K gave CO conversions above 85% at room temperature [17]. $\delta$-$Al_2O_3$ presents a little higher catalytic activity than $\gamma$-$Al_2O_3$. What is interesting is that pretreatment of the catalysts at 973 K in air containing 10% steam can enhance the catalytic activity. It is interpreted that steam treatment brings about high concentration of the surface Au-OH and/or Al-OH groups, which may act as active species for CO oxidation.

## 10.4.2 $SiO_2$ Support

Gold NPs supported on $SiO_2$ are usually not active for CO oxidation at room temperature. However, this is mainly because gold NPs are not highly dispersed and are in the size range above 10 nm in diameter. When gold is highly dispersed on fumed $SiO_2$ by using $Au(en)_2Cl_3$ as a precursor, gold NPs could be maintained in the range of 3–4 nm, even after calcination at 773 K. The reaction routes are shown in Fig. 10.10 [18]. Since $SiO_2$ is negatively charged in aqueous solution, cationic gold species can approach the surfaces of $SiO_2$ and can interact with protons.

An important trend is that on a $SiO_2$ support active base metal oxides such as $TiO_2$, $MnO_x$, $Fe_2O_3$, $Co_3O_4$, and NiO are predeposited and then gold NPs are deposited. For example, gold NPs are deposited on $TiO_2$-modified mesoporous $SiO_2$ (SBA-15) (Fig. 10.11) [19, 20]. In contrast, base metal oxides are postdeposited on Au/$SiO_2$. $MnO_x$/Au/$SiO_2$ is prepared by postdeposition of $MnO_x$ (Fig. 10.12) [19]. Although the interaction of gold NPs and $SiO_2$ is weak, causing aggregation of gold NPs, incorporation of active base metal oxides not only enhances the catalytic activity but also improves the thermal stability of gold NPs against coagulation.

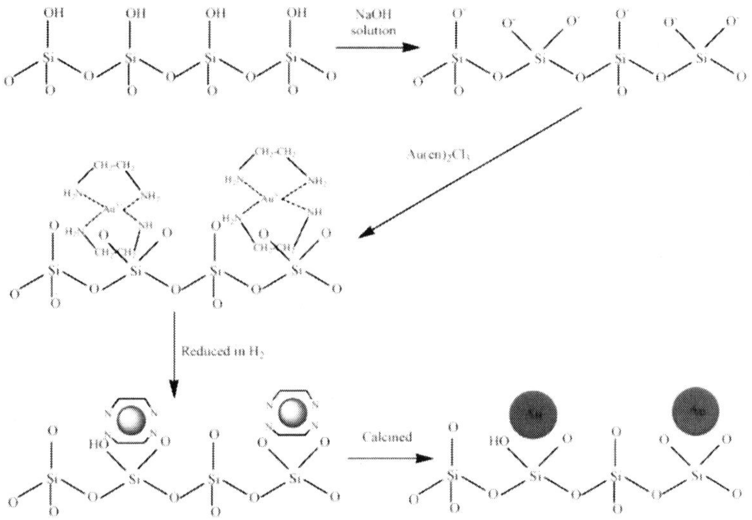

*Stability of Nanoparticulate Gold Catalysts* | **297**

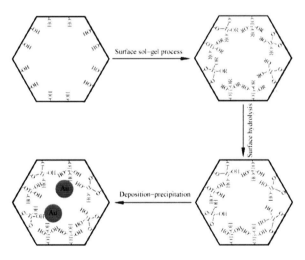

**Figure 10.10** Schematic representation of the formation of gold nanoparticles on SiO$_2$ surfaces using Au(en)$_2$Cl$_3$ as the precursor in the basic media. Reprinted from Ref. [18], Copyright (2007), with permission from Elsevier.

**Figure 10.11** The modification of mesoporous SiO$_2$ by TiO$_2$ using a surface–sol–gel approach for loading gold nanoparticles. Reproduced from Ref. [19]. Copyright © 2011, Tsinghua University Press and Springer-Verlag Berlin Heidelberg.

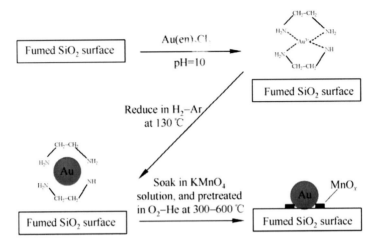

**Figure 10.12** Schematic representation of the preparation of MnO$_x$-loaded Au/SiO$_2$ by treating Au(en)$_2$Cl$_3$-derived Au/SiO$_2$ with KMnO$_4$ solution, followed by thermal activation. Reproduced from Ref. [19]. Copyright © 2011, Tsinghua University Press and Springer-Verlag Berlin Heidelberg.

### 10.4.3 TiO$_2$ Support

Titanium dioxide is a base metal oxide that has been most frequently used as a support for gold NPs because it is not active at all by itself for CO oxidation but can provide the highest catalytic activity when combined with gold NPs.

Accordingly, the reaction mechanism for CO oxidation has been most extensively and intensively studied. The role of gold NPs is first to accumulate adsorbed CO and second to activate surface lattice oxygen at Au/TiO$_2$ perimeter interfaces. Behm has discussed on CO oxidation at temperatures above 353 K and has assumed that the oxygen molecule was dissociated before reaction with CO [21] (Fig. 10.13), while Haruta assumes that molecular oxygen activated (hydroperoxide, OOH) at the perimeter reacts with CO at temperatures below 333 K [10]. He assumes that above this temperature the reaction takes place on the surfaces of gold NPs, exhibiting very low apparent activation energies.

The deactivation of gold NP catalysts is caused by the coagulation of gold NPs [22], reduction of oxidic gold species into metallic ones [23], decrease in support OH groups [23], and accumulation of

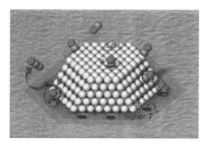

**Figure 10.13** Schematic description of the pathway for CO oxidation on Au/TiO$_2$ catalysts at $T \geq 353$ K, involving (a) CO adsorption on Au NPs, (b) reaction with activated surface lattice oxygen species at the perimeter of the Au–TiO$_2$ interface (interface sites), and (c) replenishment of these sites by dissociative adsorption of O$_2$ at the perimeter sites. (d) At higher temperatures (>353 K), migration of surface lattice oxygen and surface oxygen vacancies also gives access to neighboring surface lattice oxygen. During reaction under normal reaction conditions, only the perimeter sites are involved. Reprinted with permission from Ref. [21]. Copyright (c) 2011, John Wiley and Sons.

carbonate species [20]. Coagulation of gold NPs causes irreversible deactivation and inevitably takes place when supported gold NPs are calcined at temperatures below 573 K. Temperature for 50% conversion was 263K on Au/Na$_2$Ti$_3$O$_7$ prepared by DP and dried at 383 K ( gold diameter 1.5 nm), while it was 292 K when calcined at 673 K (gold diameter 3.4 nm). The latter catalyst is, in general, more stable against coagulation [22].

Stability against thermal sintering of gold NPs can be improved by using TiO$_2$-coated SiO$_2$ aerogel [24]. Gold is deposited by the CI method and resulted active at 273 K, even after calcination at 973 K. In the case of a Au/mesoporous TiO$_2$ whisker, a sample calcined at 873 K is more active than that calcined at 573 K for CO oxidation at 353 K. It is interpreted that the contact area of gold and interaction is increased by high-temperature calcination [25].

Veith et al. reported that gold oxide clusters grown on the hydroxylated surface of TiO$_2$ (Degussa p-25) by reactive sputtering are catalytically very active but decompose over many months [23]. The surface OH group plays a dominant role, in particular in CO oxidation at temperatures below 333 K. The loss of catalytic activity can be correlated to moisture content in the reactant gases.

The major reason for the gradual deactivation of supported gold NP catalysts is the accumulation of carbonate species at the perimeter interfaces between gold NPs and the TiO$_2$ support [15]. Therefore, this deactivation is reversible and the catalysts can be regenerated by heat treatment at 523 K in a stream of air or by irradiation of oxygen plasma.

### 10.4.4 MnO$_x$ Support

As far as the authors are concerned, manganese oxides were first used as supports for preferential CO oxidation to purify hydrogen [26]. Zuang et al. reported that $\alpha$-Mn$_2$O$_3$ produced by calcination of MnCO$_3$ was active, giving 100% conversion at 223 K with 5 wt% loading of gold [27]. As shown in Fig. 10.14, oxygen pretreatment enhanced the catalytic activity and improved the stability [28]. The superior activity was attributed to the unique redox properties and facile formation of active oxygen. Gold NPs supported on cerium-modified cryptomelane-type manganese oxide showed higher catalytic activity, high dispersion, and better stability [29].

### 10.4.5 Fe$_2$O$_3$ Support

Ferric oxide has been studied most extensively as a support for gold, owing to the potential capability for practical applications. The classical method adopted by Haruta for preparing Au/Fe$_2$O$_3$ catalysts is co-precipitation (CP). It should be noted that there are at least four procedures in the CP method: an aqueous solution of HAuCl$_4$ and Fe (NO$_3$)$_3$ is poured into an aqueous alkaline solution or vice versa, dropwise addition or instantaneous addition. The usual procedure, which may be the most reasonable one to obtain a homogeneous mixture, adopts instantaneous addition of an aqueous acidic solution into an aqueous alkaline solution. The gold and iron hydroxide mixture shows remarkably high catalytic activity when calcined at temperatures above 573 K [30].

In contrast, Hutchings used the opposite procedure, dropwise addition of an aqueous alkaline solution into an aqueous acid solution and drying at 393 K. They proposed that bilayer clusters,

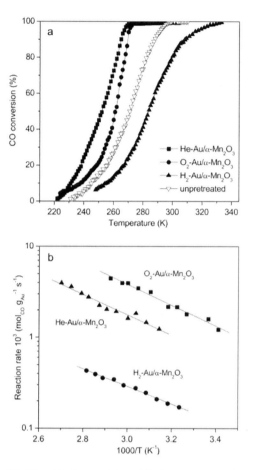

**Figure 10.14** (a) CO oxidation activity of Au/$\alpha$-Mn$_2$O$_3$ catalysts pretreated with different conditions. Reaction conditions: 20 mg catalyst, 1%CO–20%O$_2$ balanced with He (50 mL min$^{-1}$). (b) Arrhenius plots of the reaction rate vs. $1/T$ for CO oxidation over various Au/$\alpha$-Mn$_2$O$_3$ catalysts. Reprinted from Ref. [28], Copyright (2009), with permission from Elsevier.

having 0.5 nm diameter, were responsible for room-temperature CO oxidation [31]. Since then many researchers have attempted to study the catalytic components of metallic or oxidic gold supported on FeOOH, which shows fairy high catalytic activity at room temperature. They are usually more active than the catalysts calcined at 473 K [32, 33] owing to the enhancing effect of water;

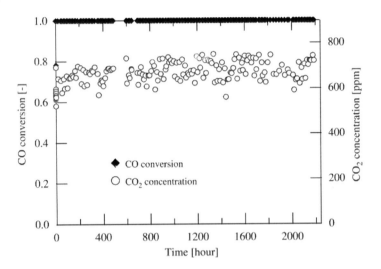

**Figure 10.15** Long-term test of Au/FeOOH prepared by two-stage successive precipitation followed by drying at 473 K in CO oxidation. CO concentration: 500 ppm; SV: 100,000 h$^{-1}$; temperature: 298 K. Reprinted from Ref. [32], Copyright (2010), with permission from Elsevier.

however, it should be kept in mind that such wet gold catalysts often change during storage in glass bottles and gradually produce relatively large gold NPs.

Mae reported that Au/FeOOH prepared by two-stage successive precipitation followed by drying at 473 K can produce a markedly long-life catalyst that maintains 100% conversion for more than 2500 h at 298 K [32] (Fig. 10.15). Figure 10.16 shows that the two-stage precipitation is more effective than instantaneous CP and HAuCl$_4^-$ dropwise CP. Zhang and coworkers prepared a catalyst composed of cationic gold, AuOOH($x$H$_2$O) and ferrihydrate, Fe$_5$HO$_8$(4H$_2$O), the synergetic combination of which yields high catalytic activity at room temperature or even below [34]. A new mechanism has been proposed where the surface lattice oxygen of the FeOOH support participates directly in CO oxidation and the reaction proceeds mainly through a redox mechanism.

Another line of approach is to develop thermally stable gold NP catalysts by using modified ferric oxides as supports. Ferric oxide doped with La calcined at 773 K gave 90% conversion at 302 K [35].

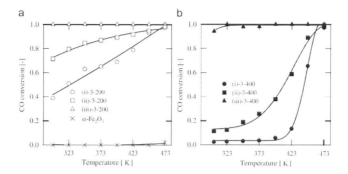

**Figure 10.16** Catalytic activities of the samples of two-step precipitation in difference conditions. CO concentration: 500 ppm; SV: 36,000 h$^{-1}$. Reprinted from Ref. [32], Copyright (2010), with permission from Elsevier.

In Fe-Ce (atomic ratio 15–85) oxide, infrared (IR) spectroscopy of adsorbed CO suggested redox couples between Au+ and Au$^0$ as an operating mechanism [36]. Spinel MgFe$_2$O$_4$ oxide yields stable gold catalysts when pretreated in oxygen at 573 K [37].

Remarkable improvement in the online stability of gold catalysts has been achieved by depositing Fe$_2$O$_3$ on Au/Fe-La oxides/Al$_2$O$_3$ [38]. Although gold NPs coagulated to form larger particles (4.1 nm from 1.6 nm) by postdeposition of Fe$_2$O$_3$, the online stability was improved and thermal regeneration treatment can totally recover the catalytic activity. This sophisticated catalyst is one of the toughest catalysts for practical applications in CO$_2$ lasers, safety gas masks, and air purifiers. An FeO$_x$ modified hydroxyapatite (HAP, Ca$_{10}$(PO$_4$)$_6$(OH)$_2$) was used as a support for gold NPs [39, 40]. Au/FeO$_x$/HAP catalyst calcined at 673 K is highly active and stable during CO oxidation at 298 K. It is sinter resistant to calcination up to 863 K, owing to a strong interaction between the HAP support (OH$^-$ and/or PO$_4^{3-}$) and the gold and FeO$_x$ NPs, and presents a longer life originated from its better prevention of carbonate accumulation.

### 10.4.6 CeO$_2$ Support

A unique low-gold-loading (below 0.1 wt%) Au/CeO$_2$ catalyst has been prepared by the DP method under different procedures [41]. Both the suspension of CeO$_2$ in water and an aqueous solution of

HAuCl$_4$ were adjusted to pH 9, warmed at 343K, and then mixed together. A sample dried at 333 K shows the highest activity and stability at ambient temperature. Base treatment of acid CeO$_2$ leads to a decrease in cationic gold and to higher Au$^0$/Au$^{\delta+}$.

## 10.5 New Attempts in the Preparation of Gold Catalysts

The impregnation method is the simplest and the most popular technique for preparing supported noble metal catalysts. However, in the case of gold, as far as we use conventional precursor HAuCl$_4$, gold is dispersed as large particles 30–100 nm in diameter. This is because chloride ions markedly enhance the coagulation of gold particles. Sakurai et al. used gold acetate instead of HAuCl$_4$ [42]. Since the solubility of gold acetate is considerably low, it is dissolved completely by refluxing the brown colloidal dispersion by adjusting the pH to 10–11 at boiling temperature. The resultant solution is transparent without any color and contains Au$^{3+}$ species like Au(OH)$_4^-$. Base metal oxide powder such as Al$_2$O$_3$, SiO$_2$, TiO$_2$,

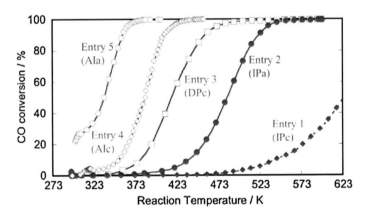

**Figure 10.17** CO conversion vs. reaction temperature curves measured on entry 1–5 Au/CeO$_2$ catalysts. AIc: alkaline impregnation using Au(OAc)$_3$, AIc: alkaline impregnation using HAuCl$_4$, DPc: deposition precipitation using HAuCl$_4$, IPa: impregnation using Au(OAc)$_3$, IPc: impregnation using HAuCl$_4$, CO(1%) + O$_2$(20%) + He(balance); SV = 300,000 h$^{-1}$ mL/g-catal. Reprinted from Ref. [42], Copyright (2013), with permission from Elsevier.

CeO$_2$, silicate like saponite clay, and Y-type zeolite was impregnated with the solution and then calcined at 350 K. In the case of Au/CeO$_2$, this technique can produce more active catalysts than other techniques. Figure 10.17 shows the conversion of CO as a function of temperature for Au/CeO$_2$ prepared by different methods. Impregnation with gold acetate alkaline solution yields higher catalytic activity, while conventional impregnation failed to exhibit activity below 523 K. The mean diameter of gold NPs was estimated to be 2.3 nm on saponite.

In the DP method, the catalytic activity of oxide-supported gold catalysts dried at room temperature appreciably depends on the pH of the alkaline solution of HAuCl$_4$. The optimum pH is 9 for TiO$_2$, 7.5 for Fe$_2$O$_3$, and 7 for SnO$_2$ and CeO$_2$ [43]. Addition of iron in the preparation lowered the rate of deactivation when TiO$_2$, SnO$_2$, and CeO$_2$ were used as supports. Improved online stability is due to gold NPs being in contact with an iron phase such as FeO(OH). Accordingly, calcination removed this stabilization. As for

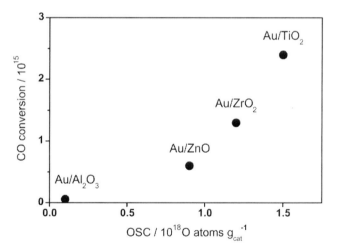

**Figure 10.18** Absolute amounts of CO converted/CO$_2$ produced in steady state during simultaneous pulses CO/Ar and O$_2$/Ar on the differently supported Au catalysts plotted against the oxygen storage capacity of the corresponding catalysts, both after pretreatment by calcination in a flow of 20 NmL min$^{-1}$ 10%O$_2$/N$_2$ at 523 K for 2 h. Reprinted from Ref. [45], Copyright (2010), with permission from Elsevier.

the influence of the surface area of the support on the catalytic activity, moderate support areas around 50 m$^2$/g give the most active catalysts as far as TiO$_2$, SnO$_2$, ZrO$_2$, and CeO$_2$ were concerned [44]. It is also interesting to note that the catalytic activity of Au/Al$_2$O$_3$, Au/TiO$_2$, Au/ZnO, and Au/ZrO$_2$, which were prepared by CI, can be correlated to the oxygen storage capacity of oxide supports [45] (Fig. 10.18).

## 10.6 Summary

CO oxidation is a representative reaction in the catalysis by gold. Mechanistic understanding at an atomic scale has been challenged by using environmental transmission electron microscopy and by using theoretical calculations. Although our literature survey was not perfect, this chapter attempted to overview recent findings in CO oxidation on supported gold catalysts in terms of thermal resistance against coagulation of gold NPs, water content, and deactivation caused by the accumulation of carbonate spectators.

A new group of supported gold catalysts has been emerging, which are used after drying without calcination in air at 573 K or above. In these catalysts crystalline base metal oxides are not formed. Gold is usually smaller than 2 nm in diameter and the supports contain much water as hydroxides. These wet catalysts usually show higher conversion of CO and better online stability owing to the co-presence of water. However, it is likely that gradual coagulation of tiny gold clusters takes place in the time range of a few to several months, causing irreversible deactivation. In contrast, conventional calcined catalysts are stable against coagulation of gold NPs at room temperature, however, are less enhanced by water. The deactivation by the accumulation of carbonate spectators is reversible and recovered by calcination.

Another new trend is that sophisticated compositions and structures have been found to be more active with longer life. Postdeposition of Fe$_2$O$_3$ on Au/La-Fe Al$_2$O$_3$ [38] and Au/Fe$_2$O$_3$/HAP [39, 40] are typical examples. Lastly, as proved by the work by F. Moreau and G. C. Bond [43, 44], the catalytic activity and life of dried catalysts can be tuned markedly by choosing optimum preparation conditions. Their attempt should be extended to calcined catalysts—for example, in the DP method, temperature, pH, concentration of the starting solution containing HAuCl$_4$, selection of alkali, agitation of dispersion, calcination temperature and atmosphere, etc.

## Acknowledgment

The authors are indebted to Ms. Mayumi Morikawa for making the list of references.

## References

1. Brody, H. (2013). Gold, *Nature*, **495**, pp. S1–S45.
2. Haruta, M., Yoshizaki, M., Cunningham, D.A.H., and Iwasaki, T. (1996). Catalysis research by use of ultra pure gas: effect of a trace amount of water, *Ultra Clean Technol.*, **8**(2), pp. 117–120 (in Japanese).
3. Cunningham, D.A.H., Kobayashi, T., Kamijo, N., and Haruta, M. (1994). Influence of dry operating conditions: observation of oscillations and low temperature CO oxidation over Co$_3$O$_4$ and Au/Co$_3$O$_4$ catalysts, *Catal. Today*, **25**, pp. 257–264.
4. Xie, X., Li, Y., Liu, Z.-Q., Haruta, M., and Shen, W. (2009). Low-temperature oxidation of CO catalysed by Co$_3$O$_4$, *Nature*, **458**, pp. 746–749.
5. Yu, Y., Takei, T., Ohashi, H., He, H., Zhang, X., and Haruta, M. (2009). Pretreatments of Co$_3$O$_4$ at moderate temperature for CO oxidation at −80°C, *J. Catal.*, **267**, pp. 121–128.
6. Freund, H.-J., Meijer, G., Scheffler, M., Schlögl, R., and Wolf, M. (2011). CO oxidation as a prototypical reaction for heterogeneous processes, *Angew. Chem., Int. Ed.*, **50**, pp. 10064–10094.
7. Tanaka, K. (2010). Unsolved problems in catalysis, *Catal. Today*, **154**, pp. 105–112.
8. Fu, Q., Li, W.-X., Yao, Y., Liu, H., Su, H.-Y., Ma, D., Gu, X.-K., Chen, L., Wang, Z., Zhang, H., Wang, B., and Bao, X. (2010). Interface-confined ferrous centers for catalytic oxidation, *Science*, **328**, pp. 1141–1144.
9. Fujitani, T., and Nakamura, I. (2011). Mechanism and active sites of the oxidation of CO over Au/TiO$_2$, *Angew. Chem., Int. Ed.*, **50**, pp. 10144–10147.

10. Haruta, M. (2011). Role of perimeter interfaces in catalysis by gold nanoparticles, *Faraday Discuss.*, **152**, pp. 11–32.
11. Green, I.X., Tang, W., Neurock, M., and Yates Jr, J.T. (2011). Spectroscopic observation of dual catalytic sites during oxidation of CO on a Au/TiO$_2$ catalyst, *Science*, **333**, pp. 736–739.
12. Uchiyama, T., Yoshida, H., Kuwauchi, Y., Ichikawa, S., Shimada, S., Haruta, M., and Takeda, S. (2011). Systematic morphology changes of gold nanoparticles supported on CeO$_2$ during CO oxidation, *Angew. Chem., Int. Ed.*, **50**, pp. 10157–10160.
13. Kuwauchi, Y., Yoshida, H., Akita, T., Haruta, M., and Takeda, S. (2012). Intrinsic catalytic structure of gold nanoparticles supported on TiO$_2$, *Angew. Chem., Int. Ed.*, **51**, pp. 7729–7733.
14. Yoshida, H., Kuwauchi, Y., Jinschek, J.R., Sun, K., Tanaka, S., Kohyama, M., Shimada, S., Haruta, M., and Takeda, S. (2012). Visualizing gas molecules interacting with supported nanoparticulate catalysts at reaction conditions, *Science*, **335**, pp. 317–319.
15. Saavedra, J., Powell, C., Panthi, B., Pursell, C.J., and Chandler, B.D. (2013). CO oxidation over Au/TiO$_2$ catalyst: pretreatment effects, catalyst deactivation, and carbonates production, *J. Catal.*, **307**, pp. 37–47.
16. Wen, L., Fu, J.-K., Gu, P.-Y., Yao, B.-X., Lin, Z.-H., and Zhou, J.-Z. (2008). Monodispersed gold nanoparticles supported on $\gamma$-Al$_2$O$_3$ for enhancement of low-temperature catalytic oxidation of CO, *Appl. Catal. B*, **79**, pp. 402–409.
17. Moroz, B.L., Pyrjaev, P.A., Zaikovskii, V.I., and Bukhtiyarov, V.I. (2009). Nanodispersed Au/Al$_2$O$_3$ catalysts for low-temperature CO oxidation: results of research activity at the Boreskov Institute of Catalysis, *Catal. Today*, **144**, pp. 292–305.
18. Zhu, H., Ma, Z., Clark, J.C., Pan, Z., Overbury, S.H., and Dai, S. (2007). Low-temperature CO oxidation on Au/fumed SiO$_2$-based catalysts prepared from Au(en)$_2$Cl$_3$ precursor, *Appl. Catal. A*, **326**, pp. 89–99.
19. Ma, Z., and Dai, S. (2011). Development of novel supported gold catalysts: a materials perspective, *Nano Res.*, **4**(1), pp. 3–32.
20. Ma, G., Binder, A., Chi, M., Liu, C., Jin, R., Jiang, D.-e., Fan, J., and Dai, S. (2012). Stabilizing gold clusters by heterostructured transition-metal oxide–mesoporous silica supports for enhanced catalytic activities for CO oxidation, *Chem. Commun.*, **48**, pp. 11413–11415.
21. Widmann, D., and Behm, R.J. (2011). Active oxygen on a Au/TiO$_2$ catalyst: formation, stability, and CO oxidation activity, *Angew. Chem., Int. Ed.*, **50**, pp. 10241–10245.

22. Tsai, J.-Y., Chao, J.-H., and Lin, C.-H. (2009). Low temperature carbon monoxide oxidation over gold nanoparticles supported on sodium titanate nanotubes, *J. Mol. Catal. A,* **298**, pp. 115–124.
23. Veith, G.M., Lupini, A.R., Pennycook, S.J., and Dudney, N.J. (2010). Influence of support hydroxides on the catalytic activity of oxidized gold clusters, *ChemCatChem,* **2**, pp. 281–286.
24. Tai, Y., and Tajiri, K. (2008). Preparation, thermal stability, and CO oxidation activity of highly loaded Au/titania-coated silica aerogel catalysts, *Appl. Catal. A,* **342**, pp. 113–118.
25. Zhu, Y., Li, W., Zhou, Y., Lu, X., Feng, X., and Yang, Z. (2009). Low-temperature CO oxidation of gold catalysts loaded on mesoporous $TiO_2$ whisker derived from potassium dititanate, *Catal. Lett.,* **127**, pp. 406–410.
26. Sanchez, R.M.T., Ueda, A., Tanaka, K., and Haruta, M. (1997). Selective oxidation of CO in hydrogen over gold supported on manganese oxides, *J. Catal.,* **168**, pp. 125–127.
27. Wang, L.-C., Huang, X.-S., Liu, Q., Liu, Y.-M., Cao, Y., He, H.-Y., Fan, K.-N., and Zhuang, J.-H. (2008). Gold nanoparticles deposited on manganese(III) oxide as novel efficient catalyst for low temperature CO oxidation, *J. Catal.,* **259**, pp. 66–74.
28. Wang, L.-C., He, L., Liu, Y.-M., Cao, Y., He, H.-Y., Fan, K.-N., and Zhuang, J.-H. (2009). Effect of pretreatment atmosphere on CO oxidation over $\alpha$-$Mn_2O_3$ supported gold catalysts, *J. Catal.,* **264**, pp. 145–153.
29. Santos, V.P., Carabineiro, S.A.C., Bakker, J.J.W., Soares, O.S.G.P., Chen, X., Pereira, X., M.F.R., Órfão, J.J.M., Figueiredo, J.L., Gascon, J., and Kapteijn, F. (2014). Stabilized gold on cerium-modified cryptomelane: highly active in low-temperature CO oxidation, *J. Catal.,* **309**, pp. 58–65.
30. Haruta, M., Yamada, N., Kobayashi, T., and Iijima, S. (1989). Gold catalysts prepared by coprecipitation for low-temperature oxidation of hydrogen and of carbon monoxide, *J. Catal.,* **115**, pp. 301–309.
31. Herzing, A.A., Kiely, C.J., Carley, A.F., Landon, P., and Hutchings, G.J. (2008). Identification of active gold nanoclusters on iron oxide supports for CO oxidation, *Science,* **321**, pp. 1331–1335.
32. Kudo, S., Maki, T., Yamada, M., and Mae, K. (2010). A new preparation method of Au/ferric oxide catalyst for low temperature CO oxidation, *Chem. Eng. Sci.,* **65,** pp. 214–219.
33. Qiao, B., Zhang, J., Liu, L., and Deng, Y. (2008). Low-temperature prepared highly effective ferric hydroxide supported gold catalysts for carbon monoxide selective oxidation in the presence of hydrogen, *Appl. Catal. A,* **340**, pp. 220–228.

34. Li, L., Wang. A., Qiao, B., Lin, J., Huang, Y., Wang, X., Zhang, T. (2013) Origin of the high activity of Au/FeO$_x$ for low-temperature CO oxidation: direct evidence for a redox mechanism, *J. Catal.*, **299**, pp. 90–100.
35. Ruihui, L., Cunman, Z., and Jianxin, M. (2010). High thermal stable gold catalyst supported on La$_2$O$_3$ doped Fe$_2$O$_3$ for low-temperature CO oxidation, *J. Rare Earths*, **28**, pp. 376–382.
36. Penkova, A., Chakarova, K., Laguna, O.H., Hadjiivanov, K., Saria, F.R., Centeno, M.A., and Odriozola, J.A. (2009). Redox chemistry of gold in a Au/FeO$_x$/CeO$_2$ CO oxidation catalyst, *Catal. Commun.*, **10**, pp. 1196–1202.
37. Jia, C.-J., Liu, Y., Schwickardi, M., Weidenthaler, C., Spliethoff, B., Schmidt, W., and Schüth, F. (2010). Small gold particles supported on MgFe$_2$O$_4$ nanocrystals as novel catalyst for CO oxidation, *Appl. Catal. A*, **386**, pp. 94–100.
38. Qi, C., Zhu, S., Su, H., Lin, H., and Guan, R. (2013). Stability improvement of Au/Fe–La–Al$_2$O$_3$ catalyst via incorporating, *Appl. Catal. B*, **138-139**, pp. 104–112.
39. Zhao, K., Qiao, B., Zhang, Y., and Wang, J. (2013). The roles of hydroxyapatite and FeO$_x$ in a Au/FeO$_x$-hydroxyapatite catalyst for CO oxidation, *Chinese J. Catal.*, **34**, pp. 1386–1394.
40. Zhao, K., Qiao, B., Wang, J., Zhang, Y., and Zhang, T. (2011). A highly active and sintering-resistant Au/FeO$_x$-hydroxyapatite catalyst for CO oxidation, *Chem. Commun.*, **47**, pp. 1779–1781.
41. Li, Q., Zhang, Y., Chen, G., Fan, J., Lan, H., and Yang, Y. (2010). Ultra-low-gold loading Au/CeO$_2$ catalysts for ambient temperature CO oxidation: Effect of preparation conditions on surface composition and activity, *J. Catal.*, **273**, pp. 167–176.
42. Sakurai, H., Koga, K., Iizuka, Y., and Kiuchi, M. (2013). Colorless alkaline solution of chloride-free gold acetate for impregnation: an innovative method for preparing highly active Au nanoparticles catalyst, *Appl. Catal. A*, **462-463**, pp. 236–246.
43. Moreau, F., and Bond, G.C. (2006). CO oxidation activity of gold catalysts supported on various oxides and their improvement by inclusion of an iron component, *Catal. Today*, **114**, pp. 362–368.
44. Moreau, F., and Bond, G.C. (2007). Influence of the surface area of the support on the activity of gold catalysts for CO oxidation, *Catal. Today*, **122**, pp. 215–221.
45. Widmann, D., Liu, Y., Schüth, F., and Behm, R.J. (2010). Support effects in the Au-catalyzed CO oxidation: correlation between activity, oxygen storage capacity, and support reducibility, *J. Catal.*, **276**, pp. 292–305.

# Chapter 11

# The Role of Gold Catalysts in C–H Bond Activation for the Selective Oxidation of Saturated Hydrocarbons

**Sarwat Iqbal, Gemma L. Brett, and Graham J. Hutchings**
*Cardiff Catalysis Institute, Cardiff University, Main Building, Park Place, Cardiff, CF10 3AT, Wales*
Hutch@cardiff.ac.uk

## 11.1 Introduction

Activation of C–H bonds during the combustion of saturated hydrocarbons is an extremely important step for the generation of power and heat. The hardest step is the activation of the first C–H bond. Once the first bond breaks, the consecutive reactions proceed more easily. Therefore, understanding the C–H bond activation in saturated hydrocarbons is important. It also improves the approach to assessing the effect of catalytic and reaction parameters on the rate and the efficiency of catalytic combustion. Many different types of catalysts are capable of oxidizing saturated hydrocarbons with varying efficiency. Among the saturated hydrocarbons methane is the most difficult compound to activate, higher hydrocarbons are

---

*Gold Catalysis: Preparation, Characterization, and Applications*
Edited by Laura Prati and Alberto Villa
Copyright © 2016 Pan Stanford Publishing Pte. Ltd.
ISBN 978-981-4669-28-3 (Hardcover), 978-981-4669-29-0 (eBook)
www.panstanford.com

easier to activate, and the hydrocarbons from $C_2$ to $C_4$ exhibit an intermediate difficulty for bond activation. The difference in activity of hydrocarbons can partly be related with the adsorption of hydrocarbons on oxide surfaces, which is considered as a prior condition to combustion. Adsorption of a larger molecule in a predissociative state is more efficient than for a smaller molecule. C–H bond activation requires harsher conditions, particularly higher temperature. The optimal temperature for selective oxidation of an alkane C–H bond until 2002 was shown by Thomas et al. [1–3] to be 100°C using transition metal ion–substituted molecular sieve catalysts, followed by use of gold catalysts by Xu et al. [4] in 2005 at 70°C.

## 11.2 Small Alkanes

### 11.2.1 Methane and Ethane

Numerous studies have been reported for the complete oxidation of alkanes, especially for methane [5–7], and these are based mainly on Pd catalysts [8, 9]. Until now, there is no general agreement on the possible mechanism(s) of C–H bond activation in saturated hydrocarbons. At higher temperatures, gas-phase contribution to combustion is reported to be an important factor, as proposed by Campbell et al. [10]. Other authors have proposed de-protonation as an important step in C–H bond activation. Choudhary et al. [11], for instance, have proposed both acidic and basic metal sites as an important factor affecting the efficiency of active sites. Palladium is considered to be an active catalyst for oxidation of methane and platinum is active for other hydrocarbons. Burch et al. [12, 13] have reported that the optimal activity of hydrocarbon occurs when the Pt surface is partially oxidized, indicating that the rate of combustion of hydrocarbon depends on the state of the catalyst surface. On the other hand, there is a proportional increase observed in combustion of methane with progressive and full oxidation of Pd to PdO. While both precious metals, Pd and Pt, are considered as active metals for the oxidation of hydrocarbons, gold was considered as an inactive/inert metal historically, in particular with respect to oxidation of alkanes. In early 1980s gold was found to be a reactive metal that can show excellent activity when supported as nanoparticles. Two independent studies were reported by Haruta et al. [14] and Hutchings [15] at about the same time, showing complete oxidation of CO and acetylene hydrochlorination, respectively, with gold catalysts. After that there was a dramatic increase in studies made with gold catalyst [16–27] (Table 11.1). In 2006, a comparison was made of $MnO_x$, $Au/MnO_x$, $CoO_x$, and $Au/CoO_x$ catalysts for methane and ethane oxidation [26]. The catalytic activity followed the trend $Au/CoO_x > CoO_x > Au/MnO_x > MnO_x$. Hashmi et al. [28] in 2005 claimed the importance of hybridization of the carbon atom in C–H activation using gold catalysts. Alkynes with $sp$ C–H bonds are easier for the insertion of a transition metal than $sp^2$ C–H bonds. The order of difficulty of substrates is reported as $sp^3$ C–H > $sp^2$ C–H > $sp$ C–H bonds. The conversion of saturated hydrocarbons, like methane and ethane, is a big challenge for both academic and industrial scientists.

There are various methods reported for direct conversion of methane (e.g., oxidative coupling or selective oxidation) and indirect conversion (e.g., synthesis gas formation) [29]. So far there is no commercial process for the direct conversion of methane to other chemicals. The current commercial technology for the synthesis of oxygenates (methanol) from methane involves an indirect process that first requires steam reforming of methane to form synthesis gas. The said technology is very successful but there are several disadvantages, particularly high cost and an intensive use of energy. The direct conversion of methane (oxidation) has been reported, with the major product being formaldehyde under very harsh conditions, by Hunter and Foster et al. [30, 31]. There is some literature available for liquid-phase oxidation of methane in a highly acidic environment [32–35]. These methods are not industrially attractive due to the difficulties in operation. Some homogeneous catalysts have been reported for methane oxidation in water (a clean solvent) [36–40]. Lin et al. [41] reported oxidation of methane and ethane at a temperature less than 100°C in aqueous phase with the $CO/O_2$ system using supported Pd and Pt catalysts. There is an in situ generation of $H_2O_2$ with hydrogen, which is used as an oxidant. A considerable amount of work has been reported

**Table 11.1** Oxidation reactions with Au-supported catalysts

| | | Oxidation reactions | | |
|---|---|---|---|---|
| Catalysts | Substrate | Reaction temperature (°C) | Conversion (%) | Ref. |
| Au/MnO$_x$ | Methane | 400 | 22 | [26] |
| Au/MnO$_x$ | Ethane | 400 | 100 | |
| Au/MnO$_x$ | Propane | 350 | 100 | |
| Au/CoO$_x$ | Methane | 350 | 95 | |
| Au/CoO$_x$ | Ethane | 250 | 100 | |
| Au/CoO$_x$ | Propane | 200 | 100 | |
| 1% Au/Pd/TiO$_2$ | Toluene | 80 | 4.3 | [42] |
| 1% Au/Pd/TiO$_2$ | Benzyl alcohol | 120 | 20 | [43] |
| 5% Au/Pd/TiO$_2$ | Methane | 90 | 0.7*, 75** | [49] |

* Turnover frequency, mole/metal/h
** Methanol selectivity

by Hutchings et al. on activation of the C–H bond in various substrates (toluene, alcohols, etc.) using gold as a catalyst [42–47] (Table 11.1). They have reported AuPd as a suitable catalyst for methane oxidation in H$_2$O$_2$, where H$_2$O$_2$ is an oxidant. Titania-supported AuPd nanoparticles are reported to be able to oxidize methane and the products are methyl hydroperoxide, methanol, and CO$_2$ using H$_2$O$_2$ in water [48]. There is an improvement in methanol production from methane oxidation with in situ generation of H$_2$O$_2$ using H$_2$ and O$_2$ added in the gas-phase reaction mixture. Further, they have reported a systematic study on the effect of reaction and catalyst variables for the oxidation of methane using supported AuPd nanoparticles. The catalysts were prepared by incipient wetness impregnation and the oxidant used was H$_2$O$_2$ [49].

The primary product was methyl hydroperoxide, which is subsequently converted to methanol with a very high selectivity, ca. 70%. The oxidation state of palladium metal is reported to be an important parameter that can influence the selectivity of methanol. The catalyst is reported to be reusable and more active in comparison to other metal catalysts (homogeneous and heterogeneous) reported in the literature. Lang et al. [50, 51] have reported the coupling of methane to ethylene with free, tiny Au$_2^+$ ions. Their studies are reported on theoretical simulations and kinetic measurements. Their observations suggest that gold catalysts are capable for methane utilization if careful optimization is performed.

In 2006 Solsona and coworkers reported catalytic oxidation of ethane for the first time with gold supported on various oxides [26]. Forde et al. [52] have reported iron and copper containing ZSM-5 catalysts as very effective for the partial oxidation of ethane with hydrogen peroxide. The products are a mixture of combined oxygenates with a selectivity up to 95.2%. High conversion of ethane to acetic acid is reported. They have shown the possibility of a complex reaction network in which the oxidationn of ethane to a range of $C_2$ oxygenates is performed. Ethene is also a by-product that is reported to be subsequently oxidized. There is some literature available on use of mixed-metal oxides [53], titanium silicate [54], Pt, and Pd supported on carbon [55] for ethane oxidation. Some biocatalytic and enzymatic systems [56, 57] are applied but they are far from implementation on a large scale.

### 11.2.2 Propane

Oxide-supported gold catalysts have been utilized for propane oxidation at temperatures above 250°C [58]. The products observed were carbon oxides and in the case of $Au/CeO_2$ catalyst, propene. A titania-supported gold catalyst was found to be the most active without pretreatment. Treating the catalysts with $H_2$ increased the activity of all catalysts, leading to higher conversions of propane. The partial oxidation of propane to propene oxide and acetone is also possible with a titanosilicate-supported gold catalyst [59]. At higher temperatures than those at which propene epoxidation takes place (443 K) with a gas mixture $C_3H_8/H_2/O_2/Ar = 2/1/1/6$, a $Au/TiO_2$ catalyst was shown to convert propane (1.4%) to propene with a selectivity of 69%, whereas a Au/titanosilicalite-1 (TS-1) catalyst under the same conditions produces acetone with a selectivity of 84% at 1.1% conversion. Using a sequential reactor, the conversion of propane is increased to 4.7% with a selectivity to propylene oxide of 89%. A possible reaction sequence has been proposed [60]. Firstly, $H_2O_2$ is formed on the gold nanoparticles from $H_2$ and $O_2$; this $H_2O_2$ then migrates to Ti sites to form Ti-hydroperoxo species, as suggested previously [61]. The propane reacts with these species to produce a 2-propoxy intermediate species, which then forms either acetone or 2-propanol. This is shown in Scheme 11.1.

$$H_2O_2 + Ti(OH) \longleftrightarrow Ti(OH)(H_2O_2)$$

$$Ti(OH)(H_2O_2) \longleftrightarrow Ti(OH_2)(OOH)$$

$$Ti(OH_2)(OOH) + CH_3CH_2CH_3 \longleftrightarrow Ti(OH_2)(OOH)(HCH(CH_3)_2)$$

$$Ti(OH_2)(OOH)(HCH(CH_3)_2) \longrightarrow Ti(OH_2)(OCH(CH_3)_2) + H_2O$$

$$Ti(OH_2)(OCH(CH_3)_2) \longleftrightarrow Ti(OH) + CH_3CH(OH)CH_3$$

$$Ti(OH_2)(OCH(CH_3)_2) + H_2O_2 \longleftrightarrow Ti(OH) + 2H_2O + CH_3COCH_3$$

**Scheme 11.1** The possible reaction sequence for the oxidation of propane.

The gold particle size has been shown to be a critical factor in the oxidation of propane over alumina-supported gold catalysts, with a particle size on 3 nm proven to be optimal [62]. Acidic additives to the catalysts, such as $VO_x$, were detrimental to propene conversion, whereas Lewis basic additives were shown to promote the $Au/Al_2O_3$ catalyst's performance. It has been proposed that the role of these promoters is to supply active oxygen as there is no clear trend between basicity and activity. Other supports tested included vanadia and molybdena, which show some activity without gold present [63]. The oxidation of propane over $V_2O_5/SiO_2$ and $Au$-$V_2O_5/SiO_2$ were compared at 420°C. The addition of gold to the support increases the conversion of propane from 7.0% to 9.6%. The same effect is seen when gold is added to $Au$-$MoO_3/SiO_2$. The addition of gold to the support also alters the selectivity from propene and CO to high concentrations of $CO_2$. It is proposed that the role of gold is to activate oxygen at high temperatures.

## 11.3 Propene

The epoxidation of propylene to propylene oxide is an important industrial process. Propylene oxide is commonly used in the synthesis of polyols and resins and in the production of polyurethane plastics. There are currently two industrial processes for the production of propene oxide from propene, the chlorohydrin process and the organic hydroperoxide process. The former has high capital costs and produces large amounts of chlorinated waste, which is environmentally detrimental. The second process involves the use of environmentally harmful organic hydroperoxides and peracids. This process produces equal amounts of propene oxide, alcohols, and acids and is therefore not ideal [64] as only around 25% of products are the desired epoxide. There is, therefore, a drive to develop new, industrially viable routes to this important chemical without the use of chlorine compounds and that yield high selectivities to propene oxide.

Hayashi et al. were the first to report the application of gold catalysts for the gas-phase epoxidation of propylene over gold catalysts in 1998 [61]. Gold supported on titania prepared by deposition-precipitation (DP) was found to achieve a selectivity to propylene oxide of >99% at a 1.1% conversion of propylene. The feed gas composition was $C_3H_6/O_2/H_2/Ar$ 10/10/10/70 with a flow rate of 2000 mL/h. The reaction temperature was 323 K with a catalyst mass of 0.5 g. Other titania-supported metals, Pd, Pt, Cu, and Ag, were also tested for the reaction. The $Cu/TiO_2$ and $Ag/TiO_2$ catalysts were relatively inactive, with conversions of propylene of <0.2%. The Pd and Pt catalysts showed higher conversions of propylene, 57% and 12%, respectively, with high selectivities >92% to the hydrogenated product, propane. No propylene oxide was detected with these catalysts; only very small amounts of oxygenated products were found.

It has been reported that three reaction pathways are possible when propene and an $O_2/H_2$ mixture are passed over a titania-supported gold catalyst: propene epoxidation, propene hydrogenation, and hydrogen combustion [65]. The optimization and control over the selectivity of this reaction lies in the size of the gold particles, support choice, and promoters.

### 11.3.1 *Effect of Support*

Gold was also supported on other metal oxide supports, such as $Al_2O_3$, $SiO_2$, $Fe_2O_3$, $Co_3O_4$, ZnO, and $ZrO_2$ but these catalysts were unsuccessful for the production of propylene oxide, with water and carbon dioxide being produced. It was therefore proposed that titanium cations at gold particle edges were intimately involved in the reaction as a site for the formation of peroxotype species. It was speculated that adsorption of propylene occurred on gold sites and that the gold present facilitated electron transfer between hydrogen and oxygen. The preparation method of the catalyst was found to be a critical factor, with the catalyst prepared by impregnation proving to be inactive.

Following this work, other support materials were developed, including Au/Ti-MCM-41. This catalyst was active for the oxidation of propene with a high selectivity to propene oxide [66]. Nijhuis et al. also endeavored to find a more active support material for the epoxidation [67]. Due to the possible peroxo nature of the mechanism, TS-1 was selected for testing. Again it was found that the presence of titanium metal was necessary for active and selective catalysts. The authors agree that oxygen is adsorbing on Ti sites and also suggested that hydrogen atoms adsorb onto the gold nanoparticles. In a hydrogen-free environment it was found that no reaction takes place, and with the apparent necessity of proximity of both oxygen and hydrogen to the propene, the authors proposed several mechanisms (Scheme 11.2).

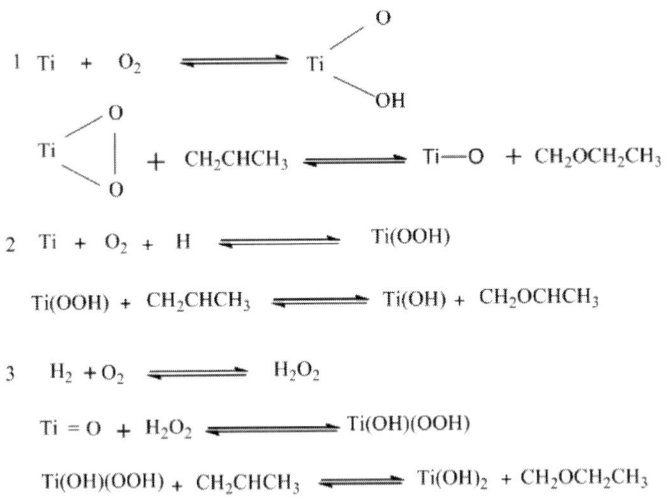

1  Ti + O$_2$ ⇌ Ti(O$_2$)OH

   Ti(O)(O) + CH$_2$CHCH$_3$ ⇌ Ti—O + CH$_2$OCH$_2$CH$_3$

2  Ti + O$_2$ + H ⇌ Ti(OOH)

   Ti(OOH) + CH$_2$CHCH$_3$ ⇌ Ti(OH) + CH$_2$OCHCH$_3$

3  H$_2$ + O$_2$ ⇌ H$_2$O$_2$

   Ti=O + H$_2$O$_2$ ⇌ Ti(OH)(OOH)

   Ti(OH)(OOH) + CH$_2$CHCH$_3$ ⇌ Ti(OH)$_2$ + CH$_2$OCH$_2$CH$_3$

**Scheme 11.2** Proposed mechanism for propylene oxidation using H$_2$O$_2$ as an oxidant.

The first mechanism involves the adsorption of molecular oxygen onto the catalyst, which then reacts with propene to produce propene oxide and water. The second mechanism proposed the presence of a hydroperoxide-like compound produced by the gold on the catalyst, which then reacts with propene. The hydroperoxide is formed by the reaction of an oxygen molecule adsorbed on a titanium site with a dissociated hydrogen atom on a gold nanoparticle. A third mechanism was proposed involving the synthesis of hydrogen peroxide from hydrogen and oxygen. The propene is then able to react with the hydrogen peroxide to form propene oxide. The adsorption of propene directly onto gold nanoparticles has been observed by X-ray absorption near-edge structure (XANES) [68], thereby supporting the reaction mechanisms proposed previously. These $\pi$-bonding species were shown to deactivate the catalyst, forming stronger bonds when both oxygen and hydrogen are present. It was further proposed that the activated propene species adsorbed on the gold particles then migrate to titania sites to form the bidentate propoxy species.

Gold supported on mixed-metal oxides containing titania has also shown good activity and selectivity for the epoxidation of propene in the vapor phase [69, 70]. Ti–silicate (Ti–$SiO_2$) and microporous TS-1 are shown to be selective to propene oxide at higher temperatures than $TiO_2$, thereby facilitating higher conversions of propene, whilst retaining high selectivities [71]. The degree of titanium present in TS-1 is an important parameter to control, as it has been shown that increasing the percentage loading leads to an increased conversion of propene. The addition of a barium promoter to this TS-1 catalyst yielded an increase in conversion of propene, from 6.6& to 9.8%, whilst maintaining a high selectivity to propene oxide (>90%). It has been observed that silylating the catalyst leads to a surface enrichment of metallic gold species [72]. Another effect of this treatment is the increase of tetrahedral titania structures.

Basic metal ions, such as potassium and sodium, have also been shown to promote the formation of propene oxide [65]. Huang et al. performed basic treatments in succession to achieve promotion effects [73]. The first step was the introduction of surface defects to the support by KOH, followed by a $HNO_3$ wash to remove the residual potassium. The gold was then deposited by a solid grinding technique to produce 1.6 nm particles. This catalyst was then treated with potassium salts, alkali hydroxides or alkaline earth-metal acetates. It was observed that catalysts impregnated with neutral potassium salt treatments such as KCl, $KNO_3$, and $K_2SO_4$ did not produce propene oxide. Basic salts of potassium produced propene oxide with selectivities of 12%–45%. Hydroxides of Na, K, Rb and Cs were also shown to promote the epoxidation reaction. It was proposed that these basic salts aided the activation and stabilization $O_2$ on the gold particles.

### 11.3.2 Gold Particle Size and Shape Effects

The role of gold particle size has also been investigated for the epoxidation of propene in the presence of oxygen and hydrogen [61]. It was found that there was an intrinsic relationship between particle size and selectivity. Nanoparticles below 2 nm preferentially hydrogenated propylene, in contrast to particles larger than 2 nm which carried out the partial oxidation reaction. It was proposed that the larger particles were metallic in nature, whereas the smaller species were oxidic in nature, and so behaved similarly to the $Pt/TiO_2$ catalyst observed previously. Gold nanoparticles above 5 nm have also been shown to favor the hydrogenation reaction; therefore a gold particle size of 2–5 nm is optimal to produce propene oxide under these conditions [65].

As shown by Hayashi et al. in their seminal work, gold particle size is crucial for high selectivities to propene oxide [61]. Therefore, the catalyst preparation method is of great importance. It was found that Au nanoparticles synthesized by the impregnation were large and spherical, producing high selectivities of $CO_2$ and water [61]. In contrast the small, hemispherical nanoparticles formed by the DP method were >99% selective to propene oxide. Recently, a solid grinding technique has been used to prepare Au/TS-1; this catalyst was shown to be active (8.8%) and yielded an 82% selectivity to propylene oxide [74]. Another reparation method employed for this reaction is the sol immobilization method. A 0.5% Au/TS-1 catalyst prepared by this method produced a conversion of 10% with a selectivity of 72% to propene oxide [75].

### 11.3.3 Promoters

In 2005, Chowdhury et al. reported the use of gold catalysts for the gas-phase epoxidation of propylene to propylene oxide under strong Lewis basic conditions [76]. Trimethylamine (TMA) was added in low concentrations to the gas reactants; this resulted in improved selectivity to propene oxide, catalyst lifetime, and $H_2$ efficiency. It was proposed that TMA partially adsorbed onto the gold sites, inhibiting the formation of $H_2O$ and oligomerization of propylene oxide. It was later observed that the regeneration of a Au/TS-1 catalyst was possible by treatment with 10–20 ppm of TMA for 2 h to produce an increase in propene conversion and propene oxide selectivity.

The introduction of CO into the gas feed controls the reaction selectivity toward propene oxide [77]. The hydrogenation reaction was performed with low-loaded (<1%) Au/Ti-SiO$_2$ in the presence of both $H_2$ and $O_2$. It was found that the introduction of CO decreased the selectivity to propane by a factor of 3. One possible reason for this selectivity change is the option that the CO binds to the gold sites usually occupied by $H_2$; therefore, less hydrogenation can take place.

### 11.3.4 Oxidation of Propene with Oxygen

It is possible to oxidize propene in hydrogen-free conditions with gold catalysts [58]. Again, it was shown that the support is critical in controlling product selectivity. At temperatures over 150°C, gold supported on main-group metal oxides yielded oxygenated products such as ethanol and propanal. In contrast, total oxidation occurs over transition metal oxide–supported catalysts. Propene oxide was observed in low selectivities (<1%) at temperatures lower than 150°C in the presence of titania-supported catalysts.

There are very few reports on the selective epoxidation of propene using oxygen alone. It has been demonstrated that the gold particle size is critical for the formation of propylene oxide from propene when oxidized in oxygen alone [78]. An alkaline-treated Au/TS-1 catalyst with a gold particle size of 1.8 nm was found to produce propene oxide (12% selectivity) at a propene conversion of 0.33%. Gold supported on other supports, including alumina, titania, and carbon, did not epoxidize propene. The products detected included high selectivities to $CO_2$ and acrolein. A study was undertaken to identify the active gold species in the reaction. Gold particles of different sizes supported on the alkaline treated titanosilicalite were tested for propene epoxidation. A gold particle size of <2 nm was discovered to be necessary to epoxidize propene. These clusters sinter to form larger nanoparticles under heat treatment, which then reduces the selectivity to propene oxide during the reaction. Further work has focused on the implementation of gold clusters for propene epoxidation.

Well-defined Au clusters supported on $Al_2O_3$ and $TiO_2$ using $O_2$ as an oxidant have been shown to produce propylene oxide under $H_2$-free conditions [79]. Lee et al. investigated the formation of propylene oxide from propylene over $Au_6$–$Au_{10}$ clusters. These clusters were shown to have high initial activities with a turnover frequency (TOF) of 0.4 molecules per site per second at 200°C; however, this activity is seen to diminish over time. It has been suggested that the nature of this deactivation stems from the loss of –OH groups on the catalyst surface. It was therefore concluded that the presence of hydroxyl groups at the edges of the gold clusters is crucial for high conversions of propene. As the amount of hydroxyl groups decreases and an increase in surface oxygen occurs there is a shift in selectivity from propene oxide toward acrolein. The selectivity toward propene oxide in the presence of oxygen alone is around 33%, which is particularly high compared to the previous literature. As has been discussed previously, the addition of hydrogen to the gas-phase reactants yields a large improvement in the selectivity to propene oxide. Water has also been shown to have a beneficial effect on the epoxidation of propene. Lee et al. observed an increase in the selectivity to propene oxide from 33% to 50% when $H_2O$ gas was introduced to the system. This selectivity was further increased by increasing the reactor temperature.

A positive effect was observed when water was substituted for $H_2$ in the reactant gas feed. The selectivity to propene oxide increased to 93%; this coincided with a slight decrease in conversion of propene. Density functional theory (DFT) calculations were carried out to probe the epoxidation reaction mechanism on these alumina- and titania-supported catalysts. It was indicated that $\pi$-bonding between the gas-phase propene occurred in a facile adsorption onto gold, resulting in a strongly bound species (0.81 and 1.09 eV). Another option for the adsorption of propene onto the gold is through the formation of C–C covalent bonds to produce a metallacyclic species. This adsorption was calculated to be favorable for the Au/Al$_2$O$_3$ catalyst; in contrast, the formation of this type of species is not favored over the Au/TiO$_2$ catalyst. A third adsorption possibility is the formation of a bidentate propoxy species through the formation of C–O bonds. Again, this species is less likely to be formed on titania- than alumina-supported catalysts due to an endothermic binding process on the former. Desorption of products from the catalysts was found to have the highest barrier to reaction at 1.6–1.7 eV. In summary, it was found that the reaction proceeds by different reaction pathways. Previously, it was thought that the presence of peroxo radicals was necessary in order the reaction to proceed via the metallacycle species; however, this is only true of the titania support. The alumina support is able to form metallacycle intermediates directly through adsorption. Therefore, the authors conclude that for Au/TiO$_2$ catalysts, the presence and dissociation of H$_2$ over the gold is required for the reaction to proceed. In the case of the Al$_2$O$_3$-supported catalyst, the presence of adsorbed hydroxyls make the reaction possible. The role of water when using a base-treated Au/TS-1 catalyst was shown experimentally to form Ti-OOH species [78]. The presence of these Ti-OOH species was monitored by in situ ultraviolet-visible (UV-Vis) spectroscopy. When the reaction proceeds without water the Ti-OOH species are not observed and the selectivity to propene oxide is considerably lower than when water is added to the reaction and the Ti-OOH species are present.

## 11.4 Effect of the Preparation Method

Various methods have been adopted for preparation of Au-supported catalysts (Table 11.2); the most commonly reported ones are impregnation and co-precipitation. A comparison of impregnation with co-precipitation using gold chloride (*metal*

Table 11.2 Effect of the preparation method on the size of gold particles

| Catalysts | Method | Au size (nm) | Substrate | Conversion (%) | Selectivity (%) | Ref. |
|---|---|---|---|---|---|---|
| Au/Fe$_2$O$_3$ | I | 0.2–0.3 | CO | 100 | | [88] |
| AuLa$_2$(OH)$_9$ | CP | 1.5–2.4 | CO | 65 | | [89] |
| Au/Fe$_2$O$_3$ | DP | >10 | CO | 10 | | [90] |
| Au/TiO$_2$ | DP | 0.1–10 | CO | 75-90 | | [91] |
| Au/Ti-silicate | ILI | >20 | Propylene | 14.6 | 70.9[a] | [86] |
| Au/Polymer | CVD | 2–3 | Benzyl alcohol | 51 | 37[b] | [87] |
| Au/ion exchange resin | DR | 5–8 | Glycerol | 90 | 60-64[c] | [92] |
| Au/C | SI | 8 | Ethane-1,2-diol | 100 | 87[d] | [24] |

I = impregnation; CP = co-precipitation; DP = deposition-precipitation; ILI = ionic liquid-enhanced immobilization; CVD = chemical vapor deposition; DR = deposition-reduction; SI = sol immobilization
[a] Propylene oxide
[b] PhCHO
[c] Glyceric acid
[d] sodium glycolate

*precursor*) shows co-precipitation as a better method of preparation [26] for oxidation of propene, ethane, and methane. The size of gold particles on the oxide support surface like CoO$_x$, MnO$_x$, CuO$_x$, and TiO$_2$ is an important parameter for activity of the catalyst. The larger the particles of gold, the lesser the activity [26], the reason being their inherently low activity and also blockage of pores on the metal support. This could be a reason of less activity with the impregnation method, which results in larger metal particles after calcination. In addition the presence of chloride is reported to be a poison for the catalyst, which can be related with the lower activity of impregnated gold catalysts [80–82]. The characterization data for gold-Cl-free and gold-Cl-containing catalysts has been reported by Solsona et al. [26] using different methods. The gold particles in impregnated catalysts are significantly larger in size and are reported to be less active for C–H activation in methane, ethane, and propene. Similar behavior has been reported by Haruta et al. [82] for the Au/TiO$_2$ catalyst. Only larger spherical particles were observed on the TiO$_2$ surface, when the impregnation method was used, resulting in combustion of propene. However, hemisphere nanoparticles were formed with the DP method, which showed higher activity for propene oxidation. The more extensive perimeter interface showed by hemispherical Au clusters (compared to spherical Au clusters) provides more active sites for $O_2$ adsorption [83].

As stated previously, the type of preparation method largely determines the catalytic activity (Table 11.2). The size of gold particles/clusters is related with $O_2$ activation. Gold clusters smaller than 2.0 nm in diameter are more active compared to larger-sized particles [79]. Corma et al. [84] have characterized gold-supported catalysts using X-ray photoelectron spectroscopy (XPS), $^{16}O_2$–$^{18}O_2$ exchange reactions. They have reported that gold clusters directly disassociate $O_2$. Recently colloid immobilization techniques have been developed for preparation of gold catalysts supported on TS-1. The mean diameter of the gold colloid was reported to be 3.6 nm and showed a very high activity for oxidation of propene at low temperature.

Solid grinding is another preparation method for getting very small gold particles (1.5–4 nm) on oxide supports [85–87]. Acetone and ethanol are commonly used solvents for mixing the gold complex in a mortar or a ball mill. During grinding the gold complex vaporizes or may absorb on the surface of the support. Gold complexes interact with the surface hydroxyl groups on the oxide surface, particularly in the case of metal-oxide supports. The solid grinding method is applicable to almost all kinds of supports, like metal oxides, carbon, and organic polymers.

The DP method is very useful when basic metal supports are the choice for gold deposition. Sodium hydroxide is a commonly used reagent in depositing gold on the support surface and can control pH over a wide range. Urea has been reported to be a good precipitating agent [93, 94] to deposit gold nanoparticles on the $TiO_2$ surface. All of the starting concentration of gold is reported to be deposited completely on the support surface when urea is used as a precipitating agent and thus is good when high loading of gold is required. These studies only report the characterization of prepared materials. The size of deposited gold metal can be varied by varying the calcination conditions (heating time, temperature, etc). The reaction between the gold precursor and urea results in the formation of a complex amorphous compound consisting of nitrogen, oxygen, and carbon when precipitated on the support, which affects the dispersion and size of gold particles. A very popular method for deposition of gold nanoparticles on carbon and polymers is deposition-reduction (DR). In this process the direct reduction of gold complex ions takes place near the support surface [92, 94]. The Au-loaded catalysts are tested for glucose and polyols oxidation. 0.1 wt% Au nanopartciles deposited on cation exchange resins (weekly acidic) and anion exchange (weakly basic) resins were found to be poorly active for glucose oxidation (activity <5%). Au deposited on strongly basic anion resins was reported to be more active (40% conversion) [96]. Gold nanoparticles supported on weak basic anion resins were tested for the oxidation of glycerol under pure oxygen, and 25% conversion was reported after 20 h of reaction time [92]. Variation of NaOH concentration in preparation showed a variation in the selectivity pattern.

### 11.4.1 Cyclohexane

The conversion of cyclohexane to cyclohexanone and cyclohexanol under mild reaction conditions is a great challenge in modern chemistry (Scheme 11.3).

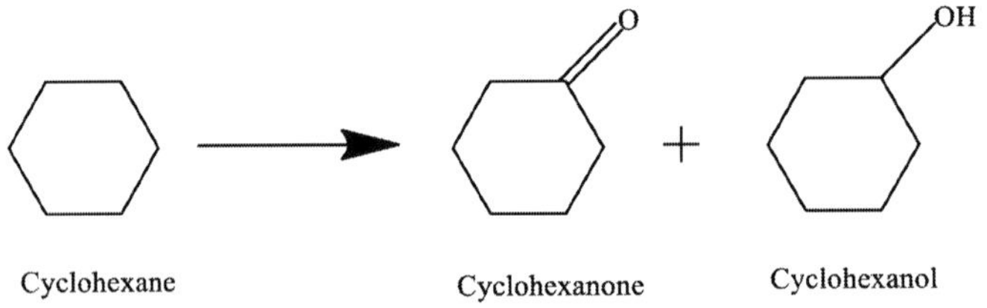

**Scheme 11.3** Cyclohexane oxidation.

Gold has been shown to catalyze the oxidation of cyclohexane. Zhao et al. [96] first demonstrated that ZSM-5-supported gold catalysts promoted C–H activation for the oxidation of cyclohexane. The reaction was carried out in solvent-free conditions at 150°C in oxygen (1 MPa). A conversion of 16% was achieved using a catalyst with 1.3% Au content on ZSM-5 with a 67% selectivity to cyclohexanone, which is an important starting material in the pharmaceutical and insecticide industries. The effects of metal [97], solvent [4], and reaction additive have been investigated. Suo et al. [96, 98] in 2004 reported the use of nanostructured gold catalysts for cyclohexane oxidation. Au/ZSM-5 and Au/MCM-41 catalysts showed a 90% selectivity to K/A-oil (a mixture of cyclohexanone and cyclohexanol) at a conversion of 10%–15%. Later various publications from different research groups reported similar observations, for example, Au/Al$_2$O$_3$ [99], Au/TiO$_2$ [99], mesoporous silica [100–102], and Au/C [4, 103]. The reaction involves a free-radical autoxidation mechanism [104, 105]. The industrial process for this reaction is limited to a 4%–12% conversion in order to have a high selectivity of the alcohol and ketone content and lower by-product formation [106, 107]. A thiol-containing organosilane, mercapto-propyltrimethoxysilane, has been used to modify the surface of the SBA-15 support for deposition of gold nanoparticles [108]. The resulting gold catalysts have been reported to be very active and stable for cyclohexane oxidation. Xu et al. [109] have reported enhanced stability of gold nanoparticles on an alumina support, achieved by silica doping via the sol–gel postmodification process, for cyclohexane oxidation in the absence of any solvent and initiator. Improved catalytic activity is attributed to decreased acidity and hydrophilicity by surface substitution of alumina.

There is a continuous discussion of the role of the Au catalyst to be that of either a true catalyst or only a promoter for cyclohexane oxidation, as observed by Della Pina et al. [110], who have debated the catalytic activity effect of gold-based materials versus gold. Xu et al. [99, 109] have reported silica-supported gold catalysts doped with TiO$_2$ to be capable of high conversion (ca. 10%) in comparison with the industrial catalyst and reported selectivity to K/A-oil of >70%. This activity was not observed in the absence of supported gold catalysts: their conclusion was that gold nanoparticles are a real catalyst for the reaction. However, there is a detailed study on the investigation of cyclohexane activity over Au/Al$_2$O$_3$, Au/TiO$_2$, and Au/SBA-15, which suggests that the reaction proceeds by a pure radical pathway [110, 111]. There is a decrease observed in the combined selectivity of cyclohexanone and cyclohexanol

(~70%) with an increase in conversion above 5%. An increase in the selectivity of adipic acid and $CO_2$ has also been reported, although the observed product evolution and by-product evolution were similar to the autoxidation process. In addition, they have also reported a complete inhibition of reaction in the presence of a radical scavenger. Through this study they have drawn a conclusion that the oxidation follows a radical-chain mechanism instead of catalytic mechanism. Later, Xiu et al. [112] showed that gold nanoclusters supported on hydroxyapatite are very active for oxidation of cyclohexane, and there is no reaction taking place in the absence of gold, although the presence of a radical initiator was necessary. Xie et al. [113] have reported an activity of 5% and a selectivity to cyclohexanol and cyclohexanone as 88% with Mn immobilized on the $Au/SiO_2$ catalyst.

Hereijgers and coworkers [114] have reported the oxidation reaction of cyclohexyl hydroxide combined with the epoxidation of cyclohexene over mesoporous titanium silicates. They have observed a very high selectivity of cyclohexanol, cyclohexanone, and epoxy-cyclohexane. The variation in surface hydrofobicity of the catalysts was found to have a profound influence on the selectivity of alicyclic oxygenates from cyclohexane oxidation.

Conte et al. [115] have shown a situation that is considered as an intermediate within the assumption of gold as a true catalyst or as a promoter in cyclohexane oxidation. They have reported gold to be capable of accelerating the reaction, without any initiator, and so it is a catalyst by definition. They have used X-band electron paramagnetic resonance (EPR) spectroscopy combined with the spin trapping technique in order to study the radical mechanism.

## 11.5 Summary

The activation of the C–H bond in small alkanes is a major challenge both for academic and for industrial researchers, taking into account the complexity of the reaction and less understanding of the mechanism involved. More effort is required, including in situ studies, to get a better understanding in order to develop an industrially viable process.

# References

1. Dugal, M., Sankar, G., Raja, R., and Thomas, J.M. (2000). Designing a heterogeneous catalyst for the production of adipic acid by aerial oxidation of cyclohexane, *Angew. Chem., Int. Ed.*, **39**, pp. 2310–2313.
2. Thomas, J.M., Raja, R., Sankar, G., and Bell, R.G. (1999). Molecular-sieve catalysts for the selective oxidation of linear alkanes by molecular oxygen, *Nature*, **398**, pp. 227–230.
3. Thomas, J.M., Raja, R., Sankar, G., and Bell, R.G. (2001). Molecular sieve catalysts for the regioselective and shape- selective oxyfunctionalization of alkanes in air, *Acc. Chem. Res.*, **34**, pp. 191–200.
4. Xu, Y.-J., Landon, P., Enache, D., Carley, A.F., Roberts, M.W., and Hutchings, G.J. (2005). Selective conversion of cyclohexane to cyclohexanol and cyclohexanone using a gold catalyst under mild conditions, *Catal. Lett.*, **101**, pp. 175–179.
5. Prasad, R., Kennedy, L.A., and Ruckenstein, E. (1984). Catalytic combustion, *Catal. Rev.: Sci. Eng.*, **26**, pp. 1–57.
6. Zwinkels, M.F.M., Jaeraas, S.G., Menon, P.G., and Griffin, T.A. (1993). Catalytic materials for high-temperature combustion, *Catal. Rev.: Sci. Eng.*, **35**, pp. 319–358.
7. Zwinkels, M.F.M., Jaras, S.G., and Menon, P.G. (1998). Catalytic fuel combustion in honeycomb monolith reactors, *Chem. Ind.*, **71**, pp. 149–177.
8. Baldwin, T.R., and Burch, R. (1990). Catalytic combustion of methane over supported palladium catalysts. I. Alumina supported catalysts, *Appl. Catal.*, **66**, pp. 337–358.
9. Hicks, R.F., Young, M.L., Lee, R.G., and Qi, H. (1990). Effect of catalyst structure on methane oxidation over palladium and alumina, *J. Catal.*, **122**, pp. 295–306.
10. Campbell, K.D., Morales, E., and Lunsford, J.H. (1987). Gas-phase coupling of methyl radicals during the catalytic partial oxidation of methane, *J. Am. Chem. Soc.*, **109**, pp. 7900–7901.
11. Choudhary, V.R., and Rane, V.H. (1991). Acidity/basicity of rare-earth oxides and their catalytic activity in oxidative coupling of methane to C2-hydrocarbons, *J. Catal.*, **130**, pp. 411–422.
12. Burch, R., Crittle, D.J., and Hayes, M.J. (1999). C-H bond activation in hydrocarbon oxidation on heterogeneous catalysts, *Catal. Today,* **47**, pp. 229–234.

13. Burch, R., Loader, P.K., and Urbano, F.J. (1996). Some aspects of hydrocarbon activation on platinum group metal combustion catalysts, *Catal. Today*, **27**, pp. 243–248.
14. Haruta, M., Yamada, N., Kobayashi, T., and Iijima, S. (1989). Gold catalysts prepared by coprecipitation for low-temperature oxidation of hydrogen and of carbon monoxide, *J. Catal.*, **115**, pp. 301–319.
15. Hutchings, G.J. (1985). Vapor phase hydrochlorination of acetylene: correlation of catalytic activity of supported metal chloride catalysts, *J. Catal.*, **96**, pp. 292–295.
16. Bianchi, C., Porta, F., Prati, L., and Rossi, M. (2000). Selective liquid phase oxidation using gold catalysts, *Top. Catal.*, **13**, pp. 231–236.
17. Bond, G.C., and Thompson, D.T. (1999). Catalysis by gold, *Catal. Rev.: Sci. Eng.*, **41**, pp. 319–388.
18. Carrettin, S., McMorn, P., Johnston, P., Griffin, K., and Hutchings, G.J. (2002) Selective oxidation of glycerol to glyceric acid using a gold catalyst in aqueous sodium hydroxide, *Chem. Commun.* (7), pp. 696–697.
19. Carrettin, S., McMorn, P., Johnston, P., Griffin, K., and Hutchings, G.J. (2003). Oxidation of glycerol using supported Pt, Pd and Au catalysts, *Phys. Chem. Chem. Phys.*, **5**, pp. 1329–1336.
20. Haruta, M. (2004). Gold as a novel catalyst in the 21st century: preparation, working mechanism and applications, *Gold Bull. (London)*, **37**, pp. 27–36.
21. Haruta, M., and Date, M. (2001). Advances in the catalysis of Au nanoparticles, *Appl. Catal. A.*, **222**, pp. 427–437.
22. Hutchings, G.J. (1996). Catalysis: a golden future, *Gold Bull. (London)*, **29**, pp. 123–130.
23. Porta, F., Prati, L., Rossi, M., Coluccia, S., and Martra, G. (2000). Metal sols as a useful tool for heterogeneous gold catalyst preparation: reinvestigation of a liquid phase oxidation, *Catal. Today*, **61**, pp. 165–172.
24. Prati, L., and Martra, G. (1999). New gold catalysts for liquid phase oxidation, *Gold Bull. (London)*, **32**, pp. 96–101.
25. Prati, L., and Rossi, M. (1998). Gold on carbon as a new catalyst for selective liquid phase oxidation of diols, *J. Catal.*, **176**, pp. 552–560.
26. Solsona, B.E., Garcia, T., Jones, C., Taylor, S.H., Carley, A.F., and Hutchings, G.J. (2006). Supported gold catalysts for the total oxidation of alkanes and carbon monoxide, *Appl. Catal. A*, **312**, pp. 67–76.

27. Thompson, D.T. (2003). Perspective on industrial and scientific aspects of gold catalysis, *Appl. Catal. A*, **243**, pp. 201–205.
28. Hashmi, A.S.K., Salathe, R., Frost, T.M., Schwarz, L., and Choi, J.-H. (2005). Homogeneous catalysis by gold: the current status of C,H activation, *Appl. Catal. A*, **291**, pp. 238–246.
29. Hutchings, G.J., Scurrell, M.S., and Woodhouse, J.R. (1989). Oxidative coupling of methane using oxide catalysts, *Chem. Soc. Rev.*, **18**, pp. 251–283.
30. Foster, N.R. (1985). Direct catalytic oxidation of methane to methanol: a review, *Appl. Catal.*, **19**, pp. 1–11.
31. Hunter, N.R., Gesser, H.D., Morton, L.A., Yarlagadda, P.S., and Fung, D.P.C. (1990). Methanol formation at high pressure by the catalyzed oxidation of natural gas and by the sensitized oxidation of methane, *Appl. Catal.*, **57**, pp. 45–54.
32. Grosse, A.V., and Snyder, J.C. (1950). *Methanol Production* (Houdry Process).
33. Periana, R.A., Taube, D.J., Gamble, S., Taube, H., Satoh, T., and Fujii, H. (1998). Platinum catalysts for the high-yield oxidation of methane to a methanol derivative, *Science*, **280**, pp. 560–564.
34. Sen, A. (1998). Catalytic functionalization of carbon-hydrogen and carbon-carbon bonds in protic media, *Acc. Chem. Res.*, **31**, pp. 550–557.
35. Stahl, S., Labinger, J.A., and Bercaw, J.E. (1998). Homogeneous oxidation of alkanes by electrophilic late transition metals, *Angew. Chem., Int. Ed.*, **37**, pp. 2181–2192.
36. Mizuno, N., Seki, Y., Nishiyama, Y., Kiyoto, I., and Misono, M. (1999). Aqueous phase oxidation of methane with hydrogen peroxide catalyzed by di-iron-substituted silicotungstate, *J. Catal.*, **184**, pp. 550–552.
37. Raja, R., and Ratnasamy, P. (1997). Oxidation of cyclohexane over copper phthalocyanines encapsulated in zeolites, *Catal. Lett.*, **48**, pp. 1–10.
38. Raja, R., and Ratnasamy, P. (1997). Direct conversion of methane to methanol, *Appl. Catal. A*, **158**, pp. L7–L15.
39. Suss-Fink, G., Nizova, G.V., Stanislas, S., and Shul'pin, G.B. (1998). Oxidations by the reagent "O2-H2O2-vanadate anion: pyrazine-2-carboxylic acid." Part 10. Oxygenation of methane in acetonitrile and water, *J. Mol. Catal. A: Chem.*, **130**, pp. 163–170.

40. Yuan, Q., Deng, W., Zhang, Q., and Wang, Y. (2007). Osmium-catalyzed selective oxidations of methane and ethane with hydrogen peroxide in aqueous medium, *Adv. Synth. Catal.*, **349**, pp. 1199–1209.
41. Lin, M., and Sen, A. (1992). A highly catalytic system for the direct oxidation of lower alkanes by dioxygen in aqueous medium. A formal heterogeneous analog of alkane monooxygenases, *J. Am. Chem. Soc.*, **114**, pp. 7307–8730.
42. bin Saiman, M.I., Brett, G.L., Tiruvalam, R., Forde, M.M., Sharples, K., Thetford, A., Jenkins, R.L., Dimitratos, N., Lopez-Sanchez, J.A., Murphy, D.M., Bethell, D., Willock, D.J., Taylor, S.H., Knight, D.W., Kiely, C.J., and Hutchings, G.J. (2012). Involvement of surface-bound radicals in the oxidation of toluene using supported Au-Pd nanoparticles, *Angew. Chem., Int. Ed.*, **51**, pp. 5981–5985.
43. Dimitratos, N., Lopez-Sanchez, J.A., Morgan, D., Carley, A.F., Tiruvalam, R., Kiely, C.J., Bethell, D., and Hutchings, G.J. (2009). Solvent-free oxidation of benzyl alcohol using Au-Pd catalysts prepared by sol immobilization, *Phys. Chem. Chem. Phys.*, **11**, pp. 5142–5153.
44. Edwards, J.K., Solsona, B., Ntainjua, N.E., Carley, A.F., Herzing, A.A., Kiely, C.J., and Hutchings, G.J. (2009). Switching off hydrogen peroxide hydrogenation in the direct synthesis process, *Science*, **323**, pp. 1037–1041.
45. Enache, D.I., Edwards, J.K., Landon, P., Solsona-Espriu, B., Carley, A.F., Herzing, A.A., Watanabe, M., Kiely, C.J., Knight, D.W., and Hutchings, G.J. (2006). Solvent-free oxidation of primary alcohols to aldehydes using Au-Pd/TiO2 catalysts, *Science*, **311**, pp. 362–365.
46. Kesavan, L., Tiruvalam, R., Ab, R.M.H., bin Saiman, M.I., Enache, D.I., Jenkins, R.L., Dimitratos, N., Lopez-Sanchez, J.A., Taylor, S.H., Knight, D.W., Kiely, C.J., and Hutchings, G.J. (2011). Solvent-free oxidation of primary carbon-hydrogen bonds in toluene using Au-Pd alloy nanoparticles, *Science*, **331**, pp. 195–199.
47. Pritchard, J.C., He, Q., Ntainjua, E.N., Piccinini, M., Edwards, J.K., Herzing, A.A., Carley, A.F., Moulijn, J.A., Kiely, C.J., and Hutchings, G.J. (2010). The effect of catalyst preparation method on the performance of supported Au-Pd catalysts for the direct synthesis of hydrogen peroxide, *Green Chem.*, **12**, pp. 915–921.
48. Pritchard, J., Kesavan, L., Piccinini, M., He, Q., Tiruvalam, R., Dimitratos, N., Lopez-Sanchez, J.A., Carley, A.F., Edwards, J.K., Kiely, C.J., and Hutchings, G.J. (2010). Direct synthesis of hydrogen peroxide and benzyl alcohol oxidation using Au-Pd catalysts prepared by sol immobilization, *Langmuir*, **26**, pp. 16568–16577.

49. Ab, R.M.H., Forde, M.M., Hammond, C., Jenkins, R.L., Dimitratos, N., Lopez-Sanchez, J.A., Carley, A.F., Taylor, S.H., Willock, D.J., and Hutchings, G.J. (2013). Systematic study of the oxidation of methane using supported gold palladium nanoparticles under mild aqueous conditions, *Top. Catal.*, **56**, pp. 1843–1857.
50. Lang, S.M., and Bernhardt, T.M. (2011). Methane activation and partial oxidation on free gold and palladium clusters: mechanistic insights into cooperative and highly selective cluster catalysis, *Faraday Discuss.*, **152**, pp. 337–351.
51. Lang, S.M., Bernhardt, T.M., Barnett, R.N., and Landman, U. (2010). Methane activation and catalytic ethylene formation on free Au2+, *Angew. Chem., Int. Ed.*, **49**, pp. 980–983.
52. Forde, M.M., Armstrong, R.D., Hammond, C., He, Q., Jenkins, R.L., Kondrat, S.A., Dimitratos, N., Lopez-Sanchez, J.A., Taylor, S.H., Willock, D., Kiely, C.J., and Hutchings, G.J. (2013). Partial oxidation of ethane to oxygenates using Fe- and Cu-containing ZSM-5, *J. Am. Chem. Soc.*, **135**, pp. 11087–11099.
53. Chen, N.F., Ueda, W., and Oshihara, K. (1999). Hydrothermal synthesis of Mo-V-M-O complex metal oxide catalysts active for partial oxidation of ethane, *Chem. Commun.*, pp. 517–518.
54. Shul'pin, G.B., Sooknoi, T., Romakh, V.B., Suess-Fink, G., and Shul'pina, L.S. (2006). Regioselective alkane oxygenation with H2O2 catalyzed by titanosilicalite TS-1, *Tetrahedron Lett.*, **47**, pp. 3071–3075.
55. Lin, M., and Sen, A. (1992). A highly catalytic system for the direct oxidation of lower alkanes by dioxygen in aqueous medium. A formal heterogeneous analog of alkane monooxygenases, *J. Am. Chem. Soc.*, **114**, pp. 7307–7308.
56. Colby, J., Stirling, D.I., and Dalton, H. (1977). The soluble methane monooxygenase of Methylococcus capsulatus (Bath). Its ability to oxygenate n-alkanes, n-alkenes, ethers, and alicyclic, aromatic and heterocyclic compounds, *Biochem. J.*, **165**, pp. 395–402.
57. Xu, F., Bell, S.G., Lednik, J., Insley, A., Rao, Z., and Wong, L.-L. (2005). The heme monooxygenase cytochrome P450cam can be engineered to oxidize ethane to ethanol, *Angew. Chem., Int. Ed.*, **44**, pp. 4029–4032.
58. Gaşior, M., Grzybowska, B., Samson, K., Ruszel, M., and Haber, J. (2004). Oxidation of CO and C3 hydrocarbons on gold dispersed on oxide supports, *Catal. Today*, **91–92**, pp. 131–135.
59. Bravo-Suárez, J.J., Bando, K.K., Lu, J., Fujitani, T., and Oyama, S.T. (2008). Oxidation of propane to propylene oxide on gold catalysts, *J. Catal.*, **255**, pp. 114–126.

60. Bravo-Suárez, J.J., Bando, K.K., Fujitani, T., and Oyama, S.T. (2008). Mechanistic study of propane selective oxidation with H2 and O2 on Au/TS-1, *J. Catal.*, **257**, pp. 32–42.
61. Hayashi, T., Tanaka, K., and Haruta, M. (1998). Selective vapor-phase epoxidation of propylene over Au/TiO2 catalysts in the presence of oxygen and hydrogen, *J. Catal.*, **178**, pp. 566–575.
62. Gluhoi, A.C., and Nieuwenhuys, B.E. (2007). Catalytic oxidation of saturated hydrocarbons on multicomponent Au/Al2O3 catalysts: effect of various promoters, *Catal. Today*, **119**, pp. 305–310.
63. Ruszel, M., Grzybowska, B., Gasior, M., Samson, K., Gressel, I., and Stoch, J. (2005). Effect of Au in V2O5/SiO2 and MoO3/SiO2 catalysts on physicochemical and catalytic properties in oxidation of C3 hydrocarbons and of CO, *Catal. Today*, **99**, pp. 151–159.
64. Monnier, J.R. (2001). The direct epoxidation of higher olefins using molecular oxygen, *Appl. Catal. A*, **221**, pp. 73–91.
65. Qi, C., Huang, J., Bao, S., Su, H., Akita, T., and Haruta, M. (2011). Switching of reactions between hydrogenation and epoxidation of propene over Au/Ti-based oxides in the presence of H2 and O2, *J. Catal.*, **281**, pp. 12–20.
66. Kalvachev, Y.A., Hayashi, T., Tsubota, S., and Haruta, M. (1999). Vapor-phase selective oxidation of aliphatic hydrocarbons over gold deposited on mesoporous titanium silicates in the co-presence of oxygen and hydrogen, *J. Catal.*, **186**, pp. 228–233.
67. Nijhuis, T.A., Huizinga, B.J., Makkee, M., and Moulijn, J.A.(1999). Direct epoxidation of propene using gold dispersed on TS-1 and other titanium-containing supports, *Ind. Eng. Chem. Res.*, **38**, pp. 884–891.
68. Nijhuis, T.A., Sacaliuc, E., Beale, A.M., van der Eerden, A.M.J., Schouten, J.C., and Weckhuysen, B.M. (2008). Spectroscopic evidence for the adsorption of propene on gold nanoparticles during the hydro-epoxidation of propene, *J. Catal.*, **25**, pp. 256–264.
69. Uphade, B.S., Yamada, Y., Akita, T., Nakamura, T., and Haruta, M. (2001). Synthesis and characterization of Ti-MCM-41 and vapor-phase epoxidation of propylene using H2 and O2 over Au/Ti-MCM-41, *Appl. Catal. A*, **215**, pp. 137–148.
70. Sinha, A.K., Seelan, S., Akita, T., Tsubota, S., and Haruta, M. (2003). Vapor phase propylene epoxidation over Au/Ti-MCM-41 catalysts prepared by different Ti incorporation modes, *Appl. Catal. A*, **240**, pp. 243–252.

71. Sinha, A.K., Seelan, S., Tsubota, S., and Haruta, M. (2004). A three-dimensional mesoporous titanosilicate support for gold nanoparticles: vapor-phase epoxidation of propene with high conversion, *Angew. Chem., Int. Ed.*, **43**, pp. 1546–1548.
72. Chowdhury, B., Bando, K.K., Bravo-Suárez, J.J., Tsubota, S., and Haruta, M. (2012). Activity of silylated titanosilicate supported gold nanoparticles towards direct propylene epoxidation reaction in the presence of trimethylamine, *J. Mol. Catal. A: Chem.*, **359**, pp. 21–27.
73. Huang, J., Takei, T., Ohashi, H., and Haruta, M. (2012). Propene epoxidation with oxygen over gold clusters: Role of basic salts and hydroxides of alkalis, *Appl. Catal. A: Gen.*, **435**, pp. 115–122.
74. Huang, J., Takei, T., Akita, T., Ohashi, H., and Haruta, M. (2010). Gold clusters supported on alkaline treated TS-1 for highly efficient propene epoxidation with $O_2$ and $H_2$, *Appl. Catal. B: Environ.*, **95**, pp. 430–438.
75. Du, M., Zhan, G., Yang, X., Wang, H., Lin, W., Zhou, Y., Zhu, J., Lin, L., Huang, J., Sun, D., Jia, L., and Li, Q. (2011). Ionic liquid-enhanced immobilization of biosynthesized Au nanoparticles on TS-1 toward efficient catalysts for propylene epoxidation, *J. Catal.*, **283**(2), pp. 192–201.
76. Chowdhury, B., Bravo-Suárez, J.J., Daté, M., Tsubota, S., and Haruta, M. (2006). Trimethylamine as a gas-phase promoter: highly efficient epoxidation of propylene over supported gold catalysts, *Angew. Chem., Int. Ed.*, **45**(3), pp. 412–415.
77. Chen, J., Halin, S.J.A., Perez Ferrandez, D.M., Schouten, J.C., and Nijhuis, T.A. (2012). Switching off propene hydrogenation in the direct epoxidation of propene over gold-titania catalysts, *J. Catal.*, **285**(1), pp. 324–327.
78. Huang, J., Akita, T., Faye, J., Fujitani, T., Takei, T., and Haruta, M. (2009). Propene epoxidation with dioxygen catalyzed by gold clusters, *Angew. Chem., Int. Ed.*, **48**(42), pp. 7862–7866.
79. Lee, S., Molina, L.M., Lopez, M., Alonso, J.A., Hammer, B., Lee, B., Seifert, S., Winans, R.E., Elam, J.W., Pellin, M.J., and Vajda S. (2009). Selective propene epoxidation on immobilized Au6–10 clusters: the effect of hydrogen and water on activity and selectivity, *Angew. Chem., Int. Ed.*, **48**(8), pp. 1467–1471.
80. Simone, D.O., Kennelly, T., Brungard, N.L., Farrauto, R.J. (1991). Reversible poisoning of palladium catalysts for methane oxidation, *Appl. Catal.*, **70**, pp. 87–100.

81. Oh, H.-S., Yang, J.H., Costello, C.K., Wang, Y.M., Bare, S.R., Kung, H.H., Kung, M.C., (2002) Selective catalytic oxidation of CO: effect of chloride on supported Au catalysts, *J. Catal.*, **210**(2), pp. 375–386.
82. Hayashi, T., Tanaka, K., and Haruta, M. (1998). Selective vapor-phase epoxidation of propylene over Au/TiO$_2$ catalysts in the presence of oxygen and hydrogen, *J. Catal.*, **178**(2), pp. 566–575.
83. Hayashi, T., Tanaka, K., and Haruta, M. (1997). Size- and support-dependency in the catalysis of gold, *Catal. Today*, **36**(1), pp. 153–166.
84. Alves, L., Ballesteros, B., Boronat, M., Cabrero-Antonino, J.R., Concepcion, P., Corma, A., Correa-Duarte, M.A., Mendoza, E. (2011). Synthesis and stabilization of subnanometric gold oxide nanoparticles on multiwalled carbon nanotubes and their catalytic activity, *J. Am. Chem. Soc.*, **133**(26), pp. 10251–10261.
85. Boronat, M., and Corma, A. (2011). Molecular approaches to catalysis, *J. Catal.*, **284**(2), pp. 138–147.
86. Du, M.-M., Zhan, G.-W., Yang, X., Wang, H.-X., Lin, W.-S., Zhou, Y., Zhu, J., Lin, L., Huang, J.-L., Sun, D.-H., Jia, L.-S., and Li, Q.-B. (2011). Ionic liquid-enhanced immobilization of biosynthesized Au nanoparticles on TS-1 toward efficient catalysts for propylene epoxidation, *J. Catal.*, **283**(2), pp. 192–201.
87. Ishida, T., Nagaoka, M., Akita, T., and Haruta, M. (2008). Deposition of gold clusters on porous coordination polymers by solid grinding and their catalytic activity in aerobic oxidation of alcohols, *Chem.: Eur. J.*, **14**(28), pp. 8456–8460.
88. Herzing, A.A., Kiely, C.J., Carley, A.F., Landon, P., and Hutchings, G.H. (2008). Identification of active gold nanoclusters on iron oxide supports for CO oxidation, *Science*, **321**(5894), pp. 1331–1335.
89. Takei T, Okuda, I., Bando, K., Akita, T., Haruta, M. (2010). Gold clusters supported on La(OH)$_3$ for CO oxidation at 193 K, *Chem. Phys. Lett.*, **493**(4–6), pp. 207–211.
90. Khoudiakov, M., Gupta, M.C., and Deevi, S. (2005). Au/Fe$_2$O$_3$ nanocatalysts for CO oxidation: a comparative study of deposition-precipitation and coprecipitation techniques, *Appl. Catal. A: Gen.*, **291**(1–2), pp. 151–161.
91. Tsubota, S., Cunningham, D.A.H., Bando, Y., Haruta, M. (1995). Preparation of nanometer gold strongly interacted with TiO$_2$ and the structure sensitivity in low-temperature oxidation of CO, *Stud. Surf. Sci. Catal.*, **91**(Preparation of Catalysts VI), pp. 227–235.

92. Villa, A., Chan-Thaw, C.E., and Prati, L. (2010). Au NPs on anionic-exchange resin as catalyst for polyols oxidation in batch and fixed bed reactor, *Appl. Catal. B: Environ.*, **96**(3–4), pp. 541–547.
93. Zanella, R., Delannoy, L., and Louis, C. (2005) Mechanism of deposition of gold precursors onto $TiO_2$ during the preparation by cation adsorption and deposition-precipitation with NaOH and urea, *Appl. Catal. A: Gen.*, **291**, pp. 62–72.
94. Zanella, R., Giorgio, S., Henry, C.R., and Louis, C. (2002). Alternative methods for the preparation of gold nanoparticles supported on $TiO_2$, *J. Phys. Chem. B*, **106**, pp. 7634–7642.
95. Ishida, T., Okamoto, S., Makiyama, R., and Haruta, M. (2009). Aerobic oxidation of glucose and 1-phenylethanol over gold nanoparticles directly deposited on ion-exchange resins, *Appl. Catal. A: Gen.*, **353**(2), pp. 243–248.
96. Zhao, R., Ji, D., Lu, G., Qian, G., Yan, L., Wang, X., and Suo, J. (2004). A highly efficient oxidation of cyclohexane over Au/ZSM-5 molecular sieve catalyst with oxygen as oxidant, *Chem. Commun.* (7), pp. 904–905.
97. Golunski, S., Rajaram, R., Hodge, N., Hutchings, G.J., and Kiely, C.J. (2002) Low-temperature redox activity in co-precipitated catalysts: a comparison between gold and platinum-group metals, *Catal. Today*, **72**(1–2), pp. 107–113.
98. Jin, G., Lu, G., Guo, Y., Guo, Y., Wang, J., Liu, X., Kong, W., and Liu, X. (2004). Modification of Ag-$MoO_3$/$ZrO_2$ catalyst with metallic chloride for propylene epoxidation by molecular oxygen, *Catal. Lett.*, **97**(3–4), pp. 191–196.
99. Xu, L.-X., He, C.-H., Zhu, M.-Q., Wu, K.-J., and Lai, Y.-L. (2007). Silica-supported gold catalyst modified by doping with titania for cyclohexane oxidation, *Catal. Lett.*, **118**(3–4), pp. 248–253.
100. Li, L., Jin, C., Wang, X., Ji, W., Pan, Y., van der Knaap, T., van der Stoel, R., and Au, C.T. (2009). Cyclohexane oxidation over size-uniform Au nanoparticles (SBA-15 hosted) in a continuously stirred tank reactor under mild conditions, *Catal. Lett.*, **129**(3–4), pp. 303–311.
101. Lue, G., Ji, D., Qian, G., Qi, Y., Wang, X., and Suo, J. (2005). Gold nanoparticles in mesoporous materials showing catalytic selective oxidation cyclohexane using oxygen, *Appl. Catal. A: Gen.*, **280**(2), pp. 175–180.
102. Zhu, K., Hu, J., and Richards, R. (2005). Aerobic oxidation of cyclohexane by gold nanoparticles immobilized upon mesoporous silica, *Catal. Lett.*, **100**(3–4), pp. 195–199.

103. Hutchings, G.J., Carrettin, S., Landon, P., Edwards, J.K., Enache, D., Knight, D.W., Xu, Y.-J., and Carley, A.F. (2006). New approaches to designing selective oxidation catalysts: Au/C a versatile catalyst, *Top. Catal.*, **38**(4), pp. 223–230.
104. Govindan, V., and Suresh, A.K. (2007). Modeling liquid-phase cyclohexane oxidation, *Ind. Eng. Chem. Res.*, **46**(21), pp. 6891–6898.
105. Masters, A.F., Beattie, J.K., and Roa, A.L. (2001). Synthesis of a CrCoAPO-5(AFI) molecular sieve and its activity in cyclohexane oxidation in the liquid phase, *Catal. Lett.*, **75**(3–4), pp. 159–162.
106. Chiusoli, G.P., and Maitlis, P. (2008). *Metal-Catalysis in Industrial Organic Processes* (RSC, Cambridge, U.K.).
107. Schuchardt, U., Cardoso, D., Sercheli, R., Pereira, R., da Cruz, R.S., Guerreiro, M.C., Mandelli, D., Spinace, E.V., and Pires, E.L. (2001) Cyclohexane oxidation continues to be a challenge, *Appl. Catal. A: Gen.*, **211**(1), pp. 1–17.
108. Wu, P., Bai, P., Lei, Z., Loh, K.P., and Zhao, X.S. (2011). Gold nanoparticles supported on functionalized mesoporous silica for selective oxidation of cyclohexane, *Microporous Mesoporous Mater.*, **141**(1–3), pp. 222–230.
109. Xu, L.-X., He, C.-H., Zhu, M.-Q., Wu, K.-J., and Lai, Y.-L. (2008). Surface stabilization of gold by sol-gel post-modification of alumina support with silica for cyclohexane oxidation, *Catal. Commun.*, **9**(5), p. 816–820.
110. Della Pina, C., Falletta, E., and Rossi, M. (2012). Update on selective oxidation using gold, *Chem. Soc. Rev.*, **41**(1), pp. 350–369.
111. Hereijgers, B.P.C., and Weckhuysen, B.M. (2010) Aerobic oxidation of cyclohexane by gold-based catalysts: new mechanistic insight by thorough product analysis, *J. Catal.*, **270**(1), pp. 16–25.
112. Liu, Y., Tsunoyama, H., Akita, T., Xie, S., and Tsukuda, T. (2011) Aerobic oxidation of cyclohexane catalyzed by size-controlled Au clusters on hydroxyapatite: size effect in the sub-2 nm regime, *ACS Catal.*, **1**(1), pp. 2–6.
113. Xie, J., Wang, Y., and Wei, Y. (2009) Immobilization of manganese tetraphenylporphyrin on Au/SiO$_2$ as new catalyst for cyclohexane oxidation with air, *Catal. Commun.*, **11**(2), pp. 110–113.

114. Hereijgers, B.P.C., Parton, R.F., and Weckhuysen, B.M. (2011). Cyclohexene epoxidation with cyclohexyl hydroperoxide: a catalytic route to largely increase oxygenate yield from cyclohexane oxidation, *ACS Catal.*, **1**, pp. 1183–1192.
115. Conte, M., Liu, X., Murphy, D.M., Whiston, K., and Hutchings, G.J. (2012). Cyclohexane oxidation using Au/MgO: an investigation of the reaction mechanism, *Phys. Chem. Chem. Phys.*, **14**(47), pp. 16279–16285.

# Chapter 12

# Liquid-Phase Oxidation Using Au-Based Catalysts

### Nikolaos Dimitratos,[a] Ceri Hammond,[a] and Peter P. Wells[b,c]

[a]*Cardiff Catalysis Institute, School of Chemistry, Cardiff University, Main Building, Park Place, Cardiff, CF10 3AT, UK*
[b]*Department of Chemistry, University College London, 20 Gordon Street, WC1H 0AJ, London, UK*
[c]*Research Complex at Harwell, Rutherford Appleton Laboratory, Harwell Oxford, Didcot, OX11 0FA, UK*
DimitratosN@cardiff.ac.uk

## 12.1 Introduction

Within the academic and industrial communities the selective oxidation of oxygen-containing organic compounds, such as alcohols, polyols, and aldehydes, primarily for the production of fine and specialty chemicals, remains a high priority. The main issue with the current industrial methods for these synthetic processes is the utilization of stoichiometric inorganic reagents. Due to the toxicity, waste, and corrosion problems associated with these inorganic reagents, these approaches are no longer acceptable from an environmental point of view. Preventing the use of

---

*Gold Catalysis: Preparation, Characterization, and Applications*
Edited by Laura Prati and Alberto Villa
Copyright © 2016 Pan Stanford Publishing Pte. Ltd.
ISBN 978-981-4669-28-3 (Hardcover), 978-981-4669-29-0 (eBook)
www.panstanford.com

toxic materials and substituting these routes with environmentally friendly processes, thus minimizing the formation of waste, remains one of the key objectives of modern-day research. The utilization of homogeneous and heterogeneous catalysts as alternatives to achieving these objectives is highly desirable. On an industrial level, it is preferable to develop heterogeneous catalysts that can be more easily recovered and recycled on a process level with a prolonged lifetime. Furthermore the use of green oxidants such as molecular oxygen or peroxides (organic and inorganic) is highly desirable since in this way the reduction of chemical waste associated with inorganic oxidants, such as $K_2Cr_2O_7$, can be achieved [1–4].

Several excellent reviews have recently been published on this topic, and have thoroughly addressed the catalytic application of homogeneous and heterogeneous catalysts in the area of liquid-phase oxidation [5–9]. The objective of this chapter is therefore to provide an updated overview of the utilization of metal nanoparticles, preferentially supported on different supports, as catalysts for the liquid-phase oxidation of oxygen-containing organic compounds.

## 12.2 Liquid-Phase Oxidation of Oxygen-Containing Organic Compounds

### 12.2.1 Supported Au Catalysts

Metal nanoparticles, and particularly Au-based unsupported and supported metal nanoparticles, have attracted much attention over the last two decades and have been widely explored as catalysts for the selective oxidation of oxygen-containing organic compounds, including alcohols, polyols, aldehydes, and sugars (Table 12.1) [10–14]. One of the first groups to demonstrate the catalytic applications of Au-based materials for liquid-phase oxidation was the group of Rossi and Prati. They demonstrated the efficient transformation of vicinal diols to $\alpha$-hydroxy carboxylates with molecular oxygen in alkaline solution using Au-based catalysts synthesized by colloidal methods [15]. High activity and high chemoselectivity of the Au catalysts toward the primary alcoholic function compared to more

**Table 12.1** Oxidation of oxygen-containing organic compounds using supported Au-based catalysts

| Substrate | Catalyst | T(°C) | Solvent | Conv. % | Sel.% (aldehyde/ketone) | Ref. |
|---|---|---|---|---|---|---|
| Propane-1,2-diol | Au/C | 90 | $H_2O$/NaOH | 78 | 100 (acid) | [15] |
| Ethane-1,2-diol | Au/C | 70 | $H_2O$/NaOH | 100 | 96 (acid) | [16] |
| D-glucose | Au/C | 50 | $H_2O$/NaOH | 99 | 99 (acid) | [19] |
| Glycerol | Au/G | 60 | $H_2O$/NaOH | 56 | 100 (acid) | [31] |
| Glycerol | Au/C | 60 | $H_2O$/NaOH | 90 | 83 (acid) | [33] |
| Glycerol | Au/$CeO_2$ | 60 | $H_2O$/NaOH | 33 | 44 (acid) | [35] |
| Glycerol | Au/$MgAl_2O_4$ | 50 | $H_2O$/NaOH | 50 | 61 (acid) | [40] |
| Glycerol | Au/N-CNF | 50 | $H_2O$/NaOH | 90 | 68 (acid) | [41] |
| Glycerol | Au/$NiO_{10}$-$TiO_{290}$ | 50 | $H_2O$/NaOH | 90 | 70 (acid) | [42] |
| Salicylic alcohol | Au/$Fe_2O_3$ | 60 | $H_2O$ | 90 | 90 | [45] |
| 3-Octanol | Au/$CeO_2$ | 80 | Solvent free | 97 | 99 | [46] |
| Ethanol | Au/$MgAl_2O_4$ | 150 | $H_2O$ | 97 | 86 | [48] |
| Benzyl alcohol | Au/GMS | 130 | $C_6H_5CH_3$ | 99 | 79 | [49] |
| 4-Methylbenzyl alcohol | Au/$Cu_5Mg_1Al_2O_x$ | 90 | $C_6H_3(CH_3)_3$ | 98 | 99 | [50] |
| 1-Phenyethanol | Au-Microgel | 60 | $H_2O$ | 75 | 100 | [52] |
| Benzyl alcohol | Au/$MnO_2$ nanorods | 120 | Solvent free | 41 | 99 | [53] |
| Benzyl alcohol | Au/$\gamma$-$Ga_2O_3$ | 130 | Solvent free | 40 | 98 | [54] |
| Benzyl alcohol | Au/$Ga_3Al_3O_9$ | 80 | $C_6H_5CH_3$ | 98 | 99 | [55] |
| Benzyl alcohol | Au/PoPD | 25 | $H_2O$ | 99 | 99(Acid) | [56] |
| 1-Phenylethanol | Au/Zn | 90 | Solvent free | 6.5 | 99 | [57] |
| Benzyl alcohol | Au/SBA-15 | 80 | $H_2O$ | 100 | 91(Acid) | [58] |
| 1-Phenylethanol | Au/PNIPAM | 80 | $H_2O$ | 99 | 99 | [59] |
| Cyclohexanol | Au/$Fe_3O_4$@$SiO_2$ | 100 | $C_6H_5CH_3$ | 42 | 100 | [60] |
| Benzyl alcohol | Au/$ZrO_2$ | 94 | Solvent free | 59.5 | 81.5 | [71] |
| 1-Phenylethanol | Au/$TiO_2$ | 90 | $H_2O$ | 99 | 100 | [75] |
| Benzyl alcohol | Au/SBA-15 | 60 | $H_2O$/$K_2CO_3$ | 99 | 87 (Acid) | [76] |

commonly use supported Pd and Pt catalysts were reported. Improvement in terms of stability was observed using Au on carbon (Au/C) catalysts, and minimum deactivation or metal leaching was observed through recycling experiments. The effect of the preparation method was subsequently investigated by Prati and Martra [16], who compared the methods of deposition-precipitation and sol immobilization for the liquid-phase oxidation of ethane-1,2-diol. The activity of Au/C catalysts prepared by the colloidal method showed an increase of catalytic activity by a factor of 2 compared to a similar catalyst prepared by the deposition-precipitation methodology. The improvement in catalytic activity was attributed to the smaller particle size and higher dispersion of Au particles on the support using the sol immobilization method. Rossi and coworkers [17] in subsequent studies demonstrated that for the interpretation of catalytic activity two parameters should be considered, (i) gold particle size and (ii) surface gold concentration. Moreover, catalytic activity was influenced considerably by the support (oxides versus carbon materials). In the former case, an increase of gold particle size led to a decrease of catalytic activity, whereas in the latter case a maximum catalytic activity was observed with gold particle mean diameters around 7–8 nm. The authors ascribed this difference to the possibility that small gold particles can lay deeper in the carbon than larger particles; therefore there is a limitation in the reagent accessibility of the deeper gold particles.

Rossi and coworkers have investigated several parameters, such as choice of stabilizer, concentration of metal, and choice of support during the synthesis of metal colloids, and have evaluated their impact on catalytic activity for the liquid-phase oxidation of polyols [18]. It was concluded that the above parameters could influence the final metal morphology and particle size. Moreover, the particular nature of the support drastically influenced the choice of the sol for maintaining, once supported, the gold particle size observed in the solution. It was concluded that the choice of stabilizer did not considerably affect final catalytic activity, since similar-sized particles showed the same activity although synthesized by different methodologies.

The catalytic and selective aerobic oxidation of D-glucose to D-gluconate was demonstrated by the same group to proceed with high selectivity (>98%) and activity with gold-based catalysts at mild reaction conditions. It was shown that Au/C could be an alternative catalyst to most of the multimetallic catalysts based on palladium and platinum metals reported in the current literature. In addition, gold-supported catalysts showed high resistance to chemical poisoning, and metal leaching was dependent on the reaction conditions [19]. In subsequent studies, Rossi and coworkers investigated the role of the support and showed that the support stabilized the gold colloids by forming a protective layer around the metal nanoparticle [20]. Furthermore, by optimizing reaction conditions and comparing with the enzymatic process reported in the literature, it was shown that turnover frequency (TOF) values of 150,000 h$^{-1}$ could be achieved at 50°C with the gold catalysts, whereas with the enzymatic catalysts TOF values of 550,000 h$^{-1}$ were reported [21]. The potential applicability of Au-based catalysts to achieve high plant productivity and future industrial utilization was therefore demonstrated. The influence of gold particle size on the activity of Au/C for the oxidation of glucose was also studied by Claus and coworkers. Au/C catalysts with a gold mean particle size in the range of 3–6 nm were prepared using a modified colloidal method. It was shown that by decreasing particle size, and therefore increasing the specific surface area of gold, the rate of glucose oxidation increased by a factor of 3. A Langmuir–Hinshelwood model was proposed whereby the overall reaction rate is limited by the surface oxidation reaction. The proposed mechanism was proposed to proceed through a dehydrogenation pathway; D-glucose first transforms in aqueous solution into the hydrate, with subsequent adsorption on the catalyst surface and dehydrogenation yielding the final product [22].

Prüße and coworkers used an alternative method for the synthesis of active Au catalysts for glucose oxidation, based on a deposition-precipitation methodology [23]. Using NaOH or urea as the precipitation agent and pH-controlling agent, and using alumina as the support, they reported highly active and selective catalysts with excellent long-term stability. A beneficial effect of using urea instead of NaOH was found to be the fact that no losses of gold were observed during the synthesis, that is, the entire Au fraction was successfully deposited onto the support material.

Haruta and coworkers [24] also reported the effect of support and gold particle size for glucose oxidation using a novel preparation method. This method is based on solid grinding by using a volatile organogold complex [Me$_2$Au(acac)] (acac = acetylacetonate). The advantage of this methodology is based on the fact that deposition of gold clusters smaller than 2 nm in diameter onto porous coordination polymers and on several kinds of metal oxides and carbon supports can be achieved. From their studies it was demonstrated that the most active catalyst was gold on $ZrO_2$ and that the control of particle size is critical for affecting activity as well as selectivity for liquid-phase oxidation reactions.

One of the first examples demonstrating the utilization of mesoporous carbon-supported Au catalysts for the aerobic oxidation of glucose under mild conditions (40°C, pH 9, bubbling with $O_2$) was reported by Hao and coworkers [25]. Mesoporous carbons can provide several advantages as a support material such as a uniform/controllable mesopore size and a high surface area, the confinement of metal nanoparticles inside the mesopore channels, and easier diffusion of reactant/product molecules. The authors synthesized mesoporous carbons controlling the mesopore size (2.3–7.6 nm) and surface chemical nature. Deposition of small gold nanoparticles (3–4 nm) was carried out by using a colloidal method. The most active supported Au catalyst was the one with a pore diameter of 5.4 nm, with supported Au catalysts with lower or higher mesopore size exhibiting lower catalytic activity. The authors speculated that there is an optimum mesopore size, where Au nanoparticles of 3.3 nm mean size can be well confined in the mesoporous channels of 5.4 nm diameter, therefore leading to enhanced catalytic activity for glucose oxidation. Their hypothesis for this enhancement was mainly due to two factors: The first factor is that reactant and substrate molecules such as glucose and gluconic acid can diffuse into the mesopores of this catalyst and achieve sufficient contact with the gold nanoparticles to react, and the second one is based on the abundant presence of active oxygen species on the catalyst surface.

Glycerol is a by-product from biodiesel production, which has recently attracted significant research interest. The main reasons are the following: Glycerol is a highly functionalized molecule, highly abundant, and a waste material that could be disposed via chemical fixation, and a large number of products can be obtained from its oxidation; therefore control of selectivity to the desired products is a significant challenge [26–30]. One of the first groups to demonstrate the utilization of Au-based catalysts for the selective oxidation of glycerol to glycerate under alkaline conditions was the group of Hutchings (Eq. 12.1). By synthesizing graphite-supported Au nanoparticles of around 25 nm mean particle size and modifying various reaction parameters, such as pressure, alkaline conditions, and catalyst loading, the authors optimized conversion of glycerol to yield glycerate at high selectivity (90%) [31]. It was proposed that sodium hydroxide is essential for the rate-determining initial dehydrogenation step, since in the presence of a base hydrogen is readily abstracted from one of the primary hydroxyl groups of glycerol. In subsequent studies the authors demonstrated the improved performance of Au catalysts in comparison to Pd- and Pt-based catalysts in terms of high selectivity to glycerate and higher catalytic activity [32].

$$\underset{\text{Glycerol}}{\text{HO}\underset{\underset{\text{OH}}{|}}{\diagup}\text{OH}} \xrightarrow[\text{Water, NaOH, 60 °C, O}_2]{\text{Au/Graphite}} \underset{\text{Glyceraldehyde}}{\text{HO}\underset{\underset{\text{OH}}{|}}{\diagup}\text{O}} \longrightarrow \underset{\text{Glyceric acid}}{\text{HO}\underset{\underset{\text{OH}}{|}}{\diagup}\underset{\underset{\text{O}}{||}}{\text{C}}\text{OH}} \quad (12.1)$$

Prati and Porta also investigated the utilization of Au catalysts synthesized by different preparation methods for glycerol oxidation, where a variation of particle sizes could be achieved. From their studies the authors concluded that well-dispersed, small gold nanoparticles of 6 nm mean diameter were responsible for high activity to glycerate [33]. However, they also observed that the initial selectivity could not be maintained at higher levels of conversion due to the consecutive oxidation of glycerate to tartronate. Large nanoparticles of around 20 nm were found to be responsible for high selectivity to glycerate and exhibited minimum overoxidation of glycerate and lower catalytic activity due to the larger gold particle size [33]. Through mechanistic studies, they also showed that the overall selectivity of the reaction is affected by a combination of factors, such as the initial intrinsic selectivity of the catalyst, base-catalyzed interconversion, and stability of the reaction products.

Claus and coworkers investigated the effect of the support and the variation of the gold particle size on glycerol oxidation. The authors concluded that under the same reaction conditions and with a similar particle size, gold nanoparticles supported on carbon were more active than those supported on oxidic materials [34]. By varying the gold particle size, they concluded that the reaction is structure sensitive, in agreement with the previous reports by Hutchings and Prati. In subsequent studies, Claus and coworkers also investigated ceria as an alternative support. Whilst the authors observed that ceria-supported Au nanoparticles were active catalysts, there was a considerable decrease in catalytic activity during recycling tests due to gold leaching [35]. The effect of gold particle size was also studied by Davis and coworkers. In this study, it was demonstrated that small gold nanoparticles are more active than large gold nanoparticles and that the selectivity to glycerate considerably decreased with the decrease of gold nanoparticle size. Important observations by Davis et al. was reported; The formation of hydrogen peroxide over all Au catalysts was observed and that the concentration of hydrogen peroxide had a direct relationship with the selectivity to glycerate; lower concentration of hydrogen peroxide was associated with higher selectivity to glycerate and lower formation of glycolate. The authors concluded that the formation of glycolic acid, which is hypothesized to be due to C–C cleavage, may be unavoidable over monometallic gold catalysts [36, 37]. In subsequent studies, Davis and coworkers reported the mechanism for glycerol oxidation to acids over various supported Au catalysts [38]. By using labeling experiments with $^{18}O_2$ and $H_2^{18}O$, they demonstrated that oxygen atoms originating from hydroxide ions are incorporated into the alcohol during the oxidation reaction, instead of molecular oxygen. Their studies were also supported by density functional theory (DFT) calculations. According to the authors, the role of oxygen is to participate in the catalytic cycle, not by dissociation to atomic oxygen, but by generating hydroxide ions via the catalytic decomposition of peroxide intermediate.

The effect of the preparation method (colloidal versus deposition-precipitation) was studied by Prati and coworkers, and it was found that a low-temperature chemical reduction of the deposited gold species enhanced activity due to the formation of gold particles in the range of 2–5 nm with a narrower particle size distribution. The use of a higher pretreatment temperature (400°C) resulted in lower activity due to the increase of gold particle size and also in higher selectivity to glycerate due to the suppression of the overoxidation at similar conversion levels [39].

The utilization of active and selective supported Au nanoparticles immobilized on $MgAl_2O_4$ spinels was demonstrated by the same group for glycerol oxidation [40]. By using a variety of preparation methods, such as flame spray pyrolysis and co-precipitation for the synthesis of spinels, and colloidal, deposition-precipitation, and magnetron sputtering methods for the deposition of Au, they explored the effect of particle size, surface area, and Al/Mg surface ratio for the selective oxidation of glycerol. It was concluded that the Al/Mg surface ratio was a significant parameter to alter selectivity with an Al-rich surface promoting C–C bond cleavage with similar particle size (2–3 nm). Selectivity to C3 products (glycerate and tartronate) was highest with Mg-rich support materials (61% selectivity at 50% conversion). Moreover, from an activity point of view, the most active catalysts were comprised of small gold particles and high support surface areas. Another important parameter that influenced activity was the atomic percentage of Au, measured by X-ray photoelectron spectroscopy (XPS) for spinels prepared by co-precipitation and flame pyrolysis methods. It was observed that with similar gold particle size (around 3 nm) for both samples the Au surface exposure was much higher on the flame pyrolysis spinel. The higher activity was therefore ascribed to the higher exposure of active sites to the substrate due to the higher concentration of surface gold species. The utilization of commercial carbon nanotubes and carbon nanofibers modified by chemical and physical methods as supports for gold nanoparticles was also investigated by Prati and coworkers [41]. Oxidative treatment with $HNO_3$ and chemical treatment with ammonia at different temperatures were used to introduce carboxylic functionalities and to produce N-containing groups for enhancing basic properties of the support, respectively. Supported Au nanoparticles on the modified supports were prepared via a colloidal method that ensures the formation of gold nanoparticles with a narrow particle size distribution. In this way, the basic and acidic properties of the supported nanoparticles on the functionalized materials were studied in the case of glycerol oxidation. It was demonstrated that an increase of basic functionalities was accompanied by a constituent increase in the activity of the catalyst, since it is well known that basic reaction conditions can promote alkoxide formation and subsequent C–H bond cleavage. Moreover, the important role of controlling the support's hydrophobicity/hydrophilicity was demonstrated, since selectivity to glycerate and tartronate as the main C3 products increased from 65% to 82% at 90% conversion with the increase of the hydrophobic surface. This result was explained due to the minimization of H$_2$O$_2$ production and therefore an overall decrease in C–C bond cleavage. Finally, the authors concluded that by using functionalized carbon materials it is possible to control both activity and selectivity in glycerol oxidation by varying not only the basic properties of the support but also hydrophobicity.

In subsequent studies Prati and coworkers investigated the role of NiO as a novel support for depositing Au nanoparticles and explored the catalytic performance of Au/NiO and Au/NiO/TiO$_2$ catalysts in the case of glycerol and ethylene-1,2-diol oxidation [42]. In this way, the influence of the support (NiO) and specifically the electronic effect of NiO on Au could be systematically studied. It was demonstrated that the dispersion of NiO nanoparticles on TiO$_2$ could significantly influence the Au nanoparticles activity and selectivity, depending on the relative ratio of NiO to TiO$_2$. Au/NiO catalysts showed high activity but lower selectivity to glycerate (55% selectivity at 90% conversion). By varying the NiO/TiO$_2$ ratio and optimizing the desired ratio (NiO$_{10}$–TiO$_{290}$), the selectivity toward glycerate increased to 75% at 90% conversion. To explain the improvement in activity by using NiO as a support, the authors performed thorough Fourier transform infrared spectroscopy (FTIR), high-resolution transmission electron microscopy (HRTEM), and XPS studies and demonstrated that the most important parameter was the interaction of the support with CO as a probe molecule. XPS analysis showed that there was no change of oxide electronic properties. On the contrary, CO adsorption studies revealed that the presence of NiO was strongly influencing the CO adsorption, which was in agreement by HRTEM analysis performed. HRTEM analysis showed that Au nanoparticles were preferentially associated with NiO either in the pure Au/NiO or in the pure Au/NiO/TiO$_2$ materials, with this interaction improving the stability of the Au nanoparticles. From these studies, the authors concluded that the interaction between NiO and Au nanoparticles is the essential factor for affecting activity and selectivity for Au/NiO and Au/NiO/TiO$_2$ catalysts.

The effect of gold particle shape at a similar gold particle size was studied by Prati and coworkers using as a reaction model the oxidation of glycerol [43]. Despite the high number of studies dealing with the effect of metal nanoparticle shape, the majority of these studies concern gas-phase reactions, and the effect of shape is not studied at similar particle sizes. Prati and coworkers showed that the variation of particle shape at a similar gold particle size has a significant effect in liquid-phase oxidation of glycerol. For generating Au nanoparticles with different shapes, the authors used two different types of commercial carbon nanofibers. The impregnation of preformed colloidal gold particles onto two different types of carbon nanofibers produced Au-supported nanoparticles with a similar particle size (3.5–3.7 nm). However, the Au particle shape was related to the surface structure of the support and therefore to the different interaction with the support. It was noted that a significant higher frequency of [111] planes of Au particles was observed to be parallel and highly favored to the one type of carbon nanofiber. In terms of activity and selectivity, performing glycerol oxidation at 50°C at 3 atm of $O_2$ under basic conditions, it was observed that similar activities were obtained with the different particle morphologies. However, in terms of selectivity, it was found that the supported Au catalysts with the highest frequency of [111] planes of Au nanoparticles in contact to the support, and therefore with the lowest exposure of [111] planes of Au nanoparticles, promoted the formation of C1 and C2 products. The authors concluded that increasing the number of Au (111) sites would favor the selectivity toward C3 products.

The role and influence of the capping agent, poly(vinyl alcohol) (PVA), in terms of activity and selectivity in glycerol oxidation were also investigated by Prati and coworkers recently [44]. The presence of the protective layer around the metal nanoparticle and the surface of the support are expected to influence catalyst properties by affecting metal–support and reactant–metal interactions. By varying the PVA/metal weight ratio and synthesizing Au/$TiO_2$ catalysts with and without the presence of the capping agent, the role of the capping agent was explored. It was observed that the presence of PVA lowered the catalytic activity of the Au/$TiO_2$ catalyst but increased the stability of the catalyst during recycling tests. Moreover, by optimizing the ratio of capping agent to metal it was found that a small amount of capping agent is beneficial for increasing activity by a factor of 2, and by varying the amount of capping agent on a preformed Au/$TiO_2$ catalyst prepared by the deposition-precipitation method, selectivity toward C3 products, such as glycerate and tartonate, was increased by approximately 7% at similar conversion levels. For elucidating the reasons for this observation, FTIR adsorption studies reveal a direct interaction between the capping agent and the glycerol molecule, thus affecting the adsorption of glycerol on the active sites of the catalyst. The authors concluded that the different adsorption mode of glycerol on the active sites in the presence or absence of a capping agent can vary activity, selectivity, and stability. Therefore a new way of designing metal-supported catalysts by the careful choice of capping agent and amount has been demonstrated.

The oxidation of alcohols with Au-based catalysts has been extensively investigated over the last decade, and several examples are illustrated in Table 12.1. Galvagno and coworkers investigated the catalytic performance of a $Au/Fe_2O_3$ catalyst synthesized by the co-precipitation method in the liquid-phase oxidation of o-hydroxybenzylalcohol under mild conditions (50°C, $P_{O2} = 1$ atm). It was reported that the catalytic activity increased with gold loading and that the selectivity to aldehydes at similar levels of conversion was higher with lower metal loading. Moreover, it was observed that the reaction is first order with respect to the organic substrate and zero order with respect to the oxygen partial pressure [45].

Corma and coworkers demonstrated the beneficial effect of combining small gold particles (2–5 nm) and nanocrystalline ceria (5 nm). Highly active, selective, and recyclable catalysts for the oxidation of alcohols into aldehydes and ketones with high turnover numbers (TONs) at solvent-free conditions were reported (Eq. 12.2). On the basis of mechanistic studies, it was demonstrated that the deposition of gold nanoparticles transforms nanocrystalline cerium oxide from a stoichiometric oxidant into a catalytic material [46].

$$\text{octan-3-ol} \xrightarrow[\text{Solvent-free, 80 °C, } O_2]{Au/CeO_2} \text{octan-3-one} \quad (12.2)$$

In subsequent studies Corma and coworkers demonstrated the catalytic efficiency of gold nanoparticles supported on nanocrystalline ceria in the aerobic oxidation of allylic alcohols. Gold catalysts were more active and chemoselective than supported palladium-based nanoparticles [47]. Yields over 90% were obtained for the oxidation of allylic alcohols in the case of 1-octen-3-ol, 2-octen-1-ol, and cinnamyl alcohol at 120°C. In the presence of supported gold nanoparticles, other by-products arising from C–C double-bond isomerization and hydrogenation were drastically reduced. The authors attributed the high chemoselectivity to the lower tendency of gold nanoparticles to form reactive gold hydride species, and therefore their lower concentration.

The efficient liquid-phase oxidation of ethanol to acetic acid was demonstrated by Christensen and coworkers (Eq. 12.3). The support was MgAl$_2$O$_4$ and the method of deposition-precipitation was used for the deposition of Au nanoparticles. High conversion of ethanol was achieved in the range of 363–473 K with a reported yield over 80% [48].

$$\text{Ethanol} \xrightarrow[\text{Water, 150 °C, O}_2]{\text{Au/MgAl}_2\text{O}_4} \text{Acetic acid} \tag{12.3}$$

Richards and coworkers reported the synthesis of Au nanoparticles confined in the walls of mesoporous silica and the catalytic performance of the synthesized materials was investigated in the aerobic oxidation of alcohols [49]. The authors showed that high activity and selectivity to aldehyde (79%–94%) was achievable using toluene as a solvent and K$_2$CO$_3$ as a base. The reusability of the catalysts was shown, indicating that the Au nanoparticles will not sinter once inside the pore channels of the mesoporous silica mesoporous silica (GMS).

Baiker and coworkers deposited Au nanoparticles (6–9 nm range) on Cu–Mg–Al mixed oxides using the deposition-precipitation methodology, and the synthesized catalysts were tested for the aerobic liquid-phase oxidation of alcohols [50]. The catalytic activity was strongly dependent on the composition of the support, especially the Cu/Mg molar ratio. X-ray absorption near-edge structure (XANES) analysis revealed the presence of reduced and oxidized gold species on the ternary mixed-oxide support before and after reaction. The best catalytic performance for a ternary mixed-oxide-supported gold catalyst was achieved using the following composition: Au/Cu$_5$Mg$_1$Al$_2$O$_x$. A range of alcohols was fully converted at high selectivity (98%). In subsequent studies, Baiker and coworkers studied in depth the oxidation state of gold during aerobic alcohol oxidation using XANES and observed that catalytic activity increased with the increase in the reduction of the gold component [51]. Therefore, they concluded that it is important that gold species in catalytic aerobic oxidation be in the metallic state.

The use of microgels for stabilizing Au nanoclusters and their use as quasi-homogeneous catalysts for the aerobic oxidation of primary and secondary alcohols in water was demonstrated by Prati and coworkers [52]. It was shown that microgels containing gold can show high activity comparable to known and well-used supported gold catalysts.

MnO$_2$ nanorods were synthesized by Cao and coworkers and used as alternative supports for depositing Au nanoparticles (4–5 nm). The authors demonstrated the high catalytic activity in the liquid-phase oxidation of benzyl alcohol and 1-phenylethanol [53]. Comparison with commercial Au/MnO$_2$ showed that the Au/MnO$_2$ nanorods were more efficient as catalysts. The improvement in activity was attributed to the collaborative effects resulting from the interaction of the Au nanoparticles and the well-defined reactive surface of the MnO$_2$ nanorods.

Au nanoparticles supported on different polymorphs of gallia ($\alpha$-$\beta$- and $\gamma$-Ga$_2$O$_3$) were synthesized and tested in the solvent-free liquid-phase oxidation of benzyl alcohol with molecular oxygen [54]. Catalytic activity varied among the gold supported on different polymorphs of gallia, indicating the impact of the support for determining catalytic behavior. The most active supported gold catalyst was obtained when $\gamma$-Ga$_2$O$_3$ was the chosen support. The reusability of the Au/$\gamma$-Ga$_2$O$_3$ was demonstrated, and its general applicability as an efficient catalyst for oxidizing a range of aromatic and aliphatic alcohols, with selectivities around 98% to their corresponding ketones and aldehydes, was also illustrated. Explanation for the improvement activity of the gold nanoparticles supported on gallia was based on spectroscopic investigation. It was reported that a significantly higher dehydrogenation capability was observed using $\gamma$-$Ga_2O_3$ in comparison to other support materials. In subsequent studies, a series of binary mesostructured Ga–Al mixed-oxide supports ($Ga_xAl_{6-x}O_9$, $x = 2, 3, 4$) were synthesized and subsequently deposited with gold [55]. The Au/$Ga_xAl_{6-x}O_9$ materials showed high catalytic performance in the aerobic oxidation of alcohols under mild conditions (80°C), and high yield to the corresponding aldehydes/ketones (Eq. 12.4). Comparison of the catalytic performance with other conventional oxide-supported systems indicated that Au/$Ga_xAl_{6-x}O_9$ was much more active due to the collaborative interaction between gold and the mixed-oxide support. Reusability tests confirmed that the catalyst is reusable and that the spinel structure of the oxide was retained during the oxidation process.

$$\text{Benzyl alcohol} \xrightarrow[\text{Mesitylene, 90 °C, } O_2]{\text{Au/}Ga_3Al_3O_9} \text{Benzaldehyde} \quad (12.4)$$

The synthesis of Au nanoparticles with various particle size (3 to 15 nm) and shape, supported on both the inner and outer surfaces of poly(o-phenylenediamine) (PoPD) hollow microspheres, was reported by Guo and coworkers [56]. The catalytic performance of the supported Au nanoparticles was investigated with the liquid-phase aerobic oxidation of alcohols using water as a solvent and $K_2CO_3$, under very mild conditions (room temperature) with yields of 91%–95%.

Hensen and coworkers demonstrated that Au nanoparticles (4–6 nm range) supported by basic hydrozincite or bismuth carbonate were efficient catalysts for liquid-phase aerobic oxidation of alcohols [57]. The catalytic performance of a series of metal (Zn, Bi, Ce, La, Zr) carbonate-supported Au catalysts was shown to depend strongly on the basicity of the supported materials. The presence of accessible strong basic sites, where the initial O–H bond cleavage is possible on the support, is an important factor for improving catalytic activity. The authors demonstrated that this approach is promising for developing heterogeneous catalysts possessing strong base sites for alcohol oxidation, because in this way replacement of soluble bases is possible.

Immobilization of approximately 1 nm Au clusters within mesoporous silicas (SBA-15, MCF, HMS) using triphenylphosphione-protected Au clusters (Au11:TPP) as precursors was reported by Tsukuda and coworkers. It was shown that the homogeneous dispersion of Au11:TPP clusters could be achieved, and by controlling the calcination procedures employed the complete removal of the protecting ligands without aggregation of the resulting Au clusters could be achieved. The resulting catalytic behavior of the gold-supported clusters was studied. Using water as a green solvent, and $K_2CO_3$ and benzyl alcohol as model substrates, a yield of 91% to benzoic acid was reported (Eq. 12.5) [58].

$$\text{Benzyl alcohol} \xrightarrow[\text{Water, } K_2CO_3, 80°C, O_2]{\text{Au/SBA-15}} \text{Benzaldehyde} \longrightarrow \text{Benzoic acid} \quad (12.5)$$

Zhang and coworkers synthesized size-controlled Au nanoparticles (2.6–6.3 nm) within a porous chelating hydrogel of poly (N-isopropylacrylamide)-co-poly[2-methacrylic acid 3-(bis-carboxyme-thylamino)-2-hydroxypropyl ester] referred to as (PNIPAM-co-PMACHE] [59]. The encapsulated Au nanoparticles were tested for the aerobic oxidation of 1-phenylethanol in the presence of water and KOH. Catalytic activity was influenced by the variation of the Au particle size. The authors claimed that the Au-encapsulated nanoparticles in the hydrogel were highly efficient and reusable catalysts.

Rossi and coworkers synthesized Au nanoparticles and immobilized them onto a magnetically recoverable support (core–shell $Fe_3O_4@SiO_2$). The catalytic performance of the synthesized supported Au catalysts was carried out in the liquid-phase oxidation of benzyl alcohol, 1-phenylethanol, and cyclohexanol. Using toluene as the solvent and $K_2CO_3$ as the base conversions of 100% were achieved at 100°C [60]. The reusability of the Au-supported catalyst was limited and the authors claimed that was due to growth of particle size and a change in the morphology of the support. The authors suggested that a magnetically recoverable support offers advantages for the potential recovery and reusability of the catalyst.

The effect of a capping agent (polyvinylpyrrolidone, PVA, poly[ethylene glycol], dodecythiol) in terms of catalytic performance was studied in the case of liquid-phase aerobic oxidation of benzyl alcohol using Au/SiO$_2$ as the model catalyst by Huang and coworkers [61]. Supported capping-ligand-free Au nanoparticles were synthesized by the deposition-precipitation method and afterward various types of capping ligands were employed. From a catalytic point of view, supported Au nanoparticles with chemisorbed capping ligands showed lower benzyl alcohol conversion than capping-ligand-free supported Au nanoparticles, demonstrating the surface-blocking effect of chemisorbed capping ligands. However, the catalytic performance depended significantly on the type of capping ligand employed. For example, by using poly(ethylene glycol) and polyvinylpyrrolidone as the capping ligands and by varying the ligand/Au molar ratio, an increase in the benzaldehyde selectivity could be observed from 43% to 99%, albeit at the expense of benzyl alcohol conversion. In the case of PVA, an increase in the capping ligand/Au molar ratio led to a large decrease in benzyl alcohol conversion, indicating a strong surface-blocking effect due to the chemisorbed capping ligand. To better understand the influence of the capping agent, XPS analysis was employed to investigate the interaction of chemisorbed capping ligands with Au nanoparticles. In the case of using polyvinylpyrrolidone, a strong decrease in the Au 4f XPS peak was detected, suggesting a strong chemisorption of the capping ligand on the Au nanoparticles. From the N 1s XPS results it was revealed that a higher N 1s binding energy was observed than for polyvinylpyrrolidone on bare SiO$_2$, suggesting a charge transfer from polyvinylpyrrolidone to the Au nanoparticles. The authors concluded that by tuning the polyvinylpyrrolidone/Au molar ratio, the selectivity to the desired products could be tuned without significantly poisoning their catalytic activity. These findings are important since parameters that influence catalytic activity by the capping ligand, such as surface-blocking effect, electronic effect, and site isolation effect, can be controlled by the proper choice and amount of capping agent, and it can serve as a novel strategy for designing active and highly selective catalysts.

Supported Au nanoparticles on ceria have shown to be a promising catalyst in the case of liquid-phase oxidation of alcohols.

Hutchings and coworkers investigated the catalytic performance of Au nanoparticles supported on different types of ceria foams [62]. By using L-asparagine to form a cerium coordination polymer foam precursor, and by varying the crystallization time from 2 to 48 h and applying a subsequent calcination treatment, various crystalline ceria materials were formed. Au nanoparticles were immobilized by using a colloidal method. The catalytic performance of Au/foam $CeO_2$ with commercial $CeO_2$ samples in the liquid-phase oxidation of benzyl alcohol was investigated under mild conditions (120 °C, 1 bar $O_2$, solvent-free conditions). It was found that Au/$CeO_2$ foams were more active by a factor of 2 than those prepared by using commercial samples of $CeO_2$. The authors performed temperature-programmed reduction (TPR) analysis to reveal the reasons for the higher catalytic activity observed with the foam $CeO_2$ samples, and it was shown that the surface of the $CeO_2$ foam catalysts was more easily reduced than the catalysts prepared using commercial $CeO_2$, suggesting that the surface oxygen availability was higher, thereby facilitating activity.

Xu and coworkers [63] investigated the effect of cerium crystallinity in the liquid-phase oxidation of benzyl alcohol under mild conditions (90 °C, chlorobenzene as solvent, 1 atm of $O_2$, and $Cs_2CO_3$) by supporting Au nanoparticles on different ceria crystal planes such as [110, 111], and [100]. Au nanoparticles supported on the ceria [110] crystal planes were more active than on the [111] and [100] crystal planes of ceria, with conversions of 96%, 79%, and 42% achieved, respectively, at benzaldehyde selectivities of 99% in all cases. To support their findings by calculating the apparent activation energy, it was found that the lowest activation energy was with the ceria [110], supporting the trend in activity and confirming that Au/$CeO_2$ with [110] crystal planes can greatly lower the C–H bond activation energy and thus increase catalytic activity. Moreover, the same trend and general applicability of using Au/$CeO_2$ with [110] crystal planes was demonstrated for a range of substituted benzyl alcohols.

A novel strategy for producing highly active and selective Au nanoparticles half-encapsulated in nano-iron-oxide particles and supported on alumina (Au–$FeO_x$–$Al_2O_3$) was reported by Ding and coworkers [64]. The synthesized supported Au nanoparticles showed much higher activity than Au nanoparticles supported on conventional bulk iron oxide (Au/FeO$_x$) and alumina (Au/Al$_2$O$_3$) in the case of liquid-phase oxidation of 1-phenylethanol. Conversion of 52% was reported in the case of the Au–FeO$_x$–Al$_2$O$_3$ catalyst, whereas for Au/FeO$_x$ and Au/Al$_2$O$_3$ conversions were lower than 6%. The promotion in catalytic activity was attributed to the cooperative effect between the Au nanoparticles and nanosized FeO$_x$ due to their increased contact boundaries and intimate electronic interactions. XPS analysis showed an increased donation from FeO$_x$ to Au, and the authors attributed this to the increase of activity. The authors presented a plausible mechanism for the catalytic reaction over Au–FeO$_x$–Al$_2$O$_3$. The 1-phenylethanol molecules are first adsorbed and activated on exposed gold nanoparticles. Subsequently, the deposition of Au on iron oxide facilitates the transfer of the lattice oxygen and the reaction with the adsorbed reactant takes place from the desired products. Finally oxygen molecules are activated at the FeO$_x$ and/or the Au–FeO$_x$ interface and then react with FeO$_x$ to reform lattice oxygen and close the catalytic cycle.

Methyl esters are important products in the chemical industry, for example, in the synthesis of flavoring agents, solvent extractants, and intermediates, as well as in the fragrance industry. Therefore, the synthesis of esters by replacing the traditional route that involves the reaction of carboxylic acid and methanol and strong acid catalysts, such as sulfuric, sulfonic, and *p*-toluenesulfonic acid or base alkoxides is desirable from an environmental standpoint [65, 66]. The selective oxidation of alcohols in methanol for the synthesis of the corresponding methyl esters has been reported by several groups and is summarized in Table 12.2. Christensen and coworkers reported the synthesis of potassium titanate nanowires and the deposition of gold nanoparticles. The authors showed gold-supported nanoparticles are efficient for the aerobic oxidation of benzyl alcohol at ambient temperature in methanol. The formation of methyl benzoate using a catalytic amount of base with 99% conversion and a yield of 93% to methyl benzoate was reported. They demonstrated the significant role of the base, since in the presence of a base deactivation of the catalyst was prevented [67]. The same group reported the production of methyl esters by oxidizing aldehydes in a primary alcohol environment and using

**Table 12.2** Oxidative esterification of alcohols/aldehydes using supported Au-based catalysts

| Substrate | Catalyst | T(°C) | Solvent | Conv. % | Sel. % | Ref. |
|---|---|---|---|---|---|---|
| Benzyl alcohol | Au/K$_2$Ti$_6$O$_{13}$ | 20 | CH$_3$OH/KOH | 99 | 93 (methyl benzoate) | [67] |
| Propane-1,2-diol | Au/Fe$_2$O$_3$ | 100 | CH$_3$OH/NaOH$_3$ | 99 | 72 (methyl lactate) | [69] |
| 5-Hydroxymethyl-furfural | Au/TiO$_2$ | 130 | CH$_3$OH/NaOH$_3$ | 100 | 98 (DMF) | [77] |
| 5-Hydroxymethyl-furfural | Au/TiO$_2$ | 30 | H$_2$O /NaOH | 100 | 71 (FDCA) | [78] |
| 5-Hydroxymethyl-furfural | Au/CeO$_2$ | 130 | CH$_3$OH/NaOH$_3$ | 100 | 100 (DMF) | [79] |
| 5-Hydroxymethyl-furfural | Au/TiO$_2$ | 130 | H$_2$O /NaOH | 100 | 95 (FDCA) | [80] |

gold-supported catalysts. They reported the selective formation of methyl benzoate at temperatures even below 0°C. The general applicability was shown by synthesizing acrylate esters at room temperature using air as the oxidant, with selectivities above 85% at 97% conversion achieved [68]. The authors proposed the following mechanism: Alcohol is first oxidized to an aldehyde, which then forms a hemiacetal with methanol. The so-formed hemiacetal is subsequently further oxidized to the corresponding ester. The aerobic oxidation of alcohols to the corresponding aldehyde has been proposed to be the rate-determining step. In subsequent studies Christensen et al. reported the utilization of polyols such as glycerol, 1,2-propanediol, and 1,3-propanediol for the synthesis of methyl esters. Using supported Au nanoparticles in the range of 3–7 nm it was reported that the synthesis of important chemicals such as methyl lactate from 1,2-propanediol, methyl acrylate from 1,3-propanediol, and dimethyl mesoxalate from glycerol can be achieved with yields in the range of 40%–90% [69].

The oxidative esterification of alcohols was also demonstrated by Rossi et al., who synthesized SiO$_2$-supported Au nanoparticles with a mean gold particle size of 5.7 nm. For the synthesis of the supported metal nanoparticles the Brust method was used. This method can form gold nanoparticles with a small particle size and a narrow particle size distribution [70]. The general applicability of the oxidation of alcohols to the corresponding methyl esters was shown, with high yields to the corresponding methyl esters (70%–95%) achieved.

The utilization of tertbutylhydroperoxide (TBHP) or hydrogen peroxide ($H_2O_2$) as an alternative oxidant has also been investigated using supported Au nanoparticles. Choudhary and coworkers have shown the use of supported Au nanoparticles in the liquid-phase oxidation of benzyl alcohol to benzaldehyde using TBHP under mild conditions 95°C [71, 72]. The effect of the support and preparation method was studied and it was concluded that the most active catalysts consist of supported Au nanoparticles with a high surface ratio of $Au^{3+}/Au^0$, whereas gold particle size seems not to be a crucial parameter in the range of 3–8 nm. Finally, the most effective supports were $TiO_2$, $ZrO_2$, and MgO. $H_2O_2$ is another alternative green oxidant [73, 74] and the utilization of $H_2O_2$ for the efficient oxidation of organic compounds could be of high interest. Cao and coworkers reported the oxidation of alcohols using $H_2O_2$ as the oxidant by gold-supported nanoparticles. 1-Phenyethanol was oxidised at 90°C using gold-supported catalysts supplied by the World Gold Council. The general applicability was demonstrated by oxidizing a range of nonactivated alcohols [75].

Tsukuda and coworkers reported the oxidation of benzyl alcohol using a combination of hydrogen peroxide and microwave irradiation. The effect of particle size was studied by synthesizing supported Au nanoparticles in the range of 0.8–1.9 nm. They concluded that there is a size dependence of the activity of Au clusters on SBA-15, with smaller gold particles exhibiting higher catalytic activity [76].

The scientific community has recently focused on the transformations of 5-hydroxymethyl-furfural (HMF) to useful organic compounds that may have a great potential in pharmaceuticals, antifungals and polymer precursors [28]. One of the most important future applications can be in the production of polymer precursors and the synthesis of 2,5-furandicarboxylic acid (FDCA) from furfural. This process can eventually replace the current industrial production of terephthalic acid, an important polymer precursor. An alternative approach could also be via the oxidative esterification of

HMF to synthesize dimethylfuroate (DMF), which can be used as a monomer for replacement of terephthalic acid in plastics and has the advantage of improved solubility in many solvents.

Christensen and coworkers reported the oxidative esterification of HMF to DMF using methanol as a solvent at 130°C and a catalytic amount of sodium methoxide (NaOCH$_3$) in the presence of a Au/TiO$_2$ catalyst with a yield of 98% (see Table 12.2) [77]. In subsequent studies the same group demonstrated the catalytic oxidation of HMF to using supported Au nanoparticles. Mild conditions (30°C) and water as a solvent were used with a reported yield of 71% [78].

Corma and coworkers reported a similar approach to synthesize DMF and FDCA. Using a Au/CeO$_2$ catalyst composed of Au nanoparticles and nanoparticulated ceria they demonstrated the efficient conversion of HMF to DMF using molecular oxygen as the oxidant and methanol as the preferred solvent. Reaction parameters, such as temperature, pressure, and substrate-to-catalyst ratio, were varied and by optimizing reaction conditions a 100% yield of DMF was obtained (Eq. 12.6). The supported Au catalyst was reusable and the authors reported a well-designed regeneration protocol [79]. In the case of FDCA synthesis the same catalyst was used and by optimizing the reaction conditions (base amount, temperature, and pressure) an FDCA yield of 95% was reported [80].

$$\text{5-hydroxymethyl-2-furfural} \xrightarrow[\text{Methanol, MeONa, 130 °C, O}_2]{\text{Au/CeO}_2} \text{2,5-Dimethylfuroate} \quad (12.6)$$

The industrially relevant oxidative transformation of carvone oxime into carvone with supported Au nanoparticles has been reported by Corma and coworkers [81]. This reaction represents a successful example for the efficient aerobic oxidation of oximes to the corresponding carbonylic compounds using gold catalysts. Carvone is an essential oil utilized in the fragrance industry. Via the utilization of supported Au nanoparticles the oxidation of the C=N bond was achieved. Several solvents were used and the most effective was aqueous solutions of ethanol and without the presence of acids. Ceria was the preferred support. The authors demonstrated that this process can replace current methods that require high amounts of acids and are environmentally unfriendly.

## 12.2.2 Supported Au–Pd- and Au–Pt-Based Catalysts

There has been an explosion of interest in the synthesis of supported bimetallic Au–Pd and Au–Pt catalysts, particularly their utilization in the aerobic liquid-phase oxidation of organic compounds such as alcohols, polyols, and sugars (Table 12.3). The main motives for the academic and industrial interest relies on the fact that (i) gold and palladium can form solid solutions in the whole range of the gold/palladium atomic ratio and (ii) the addition of a second metal can alter the electronic and geometrical properties of the synthesized nanoparticle with the formation of alloy and core–shell structures. Therefore it is expected that catalytic activity, selectivity to the desired product, and catalyst stability will be affected.

**Table 12.3** Oxidation of alcohols/polyols using supported Au–Pd-based catalysts

| Substrate | Catalyst | T (°C) | Solvent | Conv. % | Sel.% (aldehyde/ketone) | Ref. |
|---|---|---|---|---|---|---|
| D-sorbitol | Au–Pd/C | 50 | $H_2O$/NaOH | 90 | 90 (acid) | [82, 83] |
| Glycerol | Au–Pd/C | 50 | $H_2O$/NaOH | 90 | 90 (acid) | [84–87] |
| Cinnamyl alcohol | Au–Pd/C | 60 | $H_2O$ | 95 | 83 | [90] |
| Benzyl alcohol | $Au_{80}$-$Pd_{20}$/C | 60 | $H_2O$/NaOH | 90 | 99 | [91] |
| Benzyl alcohol | Au–Pd/$TiO_2$ | 100 | Solvent free | 90 | 95 | [92] |
| Benzyl alcohol | Au–Pd/C | 120 | Solvent free | 80 | 65 | [95] |
| 1,2-Propanediol | Au–Pd/$TiO_2$ | 60 | $H_2O$/NaOH | 94 | 96 (acid) | [97] |
| Benzyl alcohol | Au–Pd/PI | 100 | $C_6H_5CH_3$ | 99 | 98 | [103] |
| Benzyl alcohol | Au–Pd/SBA-15 | 80 | $H_2O$/$Na_2CO_3$ | 40 | 99 | [104] |
| Crotyl alcohol | Aucore-Pdshell | 25 | Solvent free | 61 | 87 | [106] |
| Glycerol | Au–Pt/C | 50 | $H_2O$/NaOH | 70 | 60 (glyceric acid) | [110] |
| Glycerol | AuPt/Mordenite | 100 | $H_2O$ | 70 | 83 (glyceric acid) | [111] |
| 1-Phenyethanol | Au–Pt/PI | 25 | $H_2O$/$C_6H_5CF_3$ | 99 | 99 | [113] |

Prati and coworkers studied the effect of gold with palladium or platinum and the influence of support by synthesizing bimetallic colloids and immobilizing the synthesized bimetallic colloids on supports, such as carbon and graphite [82, 83]. The catalytic performance of the synthesized materials was evaluated through the selective oxidation of polyols (sorbitol and glycerol) and alcohols (aliphatic and benzylic alcohols) under mild reaction conditions (30°C–60°C, $P_{O2}$ lower than 4 atm). In the case of polyol oxidation (sorbitol and glycerol), the bimetallic catalysts showed remarkably enhanced catalytic activity, as well as increased selectivity to the desired products with respect to the monometallic counterparts. In addition, enhanced stability due to the resistance in poisoning by molecular oxygen was observed. The variation of the Au–Pd atomic ratio was studied and a typical volcano-type relationship with catalytic behavior was observed. The most active bimetallic catalysts were those with a Au/Pd atomic ratio of 6:4. From their results the authors concluded that for enhancing the catalytic activity of a bimetallic Au–Pd system a small amount of the one metal in the presence of the other seems to be enough for the creation of active bimetallic sites.

In the case of glycerol, the choice of bimetallic system (Au–Pd or Au–Pt), preparation method, and support (carbon versus graphite) was shown not only to influence catalytic activity but also to play a role in terms of product selectivity [84, 85]. The utilization of supported Au–Pd nanoparticles showed to improve the selectivity toward the products of the oxidation of the terminal hydroxyl groups (glycerate and tartronate) with selectivities over 90% at high conversion levels (90%) (Eq. 12.7). In the case of supported Au–Pt catalysts, similar TOF values were obtained with respect to the supported Au–Pd catalysts, and an enhancement in the formation of glycolic acid was observed. These results indicated the preferred oxidation of the secondary hydroxyl group of glycerol. The effect of the Au/Pd molar ratio was studied by the same group, who observed that catalytic activity improved by increasing the Au/Pd molar ratio. The highest levels of activity were observed with a Au-rich system (Au/Pd = 9/1) [86, 87]. From their studies they concluded that surface Pd monomers in contact with Au have a promoting effect in terms of activity and lifetime of the catalyst.

Glycerol + AuPd/C, Water, NaOH, 30-50 °C, O₂ → Glyceraldehyde → Glyceric acid → Tartronic acid (12.7)

A novel methodology was used for the synthesis of single-phase supported Au–Pd catalysts. These catalysts showed a significant improvement in terms of activity and reusability than random Au–Pd catalysts during the selective oxidation of glycerol [88, 89]. By comparison of the catalytic performance of Au–Pt and Au–Pd single-phase catalysts in the selective oxidation of various primary alcohols (benzyl alcohol, cinnamyl alcohol, and 1-octanol), the catalytic behavior of Au–Pd catalysts was superior to analogue Au–Pt catalysts (Eq. 12.8). A significant improvement in catalytic activity was found when water instead of toluene was used as the desired solvent. The reported TOF values showed an increase by a factor of 1.5–6 [90]. In subsequent studies, the effect of the $Au_xPd_y$ molar ratio and the catalytic performance for a range of alcohol substrates was studied. They concluded that the most efficient Au/Pd composition was $Au_{80}/Pd_{20}$, which exhibited the highest catalyst activity [91].

Cinnamyl alcohol + PdAu/C, Water, 60 °C, O₂ → Cinnamaldehyde → 3-phenylpropan-1ol (12.8)

The general applicability of bimetallic supported Au–Pd catalysts synthesized by impregnation method was reported by Hutchings and coworkers in the case of solvent-free liquid-phase oxidation of a variety of alcohols. The synthesized catalysts (Au–Pd/TiO₂) were active and selective at solvent-free conditions using molecular oxygen as the oxidant without the use of initiators (Scheme 12.1) [92]. Reported TOF values of 269,000 h⁻¹ showed the high efficiency of the catalysts. The authors concluded that the high activity of the

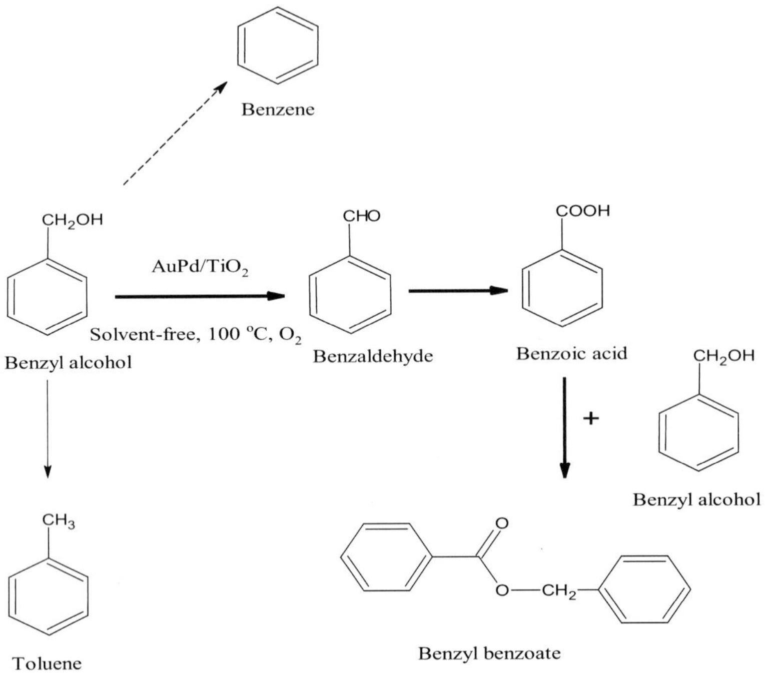

**Scheme 12.1** Solvent-free oxidation of benzyl alcohol using supported Au–Pd nanoparticles.

Au–Pd-supported catalyst was due to the formation of a gold core–palladium shell structure on the support surface and the electronic promotion of Au for Pd. In the following studies, the Au/Pd weight ratio was studied and it was found that the most active catalyst was the one with a Au/Pd weight ratio of 1/1, whereas the highest selectivity to benzaldehyde was observed with Au-rich catalysts [93]. In following studies Hutchings and coworkers investigated the synthesis of supported Au–Pd catalysts using a colloidal method and demonstrated the efficient aerobic oxidation of benzyl alcohol with very high TOFs under milder reaction conditions (120°C) [94]. Scanning transmission electron microscopy-X-ray dispersion spectroscopy (STEM-XDS) and XPS studies showed the presence of random homogeneous alloys with a metallic oxidation state for Au and the majority of Pd to be present in the metallic state. The higher activity of the supported Au–Pd catalysts synthesized by the colloidal method instead of the impregnation method was attributed mainly to the significant smaller particle size, narrow particle size distribution, and metallic oxidation state.

A preparation strategy for the synthesis of bimetallic hydrosols with the formation core–shell structures, involving the sequential addition and reduction of the metal and deposition of the bimetallic sols on carbon and titania, was used for the synthesis of supported bimetallic catalysts [95]. It was found that the catalytic activity for the aerobic oxidation of benzyl alcohol could be carried out under mild conditions (120°C, $P_{O2} = 10$ bar), and the order of metal addition has a marked effect on activity as well as selectivity. The choice of support (carbon versus titania) was shown to affect significantly the activity and distribution of products, with carbon-supported materials exhibiting an increase of activity by a factor of 2 and a lower selectivity to benzaldehyde at iso-conversion level compared to the analogous titania-supported catalysts. Mechanistic studies showed that the reaction is zero order in oxygen and the oxidation of benzaldehyde is dependent on the concentration of oxygen at the surface.

Comparison of two preparation methods (impregnation and sol immobilization methods) for the synthesis of supported Au–Pd catalysts in the liquid-phase oxidation of glycerol led to the conclusion that high activity coupled with high selectivity to the desired product (glycerate) can be achieved by gold-rich surface bimetallic nanoparticles with mean particle size of 3–5 nm and a metallic oxidation state. On the contrary, larger particles (over 6 nm) led to a significant lower catalytic activity, as it was observed with the impregnation method [96]. In subsequent studies Hutchings and coworkers demonstrated the excellent catalytic performance of supported Au–Pd nanoparticles and the strong synergistic effect of the addition of gold into the palladium metal in the liquid-phase oxidation of 1,2-propanediol to the sodium salt of lactic acid. The synthesis of lactic acid in an efficient and green way is highly desirable due to the fact that it is an important monomer for the synthesis of biodegradable polymers (Eq. 12.9) [97]. It was shown that high selectivity to lactate was possible (96%), even at 94% conversion, and that the usage of oxidants such as molecular oxygen as well as of hydrogen peroxide was sufficient, even under mild conditions (60°C and $P_{O_2} = 10$ bar or atmospheric pressure).

$$\underset{\text{1,2-propanediol}}{\text{HO}\diagdown\overset{\overset{\text{OH}}{|}}{\diagup}} \xrightarrow[\text{Water, NaOH, 60 °C, O}_2]{\text{AuPd/C}} \underset{\text{Lactaldehyde}}{\text{O}\diagdown\overset{\overset{\text{OH}}{|}}{\diagup}} \longrightarrow \underset{\text{Lactic acid}}{\text{O}\diagdown\overset{\overset{\text{OH}}{|}}{\underset{\overset{|}{\text{OH}}}{\diagup}}} \quad (12.9)$$

The effect of the Au/Pd molar ratio (1:7–7:1) in terms of catalytic activity was investigated for the liquid-phase oxidation of benzyl alcohol under solvent-free conditions using colloidal methods [98]. The optimum catalytic performance was observed with a Au/Pd 1:2 molar ratio and a selectivity to benzaldehyde of 67% at 90% conversion. Characterization by STEM-HAADF (high-angle annular dark field) analysis showed that the $Au_xPd_y$ colloidal nanoparticles were in the 3–5 nm range, and when the composition was in the 2:1–1:2 Au/Pd ratio the particles were random homogeneous Au–Pd alloys with a face-centered cubic (fcc) structure, although variations were observed at high Au/Pd and Pd/Au ratios.

Mechanistic studies were performed for the oxidation of benzyl alcohol using a Au–Pd alloy catalyst synthesized by sol immobilization method, and it was found that in the absence of oxygen benzyl alcohol was transformed into benzaldehyde and toluene at initial equal rates [99]. Introduction of oxygen significantly increased the rate of benzyl alcohol disappearance and appearance of benzaldehyde at the expense of toluene formation. It was found that at low partial pressures (below 3 bar of oxygen) rates are dependent on oxygen, suggesting that oxygen can participate in the reaction pathway as an adsorbed species.

In subsequent studies the effect in terms of catalytic performance of acidic/basic supported Au–Pd nanoparticles was studied [100]. By varying the support from titania, niobium oxide, zinc oxide, and magnesium oxide it was demonstrated the "switch off" of the disproportionation reaction and thereby the significant improvement in the selectivity to benzaldehyde (99%) at the expense of toluene could be achieved. The most suitable support to minimize the disproportionation reaction was magnesium oxide and zinc oxide.

The utilization of microstructured reactors was investigated by Gavriilidis and coworkers in the liquid-phase oxidation of benzyl alcohol under mild conditions (80°C–120°C) using Au–Pd/TiO$_2$

catalysts. The authors demonstrated that the continuous oxidation of benzyl alcohol is feasible, and by optimizing reaction parameters, 96% conversion with a selectivity to benzaldehyde of 78% was obtained at 120°C and 5 bar oxygen pressure. Moreover, they demonstrated the prospect of monitoring reactant/product species by using in situ Raman spectroscopy.

The effect of an alloy or a core–shell structure was investigated by synthesizing bimetallic Au–Pd colloids involving the sequential addition and reduction of the metal and deposition of the bimetallic colloids on carbon and titania [101]. STEM-HAADF imaging was employed for evaluating the particle size distribution, as well as to monitor the spatial distribution within individual particles, and the formation of alloy Au–Pd particles when gold and Pd were mixed together where a core–shell structure was developed when the sequential addition and reduction of metals were used was concluded from these studies. Moreover, the Au–Pd catalysts consisted of a mixture of icosahedral, decahedral, and cuboctahedral particles. In terms of catalytic activity the aerobic oxidation of benzyl alcohol could be achieved under mild conditions (120°C, $P_{O2} = 10$ bar) and the order of metal addition had a marked effect. The most effective catalysts were in the presence of a $Au_{core}$–$Pd_{shell}$ structure and TOFs up to 45,000 $h^{-1}$ were reported under mild conditions (e.g., 120°C). In subsequent studies the effect of heat treatment on the stability of these structures and catalytic performance were investigated [102]. It was found that a mild calcination treatment (e.g., 200°C) improved the catalytic activity due to the removal of the polymer layer around the Au–Pd particles, whereas a calcination treatment above 200°C affected the final morphology of the Au–Pd particles, dependent on the initial structure of the particles (alloy or core–shell) and the choice of support and led to a significant decrease in terms of activity. For example, Au–Pd particles with a random alloy structure did not show phase separation, whereas $Pd_{core}$–$Au_{shell}$ and $Au_{core}$–$Pd_{shell}$ particles started to show phase separation and the simultaneous presence of the initial structure and Au-rich particles with PdO particles. The authors concluded that in terms of optimizing catalytic performance it seems that not only an optimum particle size was required to achieve a good compromise between activity and selectivity but also an optimum ratio between PdO/Pd species was important.

The beneficial interaction of gold and palladium in bimetallic catalysts was also demonstrated by Baiker and Marx by synthesizing bimetallic catalysts using a colloidal route, where the admixing of Pd to Au resulted in the synthesis of bimetallic nanoparticles in a narrow range (2.4–3.7 nm) and a Au-rich core and a Pd-rich shell [103]. The synthesized Au–Pd nanoparticles supported on polyaniline showed high activity in the liquid-phase oxidation of benzyl alcohol and high selectivity to benzaldehyde (98%) in the presence of toluene as the solvent and aqueous solution of NaOH.

Qiao and coworkers reported the synthesis of Au–Pd/SBA-15 material by impregnation and grafting methods [104]. By using the grafting method metal nanoparticles (5 nm mean particle size) could be highly dispersed in mesoporous channels of SBA-15, showing high catalytic performance in the selective oxidation of benzyl alcohol to benzaldehyde under mild reaction conditions (80°C, water as solvent and $Na_2CO_3$). It was demonstrated that agglomeration and leaching of metal nanoparticles were avoided by restricting the nanoparticles inside the mesopores of SBA-15, therefore leading to enhanced stability and reusability of Au–Pd/SBA-15. The effective confinement of Au–Pd nanoparticles was also reported by Yang and coworkers using SBA-16 as support. Au–Pd nanoparticles were supported on SBA-16 using an adsorption method. The authors demonstrated the synergistic effect of the bimetallic Au–Pd-supported nanoparticles by showing enhanced catalytic performance of the Au–Pd catalyst compared to the Au and Pd monometallic catalysts. STEM and energy-dispersive X-ray (EDX) studies were carried out and the alloyed structure of the bimetallic nanoparticles with a Pd-rich shell and Au-core structure was shown [105].

The catalytic performance of Au–Pd core–shell colloidal nanoparticles in the aerobic oxidation of crotyl alcohol at room temperature was investigated by Scott and coworkers [106]. The synthesis of the Au–Pd core–shell nanoparticles was carried out following a sequential reduction strategy and using a colloidal methodology and polyvinylpyrrolidone as the polymer stabilizer. The sequentially reduced nanoparticles were highly active for crotyl alcohol oxidation with high selectivity to crotonaldehyde (60% conversion and 87% selectivity to crotonaldehyde). Extensive characterization by extended X-ray absorption fine structure (EXAFS) and XANES spectroscopy revealed that the Au–Pd nanoparticles were Pd enriched on the surface of the metal nanoparticles, whereas XANES spectra showed an electron transfer from Pd atoms to Au atoms. Mechanistic studies indicated that the possible mechanisms involved a redox mechanism for the core–shell nanoparticles in which Pd oxidation and re-reduction occurred as well as a $\beta$-H elimination mechanism.

A new methodology for synthesizing Au–Pd bimetallic catalysts was reported by Evangelisti and coworkers [107]. The methodology is based on the utilization of Au and Pd vapors as reagents either by simultaneous condensation or by separate evaporation of the two metals and subsequent mixing. The catalytic performance of the supported Au–Pd catalysts was tested in the aerobic oxidation of benzyl alcohol under mild conditions (60°C–100°C, 0.1–0.5 MPa of molecular oxygen) in an organic solvent and solvent-free conditions. Au–Pd bimetallic catalysts synthesized by the simultaneous condensation method showed improved activity (98% conversion) than the corresponding monometallic catalysts (2%–4% conversion), as well as the Au–Pd catalyst obtained by separate evaporation of the two metals (11%–12% conversion with selectivities around 90% toward benzaldehyde). The authors carried out extensive characterizations by means of XAFS studies and observed for the most active Au–Pd catalyst a clear charge transfer from Pd to Au by XANES data. Additional EXAFS data also showed that the Au–Pd bimetallic nanoparticles composed of an Au-rich core surrounded by a Au–Pd alloyed shell.

A new methodology for the synthesis of highly faceted, icosahedral Au–Pd core–shell nanoparticles was reported by Tilley and coworkers [108]. Their strategy was based on the utilization of a seed-mediated methodology, which allows independent control of the nucleation and growth stage. In this way, the synthesis of Au–Pd core–shell nanoparticles with a variation of Pd thickness around the Au core particle was achieved. The synthesized nanoparticles were supported onto carbon and their catalytic performance was carried out in the solvent-free aerobic oxidation of benzyl alcohol. The maximum activity occurred at a Pd shell thickness of 2.2 nm and they concluded that increasing the faceting of the particles an enhancement of activity as selectivity to benzaldehyde over 70% could be observed. The concept of this work relies on the fact that a very slow shell growth can be beneficial for future nanocrystal catalyst design.

Prati and coworkers showed the beneficial role of Pd to a Au/C catalyst for producing stable and recyclable catalysts for the selective oxidation of HMF to FDCA [109]. Au–Pd/C catalysts showed high selectivity to the desired product (>99%) at 99% conversion, and by varying the Au–Pd atomic ratio it was found that Au–Pd/C catalysts with a Au:Pd atomic ratio of 8:2 were the most promising catalysts showing high activity, selectivity, and durability. Therefore, alloying Au with Pd in an 8:2 atomic ratio produced one of the most active catalysts reported for the oxidation of HMF to FDCA to date.

The catalytic application of Au–Pt catalysts has been reported for efficient aerobic liquid-phase oxidation of alcohols and polyols [110]. Au–Pt-supported nanoparticles were immobilized on carbon using a colloidal methodology with a mean particle size in the range of 3–6 nm. The catalytic performance of these nanoparticles was investigated for the aerobic liquid-phase oxidation of glycerol. It was found that the effect of the reducing agent and the nature of the Pt precursor were affecting activity and selectivity. Au–Pt-supported nanoparticles showed a higher activity and resistance to poisoning in comparison to the monometallic Pt catalysts and a substantial increase of activity with respect to the monometallic Au- or Pt-supported nanoparticles.

In subsequent studies Prati and Hutchings demonstrated that the oxidation of polyols (glycerol, 1,2-propanediol) is possible without the use of a base under mild conditions with high selectivity to the carboxylic acid using Au–Pt nanoparticles supported on mordenite (Prati) [111] and MgO (Hutchings) [112]. A monometallic Au catalyst could not be activated without the use of a base, whereas supported Pt catalysts were efficient under base-free conditions, exhibiting high conversion but with low selectivity to glyceric acid [111]. Colloidal methods were used for the synthesis of Au–Pt nanoparticles with a mean particle size in the range of 2–4 nm.

The utilization of supported Au–Pt nanoparticles for the liquid-phase oxidation of a variety of alcohols has been presented with high yields (90%–99%) of the corresponding aldehydes or ketones at room temperature using atmospheric pressure and benzotrifluoride/water as the solvent (Eq. 12.10) [113]. A colloidal method was used for the synthesis of the active metal nanoparticles. X-ray spectroscopy confirmed the presence of both metals in each nanoparticle and the particle size distribution was in the range of 1.5–5 nm. Catalyst reusability showed that the catalysts are reusable several times without loss of activity.

The oxidation of sugars at mild conditions by using supported bimetallic Au–Pd and Au–Pt nanoparticles instead of supported monometallic Au, Pd, and Pt nanoparticles has been reported by numerous groups. Rossi and coworkers reported the catalytic performance of mono- and bimetallic catalysts based on Au, Pd, and Pt in the aerobic oxidation of glucose to gluconate in the absence and presence of a base by using colloidal methods for the synthesis of mono- and bimetallic supported nanoparticles [114]. The synthesized supported nanoparticles were in the range of 2–5 nm particle size. In the absence of a base, under acidic conditions a significant synergistic effect was observed in the case of using supported Au–Pt catalysts, and an increase of conversion by a factor of 6 was reported with respect to monometallic supported Au and Pt catalysts. Using supported Au–Pd catalysts an increase of conversion by a factor of 2 was reported. In the presence of a base no effect was observed, and the authors attributed the observed catalytic behavior to different reaction mechanisms in the presence and absence of a base.

A new strategy to design novel supported Au–Pd catalysts with high catalytic activity and selectivity in the aerobic glucose oxidation was reported by Toshima and coworkers [115]. The new strategy employs the synthesis of "crown jewel"-structured Au–Pd nanoparticles, where Au atoms are located at the top (vertex or corner sites) of Pd nanoparticles. The crown-jewel-supported Au–Pd nanoparticles showed superior activity with respect to the monometallic Au, Pd, and supported Au–Pd alloy nanoparticles, with the specific activity found to be 20–30 times higher than that of the monometallic Au and Pd catalysts and 8–10 times higher than that of Au–Pd alloy catalysts and with metal particle sizes in a similar range (1.8–2.1 nm). The authors attributed the enhancement in activity to the increase of electronegativity of the Au atoms from donation of electrons from the Pd atoms and polyvinylpyrrolidone molecules by performing XPS and DFT studies. Moreover, another factor that can explain the higher catalytic activity observed was the excess electronic charge on the Au atom that could be transferred to the absorbed oxygen molecule and it could generate a superoxo-like species that play a crucial role in the oxidation of the glucose.

$$\text{1-phenylethanol} \xrightarrow[\text{Water/Benzotrifluoride, 25 °C, O}_2]{\text{AuPt/PI}} \text{Acetophenone} \quad (12.10)$$

### 12.2.3 Supported Au–Cu- and Au–Ag-Based Catalysts

The synthesis and catalytic applications of bimetallic Au–Ag [116, 117] and Au–Cu nanocrystals [118] has recently attracted scientific community and some promising scientific results [119, 120] have been reported in the case of selective oxidation of oxygenated compounds [121]. For example, in the case of 5-hydroxymethyl-2-furfural oxidation, supported Au–Cu nanoparticles were synthesized via a colloidal route and involved the use of microwave heating. The catalytic performance of these bimetallic nanoparticles was carried out by Hutchings, Cavani, and coworkers [122]. Higher activity and stability were achieved by using bimetallic Au–Cu/TiO$_2$ catalysts with a yield of 90% to 5-hydroxymethyl-2-furancarboxylic acid after optimization of reaction conditions. Structural characterization by means of HAADF-STEM-XEDS (X-ray energy dispersive spectrometry) analysis showed that individual nanoparticles were homogeneous Au–Cu alloys with particle sizes around 4–5 nm and were a mixture of decahedral, icosahedral, and cuboctahedral morphology. In subsequent studies the same group studied the effect of the Au/Cu atomic ratio and found that the most active and reusable catalyst was with a Au/Cu atomic ratio of 1 [123]. The authors speculated that the improvement of activity, selectivity, and reusability of the Au–Cu/TiO$_2$ systems is based on the fact that Cu acts as a gold promoter and/or dispersing agent.

In the case of Au–Ag catalysts, Lu and coworkers have reported the preparation of a series of Au–Ag colloidal nanoparticles and the catalytic performance in the liquid-phase oxidation of alcohols with molecular oxygen and in the presence of $Na_2CO_3$ at ambient temperature [124]. The addition of Ag significantly enhanced catalytic activity and the highest oxidation rate was reported for a Au–Ag catalyst with a Au:Ag atomic ratio of 0.95:0.05. The synthesized Au–Ag nanoparticles were prepared by a co-reduction method using P123 as a stabilizer and $NaBH_4$ as a reducing agent. Characterization of the synthesized materials indicated the formation of alloy nanoparticles with a particle size in the range of 3–4 nm. The authors demonstrated the general applicability of the Au–Ag catalysts for the oxidation of aliphatic and aromatic alcohols with a high level of reusability and stability.

### 12.2.4 Supported Au-Based Trimetallic Catalysts

Utilization of Au-based trimetallic catalysts as efficient catalysts for the oxidation of alcohols has been recently reported. The benefit of using trimetallic catalyst is based on the fact that not only an improvement in terms of activity and selectivity but also an improvement in reusability and therefore lifetime of the catalyst could be realized.

Prati and coworkers showed the beneficial role of Bi as a promoter for Au–Pd catalysts in the case of liquid-phase oxidation of alcohols [125]. By using two different protocols Au–Pd–Bi catalysts were prepared, where Bi was deposited either prior or after the immobilization of Au–Pd colloidal nanoparticles onto the support. Characterization studies showed that the prior deposition of Bi and the Bi loading onto the support did not affect the particle size of the Au–Pd nanoparticles but did not allow a homogenous composition of the alloyed Au–Pd phase to be obtained. Conversely, the deposition of Bi on preformed Au–Pd-supported nanoparticles did not affect the alloy composition. In terms of catalysis, in the case of liquid-phase oxidation of benzyl alcohol, a suppression of the parallel reaction pathway that forms toluene was observed. In the case of glycerol oxidation, a promotion of the consecutive reaction leading to tartonate was found with a yield of 78%. The authors attributed the described catalytic performance to the selective inhibition of active Au–Pd sites by Bi.

In another example, Hutchings and coworkers reported the synthesis of supported Au–Pd–Pt nanoparticles and the catalytic performance in the aerobic liquid-phase oxidation of benzyl alcohol under mild conditions [126]. Variation of Au–Pd–Pt atomic ratios were shown to improve selectivity to benazaldehyde (95%) and decrease significantly the formation of toluene as a by-product, while still maintaining a high level of benzyl alcohol conversion. STEM-HAADF-XEDS studies showed that the mean particle size was around 2–3 nm and homogeneous trimetallic alloys were obtained. However, depending on the particle size of the individual nanoparticles a clear size-dependent composition variation was found, with the smaller (<2 nm) particles found to be Pd deficient, whereas particles larger than 10 nm tended to be Pd rich. The authors hypothesized that the beneficial effect of adding Pt atoms seems to be related to the role of Pt as an electronic or a strain modifier to the Au–Pd nanoparticles. In subsequent studies the same groups showed the beneficial role of using Au–Pd–Pt catalysts in the case of glycerol oxidation.

## 12.3 Conclusions and Future Perspectives

A great number of novel Au-based catalysts have been reported in recent years for the clean and efficient oxidation of oxygen-containing compounds with molecular oxygen or hydrogen peroxide and alkyl hydroperoxides as green oxidants. Many of the reported catalysts are based on monometallic supported nanoparticles and more recently bimetallic catalytic systems or even trimetallic systems. A closer examination of the reported bimetallic catalysts reveals that it is important for the bimetallic catalysts not only to control the particle size but also to focus on the synthesis of structures consisting of an alloy or a core–shell structure, since electronic and geometric properties are affected in this manner. In the majority of the investigated reactions, it is general accepted that the synthesis of nanoparticles of a small particle size (2–6 nm) seems to be the key for obtaining high catalytic activity. However, the control of particle size, shape, and effect of each oxidation process should be investigated more systematically. The uniformity and control of the desired structure will play a major role in the development of new and more efficient catalysts in the future, especially for industrial applications. More intense effort is needed in the area of scaling up supported gold-based nanoparticles. Unravelling the reaction pathways affecting the final morphology of metal nanoparticles and performing more intense mechanistic and characterization studies will better aid the understanding of the nature of active sites and eventually will lead to the design and synthesis of better and more efficient materials. Moreover, researchers have focused on the encapsulation of metal nanoparticles inside channels that will provide better stability; therefore this is an area that still needs attention and development. Another point is the substitution of Au catalysts with the utilization of cheaper and more abundant metals. One pathway is to use bimetallic or trimetallic catalysts and recent publications in this area have shown that this area of research is under development. Summarizing, the main challenges that researchers have to deal with in the near future are (i) understanding the formation and the role of active sites, (ii) developing more efficient, cheaper nanoparticles, particularly at lower metal loading, with a prolonged catalytic lifetime, (iii) scaling up nanoparticle synthesis, and (iv) expanding the utilization of Au-based catalysts in a larger variety of organic reactions. Finally, it is evident that gold-based catalysis still remains an area of intense research and is therefore a promising and growing area, with new materials and new reactions to be discovered and explored.

## References

1. Sheldon, R.A., Arends, I., Ten Brink, G,J., and Dijksman, A. (2002). Green catalytic oxidations of alcohols, *Acc. Chem, Res.*, **35**, p. 774781.
2. Sheldon, R.A., Arends, I., and Dijksman, A. (2000). New developments in catalytic alcohol oxidations for fine chemicals synthesis, *Catal. Today*, **57**, pp. 157–166.
3. Sheldon, R.A., Arends, I., and Hanefeld, U. (2007). *Green Chemistry and Catalysis* (Wiley-VCH, Weinheim).
4. Cavani, F. (2010). Catalytic selective oxidation faces the sustainability challenge: turning points, objectives reached, old approaches revisited

and solutions still requiring further investigation, *J. Chem. Technol. Biotechnol.*, **85**, pp. 1175–1183.

5. Mallat, T., and Baiker, A. (1994). Oxidation of alcohols with molecular-oxygen on platinum metal-catalysts in aqueous-solutions, *Catal. Today*, **19**, pp. 247–283.

6. Besson, M., and Gallezot, P. (2000). Selective oxidation of alcohols and aldehydes on metal catalysts, *Catal. Today*, **57**, pp. 127–141.

7. Matsumoto, T., Ueno, M., Wang, N., and Kobayashi, S. (2008). Recent advances in immobilized metal catalysts for environmentally benign oxidation of alcohols, *Chem. Asian J.*, **3**, pp. 196–214.

8. Mallat, T., and Baiker, A. (2004). Oxidation of alcohols with molecular oxygen on solid catalysts, *Chem. Rev.*, **104**, pp. 3037–3058.

9. Vinod, C.P., Wilson, K., and Lee, A.F. (2011). Recent advances in the heterogeneously catalysed aerobic selective oxidation of alcohols, *J. Chem. Technol. Biotechnol.*, **86**, pp. 161–171.

10. Hashmi, A.S.K., and Hutchings, G.J. (2006). Gold catalysis, *Angew. Chem., Int. Ed.*, **45**, pp. 7896–7936.

11. Della Pina, C., Falletta, E., Prati, L., and Rossi, M. (2008). Selective oxidation using gold, *Chem. Soc. Rev.*, **37**, pp. 2077–2095.

12. Campelo, J.M., Luna, D., Luque, R., Marinas, J.M., and Romero, A.A. (2009). Sustainable preparation of supported metal nanoparticles and their applications in catalysis, *ChemSusChem*, **2**, pp. 18–45.

13. Prati, L., Villa, A., Lupini, A.R., and Veith, G.M. (2012). Gold on carbon: one billion catalysts under a single label, *Phys. Chem. Chem. Phys.*, **14**, pp. 2969–2978.

14. Dimitratos, N., Lopez-Sanchez, J.A., and Hutchings, G.J. (2012). Slective liquid phase oxidation with supported metal nanoparticles, *Chem. Sci.*, **3**, pp. 20–44.

15. Prati, L., and Rossi, M. (1998). Gold on carbon as a new catalyst for selective liquid phase oxidation of diols, *J. Catal.*, **176**, pp. 552–560.

16. Prati, L., and Martra, G. (1999). New gold catalysts for liquid phase oxidation, *Gold Bull.*, **32**, pp. 96–101.

17. Bianchi, C., Porta, F., Prati, L., and Rossi, M. (2000). Selective liquid phase oxidation using gold catalysts, *Top. Catal.*, **13**, pp. 231–236.

18. Porta, F., Prati, L., Rossi, M., Coluccia, S., and Martra, G. (2000). Metal sols as a useful tool for heterogeneous gold catalyst preparation: reinvestigation of a liquid phase oxidation, *Catal. Today*, **61**, pp. 165–172.

19. Biella, S., Prati, L., and Rossi, M. (2002). Selective oxidation of D-glucose on gold catalyst, *J. Catal.*, **206**, pp. 242–247.
20. Comotti, M., Della Pina, C., Matarrese, R., and Rossi, M. (2004). The catalytic activity of "naked" gold particles, *Angew. Chem., Int. Ed.*, **43**, pp. 5812–5815.
21. Comotti, M., Della Pina, C., Falletta, E., and Rossi, M. (2006). Is the biochemical route always advantageous? The case of glucose oxidation, *J. Catal.*, **244**, pp. 122–125.
22. Onal, Y., Schimpf, S., and Claus, P. (2004). Structure sensitivity and kinetics of D-glucose oxidation to D-gluconic acid over carbon-supported gold catalysts, *J. Catal.*, **223**, pp. 122–133.
23. Baatz, C., Thielecke, N., and Prusse, U. (2007). Influence of the preparation conditions on the properties of gold catalysts for the oxidation of glucose, *Appl. Catal. B: Environ.*, **70**, pp. 653–660.
24. Ishida, T., Kinoshita, N., Okatsu, H., Akita, T., Takei, T., and Haruta, M. (2008). Influence of the support and the size of gold clusters on catalytic activity for glucose oxidation, *Angew. Chem., Int. Ed.*, **47**, pp. 9265–9268.
25. Ma, C.Y., Xue, W.J., Li, J.J., Xing, W., and Hao, Z.P. (2013) Mesoporous carbon-confined Au catalysts with superior activity for selective oxidation of glucose to gluconic acid, *Green Chem.*, **15**, pp. 1035–1041.
26. Gallezot, P. (2007). Catalytic routes from renewables to fine chemicals, *Catal. Today*, **121**, pp. 76–91.
27. Behr, A., Eilting, J., Irawadi, K., Leschinski, J., and Lindner, F. (2008). Improved utilisation of renewable resources: new important derivatives of glycerol, *Green Chem.*, **10**, pp. 13–30.
28. Corma, A., Iborra, S., and Velty, A. (2007). Chemical routes for the transformation of biomass into chemicals, *Chem. Rev.*, **107**, pp. 2411–2502.
29. Zhou, C.H.C., Beltramini, J.N., Fan, Y.X., and Lu, G.Q.M. (2008). Chemoselective catalytic conversion of glycerol as a biorenewable source to valuable commodity chemicals, *Chem. Soc. Rev.*, **37**, pp. 527–549.
30. Pagliaro, M., Ciriminna, R., Kimura, H., Rossi, M., and Della Pina, C. (2007). From glycerol to value-added products, *Angew. Chem., Int. Ed.*, **46**, pp. 4434–4440.
31. Carrettin, S., McMorn, P., Johnston, P., Griffin, K., and Hutchings, G.J. (2002). Selective oxidation of glycerol to glyceric acid using a gold catalyst in aqueous sodium hydroxide, *Chem. Commun.*, **7**, pp. 696–697.

32. Carrettin, S., McMorn, P., Johnston, P., Griffin, K., Kiely, C.J., and Hutchings, G.J. (2003). Oxidation of glycerol using supported Pt, Pd and Au catalysts, *Phys. Chem. Chem. Phys.*, **5**, pp. 1329–1336.
33. Porta, F., and Prati, L. (2004). Selective oxidation of glycerol to sodium glycerate with gold-on-carbon catalyst: an insight into reaction selectivity, *J. Catal.*, **224**, pp. 397–403.
34. Demirel-Gulen, S., Lucas, M., and Claus, P. (2005). Liquid phase oxidation of glycerol over carbon supported gold catalysts, *Catal. Today*, **102**, pp. 166–172.
35. Demirel, S., Kern, P., Lucas, M., and Claus, P. (2007). Oxidation of mono- and polyalcohols with gold: comparison of carbon and ceria supported catalysts, *Catal. Today*, **122**, pp. 292–300.
36. Ketchie, W.C., Fang, Y.L., Wong, M.S., Murayama, M., and Davis, R.J. (2007). Influence of gold particle size on the aqueous-phase oxidation of carbon monoxide and glycerol, *J. Catal.*, **250**, pp. 94–101.
37. Ketchie, W.C., Murayama, M., and Davis, R.J. (2007). Promotional effect of hydroxyl on the aqueous phase oxidation of carbon monoxide and glycerol over supported Au catalysts, *Top. Catal.*, **44**, pp. 307–317.
38. Zope, B.N., Hibbitts, D.D., Neurock, M., and Davis, R.J. (2010). Reactivity of the gold/water interface during selective oxidation catalysis, *Science*, **330**, pp. 74–78.
39. Dimitratos, N., Villa, A., Bianchi, C.L., Prati, L., and Makkee, M. (2006). Gold on titania: Effect of preparation method in the liquid phase oxidation, *Appl. Catal. A: Gen.*, **311**, pp. 185–192.
40. Villa, A., Gaiassi, A., Rossetti, I., et al. (2010). Au on $MgAl_2O_4$ spinels: The effect of support surface properties in glycerol oxidation, *J. Catal.*, **275**, pp. 108–116.
41. Prati, L., Villa, A., Chan-Thaw, C.E., Arrigo, R., Wang, D., and Su, D.S. (2011). Gold catalyzed liquid phase oxidation of alcohol: the issue of selectivity, *Faraday Discuss.*, **152**, pp. 353–365.
42. Villa, A., Veith, G.M., Ferri, D., Weidenkaff, A., Perry, K.A., Campisi, S., and Prati. L. (2013). NiO as a peculiar support for metal nanoparticles in polyols oxidation, *Catal. Sci. Technol.*, **3**, pp. 394–399.
43. Wang, D., Villa, A., Su, D.S., Prati, L., and Schlogl, R. (2013). Carbon-supported gold nanocatalysts: shape effect in the selective glycerol oxidation, *ChemCatChem*, **5**, pp. 2717–2723.
44. Villa, A., Wang, D., Veith, G.M., Vindigni, F., and Prati, L. (2013). Sol immobilization technique: a delicate balance between activity,

selectivity and stability of gold catalysts, *Catal. Sci. Technol.*, **3**, pp. 3036–3041.

45. Milone, C., Ingoglia, R., Neri, G., Pistone, A., and Galvagno, S. (2001). Gold catalysts for the liquid phase oxidation of o-hydroxybenzyl alcohol, *Appl. Catal. A: Gen.*, **211**, pp. 251–257.

46. Abad, A., Concepcion, P., Corma, A., and Garcia, H. (2005). A collaborative effect between gold and a support induces the selective oxidation of alcohols, *Angew. Chem., Int. Ed.*, **44**, pp. 4066–4069.

47. Abad, A., Almela, C., Corma, A., Garcia, H. (2006). Unique gold chemoselectivity for the aerobic oxidation of allylic alcohols, *Chem. Commun.*, **30**, pp. 3178–3180.

48. Christensen, C.H., Jorgensen, B., Rass-Hansen, J., Egeblad, K., Madsen, R., and Klitgaard, S.K. (2006). Formation of acetic acid by aqueous-phase oxidation of ethanol with air in the presence of a heterogeneous gold catalyst, *Angew. Chem., Int. Ed.*, **45**, pp. 4648–4651.

49. Hu, J.C., Chen, L.F., Zhu, K.K., Suchopar, A., and Richards, R. (2007). Aerobic oxidation of alcohols catalyzed by gold nano-particles confined in the walls of mesoporous silica, *Catal. Today*, **122**, pp. 277–283.

50. Haider, P., and Baiker, A. (2007) Gold supported on Cu-Mg-Al-mixed oxides: strong enhancement of activity in aerobic alcohol oxidation by concerted effect of copper and magnesium, *J. Catal.*, **248**, pp. 175–187.

51. Haider, P., Grunwaldt, J.D., Seidel, R., and Baiker, A. (2007). Gold supported on Cu.-Mg-Al and Cu-Ce mixed oxides: an in situ XANES study on the state of Au during aerobic alcohol oxidation, *J. Catal.*, **250**, pp. 313–323.

52. Biffis, A., Cunial, S., Spontoni, P., and Prati, L. (2007). Microgel-stabilized gold nanoclusters: powerful "quasi-homogeneous" catalysts for the aerobic oxidation of alcohols in water, *J. Catal.*, **251**, pp. 1–6.

53. Wang, L.C., Liu, Y.M., Chen, M., Cao, Y., He, H.Y., and Fan, K.N. (2008). MnO$_2$ nanorod supported gold nanoparticles with enhanced activity for solvent-free aerobic alcohol oxidation, *J. Phys. Chem. C*, **112**, pp. 6981–6987.

54. Su, F.Z., Chen, M., Wang, L.C., Huang, X.S., Liu, Y.M., and Cao, Y. (2008). Aerobic oxidation of alcohols catalyzed by gold nanoparticles supported on gallia polymorphs, *Catal. Commun.*, **9**, pp. 1027–1032.

55. Su, F.Z., Liu, Y.M., Wang, L.C., Cao, Y., He, H.Y., and Fan, K.N. (2008). Ga-Al mixed-oxide-supported gold nanoparticles with enhanced activity for aerobic alcohol oxidation, *Angew. Chem., Int. Ed.*, **47**, pp. 334–337.

56. Han, J., Liu, Y., and Guo, R. (2009). Reactive template method to synthesize gold nanoparticles with controllable size and morphology supported on shells of polymer hollow microspheres and their application for aerobic alcohol oxidation in water, *Adv. Funct. Mater.*, **19**, pp. 1112–1117.
57. Yang, J., Guan, Y.J., Verhoeven, T., van Santen, R., Li, C., and Hensen, E.J.M. (2009). Basic metal carbonate supported gold nanoparticles: enhanced performance in aerobic alcohol oxidation, *Green Chem.*, **11**, pp. 322–325.
58. Liu, Y.M., Tsunoyama, H., Akita, T., and Tsukuda, T. (2009). Preparation of similar to 1 nm gold clusters confined within mesoporous silica and microwave-assisted catalytic application for alcohol oxidation, *J. Phys. Chem. C*, **113**, pp. 13457–13461.
59. Wang, Y., Yan, R., Zhang, J.Z., and Zhang, W.Q. (2010). Synthesis of efficient and reusable catalyst of size-controlled Au nanoparticles within a porous, chelating and intelligent hydrogel for aerobic alcohol oxidation, *J. Mol. Catal. A: Chem.*, **317**, pp. 81–88.
60. Oliveira, R.L., Kiyohara, P.K., and Rossi, L.M. (2010). High performance magnetic separation of gold nanoparticles for catalytic oxidation of alcohols, *Green Chem.*, **12**, pp. 144–149.
61. Chen, K., Wu, H.T., Hua, Q., Chang, S.J., and Huang, W.X. (2013). Enhancing catalytic selectivity of supported metal nanoparticles with capping ligands, *Phys. Chem. Chem. Phys.*, **15**, pp. 2273–2277.
62. Alhumaimess, M., Lin, Z., Weng, W., et al. (2012). Oxidation of benzyl alcohol by using gold nanoparticles supported on ceria foam, *ChemSusChem*, **5**, pp. 125–131.
63. Wang, M., Wang, F., Ma, J., et al. (2014). Investigations on the crystal plane effect of ceria on gold catalysts in the oxidative dehydrogenation of alcohols and amines in the liquid phase, *Chem. Commun.*, **3**, pp. 292–294.
64. Zhao, J.B., Liu, H., Ye, S., et al. (2013). Half-encapsulated Au nanoparticles by nano iron oxide: promoted performance of the aerobic oxidation of 1-phenylethanol, *Nanoscale*, **5**, pp. 9546–9552.
65. Hudlicky, M. (1990). *Oxidations in Organic Chemistry* (American Chemical Society, Washington, DC).
66. Taft, R.W., Newman, M.S., and Verhoek, F.H. (1950). The kinetics of the base-catalyzed methanolysis of ortho-substituted, meta-substituted and para-substituted l-menthyl benzoates, *J. Am. Chem. Soc.*, **72**, pp. 4511–4519.

67. Klitgaard, S.K., DeLa Riva, A.T., Helveg, S., Werchmeister, R.M., and Christensen, C.H. (2008). Aerobic oxidation of alcohols over gold catalysts: role of acid and base, *Catal. Lett.*, **126**, pp. 213–217.
68. Marsden, C., Taarning, E., Hansen, D., et al. (2008). Aerobic oxidation of aldehydes under ambient conditions using supported gold nanoparticle catalysts, *Green Chem.*, **10**, pp. 168–170.
69. Taarning, E., Madsen, A.T., Marchetti, J.M., Egeblad, K., and Christensen, C.H. (2008). Oxidation of glycerol and propanediols in methanol over heterogeneous gold catalysts, *Green Chem.*, **10**, pp. 408–414.
70. Oliveira, R.L., Kiyohara, P.K., and Rossi, L.M. (2009). Clean preparation of methyl esters in one-step oxidative esterification of primary alcohols catalyzed by supported gold nanoparticles, *Green Chem.*, **11**, pp. 1366–1370.
71. Choudhary, V.R., Dumbre, D.K., and Bhargava, S.K. (2009). Oxidation of benzyl alcohol to benzaldehyde by tert-butyl hydroperoxide over nanogold supported on $TiO_2$ and other transition and rare-earth metal oxides, *Ind. Eng. Chem. Res.*, **48**, pp. 9471–9478.
72. Choudhary, V.R., and Dumbre, D.K. (2009). Magnesium oxide supported nano-gold: a highly active catalyst for solvent-free oxidation of benzyl alcohol to benzaldehyde by TBHP, *Catal. Commun.*, **10**, pp. 1738–1742.
73. Edwards, J.K., Ntainjua, E., Carley, A.F., Herzing, A.A., Kiely, C.J., and Hutchings, G.J. (2009). Direct synthesis of $H_2O_2$ from $H_2$ and $O_2$ over gold, palladium, and gold-palladium catalysts supported on acid-pretreated $TiO_2$, *Angew. Chem., Int. Ed.*, **48**, pp. 8512–8525.
74. Edwards, J.K., Solsona, B., N, E.N., et al. (2009). Switching off hydrogen peroxide hydrogenation in the direct synthesis process, *Science*, **323**, pp. 1037–1041.
75. Ni, J., Yu, W.J., He, L., et al. (2009). A green and efficient oxidation of alcohols by supported gold catalysts using aqueous H2O2 under organic solvent-free conditions, *Green Chem.*, **11**, pp. 756–759.
76. Liu, Y.M., Tsunoyama, H., Akita, T., and Tsukuda, T. (2010). Size effect of silica-supported gold clusters in the microwave-assisted oxidation of benzyl alcohol with $H_2O_2$, *Chem. Lett.*, **39**, pp. 159–161.
77. Taarning, E., Nielsen, I.S., Egeblad, K., Madsen, R., and Christensen, C.H. (2008). Chemicals from renewables: Aerobic oxidation of furfural and hydroxymethylfurfural over gold catalysts, *ChemSusChem*, 1, pp. 75–78.
78. Gorbanev, Y.Y., Klitgaard, S.K., Woodley, J.M., Christensen, C.H., and Riisager, A. (2009). Gold-Catalyzed Aerobic Oxidation of 5-

Hydroxymethylfurfural in Water at Ambient Temperature. *ChemSusChem*, **2**, pp. 672–675.

79. Casanova, O., Iborra, S., and Corma, A. Biomass into chemicals, pp. (2009). One pot-base free oxidative esterification of 5-hydroxymethyl-2-furfural into 2,5-dimethylfuroate with gold on nanoparticulated ceria. *J. Catal.*, **265**, pp. 109–116.

80. Casanova, O., Iborra, S., and Corma, A. Biomass into Chemicals, pp. (2009). Aerobic Oxidation of 5-Hydroxymethyl-2-furfural into 2,5-Furandicarboxylic Acid with Gold Nanoparticle Catalysts. *ChemSusChem*, **2**, pp. 1138–1144.

81. Grirrane, A., Corma, A., and Garcia, H. (2009). Gold nanoparticles supported on ceria promote the selective oxidation of oximes into the corresponding carbonylic compounds, *J. Catal.*, **268**, pp. 350–355.

82. Dimitratos, N., and Prati, L. (2005). Gold based bimetallic catalysts for liquid phase applications, *Gold Bull.*, **38**, pp. 73–77.

83. Dimitratos, N., Porta, F., Prati, L., and Villa, A. (2005). Synergetic effect of platinum or palladium on gold catalyst in the selective oxidation of D-sorbitol, *Catal. Lett.*, **99**, pp. 181–185.

84. Bianchi, C.L., Canton, P., Dimitratos, N., Porta, F., and Prati, L. (2005). Selective oxidation of glycerol with oxygen using mono and bimetallic catalysts based on Au, Pd and Pt metals, *Catal. Today*, **102**, pp. 203–212.

85. Dimitratos, N., Porta, F., and Prati, L. (2005). Au, Pd (mono and bimetallic) catalysts supported on graphite using the immobilisation method: synthesis and catalytic testing for liquid phase oxidation of glycerol, *Appl. Catal. A. Gen.*, **291**, pp. 210–214.

86. Villa, A., Campione, C., and Prati, L. (2007). Bimetallic gold/palladium catalysts for the selective liquid phase oxidation of glycerol, *Catal. Lett.*, **115**, pp. 133–136.

87. Wang, D., Villa, A., Porta, F., and Prati, L., Su, D.S. (2008). Bimetallic gold/palladium catalysts: correlation between nanostructure and synergistic effects, *J. Phys. Chem. C*, **112**, pp. 8617–8622.

88. Wang, D., Villa, A., Porta, F., Su, D.S., and Prati, L. (2006). Single-phase bimetallic system for the selective oxidation of glycerol to glycerate, *Chem. Commun.*, **18**, pp. 1956–1948.

89. Prati, L., Villa, A., Porta, F., Wang, D., and Su, D.S. (2007). Single-phase gold/palladium catalyst: the nature of synergistic effect, *Catal. Today*, **122**, pp. 386–390.

90. Dimitratos, N., Villa, A., Wang, D., Porta, F., Su, D.S., and Prati, L. (2006). Pd and Pt catalysts modified by alloying with Au in the selective oxidation of alcohols, *J. Catal.*, **244**, pp. 113–121.
91. Villa, A., Janjic, N., Spontoni, P., Wang, D., Su, D.S., and Prati, L. (2009). Au-Pd/AC as catalysts for alcohol oxidation: effect of reaction parameters on catalytic activity and selectivity, *Appl. Catal. A: Gen.*, **364**, pp. 221–228.
92. Enache, D.I., Edwards, J.K., Landon, P., et al. (2006). Solvent-free oxidation of primary alcohols to aldehydes using Au-Pd/TiO$_2$ catalysts, *Science*, **311**, pp. 362–365.
93. Enache, D.I., Barker, D., Edwards, J.K., et al. (2007). Solvent-free oxidation of benzyl alcohol using titanic-supported gold-palladium catalysts: effect of Au-Pd ratio on catalytic performance, *Catal. Today*, **122**, pp. 407–411.
94. Lopez-Sanchez, J.A., Dimitratos, N., Miedziak, P., et al. (2008). Au-Pd supported nanocrystals prepared by a sol immobilisation technique as catalysts for selective chemical synthesis, *Phys. Chem. Chem. Phys.*, **10**, pp. 1921–1930.
95. Dimitratos, N., Lopez-Sanchez, J.A., Morgan, D., et al. (2009). Solvent-free oxidation of benzyl alcohol using Au-Pd catalysts prepared by sol immobilization, *Phys. Chem. Chem. Phys.*, **11**, pp. 5142–5153.
96. Dimitratos, N., Lopez-Sanchez, J.A., Anthonykutty, J.M., et al. (2009). Oxidation of glycerol using gold-palladium alloy-supported nanocrystals, *Phys. Chem. Chem. Phys.*, **11**, pp. 4952–4961.
97. Dimitratos, N., Lopez-Sanchez, J.A., Meenakshisundaram, S., et al. (2009). Selective formation of lactate by oxidation of 1,2-propanediol using gold palladium alloy supported nanocrystals, *Green Chem.*, 11, pp. 1209–1216.
98. Pritchard, J., Kesavan, L., Piccinini, M., et al. (2010). Direct synthesis of hydrogen peroxide and benzyl alcohol oxidation using Au-Pd catalysts prepared by sol immobilization, *Langmuir*, **26**, pp. 16568–16577.
99. Meenakshisundaram, S., Nowicka, E., Miedziak, P.J., et al. (2010). Oxidation of alcohols using supported gold and gold-palladium nanoparticles, *Faraday Discuss.*, **145**, pp. 341–356.
100. Sankar, M., Nowicka, E., Tiruvalam, R., et al. (2011). Controlling the duality of the mechanism in liquid-phase oxidation of benzyl alcohol catalysed by supported Au-Pd nanoparticles, *Chem. Eur. J.*, **17**, pp. 6524–6532.

101. Tiruvalam, R.C., Pritchard, J.C., Dimitratos, N., et al. (2011). Aberration corrected analytical electron microscopy studies of sol-immobilized Au plus Pd, Au{Pd} and Pd{Au} catalysts used for benzyl alcohol oxidation and hydrogen peroxide production, *Faraday Discuss.*, **152**, pp. 63–86.
102. Pritchard, J., Piccinini, M., Tiruvalam, R., et al. (2013). Effect of heat treatment on Au-Pd catalysts synthesized by sol immobilisation for the direct synthesis of hydrogen peroxide and benzyl alcohol oxidation, *Catal. Sci. Technol.*, **3**, pp. 308–317.
103. Marx, S., and Baiker, A. (2009). Beneficial interaction of gold and palladium in bimetallic catalysts for the selective oxidation of benzyl alcohol, *J. Phys. Chem. C*, **113**, pp. 6191–6201.
104. Ma, C.Y., Dou, B.J., Li, J.J., et al. (2009). Catalytic oxidation of benzyl alcohol on Au or Au-Pd nanoparticles confined in mesoporous silica, *Appl. Catal. B: Environ.*, **92**, pp. 202–208.
105. Chen, Y.T., Lim, H.M., Tang, Q.H., et al. (2010). Solvent-free aerobic oxidation of benzyl alcohol over Pd monometallic and Au-Pd bimetallic catalysts supported on SBA-16 mesoporous molecular sieves, *Appl. Catal. A. Gen.*, **380**, pp. 55–65.
106. Balcha, T., Strobl, J.R., Fowler, C., Dash, P., and Scott, R.W.J. (2011). Selective aerobic oxidation of crotyl alcohol using AuPd core-shell nanoparticles, *ACS Catal.*, **1**, pp. 425–436.
107. Evangelisti, C., Schiavi, E., Aronica, L.A., et al. (2012). Bimetallic gold-palladium vapour derived catalysts: the role of structural features on their catalytic activity, *J. Catal.*, **286**, pp. 224–236.
108. Henning, A.M., Watt, J., Miedziak, P.J., et al. (2013). Gold-palladium core-shell nanocrystals with size and shape control optimized for catalytic performance, *Angew. Chem., Int. Ed.*, **52**, pp. 1477–1480.
109. Villa, A., Schiavoni, M., Campisi, S., Veith, G.M., and Prati, L. (2013). Pd-modified Au on carbon as an effective and durable catalyst for the direct oxidation of HMF to 2,5-furandicarboxylic acid, *ChemSusChem*, **6**, pp. 609–612.
110. Dimitratos, N., Messi, C., Porta, F., Prati, L., and Villa, A. (2006). Investigation on the behaviour of Pt(0)/carbon and Pt(0), Au(0)/carbon catalysts employed in the oxidation of glycerol with molecular oxygen in water, *J. Mol. Catal. A.: Chem.*, **256**, pp. 21–28.
111. Villa, A., Veith, G.M., and Prati, L. (2010). Selective oxidation of glycerol under acidic conditions using gold catalysts, *Angew. Chem., Int. Ed.*, **49**, pp. 4499–4502.

112. Brett, G.L., He, Q., Hammond, C., et al. (2011). Selective oxidation of glycerol by highly active bimetallic catalysts at ambient temperature under base-free conditions, *Angew. Chem., Int. Ed.*, **50**, pp. 10136–10139.

113. Miyamura, H., Matsubara, R., and Kobayashi, S. (2008) Gold-platinum bimetallic clusters for aerobic oxidation of alcohols under ambient conditions, *Chem. Commun.*, **17**, pp. 2031–2033.

114. Comotti, M., Della Pina, C., and Rossi, M. (2006) Mono- and bimetallic catalysts for glucose oxidation, *J. Mol. Catal. A.: Chem.*, **251**, pp. 89–92.

115. Zhang, H.J., Watanabe, T., Okumura, M., Haruta, M., and Toshima, N. (2013). Crown jewel catalyst: how neighboring atoms affect the catalytic activity of top Au atoms?, *J Catal.*, **305**, pp. 7–18.

116. Ma, Y.Y., Li, W.Y., Cho, E.C., et al. (2010). Au@Ag core-shell nanocubes with finely tuned and well-controlled sizes, shell thicknesses, and optical properties, *ACS Nano*, **4**, pp. 6725–6734.

117. Guan, Y., Zhao, N., Tang, B., et al. (2013). A stable bimetallic Au-Ag/TiO$_2$ nanopaper for aerobic oxidation of benzyl alcohol, *Chem. Commun.*, **49**, pp. 11524–11526.

118. Xia, Y.N., Xiong, Y.J., Lim, B., and Skrabalak, S.E. (2009). Shape-controlled synthesis of metal nanocrystals: simple chemistry meets complex physics?, *Angew. Chem., Int. Ed.*, **48**, pp. 60–103.

119. Li, W.J., Wang, A.Q., Liu, X.Y., and Zhang, T. (2012). Silica-supported Au-Cu alloy nanoparticles as an efficient catalyst for selective oxidation of alcohols, *Appl. Catal. A.: Gen.*, **433**, pp. 146–151.

120. Della Pina, C., Falletta, E., and Rossi, M. (2008). Highly selective oxidation of benzyl alcohol to benzaldehyde catalyzed by bimetallic gold-copper catalyst, *J. Catal.*, **260**, pp. 384–386.

121. Della Pina, C., Falletta, E., and Rossi, M. (2012). Update on selective oxidation using gold, *Chem. Soc. Rev.*, **41**, pp. 350–369.

122. Pasini, T., Piccinini, M., Blosi, M., et al. (2011). Selective oxidation of 5-hydroxymethyl-2-furfural using supported gold-copper nanoparticles, *Green Chem.*, **13**, pp. 2091–2099.

123. Albonetti, S., Pasini, T., Lolli, A., et al. (2012). Selective oxidation of 5-hydroxymethyl-2-furfural over TiO$_2$-supported gold-copper catalysts prepared from preformed nanoparticles: effect of Au/Cu ratio, *Catal. Today*, **195**, pp. 120–126.

124. Huang, X.M., Wang, X.G., Tan, M.W., Zou, X.J., Ding, W.Z., and Lu, X.G. (2013). Selective oxidation of alcohols on P123-stabilized Au-Ag alloy

nanoparticles in aqueous solution with molecular oxygen, *Appl. Catal. A: Gen.*, **467**, pp. 407–413.

125. Villa, A., Wang, D., Veith, G.M., and Prati, L. (2012). Bismuth as a modifier of Au-Pd catalyst: enhancing selectivity in alcohol oxidation by suppressing parallel reaction, *J Catal.*, **292**, pp. 73–80.

126. He, Q., Miedziak, P.J., Kesavan, L., et al. (2013). Switching-off toluene formation in the solvent-free oxidation of benzyl alcohol using supported trimetallic Au-Pd-Pt nanoparticles, *Faraday Discuss.*, **162**, pp. 365–378.

## Chapter 13

# Supported Gold Nanoparticles as Heterogeneous Catalysts for C–C Coupling Reactions

### Ana Primo and Hermenegildo García

*Instituto Universitario de Tecnología Química CSIC-UPV,
Universidad Politécnica de Valencia, Avda/ de Los Naranjos s/n,
46022 Valencia, Spain*
hgarcia@qim.upv.es

This chapter is focused on reported C–C bond-forming reactions catalyzed by supported gold catalysts. Initially we introduce the importance of palladium as a catalyst for these cross-coupling reactions and the fact that based on the isoelectronic configuration of the Pd(0)/Pd(II) pair with the Au(I)/Au(III) couple, there was an interest in determining the activity of supported gold catalysts for this type of reaction. In the chapter we emphasize the role of the support cooperating with the reaction mechanism, describing in detail the case of nanoparticulate ceria as a support. We have also commented on the controversy about the influence of palladium impurities on the activity of gold catalysts and how theoretical calculations indicate the notable differences in the reaction pathways between gold complexes and gold clusters and nanoparticles. Light assistance by irradiating gold nanoparticles at

---

*Gold Catalysis: Preparation, Characterization, and Applications*
Edited by Laura Prati and Alberto Villa
Copyright © 2016 Pan Stanford Publishing Pte. Ltd.
ISBN 978-981-4669-28-3 (Hardcover), 978-981-4669-29-0 (eBook)
www.panstanford.com

their characteristic surface plasmon band absorption with visible light for the Suzuki–MIyaura coupling of aryl bromides and chlorides with arylboronic acids is described and justified. The final section summarizes the state of the art in this field and how highly active supported gold catalysts for cross-couplings are still to be developed.

## 13.1 Palladium as a Catalyst for Carbon–Carbon and Carbon–Heteroatom Cross-Coupling Reactions

Carbon–carbon bond formation is certainly one of the crucial steps in any organic synthesis [1]. In classical organic synthesis, C–C bond formation is limited to a few types of reactions, many of them requiring harsh conditions and characterized by poor atom efficiency generating a considerable amount of waste. One classical example of these conventional C–C bond-forming reactions is Grignard synthesis by reaction of an organomagnesium compound with an aldehyde or ketone, leading to the formation of a secondary or tertiary alcohol [2, 3]. In this paradigmatic example, preparation of the organomagnesium compound requires an alkyl halide as a starting material and upon the formation of the alcohol, magnesium halides as well as other basic compounds are formed as wastes in equivalent amounts with respect to the target products. Also, similarly, aldolic-type reactions may need over stoichiometric amounts of strong acids or bases that have to be neutralized after the reaction during the workup. In these examples, in addition to the generation of waste and the consumption of metals, the scope of the substrates has several limitations and certain functional groups having an acid character or being susceptible to nucleophilic attacks are not compatible with the C–C bond formation reaction. In addition, a third common problem of classical C–C bond-forming reactions is related to the experimental conditions that may require dry solvents, absence of humidity, and oxygen.

Due to the above-commented drawbacks, the use of palladium as a catalyst to promote C–C bond-forming reactions has constituted one of the major achievements in organic synthesis and catalysis in the second half of the last century [4, 5]. In a certain way, the use of palladium converts organometallic synthesis from a reaction consuming stoichiometric amounts of metals to a catalytic process from the point of view of the amount of metal needed in the process, palladium in this case. Also, additional advantages making superior palladium-catalyzed reactions with respect to classical Grignard synthesis are the milder reaction conditions that do not require to exclude moisture or oxygen and the compatibility with most of the common organic functional groups, particularly OH and CO. Palladium-catalyzed reactions have also considerable flexibility with respect to the substrate that can undergo carbon–carbon formation and, therefore, in the type of product that can be obtained. Scheme 13.1 summarizes some of the palladium-catalyzed cross-coupling reactions with an indication of the reaction name.

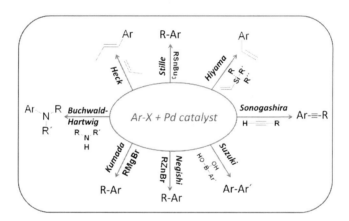

**Scheme 13.1** Summary of common palladium-catalyzed reactions with an indication of reaction names, reagents, and products.

The reaction mechanism of palladium-catalyzed carbon–carbon reactions may involve several steps but have in common the oxidative addition of Pd(0) species to the C–X or carbon–heteroatom bond to form an organopalladium intermediate that, under the reaction conditions, continues its transformation into the products generally by transmetalation in where the palladium–carbon bond is broken and closes the cycle with the reductive elimination. Scheme 13.2 illustrates the elementary steps that are common

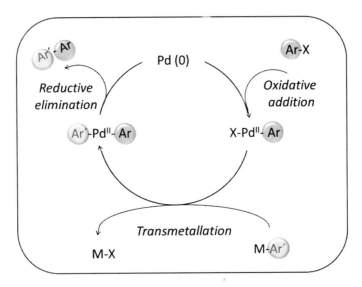

**Scheme 13.2** Elementary steps involving Pd(0) and Pd(II) species that are common in several Pd-catalyzed coupling reactions.

to many palladium-catalyzed carbon–carbon bond formation reactions.

Another feature of palladium-catalyzed reactions is that besides carbon–carbon bond formation, they can also be adapted for carbon–heteroatom bond formation (see Scheme 13.1), thus widening considerably the scope of substrates and leading to products that are difficult to obtain by other classical reactions [6].

One general trend in catalysis has been to convert a successful homogeneous reaction into a heterogeneous process [7, 8]. Initially, palladium catalysis was led by homogeneous catalysis in which the precursor of the catalytically relevant species was dissolved in the reaction media together with substrates and products. The main advantage of homogeneous catalysis is that by the use of appropriate ligands, the electron density and steric factors around the palladium can be modified in a simple manner, and in this way, highly active palladium catalyst were developed [9, 10]. However, in spite of the high activity and selectivity that some of the palladium complexes enjoy, the need to obtain products with very low metal content, together with the advantages of recovering palladium metal allowing its reuse, make it advisable to develop heterogeneous palladium catalysts. One obvious strategy for the development of heterogeneous catalysts has been to anchor or immobilize the homogeneous catalyst or a suitable derivative onto a large-surface-area insoluble solid [11]. In this way, the complex can be recovered after the reaction by filtration of the solid and reused if the catalyst has not become deactivated. The main drawback of this approach is that the stability of many palladium complexes is not so high, and after several turnovers in which the palladium complex is submitted to stress due to the change of the coordination number from di-coordinated to tetra-coordinated, decomposition of the complex may occur. This lack of stability makes the whole strategy for heterogeneization doubtful, until complete stability is not firmly demonstrated. An alternative strategy consists of anchoring on a solid support palladium metal nanoparticles (NPs) that are presumably the final palladium species after complex decomposition and agglomeration of the palladium atoms and clusters. The main problem of this strategy is that the activity of palladium NPs is much lower than the one that can be achieved by palladium complexes, and as a consequence, in addition to the diffusion limitations due to phase transfer, there are other problems associated with the lower intrinsic catalytic activity of the palladium NPs, particularly for the activation of substrates with low reactivity.

## 13.2 Gold vs. Palladium

Soon after the development of heterogeneous gold catalysis another research front associated with it was to delineate the specific catalytic activity of gold compared to other noble and transition metals [12]. In this context, Pd(0) and Pd(II) are isoelectronic to Au(I) and Au(III), and because of the interest in exploring the new opportunities offered by gold catalysis, there was an obvious interest in assessing if gold can also promote those carbon–carbon coupling reactions catalyzed by palladium previously commented.

Gold was considered up to the early 1980s as a noble metal devoid of any catalytic property. In fact, gold was considered an anomaly among the noble metals since it was the only element of this class without any catalytic activity in spite of the fact that other neighbors such as Pt, Ir, Hg, etc., were extremely efficient catalysts for many organic reactions [2]. This situation changed drastically since Haruta's seminal contribution showing that gold when forming NPs a few nanometers in size was an extremely efficient and selective catalyst for low-temperature CO oxidation [13]. Since then, part of the driving force in gold catalysts has been nanoscience, mainly because Au catalysis constitutes one of the best examples of nanoscience, that is, unique and novel properties that appear and are observed only when the solid particle has nanometric dimensions reaching the physical limit for the size of a chemical entity [14]. In fact, it was observed that the catalytic activity of Au decreases considerably as the average particle size of Au NPs increases in the range from 1 to 20 nm [15]. In this regard, recent studies have shown that clusters between 5 and 10 Au atoms are the most active gold species for acetyl chloride–assisted alkyne hydration and CO oxidation [16].

The problem is that although the optimal number of atoms for having the maximum turnover number for a catalytic reaction can correspond to subnanometric dimensions, these clusters tend to grow and are impossible to stabilize under the reaction conditions, increasing their size gradually. The most common strategy to stabilize Au NPs is by supporting them on a large-surface-area insoluble solid that establishes strong interactions with Au NPs, immobilizing them and impeding their growth. It has to be said that in addition to the role of stabilization of Au NPs, the support can also play other roles in the catalytic cycle, and for this reason, it has been experimentally observed that the nature and composition of the support play an important role on the overall catalytic activity of Au catalysts. As supports, organic polymers such poly(vinyl alcohol), polyvinylpyrrolidone and modified polystyrenes have been used, but more frequent supports for Au NPs are active carbons and specially metal oxides. The reason for this is that organic polymers exhibit generally low surface area and low accessibility and can undergo abrasion under intense stirring. As a result of the combination of these negative factors, the catalytic activity of these Au–polymer composites is typically lower than that of the materials where Au NPs are supported on other solids.

Since we have already commented that Pd(0)/Pd(II) is a good catalyst for novel C–C and C–X (X: heteroatom O, N, or S) cross-coupling reactions and considering the novelty of Au catalysis, there was a logical interest in establishing whether Au also exhibits catalytic activity for this type of reaction. In the following sections, we first will comment on the catalytic activity of Au/npCeO$_2$ for homocoupling of phenylboronic acid and later on the Suzuki–Miyaura coupling and Sonogashira reaction. We will emphasize the noninnocent role of the support on the catalytic activity and the promotional effect that light can play in Au catalysis. However, the most important issue is to establish unambiguously that Au can catalyze these reactions and the contribution that metal impurities can make to the overall activity of the material.

## 13.3 Homocoupling of Arylboronic Acids

One of the most efficient heterogeneous gold catalysts for CO oxidation [17] and alcohol oxidation [18] is constituted by Au NPs supported on CeO$_2$ NPs. The key point of this remarkable activity, besides the average particle size of Au NPs, is that the support CeO$_2$ is also in the form of NPs.

Typically, CeO$_2$ obtained by conventional synthesis is constituted of crystallites in the micrometric range. However, as the particle size of CeO$_2$ decreases, a notable change in its properties starts to appear. Thus, the 1:2 stoichiometry between Ce and O atoms decreases when ceria particle dimensions are of a few nanometers, leading to the appearance of structural defects, particularly oxygen vacancies, leading to a nonstoichiometric metal oxide. These oxygen vacancies are manifested by a deficiency in oxygen content for the solid and are present particularly on the external surface of the particle. Accompanying oxygen vacancies there is a simultaneous confusion in the oxidation state of Ce from (IV) to (III). This change in the composition of the material observed for small CeO$_2$ NPs results in an alteration in the physical and chemical properties of the solid. Perhaps the most remarkable property variation is the change from insulator to wide-bandgap semiconductor when going from micrometric, stoichiometric $CeO_2$ to nanoparticulate, nonstoichiometric $CeO_{2-x}$.

Also, from the chemical point of view, the activity of nonstoichiometric $CeO_{2-x}$ is remarkable due to the possibility to increase and decrease the oxygen content by incorporating oxygen from substrates into the solid lattice and the opposite. This exchange between the oxygen vacancies/oxygen lattice with substrates and products is part of the so-called Mars van Krevelen mechanism that can cooperate with the catalytic activity of Au NPs when they are supported on nanoparticulate ceria.

The combination of nanoparticulate, nonstoichiometric ceria with Au NPs renders a catalyst, $Au/CeO_2$, in which there is a strong interaction between Au NPs and $CeO_2$ NPs. On the one hand, Au NPs exhibit an overall positive charge compared to other supports. This positive charge on Au NPs can be experimentally evidenced by analyzing the $4f_{7/2}$ X-ray photoelectron spectroscopy (XPS) peak characteristic of Au at about 84.1 eV that upon de-convolution shows a significant contribution of up to 25% of $Au^+$. In carbon and other supports the $Au^+$ contribution to the $4f_{7/2}$ Au peak in XPS is almost absent. In addition, the presence of Au NPs on the surface of $CeO_2$ can increase apparently the deficiency in oxygen of the ceria surface, as is also demonstrated by comparison of the $Ce_{3d}$ XPS peak and by quantitative analysis of oxygen [19]. Besides XPS, CO absorption monitored by infrared (IR) spectroscopy can also detect and estimate the percentage of $Au^+$ by observing the position of the characteristic $Au^+$–CO band of carbonyl appearing at about 2140 cm$^{-1}$ that is shifted toward shorter wavenumbers compared to the typical vibration band of Au(0)–CO that appears at 2120 cm$^{-1}$ [20]. In this way, oxygen vacancies located on the periphery of Au NPs on the $CeO_2$ surface have been proposed to be the active sites for most oxidations [21]. The key point is the strong physisorption of molecular oxygen from the gas phase on the vacancies of nonstoichiometric $CeO_2$ near the Au NPs where the substrate to be oxidized is located.

$Au/CeO_2$ has also been tested for the Suzuki–Miyaura coupling of phenylboronic with iodobenzene [22]. The rationale behind the reaction is that as $Au/CeO_2$ contains a significant number of Au(I) atoms, isoelectronic with Pd(0) atoms, that could exhibit similar catalytic activity. The experimental results show, however, that this is not the case and Au/CeO$_2$ does not promote under usual reaction conditions the Suzuki cross-coupling between phenylboronic acid and iodobenzene [23]. However, homocoupling of phenylboronic acid leading to symmetric biphenyls was observed. The proposed reaction mechanism is depicted in Scheme 13.3 and has as the main steps the C–B bond breaking concomitant with the formation of the C–Au intermediate accompanied by the reduction of Au$^{3+}$ to Au$^+$. Subsequently, in the catalytic cycle, Au$^{3+}$ is formed from Au$^+$ by an oxidant and the coupling of two neighbor aryl rings.

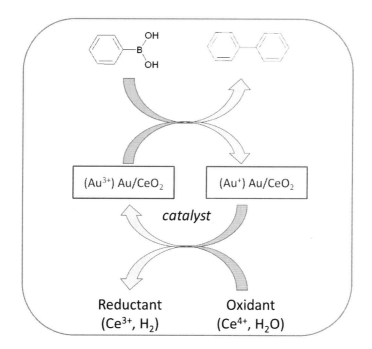

**Scheme 13.3** Homocoupling of phenylboronic acid to biphenyl by Au/CeO$_2$ catalyst with the need of an oxidizing reagent.

Other Au catalysts fail to promote phenylboronic homocoupling, and controls in the absence of Au NPs also show that the combination of these two components, Au NPs and nanoparticulate CeO$_2$, is crucial to achieve homocoupling.

## 13.4 Suzuki–Miyaura Cross-Coupling Promoted by Supported Au NPs Assisted by Light

As commented earlier, Au/CeO$_2$ fails to promote the Suzuki cross-coupling of iodobenzene and phenylboronic acid under normal reaction conditions. However, the situation changes drastically when gold catalysts are exposed to visible light irradiation [24]. The reason for this is that Au NPs exhibit a characteristic absorption band in the visible region denoted as the surface plasmon band that corresponds to the collective oxidation of electrons in a confined space [25]. This surface plasmon band is responsible for the intense pink/purple color characteristic of Au NPs. Surface plasmon bands are also present in Cu and Ag NPs, causing the visual appearance of these metal NPs to be red (Cu) or yellow (Ag). Most of the other metals, when in the form of NPs, have visually a black color, indicating that they absorb light in the whole range of visible wavelengths without exhibiting a localized surface plasmon band.

When Au NPs are supported, the exact $\lambda_{max}$ of the surface plasmon band varies depending on the average particle size, the charge density of Au NPs, and the dielectric constant of the solid support on which Au NPs are anchored. For instance, a typical $\lambda_{max}$ for 5 nm Au NPs on TiO$_2$ (P-25) is 560 nm [26].

Light excitation of Au NPs at the surface plasmon band produces three general effects [27]. On the one hand, the periodic oscillation of electrons at the Au NPs produces high, intense fluctuating electrostatic fields near these metal NPs that can affect the properties of adsorbed molecules. For instance, electronic transitions of neighbors can be enhanced by these fluctuating electrostatic fields of Au NPs. This localized field is the origin of surface-enhanced Raman spectroscopy (SERS) that makes possible recording of the Raman spectrum of a single organic molecule located at the tip of Au NPs by enhancing the Raman signal of this single molecule over 6 orders of magnitude. In SERS, preparation of the sample is crucial since this enhancement effect is only observed when the morphology of Au NPs is adequate (generally sharp tips are preferred over round particles) and the molecule is located at the positions in which the electrostatic field is reinforced.

A second defect of Au NPs can be a high temperature increase located at Au NPs for a very short period of time until the energy of the photon is dissipated to the ambient by thermalization. For instance, while gold melts at temperatures above 1500°C, Au NPs can melt at much lower temperatures of about 500°C. Laser irradiation at these Au NPs can lead to temperatures near or above the melting point of the Au NPs, and as a consequence a growth in the average particle size and a change in the Au NPs morphology can be observed due the melting of Au NPs caused by the local high temperature. Other indirect measurements of the local temperature peak that can be reached upon irradiation of Au NPs at their surface plasmon band have been based on the apparent increase of the reaction rates and Arrhenius law. Thus, the increase in the reaction rate of cumyl peroxide decomposition by irradiation at the surface plasmon band of Au NPs has led us to conclude that the temperature should be equivalent to about 300°C. Thermalization consists of converting the photon energy into heat by dissipation to the surroundings [28].

A third effect of irradiation can be the ejection of electrons from Au NPs to the surroundings, including the solvent, substrates, or even the support, if the conduction band energy of the support is accessible by these electrons [29].

Similarly, it has been observed that while under dark conditions Au NPs supported on $CeO_2$ are unable to catalyze the Suzuki cross-coupling, when the same material is illuminated with visible light, high catalytic activities for the cross-coupling between phenylboronic acid and aryl halides are observed [30]. Moreover, the system appears to be general and extremely efficiently, since besides iodobenzenes it can also promote the Suzuki coupling of bromoarenes and even the formation of *ortho*-substituted biphenyls. Table 13.1 provides a summary of some of the catalytic data reported for the light-assisted Suzuki coupling catalyzed by Au NPs.

In Suzuki cross-coupling the reactivity of the substrates depends on the electron-withdrawing or electron-donating effect of substituents, with aryl halides with electron-donating substituents being, in general, less reactive. In this regard, 4-bromoanisole is a suitable substrate to test the catalytic activity of the system and to establish the relative activity among different catalysts [31].

**Table 13.1** Selected results of the Suzuki cross-coupling of aryl boronic and aryl halides catalyzed by large Au-Pd nanostructures under the 809 nm laser illumination

| Entry | Halide | Boronic acid | Yield (%) |
|---|---|---|---|
| 1 | Ph–I | Ph–B(OH)₂ | 99 |
| 2 | Ph–Br | Ph–B(OH)₂ | 99 |
| 3 | Ph–Br | 2-CH₃-C₆H₄–B(OH)₂ | 99 |
| 4 | 2-OCH₃-C₆H₄–Br | 2-CH₃-C₆H₄–B(OH)₂ | 71 |
| 5 | 3-OCH₃-C₆H₄–Br | 2-CH₃-C₆H₄–B(OH)₂ | 43 |
| 6 | 4-H₃CO-C₆H₄–Br | 2-CH₃-C₆H₄–B(OH)₂ | 59 |

Supported Au NPs with the assistance of light can catalyze the coupling of 4-bromoanisole with phenylboronic acids.

Since several reaction mechanisms comprise an elementary step in which an electron has to be transferred to a substrate, this pathway can be promoted by the presence of Au NPs if they are illuminated. One paradigmatic example of this electron ejection by illumination of Au NPs is the promotion of the Fenton reaction at neutral pH values but by light-assisted catalysis of diamond-supported Au NPs. In the classical Fenton mechanism Fe(II) reduces hydrogen peroxide, leading to Fe(III), a hydroxide ion, and a hydroxyl radical. The Fenton reaction typically occurs at pH values below 4, this fact being related with the acid–base equilibrium of hydrogen peroxide [32, 33]. At neutral pH, no Fenton reaction leading to the generation of hydroxyl radicals takes place and a pathway involving hydroxyl peroxyl species (•OOH), a much milder oxidant than OH•, occurs. However, if the system contains Au NPs under illumination at the surface plasmon band, then a photochemical process can trigger the reduction of hydrogen peroxide by injecting the electron from Au NPs into hydrogen peroxide. The catalytic cycle, restoring the electron on positively charged Au NPs as transients, is closed because hydrogen peroxide can act also as a reducing agent, and in this way the forming oxygen restores the electron neutrality of Au NPs. Scheme 13.4 illustrates the reaction mechanism of the Au-catalyzed Fenton reaction.

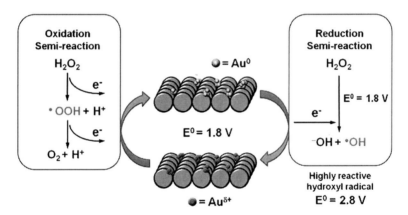

**Scheme 13.4** Reaction mechanism of the Au-catalyzed Fenton generation of OH• radicals. Notice that $H_2O_2$ acts as an oxidizing and a reducing agent. The process can take place at pH values higher than 4 if the system is illuminated with solar or visible light.

One interesting point in this process has been to determine which of the three general phenomena taking place when Au NPs are illuminated is the one responsible for enhancement of the catalytic activity. Measurements of the bulk reaction temperature as well as determination of the influence of the reaction temperature on the catalytic activity of supported Au catalysts in the dark have led us to conclude that the light-assisted enhancement of the catalytic activity of Au should be most probably due to electron ejection from Au NPs to the aryl halide photo-injection than to thermal effects. Thus, irradiation of the suspension produces only an increment of the reaction temperature of the system of a few degrees, while much higher reaction temperatures do not produce Suzuki coupling in the dark. It has to be remembered, however, that the bulk temperature may not reflect local instantaneous increases of temperature that could take place on the surface of Au NPs or in the vicinity of these NPs.

## 13.5 Sonogashira Coupling

Supported gold NPs are able to promote the Sonogashira coupling of substituted phenylacetylenes with iodobenzenes (Scheme 13.1). The yield depends on the nature and position of the substituents present on the reactants and the reaction conditions but under optimal conditions can be about 90% yield. In general *ortho*-substitution leads to lower yields due to steric encumbrance. Together with the cross-coupling product, various percentages of homocoupling by-products such as biphenyls and 1,4-diarylbutadiynes can also be present in the reaction mixture. In the next paragraphs we will comment that the product distribution is controlled in a large extent by the nature of the sites, that is, Au(I) and Au(III), and their proportion present in Au NPs.

Solid catalysts typically contain a wide distribution of metal NPs that may contain different sites, leading as result to low selectivity in the product formation. In the particular case of supported Au NPs and depending on the support, the charge density of each individual NP can vary in a certain range, depending on the size and attachment to the support. In a simplistic way some of these NPs may be considered as having some population of Au(I) and/or some percentage of Au(III) atoms. The consequence of these variations in the overall charge distribution of the Au clusters and NPs could be a difference in the selectivity, leading to variations in the product distribution.

It has to be remembered that although in metal clusters and NPs, there are no individual Au(I) and Au(III) ions since all the atoms in the NPs share the electron density, theoretical models indicate that certain atoms located at corners, edges, steps, and terraces exhibit different charge densities. In certain ways, it can be considered that the charge density of an NP is not homogeneously distributed throughout the NP and that certain atoms, particularly with low coordination, have the tendancy to concentrate. Also, Au atoms at the interface with the support exhibit a very different electron density as a consequence of the strong interaction with the support than other Au atoms.

This inhomogeneous electron density of Au atoms in a cluster or NPs can be experimentally determined by different techniques, particularly by de-convolution of the Au $4f_{7/2}$ band and by interaction by CO as a probe molecule of the electron density of Au NPs. In fact, XPS shows a primary $4f_{7/2}$ band for Au atoms accompanied by the corresponding satellites. This band appears around 84 eV and is broad, spanning many electron-volt widths, indicating that there is a distribution of Au atoms with different electron densities. It is very common to de-convolute this experimental broad band into much narrow components centered at 84, 84.5, and 86.2 eV (2 eV at half high) that would correspond to the standard values considered for Au(0), Au(I), and Au(III). Each of these individual components contributes differently to the experimental broad band and normalization of these contributions can allow one estimation of the contribution of Au(0), Au(I), and Au(III) to the total Au in the sample. According to this simple model, it can be considered that in the Au NPs there were some individual Au(I) and Au(III) ions. Figure 13.1 shows the experimental Au $4f_{7/2}$ XPS peak and the best fit to three components according to the values of Au (0), Au (I), and Au (III).

Similarly, a simple molecule like CO can interact in different ways with metal NPs. The exact position of the vibration band of CO depends on the interaction with CO with individual Au atoms in the cluster. It may also happen that the Au cluster contains a distribution of sites and accordingly CO interacts in various

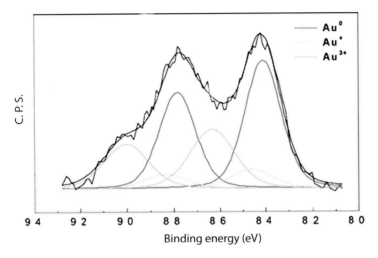

**Figure 13.1** Experimental XPS peak of $4f_{7/2}$ of Au and the best deconvolution to components corresponding to the values for Au(0), Au$^+$ and Au$^{3+}$. Reprinted with permission from Ref. [21]. Copyright 2005, WILEY.

ways with the Au NPs, leading to a broad IR band, whose deconvolution could again allow an estimation of the distribution of Au atoms in different populations of Au$^+$, Au$^{3+}$, and Au(0). More specifically, the interaction of CO with Au(0), Au(I), and Au(III) leads to maximum absorption of the vibration band at 2100 and 2150 cm$^{-1}$, respectively, and it is very common that a solid catalyst of supported Au NPs exhibits a broad band encompassing these three wavenumbers corresponding to individual interactions. As an example, Fig. 13.2 shows the CO vibration band upon absorption of CO on a Au/CeO$_2$ catalyst.

Considering that supported Au catalyst have experimentally this broad distribution of electron density in Au atoms at different positions, it is of interest to compare the catalytic activity of these solids with that of well-defined Au(I) or Au(III) complexes to determine if the product distribution observed in the heterogeneous catalysts can be rationalized as being a contribution of the activity of Au(I) and Au(III). In this regard, it should be commented that although Au(I) and/or Au(III) complexes can be prepared and characterized, and in this sense they initially are well-defined sites, during the course of the reaction, even at early stages, these

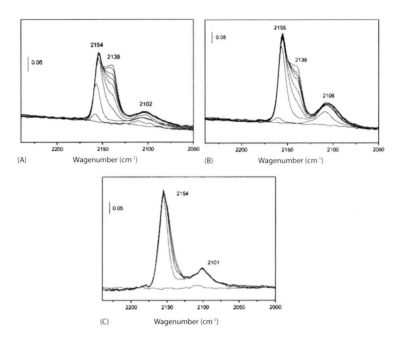

**Figure 13.2** Part of the IR spectra corresponding to the CO vibration zone together with the evolution upon CO desorption for three Au catalysts. Reprinted with permission from Ref. [34]. Copyright (2009), American Chemical Society.

complexes can undergo degradation, particularly with the formation of Au NPs. Even if these Au complexes can be considered as precatalysts due to their inherent instability, it is clear that the product distribution at low conversions and short reaction times may report on the intrinsic selectivity of Au (I) and Au (III) species.

This strategy of comparing well-defined homogeneous catalysts based on Au complexes with Au-containing solids has been applied to Sonogashira coupling, trying to understand the origin of cross- and homocoupling products and to assign each of these products to a specific site. The result of this comparison has led to the conclusion that Au(I) sites are responsible for the cross-coupling between aryl iodide and phenylacetylene, while Au(III) sites promote homocoupling of phenylacetylene to 1,4-diphenylbutadiyne. Since Au/CeO$_2$ as a catalyst leads to a mixture of these two products in a proportion of 89/11, it can be assumed that Au/CeO$_2$ behaves as having these two types of Au sites. This heterogeneity of the sites in Au/CeO$_2$ is different to the initial situation in which Au(I) and Au(III) complexes are used as catalysts. Scheme 13.5 summarizes the results obtained and the conclusion of the comparison between Au (I) and Au (III) complexes and Au/CeO$_2$ as catalysts.

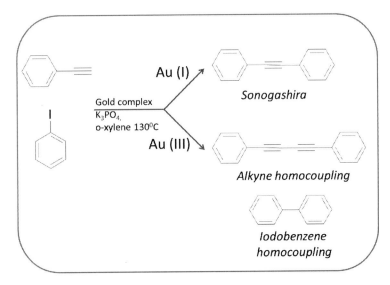

**Scheme 13.5** Specific catalytic activity of Au(I) and Au(III) complexes promoting Sonogashira cross-coupling, Au(I), or homocouplings, Au(III).

Since, as we have commented, there is a considerable influence on the nature of the support on the catalytic activity of Au, it is understandable that in the present coupling reactions those supports that, due to the stronger electronegativity with respect to the Au, increase the population of Au(III) atoms should lead to an increase in the percentage of the homocoupling product. Conversely, those supports that increase the electron density on Au NPs, decreasing the population of Au(I) and Au(III), would be inefficient as supports to promote this type of C–C coupling. In this context, and besides the use of metal oxides as supports, it has been reported that the natural biopolymer chitosan can also stabilize Au NPs by interaction of the N and O atoms of the polysaccharide with Au NPs [35]. Scheme 13.6 illustrates the preparation of Au

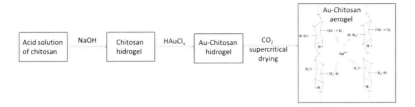

**Scheme 13.6** Preparation procedure for Au NPs embedded in chitosan fibrils.

NPs embedded into the fibrils of chitosan. Au NPs supported on a large-surface-area, porous chitosan have been found to be an active catalyst for Sonogashira coupling, leading also to a distribution of 1,2-diphenylacetylene and 1,4-diphenylbutadiyne similar to that observed for Au/CeO$_2$, indicating that in this case also there is a distribution of Au(I)- and Au(III)-like sites.

## 13.6 Role of Pd Impurities on Au-Catalyzed Sonogashira Coupling

As commented earlier, the initial hypothesis that led us to propose supported Au catalysis for C–C coupling reactions was their isoelectronic configuration of the Pd(0)–Pd(II) pair with the Au(I)–Au(III) couple. However, the reaction mechanism in Pd-catalyzed coupling consists of a series of elementary steps (see Scheme 13.2), including oxidative addition to the C–X bond, transmetallation, and reductive elimination. Each of these steps can occur with different reaction rates for the Pd(0)/Pd(II) pair compared to the Au(I)/Au(III) couple. Specifically based on the behavior of Au(I) complexes that are reluctant to undergo transmetallation with aryl alkynates, it has been proposed that Au should be unable to act as a catalyst of Sonogashira C–C coupling due to the inability of Au(I) complexes to undergo transmetallation [36]. To explain those reports in which a supported Au catalyst was found to promote Sonogashira coupling [37], it was proposed that Pd impurities present in trace amounts in Au should be the real active sites in this process, Au atoms being totally inactive to promote Sonogashira coupling or any similar C–C coupling [36].

This claim that Au-catalyzed Sonogashira coupling was due to Pd impurities present in Au was in line with other related precedents in which the claim that Fe catalysts are active for coupling was really due to Cu impurities present in this base metal [38]. Thus, many cross-coupling reactions were under suspicion and the role of impurities was established, at least in the case of Fe [39]. In this regard, it is worth remembering that, in contrast to Fe, Au is a noble metal, easy to purify and can be obtained with high purity. In fact, bulk Au metal can only react with aqua regia, a mixture of concentrated nitric and hydrochloric acids, while it can be purified from other metals by treatment with acids or bases that will dissolve impurities without attacking Au. For this reason, commercially available $HAuCl_4$ and their salts can have high purity and the Pd content in these commercial salts is in the range of a few parts per billion (ppb). Actually, the most important contaminant of Au is normally Cu that can reach a concentration in high-purity Au of 40 ppb. The high purity of Au contrast with the case of Fe for which many conventional salts have a purity of 90% or 95% and where the Fe catalyst is used in large ratios (10 mol% or higher) with respect to the substrates, making it possible to introduce as impurities a large percentage of other metals.

To address the possibility that Sonogashira coupling promoted by supported Au catalysts was really due to the presence of Pd impurities, chemical analysis of the Pd content for an active Au catalyst was provided, establishing that the Pd content was, as expected, in the range of a few parts per billion. Then, a series of supported Au catalysts was prepared contaminated, on purpose, by increasing amounts of Pd from 10 to 50 ppb, determining the influence Pd contamination on the initial reaction rates ($r_0$) of Sonogashira coupling. It was observed that even at these low Pd concentrations, $r_0$ increases linearly with the amount of Pd impurities, but extrapolation of the linear relationship to 0 ppb Pd concentration allows establishing that over 80% of the catalytic activity of supported Au NPs cannot be attributed to Pd impurities (Fig. 13.3) [40]. This extrapolation to 0 ppb Pd content is necessary

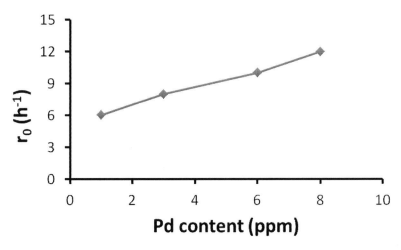

**Figure 13.3** Influence of Pd content on the initial reaction rate ($r_0$) of the Sonogashira coupling between phenylacetylene and iodobenzene catalyzed by Au/CeO$_2$.

to support the catalytic activity of Au NPs, since it is conceptually impossible to demonstrate the total absence of Pd in a sample.

These experimental data aimed at determining the intrinsic catalytic activity of a pure Au catalyst have been complemented with theoretical calculations on the reaction mechanism and activation energy barriers of each of the elementary steps of Sonogashira coupling [40]. These calculations establish clearly a difference between the behavior of monoatomic Au complexes that cannot, according to theory, promote the Sonogashira reaction and that of clusters and NPs of Au that can be active for this reaction. The key point is that the presence of several Au atoms, each of them coordinating with a substrate or reagent, makes it possible that the ensemble as a whole acts cooperatively as a single site for the reaction. Figure 13.4 illustrates the differences in the mechanism between a single Au(I) atom and a Au NP. According to these theoretical calculations, the role of the base in the process is to remove halides from the surface of Au NPs, since otherwise, these halide ions will act as poisons due to the strong interaction with Au atoms bearing a positive charge density. In conclusion, experimental and theoretical data all point out that supported Au catalysts can

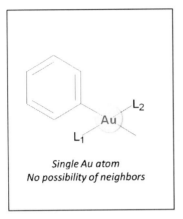 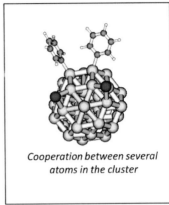

Single Au atom
No possibility of neighbors

Cooperation between several atoms in the cluster

**Figure 13.4** Outcome of the theoretical study on Au-catalyzed Sonogashira coupling, highlighting the differences between a single Au atom present in a metal complex that is inactive for these couplings and the behavior of Au clusters where two different Au atoms can bind to substrates, making possible the C–C coupling.

promote C–C coupling, even in the absence of Pd. However, from a practical point of view it appears that solid Au catalysts are still far from the performance that can exhibit Pd and particularly homogeneous Pd complexes to promote these reactions. Thus, the activity of Pd catalysts surpasses largely that of analogous Au catalysts.

## 13.7 Conclusions and Final Remarks

The earlier sections briefly summarize the current state of the art in C–C bond-forming reactions catalyzed by supported gold catalysts. We have shown that the activity of palladium complexes and solids is much higher than analogous complexes or materials with gold. This better performance of palladium is attributed to the easier reversibility between the 0 and +II oxidation states for palladium compared to the +I and +III manifold in the case of gold. Nevertheless, the intrinsic catalytic activity of Au NPs for this type of reaction has been proved on a firm experimental ground and has been rationalized by theoretical calculations.

In spite of the lower activity compared to palladium, gold catalysts may offer new possibilities for C–C and C–X coupling reactions derived from the unique features of Au NPs derived from their surface plasmon band and the possibility to assist gold-catalyzed reactions with visible or ambient light. A deeper understanding and harnessing of the process triggered by light on Au NPs can be an extremely useful tool to increase the activity and control the selectivity in gold catalysis that do not exist in the palladium analogs. Thus, considering the current interest in gold catalysis, it is clear that the field of C–C bond-forming reactions using gold is still in its infancy and that novel, more efficient heterogeneous gold catalysts will be developed in the near future, complementing those solid catalysts based on palladium for this type of reaction.

## References

1. Li, C.J. (1993). Organic reactions in aqueous media-with a focus on carbon-carbon bond formation, *Chem. Rev.*, **93**, pp. 2023–2035.
2. Smith, M.B., and March, J. (2007). *March's Advanced Organic Chemistry: Reactions, Mechanisms, and Structure* (Wiley & Sons, New Jersey).
3. Maruyama, K., and Katagiri, T. (1989). Mechanism of the Grignard reaction, *J. Phys. Org. Chem.*, **2**, pp. 205–213.
4. Ritleng, V., Sirlin, C., and Pfeffer, M. (2002). Ru-, Rh-, and Pd-catalyzed C-C bond formation involving C-H activation and addition on unsaturated substrates: reactions and mechanistic aspects, *Chem. Rev.*, **102**, pp. 1731–1769.
5. Kalyani, D., Deprez, N.R., Desai, L.V., and Sanford, M.S. (2005). Oxidative C-H activation/C-C bond forming reactions: synthetic scope and mechanistic insights, *J. Am. Chem. Soc.*, **127,** pp. 7330–7331.
6. Hartwig, J.F. (1998). Carbon-heteroatom bond-forming reductive eliminations of amines, ethers, and sulfides, *Acc. Chem. Res.*, **31**, pp. 852–860.
7. Astruc, D., Lu, F., and Ruiz Aranzaes, J. (2005). Nanoparticles as recyclable catalysts: the frontier between homogeneous and heterogeneous catalysis, *Angew. Chem., Int. Ed.,* **44**, pp. 7852–7872.
8. Corma, A., and Garcia, H. (2006). Silica-bound homogeneous catalysts as recoverable and reusable catalysts in organic synthesis, *Adv. Synth. Catal.,* **348**, pp. 1391–1412.

9. Wolfe, J.P., Singer, R.A., Yang, B.H., and Buchwald, S.L. (1999). Highly active palladium catalysts for Suzuki coupling reactions, *J. Am. Chem. Soc.*, **121**, pp. 9550–9561.
10. Old, D.W., Wolfe, J.P., and Buchwald, S.L. (1998). A highly active catalyst for palladium-catalyzed cross-coupling reactions: room-temperature Suzuki couplings and amination of unactivated aryl chlorides, *J. Am. Chem. Soc.*, **120**, pp. 9722–9723.
11. De Vos, D.E., Dams, M., Sels, B.F., and Jacobs, P.A. (2002). Ordered mesoporous and microporous molecular sieves functionalized with transition metal complexes as catalysts for selective organic transformations, *Chem. Rev.*, **102**, pp. 3615–3640.
12. Abad, A., Almela, C., Corma, A., and Garcia, H. (2006). Efficient chemoselective alcohol oxidation using oxygen as oxidant. Superior performance of gold over palladium catalysts, *Tetrahedron*, **62**, pp. 6666–6672.
13. Haruta, M., Yamada, N., Kobayashi, T., and Iijima, S. (1989). Gold catalysts prepared by coprecipitation for low-temperature oxidation of hydrogen and of carbon monoxide, *J. Catal.*, **115**, pp. 301–309.
14. Corma, A., and Garcia, H. (2008). Supported gold nanoparticles as catalysts for organic reactions, *Chem. Rev.*, **37**, pp. 2096–2126.
15. Haruta, M. (2002). *Catalysis of Gold Nanoparticles Deposited on Metal Oxides*, CATECH, **6**, pp. 102–115 (Springer).
16. Oliver-Meseguer, J., Cabrero-Antonino, J.R., Domínguez, I., Leyva-Pérez, A., and Corma, A. (2012). Small gold clusters formed in solution give reaction turnover numbers of $10^7$ at room temperature, *Science*, **338**, pp. 1452–1455.
17. Si, R., and Flytzani-Stephanopoulos, M. (2008). Shape and crystal-plane effects of nanoscale ceria on the activity of Au-CeO$_2$ catalysts for the water-gas shift reaction, *Angew. Chem., Int. Ed.*, **47**, pp. 2884–2887.
18. Corma, A., and Domine, M.E. (2005). Gold supported on a mesoporous CeO2 matrix as an efficient catalyst in the selective aerobic oxidation of aldehydes in the liquid phase, *Chem. Commun.*, **32**, pp. 4042–4044.
19. Guzman, J., Carretin, S., and Corma, A. (2005). Spectroscopic evidence for the supply of reactive oxygen during CO oxidation catalyzed by gold supported on nanocrystalline CeO$_2$, *J. Am. Chem. Soc.*, **127**, pp. 3286–3287.
20. Carretin, S., Concepción, P., Corma, A., López Nieto, J.M., and Puntes, V.F. (2004). Nanocrystalline CeO$_2$ increases the activity of Au for CO

oxidation by two orders of magnitude, *Angew. Chem., Int. Ed.*, **43**, pp. 2538–2540.

21. Abad, A., Concepción, P., Corma, A., and Garcia, H. (2005). A collaborative effect between gold and a support induces the selective oxidation of alcohols, *Angew. Chem., Int. Ed.*, **44**, pp. 4066–4069.

22. Carretin, S., Blanco, M.C., Corma, A., and Hashmi, A.S.K. (2006). Heterogeneous gold-catalysed synthesis of phenols, *Adv. Synth. Catal.*, **348**, pp. 1283–1288.

23. Carrettin, S., Guzman, J., and Corma, A. (2005). Supported gold catalyzes the homocoupling of phenylboronic acid with high conversion and selectivity, *Angew. Chem., Int. Ed.*, **44**, pp. 2242–2245.

24. Zhou, X., Liu, G., Yu, J., and Fan, W. (2012). Surface plasmon resonance-mediated photocatalysis by noble metal-based composites under visible light, *J. Mater. Chem.*, **22**, pp. 21337–21354.

25. Link, S., and El-Sayed, M.A. (1999). Size and temperature dependence of the plasmon absorption of colloidal gold nanoparticles, *J. Phys. Chem. B*, **103**, pp. 4212–4217.

26. Abad, A., Corma, A., and Garcia, H. (2007). Bridging the gap between homogeneous and heterogeneous gold catalysis: supported gold nanoparticles as heterogeneous catalysts for the benzannulation reaction, *Top. Catal.*, **44**, pp. 237–243.

27. Hallett-Tapley, G.L., D'Alfonso, C., Pacioni, N.L., McTiernan, C.D., González-Béjar, M., Lanzalunga, O., Alarcon, E.I., and Scaiano, J.C. (2013). Gold nanoparticle catalysis of the cis-trans isomerization of azobenzene, *Chem. Commun.*, **49**, pp. 10073–10075.

28. Fasciani, C., Bueno Alejo, C.J., Grenier, M., Netto-Ferreira, J.C., and Scaiano, J.C. (2011). High-Temperature organic reactions at room temperature using plasmon excitation: decomposition of dicumyl peroxide, *Org. Lett.*, **13**, pp. 204–207.

29. Alvaro, M., Aprile, C., Ferrer, B., Sastre, F., and Garcia, H. (2009). Photochemistry of gold nanoparticles functionalized with an iron(II) terpyridine complex. An integrated visible light photocatalyst for hydrogen generation, *Dalton Trans.*, **36**, pp. 7437–7444.

30. Wang, F., Li, C.C., H., Jiang, R., Sun, L.D., Li, Q., Wang, J., Yu, J.C., and Yan, C.H. (2013). Plasmonic harvesting of light energy for Suzuki coupling reactions, *J. Am. Chem. Soc.*, **135**, pp. 5588–5601.

31. Dupont, J., and Scholten, J.D. (2010). On the structural and surface properties of transition-metal nanoparticles in ionic liquids, *Chem. Soc. Rev.*, **39**, pp. 1780–1804.

32. Sempere, D., Navalon, S., Dancíková, M., Alvaro, M., and Garcia, H. (2013). Influence of pretreatments on commercial diamond nanoparticles on the photocatalytic activity of supported gold nanoparticles under natural Sunlight irradiation, *Appl. Catal. B: Environ.*, **142–143**, pp. 259–267.
33. Navalon, S., Martin, R., Alvaro, M., and Garcia, H. (2010). Gold on diamond nanoparticles as a highly efficient Fenton catalyst, *Angew. Chem., Int. Ed.*, **49**, pp. 8403–8407.
34. Boronat, M., Concepción, P., and Corma, A. (2009). Unravelling the nature of gold surface sites by combining IR spectroscopy and DFT calculations, implications in catalysis, *J. Phys. Chem. C*, **113**, pp. 16772–16784.
35. Primo, A., and Quignard, F. (2010). Chitosan as efficient porous support for dispersion of highly active gold nanoparticles: design of hybrid catalyst for carbon-carbon bond formation, *Chem. Commun.*, **46**, pp. 5593–5595.
36. Lauterbach, T., Livendahl, M., Rosellón, A., Espinet, P., and Echavarren, A.M. (2010). Unlikeliness of Pd-free gold(I)-catalyzed Sonogashira coupling reactions, *Org. Lett.*, **12**, pp. 3006–3009.
37. González-Arellano, C., Abad, A., Corma, A., Garcia, H., Iglesias, M., and Sánchez, F. (2007). Catalysis by gold(I) and gold(III): a parallelism between homo- and heterogeneous catalysts for copper-free Sonogashira cross-coupling reactions, *Angew. Chem., Int. Ed.*, **119**, pp. 1558–1560.
38. Buchwald, S.L., and Bolm, C. (2009). On the role of metal contaminants in catalyses with FeCl$_3$, *Angew. Chem., Int. Ed.*, **48**, pp. 5586–5587.
39. Thomé, I., Nijs, A., and Bolm, C. (2012). Trace metal impurities in catalysis, *Chem. Soc. Rev.*, **41**, pp. 979–987.
40. Corma, A., Juárez, R., Boronat, M., Sánchez, F., Iglesias, M., and Garcia, H. (2011). Gold catalyzes the Sonogashira coupling reaction without the requirement of palladium impurities, *Chem. Commun.*, **47**, pp. 1446–1448.

## Chapter 14

# Toward Chemoselectivity: The Case of Supported Au for Hydrogen-Mediated Reactions

**Fernando Cárdenas-Lizana and Mark A. Keane**

*Chemical Engineering, School of Engineering and Physical Sciences,*
*Heriot-Watt University, Edinburgh, EH14 4AS, Scotland*
F.CardenasLizana@hw.ac.uk, M.A.Keane@hw.ac.uk

Gold has untapped potential in terms of catalytic chemoselectivity. In this chapter, the selective action of supported gold in the hydrogenation of functionalized nitroarenes and hydrodechlorination of chloroarenes is analyzed in detail. Hydrogen/gold interactions are examined, and the crucial catalyst structural and surface properties required to achieve enhanced performance are discussed with a consideration of process variables and sustainability in terms of hydrogen utilization.

## 14.1 Introduction/Scope

The work of Haruta et al. [1] dating from 1987 is generally credited for the renaissance of gold as a catalytically active material. Haruta recorded CO oxidation activity for supported Au nanoparticles at

---

*Gold Catalysis: Preparation, Characterization, and Applications*
Edited by Laura Prati and Alberto Villa
Copyright © 2016 Pan Stanford Publishing Pte. Ltd.
ISBN 978-981-4669-28-3 (Hardcover), 978-981-4669-29-0 (eBook)
www.panstanford.com

subambient temperatures. Bond [2], however, has identified a rich and varied literature that predates Haruta's work and that first established the catalytic properties of gold. In a recent article, he has also highlighted the reactivity of gold at the nanoscale in activating small molecules [3]. Technological applications of gold now cut across automotive, electronics, medicine, nanotechnology, space, and engineering sectors, but the work conducted over the past two decades has established catalysis as a priority growth area [4]. Use of gold catalysts in oxidation is now the subject of a mature corpus of publications that goes beyond CO oxidation and encompasses the conversion of a range of volatile organic compounds (VOCs) with environmental implications and hydrogen production via the water–gas shift (WGS) reaction [5–9]. Significant oxidation activity has even been demonstrated for reaction over unsupported gold [10, 11]. While less developed, use in hydrogenation is now the subject of a burgeoning literature. This has been driven by the distinct selectivity achieved with gold when compared with conventional metals that can be exploited in the cleaner synthesis of target compounds and/or in the treatment of environmental pollutants. Chemoselectivity in the conversion of a multifunctional feedstock is the critical property in rational catalyst design for the application of green chemistry in optimizing reactant transformation to high-value products, minimizing separation operations and disposal of unwanted by-products [12]. There are a number of pertinent reviews that range from the examination of hydrogenation reactions using gold, published by Claus in 2005 [13] and that has since been supplemented by the report of McEwan et al. [14], to the consideration of hydrogen–gold interactions [15], liquid-phase hydrogenation [16], reduction of nitrocompounds [17], and information gleaned from model systems and theoretical calculations [18].

There are few reported examples of hydrogenation promoted by homogenous gold catalysts. We can flag studies by Corma et al. [19, 20] on Au(I) and Au(III) complexes in the conversion of nitrocompounds. We should also note work on gold complexes in enantioselective hydrogenation, which has been reviewed by Widenhoefer [21] and Pradal et al. [22]. Practical application of homogeneous catalysis has associated sustainability constraints due to the additional use of chemicals as solvents/hydrogen donors and the, often difficult, downstream product/solvent/catalyst separation steps. This chapter is focused on solid supported Au catalysts in continuous gas-phase processes. Where applicable, we relate gas-phase data to trends observed in liquid-phase operation. Use of gold catalysts in hydrogenation has largely been conducted in batch liquid-phase operations at elevated $H_2$ pressures, resulting in energy-intensive processes with safety implications for large-scale production. Incomplete mixing in batch reactors results in significant mass/heat transfer gradients, which extend reaction time, leading to by-product formation with the requirement for subsequent separation/purification stages to extract the desired product. A move from batch liquid to continuous gas flow operation circumvents downtime and use of solvents and derivatization agents, while facilitating high throughput. Catalytic action in terms of activity, selectivity, and stability can show a critical dependence on Au dispersion where the Au–support interface and Au electronic structure have crucial roles to play. We discuss these effects, with particular emphasis on the role of the support and the modifying effect of a second metal to govern and even tune product distribution.

This chapter is organized to consider the use of gold in two topical areas of catalysis that we illustrate with two case studies: (i) catalytic hydrodechlorination (HDC) as an environmental remediation strategy and (ii) selective nitroarene hydrogenation as a clean and sustainable route to high-value functionalized aromatic amine products. We address process variables and examine key catalyst structural features that determine performance with a consideration of sustainability in practical application. The chapter ends with a consideration of possible future research directions.

## 14.2 Application of Gold in Hydrogen-Mediated Reactions

### 14.2.1 *Hydrogen–Gold Interaction*

A prerequisite for catalytic hydrogenation is dissociative adsorption to generate surface-reactive hydrogen. In general terms, gold has found limited application in hydrogenation reactions due to its low capacity for hydrogen chemisorption. This can be attributed to the absence of unpaired $d$-electrons but temperature-induced $5d \rightarrow 6s$ transition generates $d$-electrons that can facilitate hydrogen adsorption/dissociation [23, 24]. Molecular hydrogen does not chemisorb on bulk gold [13, 25], weakly interacting at 78 K (hydrogen coverage <0.015 at $P = 0.3$ Pa) and desorbing at 125 K (desorption activation energy = 12 kJ mol$^{-1}$) [26] where low-coordinated gold atoms on film surfaces were identified as adsorption sites. The use of a (carbonaceous or oxide) support on which to anchor Au particles can ensure a well-dispersed metal phase at the *nano*scale, and Lin and Vannice [27] detected weak, reversible adsorption of hydrogen (ca. 1% coverage of surface gold atoms) on (25 nm) gold particles supported on $TiO_2$ at 300–473 K. The authors proposed that hydrogen adsorption is an activated process that is enhanced at higher temperatures. Jia et al. [28] reported that 14 ± 2% of surface atoms in (3.8 nm) Au/Al$_2$O$_3$ adsorb irreversibly held hydrogen at 273 K, while Bus et al. [29] studying the hydrogen capacity of Au/Al$_2$O$_3$ and Au/SiO$_2$ (1–1.5 nm) provided evidence that 10%–30% of the total adsorbed hydrogen did not desorb under evacuation at 473 K. The same authors demonstrated (by in situ X-ray absorption spectroscopy, chemisorption, and hydrogen–deuterium [H$_2$–D$_2$] exchange experiments) dissociative hydrogen adsorption on Au/Al$_2$O$_3$ with a greater uptake on smaller metal particles and proposed that corner and edge atoms were responsible for hydrogen dissociation. Further evidence for hydrogen chemisorption on defect sites associated with small Au nanocrystals was later provided by experimental [30] and theoretical [31] (Fig. 14.1) work. Temperature dependence of hydrogen uptake has been confirmed [27, 32] with a threefold increase from 298 to 373 K [29]. Moreover, there is evidence [33, 34], based on electron energy loss spectroscopy (EELS) analysis, for an induced dipole due to hydrogen adsorption, while changes in the X-ray absorption near-edge structure (XANES) response [35, 36] suggest (slight) variations in gold electronic structure. Boronat et al. [30] have proposed that the net charge on supported gold can impact on hydrogen interaction where dissociation is favored on sites that are neutral or with a net charge close to zero.

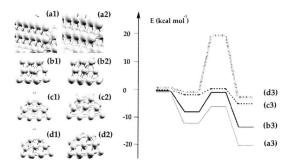

**Figure 14.1** Model for hydrogen (1) adsorption and (2) dissociation with (3) calculated energy profile for (a) a monoatomic row on a defective Au(111) surface and (b and d) corner and (c) top of an isolated Au$_{25}$ with Au atoms directly bonded to four (b) or five (c and d) gold atoms. Adapted from Ref. [31].

In summary, although the nature of H$_2$–Au interactions in supported systems requires further fundamental analysis, the consensus that emerges from the literature suggests a high activation energy barrier for dissociative adsorption where uptake capacity is dependent on gold coordination, with dissociative chemisorption favored on low-coordination (defects, edge, and corners) sites associated with smaller (<10 nm) metal nanoparticles. Gold electronic density is also critical in determining hydrogen uptake, which is enhanced at higher temperatures.

### 14.2.2 Hydrogen-Mediated Reactions Catalyzed by Gold

H$_2$–D$_2$ isotopic exchange is an effective means of probing hydrogen/surface interactions. In a series of publications [37–39] Kislyuk and Tretyakov demonstrated the formation of atomic hydrogen via a first-order reaction and activation energies of 71–75 kJ mol$^{-1}$ with hydrogen atom recombination on adjacent gold sites at 300 K that followed a Langmuir–Hinshelwood mechanism. These findings confirmed H$_2$–D$_2$ isotopic exchange on gold. Gluhoi et al. [40] studying H$_2$–D$_2$ exchange on Au/Al$_2$O$_3$ (3 nm) reported activity consistent with activation-dissociation of H$_2$ (D$_2$) at Au defect sites (e.g., steps, kinks, and corners) and isotopic equilibration at the Au–oxide support interface. This is in line with the work of Boronat et al.

[41], who combining infrared/density functional theory (IR/DFT) analysis with H/D isotopic exchange experiments demonstrated that low-coordinated neutral gold atoms located at corners and edges (i.e., not involved in Au–O–Ti linkages) were responsible for $H_2$ dissociation on Au/TiO$_2$.

The hydrogenation of unsaturated hydrocarbons has been typically conducted over noble metal–based (e.g., Pd, Pt, Rh, and/or Ru) catalysts. In the early 1970s, Bond et al. [42] established the requirement for smaller Au particles and surface defects in order to achieve significant conversion in the hydrogenation of buta-1,3-diene and but-2-yne. The lower catalytic activity of Au in hydrogen-mediated reactions relative to conventional transition metals is generally attributed to the less effective activation/dissociation of hydrogen [43–45]. Nevertheless, supported gold has shown promising results in terms of hydrogenation chemoselectivity [16, 17, 46]. The early studies largely focused on hydrogen adsorption/activation, hydrogen exchange reactions, and conversion of "simple" molecules such as CO, $CO_2$, alkenes, and alkadienes. The development of methodologies for the preparation of gold as active supported nanoparticles extended applications to consider multifunctional organic reactants, for example, $\alpha$, $\beta$-unsaturated aldehydes, ketones, and functionalized nitroarenes. This is illustrated by the increasing number of publications and filed patents in the field. In contrast to conventional hydrogenation catalysts, which often require complex modifications and incorporation of promoters/additives, supported gold can deliver enhanced chemoselectivity in the targeted hydrogenation of multifunctional reactants. This property can be exploited in the clean production of commodity chemicals.

In terms of environmental remediation, gold catalysis has been employed in the conversion of CO/$CO_2$ (to methanol) and reduction of nitric oxide [43]. Shibata et al. [47] demonstrated high activity for the hydrogenation of CO over an amorphous $Au_{25}Zr_{75}$ alloy, which was oxidized to $Au^0$ and $ZrO_2$ under reaction conditions (6.0 MPa; 523 K). Koeppel et al. [48] reported the formation of metallic gold particles (8.5 nm) and $ZrO_2$ by exposing the same alloy to $CO_2$ hydrogenation conditions to generate methanol and CO as principal products. The nature of the carrier is critical in methanol production where methane formation is a feature of reaction over

TiO$_2$- and Fe$_2$O$_3$-supported gold [49, 50] (relative to Au/ZrO$_2$ [51]). Gold supported on ZnO or ZnFe$_2$O$_4$ (as strongly basic carriers) outperformed conventional Cu–ZnO/Al$_2$O$_3$ in this reaction [49, 50]. Strunk et al. [52] have proposed that oxygen vacancies on ZnO are active sites for methanol production where gold incorporation can increase the surface density of these vacancies. Product distribution in the hydrogenation of CO over Au/ZnO can be effectively tuned by co-incorporation of alumina or zeolite Y [53]. Enhanced methanol yield was reported by Sakurai and Haruta [49] in the hydrogenation of CO$_2$ over Au/ZnO–TiO$_2$ and attributed to an increased Au–ZnO interface. Gold also shows advantages in the conversion of environmental toxins, notably CO, unburnt hydrocarbons, and NO$_x$ in air pollution abatement [54]. Reduction of NO (to N$_2$) with CO from exhaust gas has been demonstrated for Au on $\alpha$-Fe$_2$O$_3$ and NiFe$_2$O$_4$ at $T < 373$ K. Activity is strongly dependant on the carrier [55, 56] with elevated NO conversion resulting from the introduction of Mn$_2$O$_3$ [57] and/or the presence of O$_2$ and moisture [58]. In contrast, increased reaction temperatures are required for Pt group metals, which promote preferential formation of N$_2$O [55, 56] and NO$_2$ [59]. Recent studies have demonstrated effective gold application in catalytic HDC where hydrogenolytic scission of C–Cl (with HCl release) is a viable nondestructive approach for the treatment of toxic chlorinated waste. This application will be discussed in detail in Section 14.3.

Taking an overview of hydrogenation applications of gold in the manufacture of intermediates/products for the chemical industry (e.g., production of pesticides, herbicides, pigments, pharmaceuticals, or polymers) it is clear that research has been directed primarily at three groups of reactions: (i) partial hydrogenation of C≡C (in alkynes) and C=C (in alkenes and alkadienes); (ii) selective hydrogenation of C=O (in $\alpha$, $\beta$-unsaturated carbonyl compounds); and (iii) –NO$_2$ reduction in functionalized nitroarenes. We examine the chemoselective formation of functionalized anilines in Section 14.4 and provide here a general perspective on the other processes. Partial hydrogenation of C≡C and C=C is important in the purification of industrial (alkyne or alkadiene+alkene) mixtures, that is, upgrading C$_2$–C$_6$ fractions from steam crackers. Because of the detrimental (poisoning) effect of these unsaturated components in downstream polymerization units, there is a requirement for selective removal of trace amounts from the feed. Palladium-based systems have been widely applied in industry [60], notably the Lindlar catalyst (5% w/w Pd/CaCO$_3$ modified by lead acetate) [61, 62]. Nonetheless, limited selectivity and deactivation with time on-stream are decided drawbacks [63] that need to be addressed. There is evidence in the literature of enhanced activity [64], selectivity [28], and stability [64] for gold catalysts. Increased selectivity has been attributed to preferential adsorption on gold edge sites [65] but the genesis of carbon deposits results in loss of activity [64, 66]. Metallic (Au$^0$) [67, 68] and cationic (Au$^{3+}$) [69, 70] species have been suggested as active sites where (i) particle size [71] (with ca. 3 nm as optimum [28, 72]), (ii) choice of support (e.g., TiO$_2$ vs. Fe$_2$O$_3$ [73]), (iii) preparation method [64], and/or (iv) formation of bimetallic ensembles by incorporation of a second metal (e.g., Au$_{core}$–Pd$_{shell}$ [74], Au–Fe [75]) can impact the catalytic response.

The synthesis of unsaturated alcohols by selective hydrogenation of $\alpha$, $\beta$-unsaturated aldehydes, that is, preferential C=O→C–OH over the thermodynamically favored C=C→C–C reduction step, represents an important area of investigation. The formation of saturated aldehydes has been reported for reaction over supported Pt, Rh, and Pd catalysts [13, 76, 77], typically in liquid-phase operation. Higher conversions in the hydrogenation of crotonaldehyde have been achieved using (S [78], Fe, Mo, or W [79]) promoted gold systems. In the hydrogenation of biphenyl, a linear correlation between activity and Au$^{\delta+}$ content was established [81]. You et al. [82] using a series of Mg$_x$AlO hydrotalcite–supported Au catalysts reported increased activity and selectivity to $\alpha$, $\beta$-unsaturated alcohols (from the corresponding $\alpha$, $\beta$-unsaturated aldehyde) at higher Au$^{3+}$/Au$^0$ ratios (achieved by varying calcination temperature). Claus et al. [44, 80, 83, 84] have studied the selective hydrogenation of acrolein over ZrO$_2$-, ZnO-, TiO$_2$-, and SiO$_2$-supported Au catalysts and established higher allyl alcohol selectivity over Au/ZrO$_2$ relative to Pt/ZrO$_2$ [83]. Lenz et al. [85] studying the liquid-phase hydrogenation of crotonaldehyde and cinnamaldehyde over gold supported on iron oxides proposed (on the basis of transmission electron microscopy [TEM], temperature-programmed reduction [TPR], and XANES analyses) that high selectivity to the unsaturated alcohol was related to the morphology (but not the size) of the supported Au nanoparticles, with no apparent dependence on support redox properties or charge on the gold particles. Mohr et al. [44] reported improved activity with the formation of "rounded" gold nanoparticles that bear a preponderance of electron-deficient and low-coordination sites. Coordinatively unsaturated (edge) sites were taken to be responsible for the selective reduction of the carbonyl functionality [44] consistent with preferential ally alcohol formation over Au–In [80] where In located in Au crystal faces, leaving edges free (Fig. 14.2). Nevertheless, extremely small metal particles (1.1–2.0

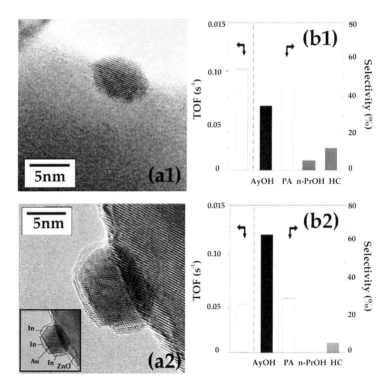

**Figure 14.2** (a) HRTEM images with inset showing location of individual elements within the metal nanoparticle and support; (b) catalytic results in the hydrogenation of acrolein (593 K) in terms of turnover frequency (TOF) and selectivity to propenol (AyOH), propionaldehyde (PA), n-propanol (n-PrOH), and $C_2$ and $C_3$ hydrocarbons (HC) over (1) Au/ZnO and (2) Au–In/ZnO. Adapted from Ref. [80].

nm) exhibited antipathetic structure sensitivity, that is, increased activity (2→400 mmol $g_{Au}^{-1}$ $s^{-1}$) and allyl alcohol selectivity (19%→26%) with increasing size. This was explained on the basis of a preferential adsorption of the C=O function on face atoms in (111) planes and attributed (on the basis of TEM/high-resolution TEM [HRTEM] and electron paramagnetic resonance) to quantum sizes effects that modify the electronic properties of very small gold nanoparticles [86]. A preferential C=O attack in the hydrogenation of $\alpha$, $\beta$-unsaturated ketones was shown for the first time by Milone and coworkers [87] who determined that the catalytic response was strongly influenced by the support [88], where the formation of $Au^{\delta-}$ favored C=O activation and hydrogenation [89]. The versatility of gold in the selective hydrogenation of $\alpha$, $\beta$-unsaturated aldehydes and ketones to allyl alcohols has been demonstrated by Mertens et al. [90], who, using polymer-stabilized metal (Ag, Au, Co, Cu, Ir, Ni, Pt, Ru) clusters, achieved the highest yields (25%–88%) to the target alcohol over $Au^0$ nanocolloids for a series of 24 reactants.

## 14.3 Case Study 1: Environmental Pollution Control; Hydrodechlorination of Chloroaromatics

### 14.3.1 Background

Halogen-containing waste is xenobiotic and there is no natural means of ameliorating the negative environmental impact. The presence of chloroarenes, the subject of this case study, in effluent discharges is of increasing concern due to mounting evidence of adverse stratospheric ozone, ecological, and public health impacts [91]. Catalytic HDC is an emerging green chlorowaste treatment with many advantages over standard control methodologies (e.g., incineration, photolysis, ozonation, adsorption, and thermal [non-catalytic] de-chlorination). HDC involves hydrogen cleavage of one or more C–Cl bonds, lowering of toxicity, and generation of reusable raw material. The application of catalytic HDC in environmental pollution abatement has been shown to lower the ecotoxicity of chlorinated waste streams by over 80% [92].

Gas-phase catalytic HDC has been reported for the conversion of a range of aliphatic [93–101] and aromatic [102–111] feedstock. In terms of chloroarene HDC, chlorobenzene has been the overwhelming choice as a model reactant and has been investigated over an array of supported metal catalysts, notably Pd [94, 109, 110] and Ni [103, 108, 111, 112] where the supports that have been used include carbon [109, 110, 112], $Al_2O_3$ [94, 112], $SiO_2$ [112, 113], and MgO [112, 114]. HDC efficiency has been assessed in terms of the response to reaction temperature [115, 116], the reactants' partial pressure [103, 115], catalyst support [116, 117], metal dispersion [103, 118], and use of additives [119, 120] in the treatment of chlorinated dioxins/furans [119, 120], toluenes [103], and phenols [103, 116, 118].

## 14.3.2 Gold-Promoted Gas-Phase Catalytic Hydrodechlorination of Chlorophenols

The lower activity exhibited by supported gold relative to traditional transition metals in hydrogenation extends to catalytic HDC. Indeed, the gas-phase conversion of 2,4-dichlorophenol (2,4-DCP) over $Au/SiO_2$ and $Au/TiO_2$ was negligible ($\leq 3\%$), generating 2-chlorophenol (2-CP) as the sole product under conditions where near-complete de-chlorination to phenol was achieved over supported Ni [121]. Nevertheless, gold has found effective use as a promoter of unsupported [122–128] and ($Al_2O_3$ [123, 129] carbon [130–132]) supported Pd in the hydrotreatment of trichloroethene [122, 123, 125, 126], perchloroethene [124], $CCl_2F_2$ [129–132], diclofenac [127], and cloropyralid [128]. The HDC response exhibited by Au–Ni bimetallic combinations has also been considered [121, 133–135]. We have compared the catalytic action of $Ni/SiO_2$ with that of Au–Ni/SiO (prepared by co-impregnation) [121]. The Ni particles in activated $Ni/SiO_2$ were well dispersed (1–6 nm diameter range) with a mean size = 1.4 nm. The metal phase associated with Au–Ni/SiO$_2$ exhibited a wider size range, which can be attributed in part to the use of a Cl-containing Au precursor [136, 137]. Energy-dispersive X-ray (EDX) analysis has revealed that the particles possessed a variable Ni–Au composition with a range of Ni/Au atomic ratios (<1–40). Hydrogen chemisorption

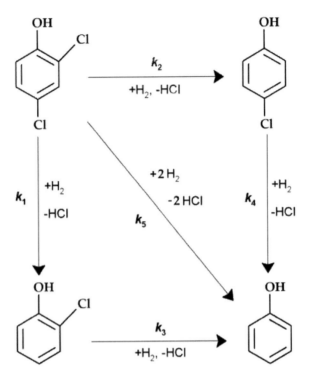

**Figure 14.3** 2,4-Dichlorophenol (2,4-DCP) HDC reaction scheme. Reprinted from Ref. [138], Copyright (2013), with permission from Elsevier.

on Au–Ni/SiO$_2$ was significantly lower relative to Ni/SiO$_2$ (18 vs. 843 μmol g$_{Ni}$$^{-1}$) diagnostics of surface Ni–Au interaction that serve to inhibit H$_2$ chemisorption on Ni. The HDC of 2,4-DCP over two reaction cycles generated 2-CP and phenol as partially and fully de-chlorinated products.

In the HDC of polychlorinated phenols, de-chlorination is inhibited due to steric hindrance with respect to *ortho*-positioned Cl substituent(s) [102, 139]. This effect is well illustrated in case of 2,4-DCP (see reaction pathway in Fig. 14.3) where 2-CP has been consistently isolated as the predominant product of partial de-chlorination [121, 133, 140]. A temporal decline in activity with time on-stream was recorded for both (Ni/SiO$_2$ and Au–Ni/SiO$_2$) catalysts. Time dependent loss of activity is a general feature of gas-phase HDC [93, 141–144] and has been ascribed to coke deposition

[93, 143, 144] and/or the formation of surface metal chlorides [141, 142] and/or metal sintering [141]. A reaction over Ni/SiO$_2$ and Au–Ni/SiO$_2$ in the first reaction cycle delivered similar levels of de-chlorination. A resumption of the reaction after thermal treatment in H$_2$ delivered, in the case of Ni/SiO$_2$, a fractional de-chlorination that was comparable to that obtained at the conclusion of the previous reaction cycle. In contrast, the thermal treatment resulted in a greater than twofold increase in 2,4-DCP conversion over Au–Ni/SiO$_2$ to outperform Ni/SiO$_2$. In terms of selectivity, reaction over Ni/SiO$_2$ resulted in preferential phenol formation, whereas the incorporation of Au favored partial de-chlorination to 2-CP. Reaction selectivity in the second cycle for Ni/SiO$_2$ essentially coincided with that attained at the completion of the first reaction. The increase in fractional de-chlorination over Au–Ni/SiO$_2$ postthermal treatment was accompanied by a significant shift to complete de-chlorination (to phenol).

This Au–Ni synergism extended to alumina-supported bimetallics prepared by the reductive deposition of Au onto Ni. The representative TEM images provided in Fig. 14.4a illustrate metal dispersion in Au–Ni/Al$_2$O$_3$ and the particle size distribution is given in Fig. 14.4b. The Ni/Au atomic ratios from EDX analysis of the encircled areas are included in Fig. 14.4a and exhibit significant variability (Ni/Au = 0.1–8.8). As a general observation, the gold content was distinctly higher for larger cluster sizes (>15 nm). In catalyst tests, there was again no detectable 4-CP production or subsequent phenol hydrogenation to cyclohexanol and cyclohexanone. In the first reaction cycle, initial HDC activity was equivalent for Ni/Al$_2$O$_3$ and Au–Ni/Al$_2$O$_3$ and both catalysts exhibited a temporal decrease in fractional de-chlorination. A resumption of the reaction over Ni/Al$_2$O$_3$, after thermal treatment in hydrogen, delivered (reaction cycle 2) activity values that showed continual decline with low HDC activity in reaction cycle 3. As observed for the silica-based system, the same thermal treatment resulted in an appreciable enhancement of HDC activity over Au–Ni/Al$_2$O$_3$, which greatly exceeded Ni/Al$_2$O$_3$. A temporal (over 8 h on-stream) HDC decrease was again in evidence but thermal treatment restored activity so that the profile for reaction cycle 3 was equivalent to that in reaction cycle 2 (Fig. 14.5).

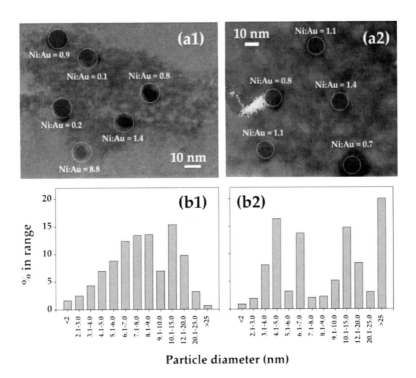

**Figure 14.4** (a) Representative TEM images with (b) associated particle size distributions of (1) activated unused and (2) used Au–Ni/Al$_2$O$_3$ catalysts. Note: EDX results (Ni/Au) for the encircled areas are included. Adapted from Ref. [134].

Initial phenol selectivity was lower over Au–Ni/Al$_2$O$_3$ but full HDC was increasingly favored in subsequent reaction cycles [134]. The initial equivalency of HDC performance for Ni/Al$_2$O$_3$ and Au–Ni/Al$_2$O$_3$ is to be expected, given the low activity associated with Au/Al$_2$O$_3$, that is, in terms of a solely additive effect. The enhancement of HDC performance for Au–Ni/Al$_2$O$_3$ postreaction/postthermal treatment suggests some surface modification that results in more effective HDC. To further probe this effect, the activated Au–Ni/Al$_2$O$_3$ was subjected to a thermal treatment with water vapor and HCl. Contact with water vapor had no effect on HDC behavior but HCl treatment delivered a HDC response that was comparable to that obtained in reaction cycles 2 and 3.

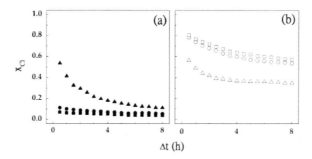

**Figure 14.5** Variation of fractional 2,4-dichlorophenol (2,4-DCP) dechlorination ($x_{Cl}$) with time on-stream ($\Delta t$) for reaction over (a) Ni/Al$_2$O$_3$ (solid symbols) and (b) Au–Ni/Al$_2$O$_3$ over three consecutive cycles (1$^{st}$ cycle: ▲, △; 2$^{nd}$ cycle: ●, ○; 3$^{rd}$ cycle: ■, □). Adapted from Ref. [134].

In contrast, HCl treatment of Ni/Al$_2$O$_3$ lowered initial fractional conversion (from 0.70 to 0.16) [134]. Improved HDC performance can then be arrived at by a direct thermal treatment of Au–Ni/Al$_2$O$_3$ with HCl. The particle size distribution for Au–Ni/Al$_2$O$_3$ was significantly wider after reaction cycle 3 (Fig. 14.4b [2]). There was also a measurable broadening of the Ni particle size distribution in Ni/Al$_2$O$_3$ postreaction, which can be ascribed to a halide-induced agglomeration of Ni particles due to the surface mobility of Ni–Cl species [145]. However, the thermal treatment following the reaction resulted in a more uniform Ni/Au atom ratio (Fig. 14.4a [2]).

We can attribute enhanced HDC performance to surface restructuring leading to a more homogeneous combination of Ni and Au. Molenbroek et al. [146, 147] have reported improved time on-stream performance for Ni–Au/SiO$_2$ relative to Ni/SiO$_2$ in butane steam reforming where the formation of a Ni–Au surface alloy was proposed to limit coke deposition and prolong activity. Moreover, Triantafyllopoulos and Neophytides [148] noted that the addition of gold to Ni/YSZ served to suppress carbon deposition during methane steam reforming. Surface Au/Ni interactions have been invoked to account for higher WGS activities for Au–Ni/Fe$_2$O$_3$ [149], increased isomerization rate [150], and enhanced propane de-hydrogenation [151] over Au–Ni/SiO$_2$ when compared with the corresponding monometallic systems. While Au/Al$_2$O$_3$ exhibited negligible activity, the Au component can have a direct contribution to HDC through activation of the C–Cl bond with subsequent attack from reactive hydrogen dissociated at the Ni centers. A similar rationale has been proposed by Venezia et al. [152] for the hydrodesulfurization of thiophene over Pd–Au/SiO$_2$. Moreover, Heinrichs et al. [153] have explained the selectivity response that they observed in the de-chlorination of 1,2-dichloroethane over Pd–Ag/SiO$_2$ by assuming Cl interaction at Ag sites with Pd as a source of dissociated hydrogen.

Taking 2,4-DCP as feed, controlled removal of the *ortho*-Cl substituent represents a significant advance in chlorowaste treatment, allowing the recycle of 4-CP as an important raw material in the production of fungicides [154], molluskicides [155], dichlorophene (used as antimicrobial finishing for textiles/paper), and fenticlor (for algae control in water-circulating systems) [156]. The catalytic behavior of nanoscale Au is sensitive to support redox properties, which govern metal–support electron transfer [157, 158]. We can flag work by Hildebrand et al. [159, 160] where Fe$_3$O$_4$ was used to support Pd and Au–Pd in the liquid-phase HDC of trichloroethene and chlorobenzene. Gas-phase conversion of 2,4-DCP over Au/Fe$_2$O$_3$ generated 4-CP as the principal product. The rate constants for each step (Fig. 14.3) confirmed that 2,4-DCP HDC proceeded via a predominant stepwise mechanism, that is, $k_1 + k_2 + k_3 + k_4 = 58$ mol$_{Cl}$ h$^{-1}$ m$_{Au}$$^{-2}$ >> $k_5 = 0.7$ mol$_{Cl}$ h$^{-1}$ m$_{Au}$$^{-2}$ (at 423 K), where removal of the *ortho*-positioned Cl proceeded at a greater rate than *para*-Cl ($k_2/k_1$ and $k_3/k_4$ >>1) [157, 158]. The selective scission of *ortho*-Cl suggests that this C–Cl bond is preferentially activated on Au/Fe$_2$O$_3$. Under identical reaction conditions, Au/Al$_2$O$_3$ only generated 2-CP as a product. The use of a reducible support must play a critical role in facilitating *ortho*-attack. Partial reduction of Fe$_2$O$_3$ during TPR (promoted by Au) results in the removal of surface oxygen to form vacancies, which have been proposed to generate electron-deficient Au species [138, 161, 162]. 2,4-DCP adsorption on (positively) charged Au clusters via the lone pair of electrons associated with –OH facilitates polarization (and activation) of the *ortho*- C–Cl bond as a result of inductive effects. This renders the *ortho*-Cl substituent more susceptible to electrophilic attack than the *para*-counterpart. It should be noted that the overall specific

**Table 14.1** Specific HDC rate constant ($k$) and selectivity to phenol, 2-CP, and 4-CP in the conversion of 2,4-DCP in different solvents/carriers and a series of chlorophenols (in ethanol) over Au/Fe$_2$O$_3$

|  |  | $k \times 10^4$ ($mol_{Cl}h^{-1}m_{Au}^{-2}$) | Phenol (%) | 2-CP (%) | 4-CP (%) |
|---|---|---|---|---|---|
| Solvent/Carrier (HDC of 2,4-DCP) | Water | 10.8 | 5 | 3 | 92 |
|  | Methanol | 1.3 | 7 | 4 | 89 |
|  | Ethanol | 5.1 | 5 | 2 | 93 |
|  | n-Propanol | 5.6 | 3 | 4 | 93 |
|  | n-Butanol | 6.7 | 6 | 1 | 93 |
|  | n-Hexane | 2.5 | 5 | 4 | 91 |
|  | Cyclohexane | 2.4 | 8 | 4 | 88 |
| Reactant (in ethanol) | 2-CP | 11.6 | 100 | – | – |
|  | 3-CP | 4.3 | 100 | – | – |
|  | 4-CP | 4.2 | 100 | – | – |
|  | 2,3-DCP | 4.4 | 0 | 8 | 92 |
|  | 2,4-DCP | 5.1 | 6 | 0 | – |
|  | 2,5-DCP | 8.6 | 0 | 17 | 83 |
|  | 2,6-DCP | 10.0 | 56 | 44 | – |
|  | 3,4-DCP | 2.8 | 27 | – | 18 |
|  | 3,5-DCP | 3.7 | 50 | – | 50 |
|  | 2,4,6-TCP[a] | 8.5 | 29 | – | – |

[a] Selectivity to 2,4-DCP = 34%.
Source: Adapted from Ref. [138].

2,4-DCP HDC rate delivered by Au/Fe$_2$O$_3$ ([$k_1 + k_2 + k_5$] = 11 × 10$^{-4}$ mol$_{Cl}$ h$^{-1}$ m$_{Au}$$^{-2}$) is close to that (14 × 10$^{-4}$ mol$_{Cl}$ h$^{-1}$ m$_{Ni}$$^{-2}$) reported elsewhere [118] for reaction over Ni nanoparticles (mean diameter = 1.4 nm) supported on SiO$_2$.

The effect of the carrier on HDC performance was considered by examining the action of protonated (water and alcohols) and nonprotonated (n-hexane and cyclohexane) solvents. Competitive adsorption can result in rate inhibition due to occlusion of active sites, as has been reported for gas-phase hydrogenation [163–165]. The results presented in Table 14.1 show that 2,4-DCP HDC activity using organic carriers was significantly lower than that recorded for the aqueous 2,4-DCP feed. This can be ascribed to water dissociation on Au/Fe$_2$O$_3$ to produce spillover hydrogen that has been shown to contribute to WGS reaction activity [166, 167]. Taking the organic carriers, th eHDC rate was higher with alcohols, increasing from methanol to *n*-butanol. Alcohol chemisorption on metal oxides proceeds via interaction between the oxygen electron lone pair and a Lewis acid site on the oxide [168, 169] with H abstraction and alcoxide formation that is enhanced with increasing alcohol chain length [170] with a resultant improved reaction rate. Hydrogen abstraction is not possible for alkane solvents with a consequent lower HDC activity. The use of different carriers had little effect on product selectivity where, at the same 2,4-DCP conversion, product distribution was essentially equivalent. Enhanced hydrogenolytic cleavage of sterically constrained Cl is further demonstrated for a range of range of mono-, di-, and trichlorophenols (TCPs) in Table 14.1. We can flag the higher HDC rate exhibited by those reactants bearing Cl substituents in the *ortho*-position, notably 2-CP and 2,6-DCP. The unique HDC selectivity is exemplified by the preferential formation of 3-CP (relative to 2-CP) in the HDC of 2,3- and 2,5-DCP and in the conversion of 2,4,6-TCP where 2,6-DCP was not isolated as partial HDC product. In contrast, 2-CP [140, 171] and 2,6-DCP [102, 172] have been identified as principal products in the hydrotreatment of polychlorinated phenols over Ni- and Pd-based catalysts. The facility of gold on iron oxide to selectively cleave sterically constrained Cl can serve to extend the range of possible products generated in the treatment of chlorinated streams, allowing controlled recycle of target product(s).

## 14.4 Case Study 2: Production of Fine Chemicals; Hydrogenation of Nitroaromatics

### 14.4.1 *Background*

Aromatic amines are widely used in the fine-chemical sector for the manufacture of a diversity of pharmaceuticals, insecticides, dyes/pigments, and cosmetics [173, 174]. Conventional production routes are based on batch processes either by Fe-promoted reduction in acid media (Béchamp reaction), which generates appreciable toxic Fe/FeO sludge waste, or hydrogenation over transition metal catalysts [175, 176]. Catalytic hydrogenation represents a

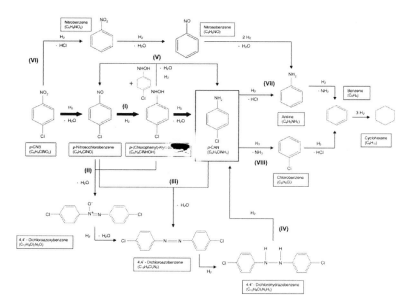

**Figure 14.6** Reaction pathways in the hydrogen-mediated conversion of p-CNB. Targeted route (I) to p-chloroaniline (p-CAN) is represented by bold arrows. Reprinted from Ref. [180], Copyright (2008), with permission from Elsevier.

cleaner option but the formation of toxic azo- [177] and azoxy- [178] derivatives is still a decided drawback. High selectivity in the conversion of substituted nitroarenes to the target amine remains challenging. By way of illustration, the hydrogenation of p-chloronitrobenzene (p-CNB) generates a range of intermediates and by-products [174], as shown in Fig. 14.6. The substantial waste generated in the transformation of substituted nitroarenes led Sheldon to introduce the concept of the **e**(nvironmental) **f**actor ($kg_{waste}\ kg_{product}^{-1}$), which highlighted the severe ecological impact of amine production [179].

The chemoselective reduction of substituted nitroarenes to the corresponding functionalized aniline is possibly one of the most widely studied catalytic applications of gold in hydrogenation [17, 174, 176, 181, 182]. Gold has been demonstrated to effectively activate the $-NO_2$ function, circumventing the formation (and exothermic decomposition) of reaction intermediates (e.g., nitrosobenzenes and hydroxylamine).

In 2006, Chen et al. [183] and Corma and Serna [184] reported breakthrough work demonstrating the potential of gold for the chemoselective hydrogenation of aromatic nitrocompounds. They established selective $-NO_2$ reduction in the presence of other reactive functionalities ($-CH_3$, $-OH$, $-COOCH_3$, $-OC_2H_5-C=C$, $-C\equiv C$, $-C=O$, and $-C\equiv N$) for liquid-phase reaction over oxide-supported gold ($SiO_2$ [183], $TiO_2$ [184], and $Fe_2O_3$ [184]). Preferential nitrogroup adsorption and activation on Au/TiO$_2$ in the conversion of 3-nitrostyrene [157] contrasted with *non*selective reaction over supported Pt [185]. This work was extended to encompass high selectivity in the hydrogenation of chloronitrobenezenes over Au/ZrO$_2$ [186] and production of oximes from $\alpha, \beta$-unsaturated nitroreactants [187]. Recent studies using different carriers (Al$_2$O$_3$ [188], TiO$_2$ [189], CeO$_2$ [190], Fe$_2$O$_3$ [191], Fe(OH)$_x$ [192, 193], ZrO$_2$ [186], carbon [194], Mo$_2$N [195], and polymer [196]) have confirmed the general applicability of gold catalysts for the liquid-phase chemoselective hydrogenation of a series of nitroarenes (nitrophenol [190], halonitrobenzenes [186, 195], nitrostyrenes [188, 196], nitroarylalkynes [191], nitrobenzaldehyde [192, 193], nitroacetophenone [192, 193], and nitroanisole [189]). Catalytic activity and/or selectivity in nitroarene reduction has shown a dependence on metal/support interaction(s) and metal dispersion [197] and variations in process variables such as temperature and/or pressure [184, 186], reactant concentration [198, 199], the nature of the reaction medium [200], and solvent polarity [201, 202].

### 14.4.2 Gold-Promoted Gas-Phase Catalytic Hydrogenation of Nitrocompounds

It has been reported [161, 163, 203–206] that oxide-supported gold is fully selective in the gas-phase continuous hydrogenation of halonitroarenes to the corresponding haloaniline. Reaction exclusivity to *p*-CAN (from *p*-CNB) was achieved over Au/Al$_2$O$_3$ (2–8 nm Au) with no detectable undesired by-products from condensation or HDC (see Fig. 14.6) [203, 206, 207]. Stable conversion with

**Table 14.2** Pseudo-first-order rate constants ($k$) and reaction products obtained in the hydrogenation of a range of nitroarenes over Au/Al$_2$O$_3$

| Reactant | Rate constant k (h$^{-1}$) | Product(s) |
|---|---|---|
| p-Nitrotoluene (p-NT) | No conversion | – |
| Nitrobenzene (NB) | 2 | AN |
| m-Chloronitrobenzene (m-CNB) | 2 | m–CAN |
| p-Bromonitrobenzene (p-BNB) | 4 | p–BAN |
| p-Chloronitrobenzene (p-CNB) | 4 | p–CAN |
| o-Chloronitrobenzene (o-CNB) | 9 | o–CAN |
| 2,4-Dichloronitrobenzene (2,4-DNB) | 90 | 2,4-DCAN |
| 3,4-Dichloronitrobenzene (3,4-DNB) | 90 | 3,4-DCAN |
| 3,5-Dichloronitrobenzene (3,5-DNB) | 90 | 3,5-DCAN |

*Source*: Reprinted from Ref. [203], Copyright (2008), with permission from Elsevier.

full p-CAN selectivity was maintained over the temperature range of 393–523 K ($\Delta E_a = 49$ kJ mol$^{-1}$) for up to 80 h on-stream [203]. Such stable and exclusive p-CAN formation is noteworthy in light of the published work on gas-phase nitroarene hydrogenation where a temporal decline in activity has been reported (over supported Pd [208–210] and Cu [211, 212]) and attributed to deleterious effects of water as a by-product [210], metal leaching [209], and coking [208, 211, 212]. Exclusivity extended to a range of mono- and disubstituted halonitroarenes (see Table 14.2) where the activity sequence (NT <NB ≤ CNB/BNB <DNB) suggests electron-withdrawing substituent activation, consistent with the participation of a negatively charged reaction intermediate [213]. The role of gold particle size in hydrogenation is sensitive to the nature of the reactant [214], support [89], and even catalyst preparation [215]. Exclusive –NO$_2$ reduction was established in the conversion of p-CNB and m−dinitrobenzene (m−DNB) over a series of oxide-supported gold catalysts (1–10 mol% Au) with different mean particle sizes (1.5–10.0 nm) [204, 205, 216]. An increase in turnover frequency (TOF) with decreasing gold size (from 10 to 3 nm) was taken to be diagnostic of structure sensitivity with enhanced intrinsic –NO$_2$ hydrogenation efficiency for smaller gold particles. However, the observed specific rate was lower for gold particles <2 nm, a result that was attributed to quantum size effects

[217] with a transition in the electronic state of gold from metal to nonmetal [218, 219].

The primary function of the catalyst support is to provide a surface on which the metal phase is dispersed. This facilities formation of smaller metal particle sizes, which, in turn, provides a higher metal surface area that should favorably impact on catalytic activity. The redox character of the carrier is significant as partial support reduction can induce geometric variations in the supported Au phase with the exposure of a specific crystallographic plane with distinct catalytic properties [220]. Reducible iron oxides have served as effective gold supports with a superior catalytic response for gold on $Fe_2O_3$ relative to $TiO_2$ [157] and $Fe_3O_4$ [75] in nitroarene and acetylene hydrogenation, respectively. The reducibility of hematite ($\alpha$-$Fe_2O_3$) can be assessed from the TPR profile in Fig. 14.7a, where structural modifications can be assessed from X-ray diffraction (XRD) analysis (see insets). Hydrogen consumption during TPR exhibited four temperature-related maxima (at 562, 666, 965, and 1183 K). The first two signals can be attributed to the reduction of $Fe_2O_3$ to $Fe_3O_4$ (as confirmed by XRD, coincidence of profile C and the JCPDS-ICDD reference for $Fe_3O_4$, profile F). The associated hydrogen consumption (0.17 mol $mol_{Fe}^{-1}$) matches that required for $Fe_2O_3 \rightarrow Fe_3O_4$. The occurrence of a two-stage reduction of $Fe_2O_3$ to $Fe_3O_4$ has been reported elsewhere [221, 222] and linked to support de-hydroxylation with concomitant formation of surface oxygen vacancies. The broader hydrogen consumption peaks at 965 and 1183 K are due to further reduction of $Fe_3O_4$ to metallic iron, as confirmed by the XRD diffractogram of the sample taken to 1273 K (match of profile D with the JCPDS-ICDD reference for Fe, profile G). Hematite-supported gold was prepared by deposition-precipitation, where the resultant TPR profile (Fig. 14.7b) deviates from that recorded for $Fe_2O_3$ in that the peaks are shifted to lower temperatures by up to 200 K. This suggests a more facile $Fe_2O_3$ reduction in the presence of gold and is consistent with previous reports [166, 223]. A comparison of the XRD patterns of as-prepared Au/$Fe_2O_3$ (profile A) with that obtained post-TPR to 423 K (profile B) can be made from the $2\theta$ range outlined in the dashed boxes. The relative intensity of the two main peaks at $2\theta = 33.2°$ and $35.6°$ in profile B (3/4) differs from that which characterizes profile

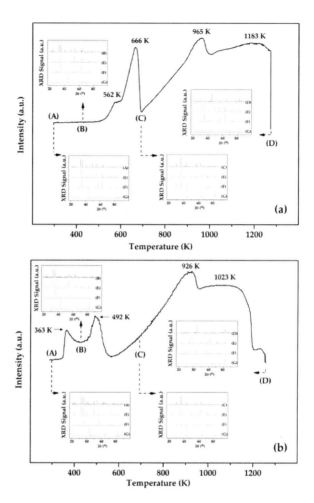

**Figure 14.7** TPR and XRD patterns (insets) for (a) Fe$_2$O$_3$ support and (b) Au/Fe$_2$O$_3$; samples (A) as prepared and reduced at 423K (B), 673K (C), and 1273K (D). Reference (JCPDS-ICDD) diffractograms for $\alpha$-Fe$_2$O$_3$ (33-0664 (E)), Fe$_3$O$_4$ (19-0629 (F)) and Fe (06-0696 (G)). Adapted from Ref. [138].

A (4/3) and the reference $\alpha$-Fe$_2$O$_3$ (profile E). As the XRD peak at $2\theta = 35.6°$ for $\alpha$-Fe$_2$O$_3$ coincides with the main characteristic peak for Fe$_3$O$_4$ ($2\theta = 35.4°$, see profile F), this variation in relative intensity suggests a partial reduction of the support. Similarly, a consideration of the XRD signals contained within the dashed box at

$2\theta = 62.4°$ and $64.0°$ in Fig. 14.7b reveals differences in the relative intensities of the signals for $\alpha$-Fe$_2$O$_3$ (profiles A and E) with the simultaneous appearance of a signal characteristic of Fe$_3$O$_4$ at $2\theta = 57.0°$ (see profile F). These results demonstrate partial reduction of the support after TPR of Au/Fe$_2$O$_3$ to 423 K. The absence of reflections due to gold in the XRD patterns is indicative of the occurrence of small clusters. This was confirmed by TEM analyses, where the quasi-spherical gold particles exhibited a narrow size distribution and surface area–weighted mean of 2 nm [138]. The catalytic action of nanosized gold on hematite and magnetite (Fe$_3$O$_4$) was compared in the gas-phase hydrogenation of p-CNB and m-DNB [161]. HRTEM and X-ray photoelectron spectroscopy (XPS) analysis established an encapsulation of Au in the Fe$_3$O$_4$ matrix following TPR to 423 K, which inhibited hydrogenation activity. Encapsulation or "decoration" [224] is a documented feature of gold on reducible carriers such as TiO$_2$ [225] and CeO$_2$ [226]. TPR to 673 K resulted in gold segregation on the Fe$_3$O$_4$ surface and the formation of particles with mean diameter = 4 nm. A similar TOF was recorded for Au/Fe$_2$O$_3$ and Au/Fe$_3$O$_4$ with exclusive formation of p-chloroaniline and m-nitroaniline.

Support acid–base properties are also important in terms of metal$\leftrightarrow$support electron transfer, resulting in the formation of positively (electron-deficient) or negatively charged gold particles [89, 227, 228]. We can flag a recent study where catalyst performance (gold on TiO$_2$, ZrO$_2$, and Al$_2$O$_3$) in 4-nitrobenzaldehyde hydrogenation was correlated with support Lewis acidity [229]. Taking the ZrO$_2$ carrier, synthesis by precipitation of ZrOCl$_2$ generates a mixed (tetragonal and monoclinic) phase with associated pH point of zero charge (pH$_{PZC}$) = 7.4 [230 231]. Gold incorporation by deposition-precipitation enhanced the monoclinic ZrO$_2$ content, which is sensitive to solution pH during synthesis and postsynthesis treatment [232] TPR (Au$^{3+}$ $\rightarrow$Au) generated a mean particle size = 7 nm with greater hydrogen release (from TPD analysis) relative to chemisorption [229], which can be ascribed to spillover species [233]. Hydrogenation of 4-nitrobenzaldehyde over Au/ZrO$_2$ generated 4-aminobenzaldehyde as the sole product [229]. Au/TiO$_2$, synthesized and activated in a similar manner, resulted in a narrower distribution of smaller gold particles (mean = 4.7 nm) that again exhibited 100% selectivity to 4-aminobenzaldehyde but with a higher TOF relative to Au/ZrO$_2$. In contrast, reaction over Au/Al$_2$O$_3$ generated 4-nitrobenzyl alcohol attributable to greater surface Lewis acidity (as demonstrated by Fourier transform infrared spectroscopy [FTIR] analysis following pyridine adsorption), which favors formation of an intermediate benzoate that is reduced to the alcohol [229].

One of the major challenges in hydrogenation over supported gold is the low associated reaction rates. Catalyst development to achieve higher product yields is a crucial requirement for application at an industrial scale [234, 235]. Given that the limiting factor is hydrogen activation, the use of a (i) second metal and the use of a (ii) support that can activate hydrogen and increase surface concentration represent two feasible routes to higher selective hydrogenation rates. The incorporation of palladium with gold (on an alumina carrier, Pd–Au/Al$_2$O$_3$) was examined as a means of enhancing hydrogen dissociation [236]. The rate of *p*-CNB hydrogenation was elevated where Au/Pd $\geq$ 20 with 100% selectivity to *p*-CAN. Reaction at higher palladium content (Au/Pd = 8) resulted in aniline formation. The combination of Au/Al$_2$O$_3$ and Pd/Al$_2$O$_3$ as a physical mixture (Au/Pd = 20) promoted the formation of nitrobenzene and aniline via composite HDC/hydrogenation (see Fig. 14.6). The Au/Al$_2$O$_3$ component in the mixture had little effect on reaction selectivity, which was governed by Pd/Al$_2$O$_3$. The distinct catalytic response observed for the supported bimetallic suggests that the contribution of both metals is not merely an additive effect but the result of a surface synergism in Pd–Au/Al$_2$O$_3$. This was investigated by FTIR–diffuse reflectance infrared Fourier transform spectroscopy (DRIFTS) analysis using CO as a probe molecule [236]. The spectra in the carbonyl region for the bimetallic catalyst exhibited carbonyl bands at high (2150–2050 cm$^{-1}$) and low (2000–1900 cm$^{-1}$) wavenumbers that deviated significantly from that expected on the basis of a simple additive contribution due to spectra for Pd/Al$_2$O$_3$ and Au/Al$_2$O$_3$. The band at lower wavenumbers can be assigned to CO-bridged species on Pd crystal planes [237, 238]. Two resolved peaks at 1955 and 1935 cm$^{-1}$ for Pd–Au/Al$_2$O$_3$ suggest the predominance of twofold bridged surface species, which are diagnostic of bimetallic particles

[239, 240]. Spectroscopic analysis of CO adsorption on Au–Pd films [239, 241] and Au–Pd clusters [240] suggests that bands at higher wavenumbers (2110–2135 cm$^{-1}$) are due to CO linearly adsorbed on Au$^0$ sites in monometallic or bimetallic particles. The DRIFTS results indicate Pd–Au surface interaction that serves to elevate the selective hydrogenation rate and that is sensitive to the Au/Pd ratio [236]. We note that Au–Pt supported on functionalized Si-MCM-41 [242] and chloride dendrons [243] have shown higher activity in the hydrogenation of nitrotoluene (to toluidine) compared to monometallic platinum and gold nanoparticles. This was linked to preferential adsorption on the electron-deficient platinum shell (due to electron withdrawal from the gold core) in Au$_{core}$Pt$_{shell}$ nanoparticles.

As Mo$_2$N [244] and Mo$_2$C [245] can chemisorb hydrogen, use as gold supports should serve to increase the available hydrogen and elevate the hydrogenation rate. This was examined in a series of studies [195, 246, 247], taking p-CNB as a test reactant where Au/Al$_2$O$_3$ served as the benchmark [236]. The work was extended to consider palladium–nitride formulations (Pd/Mo$_2$N) and the results obtained are summarized in Fig. 14.8). Au/Mo$_2$N outperformed Au/Al$_2$O$_3$ that exhibited an equivalent mean Au particle size (= 8 nm).

Under the same reaction conditions, Pd/Mo$_2$N was *non*selective and promoted a combined HDC/hydrogenation to give nitrobenzene and aniline. The data support the viability of Au/Mo$_2$N for the selective reduction of nitroarenes. The influence of the nitride crystallographic phase (tetragonal $\beta$-Mo$_2$N vs. cubic $\gamma$-Mo$_2$N) and surface area (7–66 m$^2$ g$^{-1}$) was also considered [246]. Hydrogen TPD analysis established a higher specific (per unit area) H$_2$ content for $\beta$-Mo$_2$N relative to $\gamma$-Mo$_2$N. This was attributed to a greater number of nitrogendeficient sites on $\beta$-Mo$_2$N (as established by XPS), which translated into a higher rate. Specific activity was independent of the Brunauer–Emmett–Teller (BET) surface area. Incorporation of Au served to enhance H$_2$ uptake capacity and elevated the hydrogenation rate where p-CAN was again the sole product; Au/$\beta$-Mo$_2$N outperformed Au/$\gamma$-Mo$_2$N.

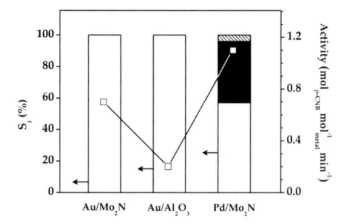

**Figure 14.8** Specific activity with selectivity to p-CAN (open bars), aniline (hatched bars), and nitrobenzene (solid bars) under conditions of equal p-CNB conversion (20%) over Au/Mo$_2$N, Au/Al$_2$O$_3$, and Pd/Mo$_2$N. Reprinted from Ref. [195], Copyright (2012), with permission from Elsevier.

## 14.5 Concluding Remarks and a Look to the Future

The study of heterogenous gold catalysis is a fast-moving field of research, characterized by changes of emphasis and often of perception. The selectivity of supported nanoscale gold is now well established but with as yet untapped potential in the production of a diversity of high-value chemicals. Gold-promoted hydrogenation is still at the formative stages but there is increasing evidence of commercial possibilities that can build on reaction selectivity. In this chapter, we have highlighted the capability of gold catalysts to reduce the nitrogroup in functionalized nitrobenzenes, with multiple applications in the fine-chemical industry. This work can be exploited further by targeting reactions of increasing complexity in terms of selectivity, leading to clean production of commercially important multifunctional aromatic amines where existing methodologies generate low product yields and appreciable toxic waste. We have also identified an emerging application of supported gold in catalytic HDC as a progressive means of detoxifying chlorinated waste and facilitating raw material recycle. The viability of selective catalytic hydrogenolysis of sterically constrained Cl substituents represents a critical advancement in the transformation of chlorinated waste. Future work should be directed at extending the database of catalytic HDC to consider a wider range of reaction conditions and alternative gold catalyst formulation with a full assessment of structure sensitivity in terms of product distribution.

A critical feature of future work on gold catalysts will be the development of unified synthesis methodologies that draws on the existing procedures but with a theoretical basis. Gold catalyst development at the *nano*scale will require an explicit knowledge of structure/performance relationships. Progress in catalyst preparation must be directed at rational design with a targeted function rather than trial-and-error catalyst preparation. Researchers must move away from conventional cycles of catalyst preparation with testing where serendipitous incremental improvements are used to inform subsequent catalyst synthesis. Fundamental studies are called for that combine surface imaging and in situ spectroscopy with analysis of reaction kinetics to establish how local bonding and formation of surface species and reaction intermediates determine catalyst performance. DFT calculations are significant in this respect and can support experimental measurements and elucidate reactant/surface dynamics. Molecular-level mechanistic information gleaned from surface science methodologies must be coupled with catalyst testing in practical operation. This requires surface science measurements that bridge the pressure gap that exists between the analysis of model surfaces and practical heterogeneous catalysts.

The results presented in this chapter have illustrated the critical role of gold particle size and electronic character in determining performance, which is influenced by the support either indirectly in terms of interactions that modify the gold active site or that can directly, notably in terms of surface acidity, influence adsorption and reactive intermediate formation. Hydrogen activation by gold represents a limiting step due to the high-activation-energy barrier for dissociative adsorption. Reactions are conducted in excess hydrogen where the vast majority is unreacted, resulting in fundamental process inefficiency. Hydrogen is not a naturally occurring feedstock and production, storage, and transport are sustainability constraints. Effective exploitation of the unique selectivity of gold catalysis in hydrogen-mediated reactions must address hydrogen utilization. We have shown improvements resulting from the introduction of a second metal (palladium) and the use of molybdenum nitrides or carbides as supports with hydrogen uptake capacity. There is, however, considerable scope for further enhancement. A possible innovative approach is the coupling of catalytic de-hydrogenation as a source of reactive hydrogen for hydrogenation.

A switch from batch liquid-phase to continuous gas-phase operation is an important sustainability consideration. Economies of scale favor continuous processes for high throughput. In principle, the catalytic trends observed in batch liquid-phase hydrogenation should largely apply to continuous gas-phase operation, although reaction conditions in terms of contact time, pressure, and temperature can vary significantly and considerations with respect to heat and mass transport effects are quite different. An examination of catalysts to promote hydrogenation in both (continuous) gas and (batch/continuous) liquid modes should be conducted in parallel. This should incorporate a comprehensive evaluation of catalyst deactivation and regeneration. A full sustainability analysis that considers energy requirements and overall throughput is required to arrive at the best practicable option for the effective delivery of next-generation fine chemicals to the marketplace. Use of biomass-derived reactants as alternatives to petroleum-based feedstock is certain to become a critical theme in the coming years. Gold catalysis will have a critical role to play in facilitating cleaner turnover of bioderived feedstock to target products.

## References

1. Haruta, M., Kobayashi, T., Sano, H., and Yamada, N. (1987). Novel gold catalysts for the oxidation of carbon monoxide at a temperature far below 0°C, *Chem. Lett.*, **2**, pp. 405–408.
2. Bond, G. (2008). The early history of catalysis by gold, *Gold Bull.*, **41**, pp. 235–241.
3. Bond, G.C. (2012). Chemisorption and reactions of small molecules on small gold particles, *Molecules*, **17**, pp. 1716–1743.

4. Corti, C.W., Holliday, R.J., and Thompson, D.T. (2007). Progress towards the commercial application of gold catalysts, *Top. Catal.*, **44**, pp. 331–343.
5. Barakat, T., Rooke, J.C., Genty, E., Cousin, R., Siffert, S., and Su, B.-L. (2013). Gold catalysts in environmental remediation and water-gas shift technologies, *Energy Environ. Sci.*, **6**, pp. 371–391.
6. Scirè, S., and Liotta, L.F. (2012). Supported gold catalysts for the total oxidation of volatile organic compounds, *Appl. Catal. B: Environ.*, **125**, pp. 222–246.
7. Gaálová, J., Topka, P., Kaluža, L., and Šolcová, O. (2011). Gold versus platinum on ceria-zirconia mixed oxides in oxidation of ethanol and toluene, *Catal. Today*, **175**, pp. 231–237.
8. Tabakova, T., Ilieva, L., Ivanov, I., Zanella, R., Sobczak, J.W., Lisowski, W., Kaszkur, Z., and Andreeva, D. (2013). Influence of the preparation method and dopants nature on the WGS activity of gold catalysts supported on doped by transition metals ceria, *Appl. Catal. B: Environ.*, **136–137**, pp. 70–80.
9. Tao, F., and Ma, Z. (2013). Water-gas shift on gold catalysts: catalyst systems and fundamental studies, *Phys. Chem. Chem. Phys.*, **15**, pp. 15260–15270.
10. Angelici, R.J. (2013). Bulk gold (non-nanogold) catalysis of aerobic oxidations of amines, isocyanides, carbon monoxide, and carbene precursors, *Catal. Sci. Technol.*, **3**, pp. 279–296.
11. Mikami, Y., Dhakshinamoorthy, A., Alvaro, M., and Garcia, H. (2013). Catalytic activity of unsupported gold nanoparticles, *Catal. Sci. Technol.*, **3**, pp. 58–69.
12. Sheldon, R.A., Arends, I., and Hanefeld, U. (2007). *Green Chemistry and Catalysis* (Willey-VCH, Weinheim, Germany).
13. Claus, P. (2005). Heterogeneously catalyzed hydrogenation using gold catalysts, *Appl. Catal. A: Gen.*, **291**, pp. 222–229.
14. McEwan, L., Julius, M., Roberts, S., and Fletcher, J.C.Q. (2010). A review of the use of gold catalysts in selective hydrogenation reactions, *Gold Bull.*, **43**, pp. 298–306.
15. Kartusch, C., and van Bokhoven, J.A. (2009). Hydrogenation over gold catalysts: the interaction of gold with hydrogen, *Gold Bull.*, **42**, pp. 343–348.
16. Mitsudome, T., and Kaneda, K. (2013). Gold nanoparticle catalysts for selective hydrogenations, *Green Chem.*, **15**, pp. 2636–2654.

17. Cárdenas-Lizana, F., and Keane, M.A. (2013). The development of gold catalysts for use in hydrogenation reactions, *J. Mater. Sci.*, **48**, pp. 543–564.
18. Pan, M., Brush, A.J., Pozun, Z.D., Ham, H.C., Yu, W.-Y., Henkelman, G., Hwang, G.S., and Mullins, C.B. (2013). Model studies of heterogeneous catalytic hydrogenation reactions with gold, *Chem. Soc. Rev.*, **42**, pp. 5002–5013.
19. Corma, A., González-Arellano, C., Iglesias, M., and Sánchez, F. (2009). Gold complexes as catalysts: chemoselective hydrogenation of nitroarenes, *Appl. Catal. A: Gen.*, **356**, pp. 99–102.
20. Comas-Vives, A., González-Arellano, C., Corma, A., Iglesias, M., Sánchez, F., and Ujaque, G. (2006). Single-site Homogeneous And Heterogenized Gold(III) hydrogenation catalysts: mechanistic implications, *J. Am. Chem. Soc.*, **128**, pp. 4756–4765.
21. Widenhoefer, R.A. (2008). Recent developments in enantioselective gold(I) catalysis, *Chem. Eur. J.*, **14**, pp. 5382–5391.
22. Pradal, A., Toullec, P.Y., and Michelet, V. (2011). Recent developments in asymmetric catalysis in the presence of chiral gold complexes, *Synthesis*, pp. 1501–1514.
23. Bond, G.C., and Sermon, P.A. (1973). Gold catalysts for olefin hydrogenation, *Gold Bull.*, **6**, pp. 102–105.
24. Schwank, J. (1983). Catalytic gold. Applications of elemental gold in heterogeneous catalysis, *Gold Bull.*, **16**, pp. 103–110.
25. Barrio, L., Liu, P., Rodríguez, J.A., Campos-Martín, J.M., and Fierro, J.L. G. (2006). A density functional theory study of the dissociation of $H_2$ on gold clusters: importance of fluxionality and ensemble effects, *J. Chem. Phys.*, **125**, pp. 164715–/(1–5).
26. Stobiński, L., Zommer, L., and Duś, R. (1999). Molecular hydrogen interactions with discontinuous and continuous thin gold films, *Appl. Surf. Sci.*, **141**, pp. 319–325.
27. Lin, S., and Vannice, M.A. (1991). Gold dispersed on $TiO_2$ and $SiO_2$: adsorption properties and catalytic behavior in hydrogenation reactions, *Catal. Lett.*, **10**, pp. 47–62.
28. Jia, J., Haraki, K., Kondo, J.N., Domen, K., and Tamaru, K. (2000). Selective hydrogenation of acetylene over $Au/Al_2O_3$ catalyst, *J. Phys. Chem. B*, **104**, pp. 11153–11156.
29. Bus, E., Miller, J.T., and van Bokhoven, J.A. (2005). Hydrogen chemisorption on $Al_2O_3$-supported gold catalysts, *J. Phys. Chem. B*, **109**, pp. 14581–14587.

30. Boronat, M., Illas, F., and Corma, A. (2009). Active sites for $H_2$ adsorption and activation in $Au/TiO_2$ and the role of the support, *J. Phys. Chem. A*, **113**, pp. 3750–3757.
31. Corma, A., Boronat, M., González, S., and Illas, F. (2007). On the activation of molecular hydrogen by gold: a theoretical approximation to the nature of potential active sites, *Chem. Commun.*, **32**, pp. 3371–3373.
32. Berndt, H., Pitsch, I., Evert, S., Struve, K., Pohl, M.-M., Radnik, J., and Martin, A. (2003). Oxygen adsorption on $Au/Al_2O_3$ catalysts and relation to the catalytic oxidation of ethylene glycol to glycolic acid, *Appl. Catal. A: Gen.*, **244**, pp. 169–179.
33. Svensson, K., Bellman, J., Hellman, A., and Andersson, S. (2005). Dipole active rotations of physisorbed $H_2$ and $D_2$, *Phys. Rev. B*, **71**, pp. 245402-/(1–6).
34. Bellman, J., Svensson, K., and Andersson, S. (2006). Molecular hydrogen adsorption at surface adatoms, *J. Chem. Phys.*, **125**, pp. 064704-/(1–4).
35. Lytle, F.W. (1976). Determination of $d$-band occupancy in pure metals and supported catalysts by measurement of L3 x-ray absorption threshold, *J. Catal.*, **43**, pp. 376–379.
36. Hammer, B., and Nørskov, J.K. (1995). Why gold is the noblest of all the metals, *Nature*, **376**, pp. 238–240.
37. Kislyuk, M.U., and Tretyakov, I.I. (1976). Kinetics of thermal atomization of hydrogen on gold, *Kinet. Catal.*, **17**, pp. 1302–1305.
38. Kislyuk, M.U., and Tretyakov, I.I. (1977). Effect of additions of platinum on heterogeneous atomization of hydrogen on gold, *Kinet. Catal.*, **18**, pp. 1108–1110.
39. Kislyuk, M.U., and Tretyakov, I.I. (1975). Relative roles of surface and impact mechanisms in heterogeneous recombinaton of hydrogenatoms on metals of copper subgroup, *Kinet. Catal.*, **16**, pp. 681–683.
40. Gluhoi, A.C., Vreeburg, H.S., Bakker, J.W., and Nieuwenhuys, B.E. (2005). Activation of CO, $O_2$ and $H_2$ on gold-based catalysts, *Appl. Catal. A: Gen.*, **291**, pp. 145–150.
41. Boronat, M., Concepción, P., and Corma, A. (2009). Unravelling the nature of gold surface sites by combining IR spectroscopy and DFT calculations. implications in catalysis, *J. Phys. Chem. C*, **113**, pp. 16772–16784.
42. Bond, G.C., Sermon, P.A., Webb, G., Buchanan, D.A., and Wells, P.B. (1973). Hydrogenation over supported gold catalysts, *J. C. S. Chem. Commun.*, **13**, pp. 444–445.

43. Bond, G.C., Louis, C., and Thompson, D.T. (2006). *Catalysis by Gold* (Imperial College Press, London).
44. Mohr, C., Hofmeister, H., and Claus, P. (2003). The influence of real structure of gold catalysts in the partial hydrogenation of acrolein, *J. Catal.*, **213**, pp. 86–94.
45. Zanella, R., Louis, C., Giorgio, S., and Touroude, R. (2004). Crotonaldehyde hydrogenation by gold supported on TiO$_2$: structure sensitivity and mechanism, *J. Catal.*, **223**, pp. 328–339.
46. Zhang, Y., Cui, X., Shi, F., and Deng, Y. (2012). Nano-gold catalysis in fine chemical synthesis, *Chem. Rev.*, **112**, pp. 2467–2505.
47. Shibata, M., Kawata, N., Masumoto, T., and Kimura, H. (1985). CO hydrogenation over an amorphous gold-zirconium alloy, *Chem. Lett.*, **14**, pp. 1605–1608.
48. Koeppel, R.A., Baiker, A., Schild, C., and Wokaun, A. (1991). Carbon dioxide hydrogenation over Au/ZrO$_2$ catalysts from amorphous precursors: catalytic reaction mechanism, *J. Chem. Soc., Faraday Trans.*, **87**, pp. 2821–2828.
49. Sakurai, H., and Haruta, M. (1996). Synergism in methanol synthesis from carbon dioxide over gold catalysts supported on metal oxides, *Catal. Today,* **29**, pp. 361–365.
50. Sakurai, H., Tsubota, S., and Haruta, M. (1993). Hydrogenation of CO$_2$ over gold supported on metal-oxides, *Appl. Catal. A: Gen.*, **102**, pp. 125–136.
51. Baiker, A., Kilo, M., Maciejewski, M., Menzi, S., Wokaun, A., Waugh, K.C., Krauss, H.L., Prins, R., Kochloefl, K., Holderich, W., Hojlundnielsen, P.E., Trimm, D.L., Haruta, M., and Rooney, J.J. (1993). Hydrogenation of CO$_2$ over copper, silver and gold zirconia catalysts: comparative study of catalyst properties and reaction pathways, in *New Frontiers in Catalysis: Proceedings of the 10th International Congress on Catalysis, Budapest, 19–24 July 1992*, pp. 1257–1272, eds. Guczi, L., Solymosi, F., and Tétényi, P. (Elsevier, Budapest).
52. Strunk, J., Käehler, K., Xia, X., Comotti, M., Schüth, F., Reinecke, T., and Muhler, M. (2009). Au/ZnO as catalyst for methanol synthesis: the role of oxygen vacancies, *Appl. Catal. A: Gen.*, **359**, pp. 121–128.
53. Zhao, Y., Mpela, A., Enache, D.I., Taylor, S.H., Hildebrandt, D., Glasser, D., Hutchings, G.J., Atkins, M.P., and Scurrell, M.S. (2007). Study of carbon monoxide hydrogenation over supported Au catalysts, in *Studies in Surface Science and Catalysis*, pp. 141–151, eds. Davis, B.H., and Occelli, M.L. (Elsevier, Amsterdam).

54. Zhang, Y.L., Cattrall, R.W., McKelvie, I.D., and Kolev, S.D. (2011). Gold, an alternative to platinum group metals in automobile catalytic converters, *Gold Bull.*, **44**, pp. 145–153.
55. Haruta, M. (1997). Size- and Support-dependency in the catalysis of gold, *Catal. Today*, **36**, pp. 153–166.
56. Haruta, M. (1997). Novel catalysis of gold deposited on metal oxides, *Catal. Surv. Jpn.*, **1**, pp. 61–73.
57. Ueda, A., and Haruta, M. (1998). Reduction of nitrogen monoxide with propene over Au/Al$_2$O$_3$ mixed mechanically with Mn$_2$O$_3$, *Appl. Catal. B: Environ.*, **18**, pp. 115–121.
58. Ueda, A., Oshima, T., and Haruta, M. (1997). Reduction of nitrogen monoxide with propene in the presence of oxygen and moisture over gold supported on metal oxides, *Appl. Catal. B: Environ.*, **12**, pp. 81–93.
59. Amiridis, M.D., Roberts, K.L., and Pereira, C.J. (1997). The selective catalytic reduction of NO by propylene over Pt supported on dealuminated Y zeolite, *Appl. Catal. B: Environ.*, **14**, pp. 203–209.
60. Molnár, A., Sárkány, A., and Varga, M. (2001). Hydrogenation of carbon-carbon multiple bonds: chemo-, regio- and stereo-selectivity, *J. Mol. Catal. A: Chem.*, **173**, pp. 185–221.
61. Ulan, J.G., Kuo, E., Maier, W.F., Rai, R.S., and Thomas, G. (1987). Effect of lead acetate in the preparation of the lindlar catalyst, *J. Org. Chem.*, **52**, pp. 3126–3132.
62. Ghosh, A.K., and Krishnan, K. (1998). Chemoselective catalytic hydrogenation of alkenes by lindlar catalyst, *Tetrahedron Lett.*, **39**, pp. 947–948.
63. Jung, A., Jess, A., Schubert, T., and Schütz, W. (2009). Performance of carbon nanomaterial (nanotubes and nanofibres) supported platinum and palladium catalysts for the hydrogenation of cinnamaldehyde and of 1-octyne, *Appl. Catal. A: Gen.*, **362**, pp. 95–105.
64. Choudhary, T.V., Sivadinarayana, C., Datye, A.K., Kumar, D., and Goodman, D.W. (2003). Acetylene hydrogenation on Au-based catalysts, *Catal. Lett.*, **86**, pp. 1–8.
65. Segura, Y., López, N., and Pérez-Ramírez, J. (2007). Origin of the superior hydrogenation selectivity of gold nanoparticles in alkyne plus alkene mixtures: triple- versus double-bond activation, *J. Catal.*, **247**, pp. 383–386.
66. Azizi, Y., Petit, C., and Pitchon, V. (2008). Formation of polymer-grade ethylene by selective hydrogenation of acetylene over Au/CeO$_2$ catalyst, *J. Catal.*, **256**, pp. 338–344.

67. Hugon, A., Delannoy, L., and Louis, C. (2008). Supported gold catalysts for the selective hydrogenation of 1,3-butadiene in the presence of an excess of alkenes, *Gold Bull.,* **41**, pp. 127–138.
68. Hugon, A., Delannoy, L., and Louis, C. (2009). Influence of the reactant concentration in selective hydrogenation of 1,3-butadiene over supported gold catalysts under alkene rich conditions: a consideration of reaction mechanism, *Gold Bull.,* **42**, pp. 310–320.
69. Zhang, X., Shi, H., and Xu, B.-Q. (2005). Catalysis by gold: isolated surface $Au^{3+}$ ions are active sites for selective hydrogenation of 1,3-butadiene over $Au/ZrO_2$ catalysts, *Angew. Chem., Int. Ed.,* **44**, pp. 7132–7135.
70. Zhang, X., Shi, H., and Xu, B.-Q. (2007). Comparative study of $Au/ZrO_2$ catalysts in CO oxidation and 1,3-butadiene hydrogenation, *Catal. Today,* **122**, pp. 330–337.
71. Chou, J., Franklin, N.R., Baeck, S.-H., Jaramillo, T.F., and McFarland, E.W. (2004). Gas-phase catalysis by micelle derived Au nanoparticles on oxide supports, *Catal. Lett.,* **95**, pp. 107–111.
72. Nikolaev, S.A., and Smirnov, V.V. (2009). Selective hydrogenation of phenylacetylene on gold nanoparticles, *Gold Bull.,* **42**, pp. 182–189.
73. Lopez-Sanchez, J.A., and Lennon, D. (2005). The use of titania- and iron oxide-supported gold catalysts for the hydrogenation of propyne, *Appl. Catal. A: Gen.,* **291**, pp. 230–237.
74. Okitsu, K., Murakami, M., Tanabe, S., and Matsumoto, H. (2000). Catalytic behavior of Au core / Pd shell bimetallic nanoparticles on silica prepared by sonochemical and sol-gel processes, *Chem. Lett.,* **29**, pp. 1336–1337.
75. Sárkány, A., Schay, Z., Frey, K., Széles, É., and Sajó, I. (2010). Some features of acetylene hydrogenation on Au-iron oxide catalyst, *Appl. Catal. A: Gen.,* **380**, pp. 133–141.
76. Bus, E., Prins, R., and van Bokhoven, J.A. (2007). Origin of the cluster-size effect in the hydrogenation of cinnamaldehyde over supported Au catalysts, *Catal. Commun.,* **8**, pp. 1397–1402.
77. Milone, C., Trapani, M.C., and Galvagno, S. (2008). Synthesis of cinnamyl ethyl ether in the hydrogenation of cinnamaldehyde on $Au/TiO_2$ catalysts, *Appl. Catal. A: Gen.,* **337**, pp. 163–167.
78. Bailie, J.E., and Hutchings, G.J. (1999). Promotion by sulfur of gold catalysts for crotyl alcohol formation from crotonaldehyde hydrogenation, *Chem. Commun.,* **21**, pp. 2151–2152.

79. Chen, H.-Y., Chang, C.-T., Chiang, S.-J., Liaw, B.-J., and Chen, Y.-Z. (2010). Selective hydrogenation of crotonaldehyde in liquid-phase over Au/Mg$_2$AlO hydrotalcite catalysts, *Appl. Catal. A: Gen.*, **381**, pp. 209–215.
80. Mohr, C., Hofmeister, H., Radnik, J., and Claus, P. (2003). Identification of active sites in gold-catalyzed hydrogenation of acrolein, *J. Am. Chem. Soc.*, **125**, pp. 1905–1911.
81. Castaño, P., Zepeda, T.A., Pawelec, B., Makkee, M., and Fierro, J.L. G. (2009). Enhancement of biphenyl hydrogenation over gold catalysts supported on Fe-, Ce- and Ti-modified mesoporous silica (HMS), *J. Catal.*, **267**, pp. 30–39.
82. You, K.-J., Chang, C.-T., Liaw, B.-J., Huang, C.-T., and Chen, Y.-Z. (2009). Selective hydrogenation of $\alpha$, $\beta$-unsaturated aldehydes over Au/Mg$_x$AlO hydrotalcite catalysts, *Appl. Catal. A: Gen.*, **361**, pp. 65–71.
83. Mohr, C., Hofmeister, H., Lucas, M., and Claus, P. (2000). Gold catalysts for the partial hydrogenation of acrolein, *Chem. Eng. Technol.*, **23**, pp. 324–328.
84. Radnik, J., Mohr, C., and Claus, P. (2003). On the origin of binding energy shifts of core levels of supported gold nanoparticles and dependence of pretreatment and material synthesis, *Phys. Chem. Chem. Phys.*, **5**, pp. 172–177.
85. Lenz, J., Campo, B.C., Alvarez, M., and Volpe, M.A. (2009). Liquid phase hydrogenation of $\alpha$, $\beta$-unsaturated aldehydes over gold supported on iron oxides, *J. Catal.*, **267**, pp. 50–56.
86. Claus, P., Brückner, A., Mohr, C., and Hofmeister, H. (2000). Supported gold nanoparticles from quantum dot to mesoscopic size scale: effect of electronic and structural properties on catalytic hydrogenation of conjugated functional groups, *J. Am. Chem. Soc.*, **122**, pp. 11430–11439.
87. Milone, C., Ingoglia, R., Tropeano, M.L., Neri, G., and Galvagno, S. (2003). First example of selective hydrogenation of unconstrained $\alpha$, $\beta$-unsaturated ketone to $\alpha$, $\beta$-unsaturated alcohol by molecular hydrogen, *Chem. Commun.*, pp. 868–869.
88. Milone, C., Crisafulli, C., Ingoglia, R., Schipilliti, L., and Galvagno, S. (2007). A comparative study on the selective hydrogenation of $\alpha$, $\beta$−unsaturated aldehyde and ketone to unsaturated alcohols on Au supported catalysts, *Catal. Today*, **122**, pp. 341–351.
89. Milone, C., Ingoglia, R., Schipilliti, L., Crisafulli, C., Neri, G., and Galvagno, S. (2005). Selective hydrogenation of $\alpha$, $\beta$-unsaturated ketone to $\alpha$, $\beta$-

unsaturated alcohol on gold-supported iron oxide catalysts: role of the support, *J. Catal.*, **236**, pp. 80–90.

90. Mertens, P.G. N., Vandezande, P., Ye, X., Poelman, H., Vankelecom, I.F. J., and D.E. De_Vos, D.E. (2009). Recyclable Au$^0$, Ag$^0$ and Au$^0$–Ag$^0$ nanocolloids for the chemoselective hydrogenation of $\alpha$, $\beta$-unsaturated aldehydes and ketones to allylic alcohols, *Appl. Catal. A: Gen.*, **355**, pp. 176–183.

91. Hodnebrog, Ø., Etminan, M., Fuglestvedt, J.S., Marston, G., Myhre, G., Nielsen, C.J., Shine, K.P., and Wallington, T.J. (2013). Global warming potentials and radiative efficiencies of halocarbons and related compounds: a comprehensive review, *Rev. Geophys.*, **51**, pp. 300–378.

92. Gómez-Quero, S., Cárdenas-Lizana, F., and Keane, M.A. (2008). Effect of metal dispersion on the liquid phase hydrodechlorination of 2,4-dichlorophenol over Pd/Al$_2$O$_3$, *Ind. Eng. Chem. Res.*, **47**, pp. 6841–6853.

93. Ordóñez, S., Sastre, H., and Díez, F.V. (2003). Hydrodechlorination of tetrachloroethene over Pd/Al$_2$O$_3$: influence of process conditions on catalyst performance and stability, *Appl. Catal. B: Environ.*, **40**, pp. 119–130.

94. López, E., Ordóñez, S., Sastre, H., and Díez, F.V. (2003). Kinetic study of the gas-phase hydrogenation of aromatic and aliphatic organochlorinated compounds using a Pd/Al$_2$O$_3$ catalyst, *J. Hazard. Mater.*, **97**, pp. 281–294.

95. Kulkarni, P.P., Kovalchuk, V.I., and d'Itri, J.L. (2002). Oligomerization pathways of dichlorodifluoromethane hydrodechlorination catalyzed by activated carbon supported Pt-Cu, Pt-Ag, Pt-Fe, and Pt-Co, *Appl. Catal. B: Environ.*, **36**, pp. 299–309.

96. Tavoularis, G., and Keane, M.A. (1999). Gas phase catalytic dehydrochlorination and hydrodechlorination of aliphatic and aromatic systems, *J. Mol. Catal. A: Chem.*, **142**, pp. 187–199.

97. Barrabes, N., Cornado, D., Foettinger, K., Dafinov, A., Llorca, J., Medina, F., and Rupprechter, G. (2009). Hydrodechlorination of trichloroethylene on noble metal promoted Cu-hydrotalcite-derived catalysts, *J. Catal.*, **263**, pp. 239–246.

98. Barrabés, N., Föttinger, K., Dafinov, A., Medina, F., Rupprechter, G., Llorca, J., and Sueiras, J.E. (2009). Study of Pt–CeO$_2$ interaction and the effect in the selective hydrodechlorination of trichloroethylene, *Appl. Catal. B: Environ.*, **87**, pp. 84–91.

99. Álvarez-Montero, M.A., Gómez-Sainero, L.M., Martín-Martínez, M., Heras, F., and Rodriguez, J.J. (2010). Hydrodechlorination of

chloromethanes with Pd on activated carbon catalysts for the treatment of residual gas streams, *Appl. Catal. B: Environ.*, **96**, pp. 148–156.

100. Ordóñez, S., Vivas, B.P., and Díez, F.V. (2010). Minimization of the deactivation of palladium catalysts in the hydrodechlorination of trichloroethylene in wastewaters, *Appl. Catal. B: Environ.*, **95**, pp. 288–296.

101. Martin-Martinez, M., Gómez-Sainero, L.M., Alvarez-Montero, M.A., Bedia, J., and Rodriguez, J.J. (2013). Comparison of different precious metals in activated carbon-supported catalysts for the gas-phase hydrodechlorination of chloromethanes, *Appl. Catal. B: Environ.*, **132–133**, pp. 256–265.

102. Shin, E.-J., and Keane, M.A. (2000). Gas phase catalytic hydroprocessing of trichlorophenols, *J. Chem. Technol. Biotechnol.*, **75**, pp. 159–167.

103. Menini, C., Park, C., Shin, E.-J., Tavoularis, G., and Keane, M.A. (2000). Catalytic hydrodehalogenation as a detoxification methodology, *Catal. Today*, **62**, pp. 355–366.

104. Keane, M.A., and Murzin, D.Y. (2001). A kinetic treatment of the gas phase hydrodechlorination of chlorobenzene over nickel/silica: beyond conventional kinetics, *Chem. Eng. Sci.*, **56**, pp. 3185–3195.

105. Keane, M.A., Pina, G., and Tavoularis, G. (2004). The catalytic hydrodechlorination of mono-, di- and trichlorobenzenes over supported nickel, *Appl. Catal. B: Environ.*, **48**, pp. 275–286.

106. Amorim, C., Yuan, G., Patterson, P.M., and Keane, M.A. (2005). Catalytic hydrodechlorination over Pd supported on amorphous and structured carbon, *J. Catal.*, **234**, pp. 268–281.

107. Babu, N.S., Lingaiah, N., Kumar, J.V., and Prasad, P.S. S. (2009). Studies on alumina supported Pd–Fe bimetallic catalysts prepared by deposition–precipitation method for hydrodechlorination of chlorobenzene, *Appl. Catal. A: Gen.*, **367**, pp. 70–76.

108. Chen, J., Zhou, J., Wang, R., and Zhang, J. (2009). Preparation, characterization, and performance of HMS-supported Ni catalysts for hydrodechlorination of chorobenzene, *Ind. Eng. Chem. Res.*, **48**, pp. 3802–3811.

109. Amorim, C., and Keane, M.A. (2008). Effect of surface acid groups associated with amorphous and structured carbon on the catalytic hydrodechlorination of chlorobenzenes, *J. Chem. Technol. Biotechnol.*, **83**, pp. 662–672.

110. Amorim, C., and Keane, M.A. (2008). Palladium supported on structured and nonstructured carbon: a consideration of Pd particle size and the nature of reactive hydrogen, *J. Colloid Interf. Sci.,* **322**, pp. 196–208.
111. Cecilia, J.A., Jiménez-Morales, I., Infantes-Molina, A., Rodríguez-Castellón, E., and Jiménez-López, A. (2013). Influence of the silica support on the activity of Ni and Ni$_2$P based catalysts in the hydrodechlorination of chlorobenzene. Study of factors governing catalyst deactivation, *J. Mol. Catal. A: Chem.,* **368–369**, pp. 78–87.
112. Keane, M.A., Park, C., and Menini, C. (2003). Structure sensitivity in the hydrodechlorination of chlorobenzene over supported nickel, *Catal. Lett.,* **88**, pp. 89–94.
113. Jujjuri, S., Ding, E., Shore, S.G., and Keane, M.A. (2003). Gas-phase hydrodechlorination of chlorobenzenes over silica-supported palladium and palladium-ytterbium, *Appl. Organometal. Chem.,* **17**, pp. 493–498.
114. Lingaiah, N., Prasad, P.S. S., Rao, P.K., Smart, L.E., and Berry, F.J. (2001). Studies on magnesia supported mono- and bimetallic Pd-Fe catalysts prepared by microwave irradiation method, *Appl. Catal. A: Gen.,* **213**, pp. 189–196.
115. Tavoularis, G., and Keane, M.A. (1999). The gas phase hydrodechlorination of chlorobenzene over nickel/silica, *J. Chem. Technol. Biotechnol.,* **74**, pp. 60–70.
116. Shin, E.-J., and Keane, M.A. (1999). Detoxifying chlorine rich gas streams using solid supported nickel catalysts, *J. Hazard. Mater. B,* **66**, pp. 265–278.
117. Wu, W., Xu, J., and Ohnishi, R. (2005). Complete hydrodechlorination of chlorobenzene and its derivatives over supported nickel catalysts under liquid phase conditions, *Appl. Catal. B: Environ.,* **60**, pp. 129–137.
118. Shin, E.-J., and Keane, M.A. (2000). Structure sensitivity in the hydrodechlorination of chlorophenols, *React. Kinet. Catal. Lett.,* **69**, pp. 3–8.
119. Mitoma, Y., Tasaka, N., Takase, M., Masuda, T., Tashiro, H., Egashira, N., and Oki, T. (2006). Calcium-promoted catalytic degradation of PCDDs, PCDFs, and coplanar PCBs under a mild wet process, *Environ. Sci. Technol.,* **40**, pp. 1849–1854.
120. Cobo, M.I., Conesa, J.A., and Montes_de_Correa, C. (2008). The effect of NaOH on the liquid-phase hydrodechlorination of dioxins over Pd/$\gamma$-Al$_2$O$_3$, *J. Phys. Chem. A,* **112**, pp. 8715–8722.

121. Yuan, G., Louis, C., Delannoy, L., and Keane, M.A. (2007). Silica- and titania-supported Ni-Au: application in catalytic hydrodechlorination, *J. Catal.*, **247**, pp. 256–268.
122. Wong, M.S., Alvarez, P.J. J., Fang, Y.-L., Akçin, N., Nutt, M.O., Miller, J.T., and Heck, K.N. (2009). Cleaner water using bimetallic nanoparticle catalysts, *J. Chem. Technol. Biotechnol.*, **84**, pp. 158–166.
123. Nutt, M.O., Hughes, J.B., and Wong, M.S. (2005). Designing Pd-on-Au bimetallic nanoparticle catalysts for trichloroethene hydrodechlorination, *Environ. Sci. Technol.*, **39**, pp. 1346–1353.
124. Zhao, Z., Fang, Z.-L., Alvarez, P.J.J., and Wong, M.S. (2013). Degrading perchloroethene at ambient conditions using Pd and Pd-on-Au reduction catalysts, *Appl. Catal. B: Environ.*, **140–141**, pp. 468–477.
125. Pretzer, L.A., Song, H.J., Fang, Y.-L., Zhao, Z., Guo, N., Wu, T., Arslan, I., Miller, J.T., and Wong, M.S. (2013). Hydrodechlorination catalysis of Pd-on-Au nanoparticles varies with particle size, *J. Catal.*, **298**, pp. 206–217.
126. Andersin, J., Parkkinen, P., and Honkala, K. (2012). Pd-catalyzed hydrodehalogenation of chlorinated olefins: theoretical insights to the reaction mechanism, *J. Catal.*, **290**, pp. 118–125.
127. De_Corte, S., Sabbe, T., Hennebel, T., Vanhaecke, L., De_Gusseme, B., Verstraete, W., and Boon, N. (2012). Doping of biogenic Pd catalysts with Au enables dechlorination of diclofenac at environmental conditions, *Water Res.*, **46**, pp. 2718–2726.
128. Teevs, L., Vorlop, K.-D., and Prüße, U. (2011). Model study on the aqueous-phase hydrodechlorination of clopyralid on noble metal catalysts, *Catal. Commun.*, **14**, pp. 96–100.
129. Legawiec-Jarzyna, M., Srebowata, A., and Karpinski, Z. (2003). Hydrodechlorination of dichlorodifluoromethane (CFC-12) over Pd/Al$_2$O$_3$ and Pd-Au/Al$_2$O$_3$ catalysts, *React. Kinet. Catal. Lett.*, **79**, pp. 157–163.
130. Bonarowska, M., Burda, B., Juszczyk, W., Pielaszek, J., Kowalczyk, Z., and Karpiński, Z. (2001). Hydrodechlorination of CCl$_2$F$_2$ (CFC-12) over Pd-Au/C catalysts, *Appl. Catal. B: Environ.*, **35**, pp. 13–20.
131. Bonarowska, M., Pielaszek, J., Semikolenov, V.A., and Karpiński, Z. (2002). Pd-Au/Sibunit carbon catalysts: characterization and catalytic activity in hydrodechlorination of dichlorodifluoromethane (CFC-12), *J. Catal.*, **209**, pp. 528–538.
132. Bonarowska, M., Pielaszek, J., Juszczyk, W., and Karpiński, Z. (2000). Characterization of Pd–Au/SiO$_2$ catalysts by x-ray diffrac-

tion, temperature-programmed hydride decomposition, and catalytic probes, *J. Catal.,* **195**, pp. 304–315.

133. Yuan, G., Lopez, J.L., Louis, C., Delannoy, L., and Keane, M.A. (2005). Remarkable hydrodechlorination activity over silica supported nickel/gold catalysts, *Catal. Commun.,* **6**, pp. 555–562.

134. Keane, M.A., Gómez-Quero, S., Cárdenas-Lizana, F., and Shen, W. (2009). Alumina-supported Ni–Au: surface synergistic effects in catalytic hydrodechlorination, *ChemCatChem,* **1**, pp. 270–278.

135. Juszczyk, W., Colmenares, J.C., Śrębowata, A., and Karpiński, Z. (2011). The effect of copper and gold on the catalytic behavior of nickel/alumina catalysts in hydrogen-assisted dechlorination of 1,2-dichloroethane, *Catal. Today,* **169**, pp. 186–191.

136. Haruta, M. (2002). Catalysis of gold nanoparticles deposited on metal oxides, *CATTECH,* **6**, pp. 102–115.

137. Kung, H.H., Kung, M.C., and Costello, C.K. (2003). Supported Au catalysts for low temperature CO oxidation, *J. Catal.,* **216**, pp. 425–432.

138. Gómez-Quero, S., Cárdenas-Lizana, F., and Keane, M.A. (2013). Unique selectivity in the hydrodechlorination of 2,4-dichlorophenol over hematite-supported Au, *J. Catal.,* **303**, pp. 41–49.

139. Pozan, G.S., and Boz, I. (2008). Hydrodechlorination of 2,3,5-trichlorophenol in methanol/water on carbon supported Pd-Rh catalysts, *Environ. Eng. Sci.,* **25**, pp. 1197–1202.

140. Pina, G., Louis, C., and Keane, M.A. (2003). Nickel particle size effects in catalytic hydrogenation and hydrodechlorination: phenolic transformations over nickel/silica, *Phys. Chem. Chem. Phys.,* **5**, pp. 1924–1931.

141. Gampine, A., and Eyman, D.P. (1998). Catalytic hydrodechlorination of chlorocarbons. 2. Ternary oxide supports for catalytic conversions of 1,2-dichlorobenzene, *J. Catal.,* **179**, pp. 315–325.

142. Gopinath, R., Lingaiah, N., Sreedhar, B., Suryanarayana, I., Prasad, P.S.S., and Obuchi, A. (2003). Highly stable Pd/CeO$_2$ catalyst for hydrodechlorination of chlorobenzene, *Appl. Catal. B: Environ.,* **46**, pp. 587–594.

143. González, C.A., and Montes_de_Correa, C. (2009). Catalytic hydrodechlorination of tetrachloroethylene over Pd/TiO$_2$ minimonoliths, *Ind. Eng. Chem. Res.,* **49**, pp. 490–497.

144. López, E., Ordóñez, S., and Díez, F.V. (2006). Deactivation of a Pd/Al$_2$O$_3$ catalyst used in hydrodechlorination reactions: influence of the nature

of organochlorinated compound and hydrogen chloride, *Appl. Catal. B: Environ.*, **62**, pp. 57–65.

145. Murthy, K.V., Patterson, P.M., Jacobs, G., Davis, B.H., and Keane, M.A. (2004). An exploration of activity loss during hydrodechlorination and hydrodebromination over Ni/SiO$_2$, *J. Catal.*, **223**, pp. 74–85.

146. Besenbacher, F., Chorkendorff, I., Clausen, B.S., Hammer, B., Molenbroek, A.M., Nørskov, J.K., and Stensgaard, I. (1998). Design of a surface alloy catalyst for steam reforming, *Science*, **279**, pp. 1913–1915.

147. Molenbroek, A.M., and Nørskov, J.K. (2001). Structure and reactivity of Ni-Au nanoparticle catalysts, *J. Phys. Chem. B*, **105**, pp. 5450–5458.

148. Triantafyllopoulos, N.C., and Neophytides, S.G. (2006). Dissociative adsorption of CH$_4$ on NiAu/YSZ: the nature of adsorbed carbonaceous species and the inhibition of graphitic C formation, *J. Catal.*, **239**, pp. 187–199.

149. Venugopal, A., Aluha, J., and Scurrell, M.S. (2003). The water-gas shift reaction over Au-based, bimetallic catalysts. The Au-M (M=Ag, Bi, Co, Cu, Mn, Ni, Pb, Ru, Sn, Tl) on iron(III) oxide system, *Catal. Lett.*, **90**, pp. 1–6.

150. Vasil'kov, A.Y., Nikolaev, S.A., Smirnov, V.V., Naumkin, A.V., Volkov, I.O., and Podshibikhin, V.L. (2007). An XPS study of the synergetic effect of gold and nickel supported on SiO$_2$ in the catalytic isomerization of allylbenzene, *Mendeleev Commun.*, **17**, pp. 268–270.

151. Yan, Z., Yao, Y., and Goodman, D.W. (2012). Dehydrogenation of propane to propylene over supported model Ni–Au catalysts, *Catal. Lett.*, **142**, pp. 714–717.

152. Venezia, A.M., Parola, V.L., Deganello, G., Pawelec, B., and Fierro, J.L. G. (2003). Synergetic effect of gold in Au/Pd catalysts during hydrodesulfurization reactions of model compounds, *J. Catal.*, **215**, pp. 317–325.

153. Heinrichs, B., Delhez, P., Schoebrechts, J.-P., and Pirard, J.-P. (1997). Palladium–silver sol-gel catalysts for selective hydrodechlorination of 1,2-dichloroethane into ethylene, *J. Catal.*, **172**, pp. 322–335.

154. Müller, F., Ackermann, P., and Margot, P. (2011). *Fungicides, Agricultural, 1. Fundamentals* (Wiley-VCH Verlag, Weinheim).

155. Schnorbach, H.-J., Matthaei, H.-D., and Müller, F. (2008). *Molluskicides* (Wiley-VCH Verlag, Weinheim).

156. Paulus, W. (2005). Phenol derivates, in *Ullmann's Encyclopedia of Industrial Chemistry* (Wiley-VCH Verlag, Weinheim).

157. Boronat, M., Concepción, P., Corma, A., González, S., Illas, F., and Serna, P. (2007). A molecular mechanism for the chemoselective hydrogenation of substituted nitroaromatics with nanoparticles of gold on $TiO_2$ catalysts: a cooperative effect between gold and the support, *J. Am. Chem. Soc.,* **129**, pp. 16230–16237.

158. Sandoval, A., Gómez-Cortés, A., Zanella, R., Díaz, G., and Saniger, J.M. (2007). Gold nanoparticles: support effects for the WGS reaction, *J. Mol. Catal. A: Chem.,* **278**, pp. 200–208.

159. Hildebrand, H., Mackenzie, K., and Kopinke, F.-D. (2009). Highly active Pd-on-magnetite nanocatalysts for aqueous phase hydrodechlorination reactions, *Environ. Sci. Technol.,* **43**, pp. 3254–3259.

160. Hildebrand, H., Mackenzie, K., and Kopinke, F.-D. (2008). Novel nanocatalysts for wastewater treatment, *Global NEST J.,* **10**, pp. 47–53.

161. Cárdenas-Lizana, F., Gómez-Quero, S., Kiwi-Minsker, L., and Keane, M.A. (2012). Gold nano-particles supported on hematite and magnetite as highly selective catalysts for the hydrogenation of nitro-aromatics, *Int. J. Nanotech.,* **9**, pp. 92–112.

162. Hao, Z., An, L., Wang, H., and Hu, T. (2000). Mechanism of gold activation in supported gold catalysts for CO oxidation, *React. Kinet. Catal. Lett.,* **70**, pp. 153–160.

163. Cárdenas-Lizana, F., Gómez-Quero, S., and Keane, M.A. (2008). Exclusive production of chloroaniline from chloronitrobenzene over $Au/TiO_2$ and $Au/Al_2O_3$, *ChemSusChem,* **1**, pp. 215–221.

164. Shin, E.-J., and Keane, M.A. (1998). Catalytic hydrogen treatment of aromatic alcohols, *J. Catal.,* **173**, pp. 450–459.

165. Guevara, A., Bacaud, R., and Vrinat, M. (2003). Solvent effect in gas–liquid hydrotreatment reactions, *Appl. Catal. A: Gen.,* **253**, pp. 515–526.

166. Silberova, B.A. A., Mul, G., Makkee, M., and Moulijn, J.A. (2006). DRIFTS study of the water–gas shift reaction over $Au/Fe_2O_3$, *J. Catal.,* **243**, pp. 171–182.

167. Boccuzzi, F., Chiorino, A., Manzoli, M., Andreeva, D., and Tabakova, T. (1999). FTIR study of the low-temperature water–gas shift reaction on $Au/Fe_2O_3$ and $Au/TiO_2$ catalysts, *J. Catal.,* **188**, pp. 176–185.

168. Cai, S., and Sohlberg, K. (2003). Adsorption of alcohols on $\gamma$-alumina (1 1 0 C), *J. Mol. Catal. A: Chem.,* **193**, pp. 157–164.

169. Shi, B., Dabbagh, H.A., and Davis, B.H. (1999). Alcohol dehydration. Isotope studies of the conversion of 3-pentanol, *J. Mol. Catal. A: Chem.,* **141**, pp. 257–262.

170. Carey, F.A. (2003). *Organic Chemistry*, 5th ed. (Mc Graw-Hill, New York).
171. Calvo, L., Mohedano, A.F., Casas, J.A., Gilarranz, M.A., and Rodríguez, J.J. (2004). Treatment of chlorophenols-bearing wastewaters through hydrodechlorination using Pd/activated carbon catalysts, *Carbon*, **42**, pp. 1377–1381.
172. Bovkun, T.T., Sasson, Y., and Blum, J. (2005). Conversion of chlorophenols into cyclohexane by a recyclable Pd-Rh catalyst, *J. Mol. Catal. A: Chem.*, **242**, pp. 68–73.
173. Vogt, P.F., and Gerulis, J.J. (2005). Aromatic amines, in *Ullmann's Encyclopedia of Industrial Chemistry* (Wiley-VCH, Weinheim).
174. Blaser, H.U., Steiner, H., and Studer, M. (2009). Selective catalytic hydrogenation of functionalized nitroarenes: an update, *ChemCatChem*, **1**, pp. 210–221.
175. Maxwell, G.R. (2004). *Synthetic Nitrogen Products: A Practical Guide to the Products and Processes* (Kluwer Academic/Plenum, New York).
176. Blaser, H.U., Siegrist, U., Steiner, H., and Studer, M. (2001). Aromatic nitro compounds, in *Fine Chemicals through Heterogeneous Catalysis* (Wiley-VCH, Weinheim).
177. Zhao, F., Zhang, R., Chatterjee, M., Ikushima, Y., and Araib, M. (2004). Hydrogenation of nitrobenzene with supported transition metal catalysts in supercritical carbon dioxide, *Adv. Synth. Catal.*, **346**, pp. 661–668.
178. Li, C.-H., Yu, Z.-X., Yao, K.-F., Jib, S.-F., and Liang, J. (2005). Nitrobenzene hydrogenation with carbon nanotube-supported platinum catalyst under mild conditions, *J. Mol. Catal. A: Chem.*, **226**, pp. 101–105.
179. Sheldon, R.A. (2007). The E factor: fifteen years on, *Green Chem.*, **9**, pp. 1273–1283.
180. Cárdenas-Lizana, F., Gómez-Quero, S., and Keane, M.A. (2008). Clean production of chloroanilines by selective gas phase hydrogenation over supported Ni catalysts, *Appl. Catal. A: Gen.*, **334**, pp. 199–206.
181. Xiao, C., Wang, X.D., Lian, C., Liu, H.Q., Liang, M.H., and Wang, Y. (2012). Selective hydrogenation of halonitrobenzenes, *Curr. Org. Chem.*, **16**, pp. 280–296.
182. Pietrowski, M. (2012). Recent developments in heterogeneous selective hydrogenation of halogenated nitroaromatic compounds to halogenated anilines, *Curr. Org. Synth.*, **9**, pp. 470–487.
183. Chen, Y., Qiu, J., Wang, X., and Xiu, J. (2006). Preparation and application of highly dispersed gold nanoparticles supported on silica for catalytic

hydrogenation of aromatic nitro compounds, *J. Catal.,* **242**, pp. 227–230.

184. Corma, A., and Serna, P. (2006). Chemoselective hydrogenation of nitro compounds with supported gold catalysts, *Science,* **313**, pp. 332–334.

185. Corma, A., Concepción, P., and Serna, P. (2007). A different reaction pathway for the reduction of aromatic nitro compounds on gold catalysts, *Angew. Chem., Int. Ed.,* **46**, pp. 7266–7269.

186. He, D., Shi, H., Wu, Y., and Xu, B.-Q. (2007). Synthesis of chloroanilines: selective hydrogenation of the nitro in chloronitrobenzenes over zirconia-supported gold catalyst, *Green Chem.,* **9**, pp. 849–851.

187. Corma, A., Serna, P., and García, H. (2007). Gold catalysts open a new general chemoselective route to synthesize oximes by hydrogenation of $\alpha$, $\beta$-unsaturated nitrocompounds with $H_2$, *J. Am. Chem. Soc.,* **129**, pp. 6358–6359.

188. Shimizu, K., Miyamoto, Y., Kawasaki, T., Tanji, T., Tai, Y., and Satsuma, A. (2009). Chemoselective hydrogenation of nitroaromatics by supported gold catalysts: mechanistic reasons of size- and support-dependent activity and selectivity, *J. Phys. Chem. C,* **113**, pp. 17803–17810.

189. Gkizis, P.L., Stratakis, M., and Lykakis, I.N. (2013). Catalytic activation of hydrazine hydrate by gold nanoparticles: chemoselective reduction of nitro compounds into amines, *Catal. Commun.,* **36**, pp. 48–51.

190. Zhang, J., Chen, G., Chaker, M., Rosei, F., and Ma, D. (2013). Gold nanoparticle decorated ceria nanotubes with significantly high catalytic activity for the reduction of nitrophenol and mechanism study, *Appl. Catal. B: Environ.,* **132–133**, pp. 107–115.

191. Yamane, Y., Liu, X., Hamasaki, Y., Ishida, T., Haruta, M., Yokoyama, T., and Tokunaga, M. (2009). One-pot synthesis of indoles and aniline derivatives from nitroarenes under hydrogenation condition with supported gold nanoparticles, *Org. Lett.,* **11**, pp. 5162–5165.

192. Liu, L., Qiao, B., Ma, Y., Zhang, J., and Deng, Y. (2008). Ferric hydroxide supported gold subnano clusters or quantum dots: enhanced catalytic performance in chemoselective hydrogenation, *Dalton Trans.,* **19**, pp. 2542–2548.

193. Liu, L., Qiao, B., Chen, Z., Zhang, J., and Deng, Y. (2009). Novel chemoselective hydrogenation of aromatic nitro compounds over ferric hydroxide supported nanocluster gold in the presence of CO and $H_2O$, *Chem. Commun.,* **6**, pp. 653–655.

194. Wang, H., Dong, Z., and Na, C. (2013). Hierarchical carbon nanotube membrane-supported gold nanoparticles for rapid catalytic reduction of *p*-nitrophenol, *ACS Sust. Chem. Eng.,* **1**, pp. 746–752.

195. Cárdenas-Lizana, F., Lamey, D., Perret, N., Gómez-Quero, S., Kiwi-Minsker, L., and Keane, M.A. (2012). Au/Mo$_2$N as a new catalyst formulation for the hydrogenation of p-chloronitrobenzene in both liquid and gas phases, *Catal. Commun.*, **21**, pp. 46–51.

196. Matsushima, Y., Nishiyabu, R., Takanashi, N., Haruta, M., Kimura, H., and Kubo, Y. (2012). Boronate self-assemblies with embedded Au nanoparticles: preparation, characterization and their catalytic activities for the reduction of nitroaromatic compounds, *J. Mater. Chem.*, **22**, pp. 24124–24131.

197. Hartfelder, U., Kartusch, C., Makosch, M., Rovezzi, M., Sa, J., and van Bokhoven, J.A. (2013). Particle size and support effects in hydrogenation over supported gold catalysts, *Catal. Sci. Technol.*, **3**, pp. 454–461.

198. Han, X.-X., Zhou, R.-X., Zheng, X.-M., and Jiang, H. (2003). Effect of rare earths on the hydrogenation properties of p-chloronitrobenzene over polymer-anchored platinum catalysts, *J. Mol. Catal. A: Chem.*, **193**, pp. 103–108.

199. Coq, B., Tijani, A., Dutartre, R., and Figuéras, F. (1993). Influence of support and metallic precursor on the hydrogenation of p-chloronitrobenzene over supported platinum catalysts, *J. Mol. Catal.*, **79**, pp. 253–264.

200. Xu, D.-Q., Hu, Z.-Y., Li, W.-W., Luo, S.-P., and Xu, Z.-Y. (2005). Hydrogenation in ionic liquids: an alternative methodology toward highly selective catalysis of halonitrobenzenes to corresponding haloanilines, *J. Mol. Catal. A: Chem.*, **235**, pp. 137–142.

201. Liu, Y.-C., Huang, C.-Y., and Chen, Y.-W. (2006). Liquid-phase selective hydrogenation of p-chloronitrobenzene on Ni-P-B nanocatalysts, *Ind. Eng. Chem. Res.*, **45**, pp. 62–69.

202. Uflyand, I.E., Ilchenko, I.A., Sheinker, V.N., and Bulatov, A.V. (1991). Heterogenization of palladium (II) chelates on a sibunite, *Trans. Met. Chem.*, **16**, pp. 293–295.

203. Cárdenas-Lizana, F., Gómez-Quero, S., and Keane, M.A. (2008). Ultra-selective gas phase catalytic hydrogenation of aromatic nitro compounds over Au/Al$_2$O$_3$, *Catal. Commun.*, **9**, pp. 475–481.

204. Cárdenas-Lizana, F., Gómez-Quero, S., Perret, N., and Keane, M.A. (2009). Support effects in the selective gas phase hydrogenation of p-chloronitrobenzene over gold, *Gold Bull.*, **42**, pp. 124–132.

205. Cárdenas-Lizana, F., Gómez-Quero, S., Perret, N., and Keane, M.A. (2011). Gold catalysis at the gas-solid interface: role of the support

in determining activity and selectivity in the hydrogenation of *m*-dinitrobenzene, *Catal. Sci. Technol.,* **1**, pp. 652–661.

206. Wang, X., Perret, N., Delgado, J.J., Blanco, G., Chen, X., Olmos, C.M., Bernal, S., and Keane, M.A. (2013). Reducible support effects in the gas phase hydrogenation of *p*-chloronitrobenzene over gold, *J. Phys. Chem. C,* **117**, pp. 994–1005.

207. Wang, X., Perret, N., and Keane, M.A. (2012). The role of hydrogen partial pressure in the gas phase hydrogenation of *p*-chloronitrobenzene over alumina supported Au and Pd: a consideration of reaction thermodynamics and kinetics, *Chem. Eng. J.,* **210**, pp. 103–113.

208. Vishwanathan, V., Jayasri, V., Basha, P.M., Mahata, N., Sikhwivhilu, L.M., and Coville, N.J. (2008). Gas phase hydrogenation of *ortho*-chloronitrobenzene (*o*-CNB) to *ortho*-chloroaniline (*o*-CAN) over unpromoted and alkali metal promoted-alumina supported palladium catalysts, *Catal. Commun.,* **9**, pp. 453–458.

209. Yeong, K.K., Gavriilidis, A., Zapf, R., and Hessel, V. (2003). Catalyst preparation and deactivation issues for nitrobenzene hydrogenation in a microstructural falling film reactor, *Catal. Today,* **81**, pp. 641–651.

210. Sangeetha, P., Seetharamulu, P., Shanthi, K., Narayanan, S., and Raob, K.S. R. (2007). Studies on Mg-Al oxide hydrotalcite supported Pd catalysts for vapor phase hydrogenation of nitrobenzene, *J. Mol. Catal. A: Chem.,* **273**, pp. 244–249.

211. Diao, S., Qian, W., Luo, G., Wei, F., and Wang, Y. (2005). Gaseous catalytic hydrogenation of nitrobenzene to aniline in a two-stage fluidized bed reactor, *Appl. Catal. A: Gen.,* **286**, pp. 30–35.

212. Petrov, L., Kumbilieva, K., and Kirkov, N. (1990). Kinetic model of nitrobenzene hydrogenation to aniline over industrial copper catalysts considering the effects of mass transfer and deactivation, *Appl. Catal.,* **59**, pp. 31–43.

213. Coq, B., and Figuéras, F. (1998). Structure-activity relationships in catalysis by metals: some aspecs of particle size, bimetallic and supports effects, *Coord. Chem. Rev.,* **178–180**, pp. 1753–1783.

214. Sakurai, H., and Haruta, M. (1995). Carbon dioxide and carbon monoxide hydrogenation over gold supported on titanium, iron, and zinc oxides, *Appl. Catal. A: Gen.,* **127**, pp. 93–105.

215. Schimpf, S., Lucas, M., Mohr, C., Rodemerck, U., Brückner, A., Radnick, J., Hofmeister, H., and Claus, P. (2002). Supported gold nanoparticles: in-depth catalyst characterization and application in hydrogenation and oxidation reactions, *Catal. Today,* **72**, pp. 63–78.

216. Cárdenas-Lizana, F., Gómez-Quero, S., Idriss, H., and Keane, M.A. (2009). Gold particle size effects in the gas-phase hydrogenation of m-dinitrobenzene over Au/TiO$_2$, *J. Catal.*, **268**, pp. 223–234.
217. Valden, M., Lai, X., and Goodman, D.W. (1998). Onset of catalytic activity of gold clusters on titania with the appearance of nonmetallic properties, *Science*, **281**, pp. 1647–1650.
218. Okazaki, K., Ichikawa, S., Maeda, Y., Haruta, M., and Kohyama, M. (2005). Electronic structures of Au supported on TiO$_2$, *Appl. Catal. A: Gen.*, **291**, pp. 45–54.
219. Vinod, C.P., Kulkarni, G.U., and Rao, C.N. R. (1998). Size-dependent changes in the electronic structure of metal clusters as investigated by scanning tunneling spectroscopy, *Chem. Phys. Lett.*, **289**, pp. 329–333.
220. Min, B.K., Wallace, W.T., and Goodman, D.W. (2006). Support effects on the nucleation, growth, and morphology of gold nano-clusters, *Surf. Sci.*, **600**, pp. L7–L11.
221. Jozwiak, W.K., Kaczmarek, E., Maniecki, T.P., Ignaczak, W., and Maniukiewicz, W. (2007). Reduction behavior of iron oxides in hydrogen and carbon monoxide atmospheres, *Appl. Catal. A: Gen.*, **326**, pp. 17–27.
222. Munteanu, G., Ilieva, L., and Andreeva, D. (1997). Kinetic parameters obtained from TPR data for $\alpha$-Fe$_2$O$_3$ and Au/$\alpha$-Fe$_2$O$_3$ systems, *Thermochim. Acta*, **291**, pp. 171–177.
223. Venugopal, A., and Scurrell, M.S. (2004). Low temperature reductive pretreatment of Au/Fe$_2$O$_3$ catalysts, TPR/TPO studies and behaviour in the water–gas shift reaction, *Appl. Catal. A: Gen.*, **258**, pp. 241–249.
224. Bernal, S., Calvino, J.J., Cauqui, M.A., Cifredo, G.A., Jobacho, A., and Rodríguez-Izquierdo, J.M. (1993). Metal-support interaction phenomena in rhodium/ceria and rhodium/titania catalysts: comparative study by high-resolution transmission electron spectroscopy, *Appl. Catal. A: Gen.*, **99**, pp. 1–8.
225. Cárdenas-Lizana, F., de_Pedro, Z.M., Gómez-Quero, S., and Keane, M.A. (2010). Gas phase hydrogenation of nitroarenes: a comparison of the catalytic action of titania supported gold and silver, *J. Mol. Catal.*, **326**, pp. 48–54.
226. Akita, T., Okumura, M., Tanaka, K., Kohyama, M., and Haruta, M. (2006). Analytical TEM observation of Au nano-particles on cerium oxide, *Catal. Today*, **117**, pp. 62–68.
227. Stakheev, A.Y., and Kustov, L.M. (1999). Effects of the support on the morphology and electronic properties of supported metal clusters:

modern concepts and progress in 1990s, *Appl. Catal. A: Gen.,* **188**, pp. 3–35.

228. Centeno, M.A., Carrizosa, I., and Odriozola, J.A. (2003). Deposition–precipitation method to obtain supported gold catalysts: dependence of the acid–base properties of the support exemplified in the system $TiO_2$–$TiO_xN_y$–TiN, *Appl. Catal. A: Gen.,* **246**, pp. 365–372.

229. Perret, N., Wang, X., Onfroy, T., Calers, C., and Keane, M.A. (2014). Selectivity in the gas-phase hydrogenation of 4-nitrobenzaldehyde over supported Au catalysts, *J. Catal.,* **309**, pp. 333–342.

230. Chuah, G.K., Jaenicke, S., Cheong, S.A., and Chan, K.S. (1996). The influence of preparation conditions on the surface area of zirconia, *Appl. Catal. A: Gen.,* **145**, pp. 267–284.

231. Ardizzone, S., and Bianchi, C.L. (1999). Electrochemical features of zirconia polymorphs. The interplay between structure and surface OH species, *J. Electroanal. Chem.,* **465**, pp. 136–141.

232. Xie, S., Iglesia, E., and Bell, A.T. (2000). Water-assisted tetragonal-to-monoclinic phase transformation of $ZrO_2$ at low temperatures, *Chem. Mater.,* **12**, pp. 2442–2447.

233. Jung, K.-D., and Bell, A.T. (2000). Role of hydrogen spillover in methanol synthesis over $Cu/ZrO_2$, *J. Catal.,* **193**, pp. 207–223.

234. Blaser, H.-U. (2006). A golden boost to an old reaction, *Science,* **313**, pp. 312–313.

235. Boronat, M., and Corma, A. (2010). Origin of the different activity and selectivity toward hydrogenation of single metal Au and Pt on $TiO_2$ and bimetallic Au-Pt/$TiO_2$ catalysts, *Langmuir,* **26**, pp. 16607–16614.

236. Cárdenas-Lizana, F., Gómez-Quero, S., Hugon, A., Delannoy, L., Louis, C., and Keane, M.A. (2009). Pd-promoted selective gas phase hydrogenation of *p*-chloronitrobenzene over alumina supported Au, *J. Catal.,* **262**, pp. 235–243.

237. Hadjiivanov, K.I., and Vayssilov, G.N. (2002). Characterization of oxide surfaces and zeolites by carbon monoxide as an IR probe molecule, *Adv. Catal.,* **47**, pp. 307–511.

238. Sales, E.A., Jove, J., Mendes, M. d. J., and Bozon-Verduraz, F. (2000). Palladium, palladium–tin, and palladium–silver catalysts in the selective hydrogenation of hexadienes: TPR, Mössbauer, and infrared studies of adsorbed CO, *J. Catal.,* **195**, pp. 88–95.

239. Wei, T., Wang, J., and Goodman, D.W. (2007). Characterization and chemical properties of Pd-Au alloy surfaces, *J. Phys. Chem. C,* **111**, pp. 8781–8788.

240. Luo, K., Wei, T., Yi, C.-W., Axnanda, S., and Goodman, D.W. (2005). Preparation and characterization of silica supported Au-Pd model catalysts, *J. Phys. Chem. B,* **109**, pp. 23517–23522.
241. Yi, C.W., Luo, K., Wei, T., and Goodman, D.W. (2005). The composition and structure of Pd-Au surfaces *J. Phys. Chem. B,* **109**, pp. 18535–18540.
242. Joseph, T., Kumar, K.V., Ramaswamy, A.V., and Halligudi, S.B. (2007). Au-Pt nanoparticles in amine functionalized MCM-41: catalytic evaluation in hydrogenation reactions, *Catal. Commun.,* **8**, pp. 629–634.
243. Zhang, W., Li, L., Du, Y., Wang, X., and Yang, P. (2009). Gold/Platinum bimetallic core/shell nanoparticles stabilized by a Fréchet-type dendrimer: preparation and catalytic hydrogenations of phenylaldehydes and nitrobenzenes, *Catal. Lett.,* **127**, pp. 429–436.
244. Cárdenas-Lizana, F., Gómez-Quero, S., Perret, N., Kiwi-Minsker, L., and Keane, M.A. (2011). $\beta$-Molybdenum nitride: synthesis mechanism and catalytic response in the gas phase hydrogenation of *p*-chloronitrobenzene, *Catal. Sci. Technol.,* **1**, pp. 794–801.
245. De‗Lucas‗Consuegra, A., Patterson, P.M., and Keane, M.A. (2006). Use of unsupported and silica supported molybdenum carbide to treat chloroarene gas streams *Appl. Catal. B: Environ.,* **65**, pp. 227–239.
246. Perret, N., Cárdenas-Lizana, F., Lamey, D., Laporte, V., Kiwi-Minsker, L., and Keane, M.A. (2012). Effect of crystallographic phase ($\beta$ vs. $\gamma$) and surface area on gas phase nitroarene hydrogenation over $Mo_2N$ and $Au/Mo_2N$, *Top. Catal.,* **55**, pp. 955–968.
247. Perret, N., Wang, X., Delannoy, L., Potvin, C., Louis, C., and Keane, M.A. (2012). Enhanced selective nitroarene hydrogenation over Au supported on $\beta$-$Mo_2C$ and $\beta$-$Mo_2C/Al_2O_3$, *J. Catal.,* **286**, pp. 172–183.

# Chapter 15

# Homogenous Gold Catalysis

**David Zahner, Matthias Rudolph, and A. Stephen K. Hashmi**

*Organisch-Chemisches Institut, Universität Heidelberg, Im Neuenheimer Feld 270,
69120 Heidelberg, Germany*
hashmi@hashmi.de

## 15.1 Introduction

For a long time there existed only singular, scattered investigations on homogeneous gold-catalyzed reactions. Following a prepeak of activity between 1986 and 1998 on a very specific reactivity pattern, from the year 2000 onward an exponential growth of the field was observed. Now, 14 years later, the field has stabilized at high numbers of yearly publications, but most important, still many new discoveries are being published. These discoveries do not just represent minor variations, but sometimes entirely new reactivity patterns, unprecedented in the field of organometallic chemistry, are still found. A shift that was observed only in the last years is that after a strong focus on methodology development and mechanistic studies in the early phase, now increasingly applications in synthesis (both total synthesis of natural products and synthetic applications in the field of compounds for material science) are reported. Another

---

*Gold Catalysis: Preparation, Characterization, and Applications*
Edited by Laura Prati and Alberto Villa
Copyright © 2016 Pan Stanford Publishing Pte. Ltd.
ISBN 978-981-4669-28-3 (Hardcover), 978-981-4669-29-0 (eBook)
www.panstanford.com

new focus is the development of ligands for highly active catalysts [1].

Until 2000 homogeneous gold catalysis was the "small brother" of heterogeneous gold catalysis, but now the "younger" discipline has grown to the same size. In the subsequent sections we present the most common reactivity patterns of the field and provide key references to early publications and latest reviews on the individual conversions.

## 15.2 The First Methodology: Asymmetric Gold Catalysis

The first homogenous gold catalysis was an asymmetric aldol reaction (Scheme 15.1) discovered by Ito, Sawamura, and Hayashi in 1986 [2]. While many Lewis acidic metal centers were investigated, gold proved to be superior in this case regarding yield and selectivity. Although this reaction was later further investigated and reviewed [3], gold catalysis itself was overlooked at that time and not pursued any further.

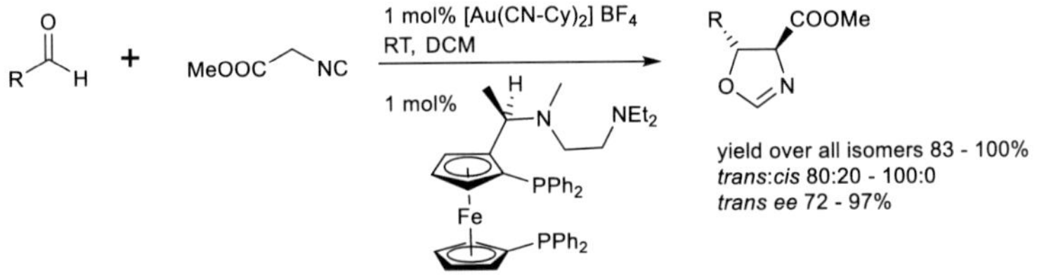

**Scheme 15.1** First catalytic asymmetric aldol reaction of aldehydes and isocycanoacetates.

It is worth to note that despite the fact that even though this first homogeneous gold-catalyzed methodology was performed enantioselectively, later with the rising interest in gold catalysis asymmetric gold catalysis was only sparely investigated in the initial phase. The first reviews on this topic, published in 2008 [4], summarized that chirality can be achieved by asymmetric ligands, counterions, or chirality transfer. Since then all of these principles were applied to (mostly) previously known reactions [4, 5] whose general reaction principles shall be discussed in the following sections.

## 15.3 The Most Basic Reactivity Pattern: Nucleophilic Attack on Carbon–Carbon Multiple Bonds

One of the most basic principles in homogenous gold catalysis is the addition of a nucleophile onto a carbon–carbon multiple bond that is activated by a gold catalyst. The resulting organogold compound is then protodeaurated, that is, the gold is replaced by the proton that in most cases was formerly attached to the incoming nucleophile (Scheme 15.2).

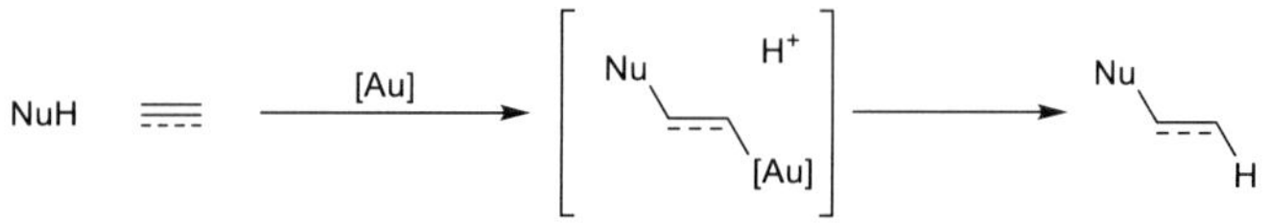

**Scheme 15.2** Activation of carbon–carbon double and triple bonds for the addition of nucleophiles is the most basic reactivity pattern.

Mostly alkynes or also the isomeric allenes are used for simple nucleophilic attacks; sometimes they can even be interconverted during the reaction, while still, in contrast to alkynes (and allenes), alkenes as an industrially very important substrate class are not frequently investigated, since they often require activated substrates or an intramolecular reaction [6].

A broad range of nucleophiles can serve for this general reaction mode; in the following sections the most frequently used nucleophiles will be reviewed.

### 15.3.1 Nitrogen Nucleophiles

Amines as nucleophiles have often been used in gold chemistry. However, often only anilines or (sulfon)amides are used in intermolecular hydroamination reactions in order to prevent catalyst poisoning [7].

The first historical example was reported by Utimoto and coworkers in 1987 (Scheme 15.3), where an yne amine cyclizes intramoleculary at first to an enamine, which then tautomerizes to an imine [8].

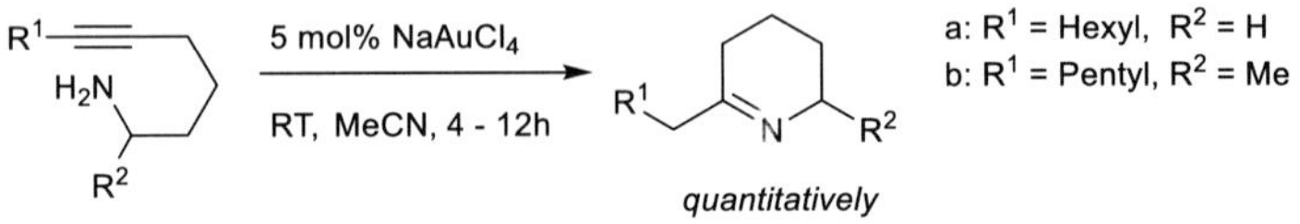

$$R^1-\!\!\!\equiv\!\!\!-CH_2-NH_2-CHR^2 \xrightarrow[\text{RT, MeCN, 4 - 12h}]{\text{5 mol\% NaAuCl}_4} R^1\text{-ring-}N\text{-}R^2$$

a: $R^1$ = Hexyl, $R^2$ = H
b: $R^1$ = Pentyl, $R^2$ = Me

quantitatively

**Scheme 15.3** Nitrogen heterocycles are accessible by intramolecular nucleophilic additions.

As previously discussed, the interest in homogenous gold catalysis faded soon after the discoveries in the 1980s, so it was not until 2003 that the first intermolecular hydroamination was published by Tanaka and coworkers (Scheme 15.4). In this example an alkyne was reacted with an aniline derivative or phenylhydrazine. The alkyne could be aromatic or aliphatic; however, selectivity problems occurred for nonterminal unsymmetrical alkynes [9].

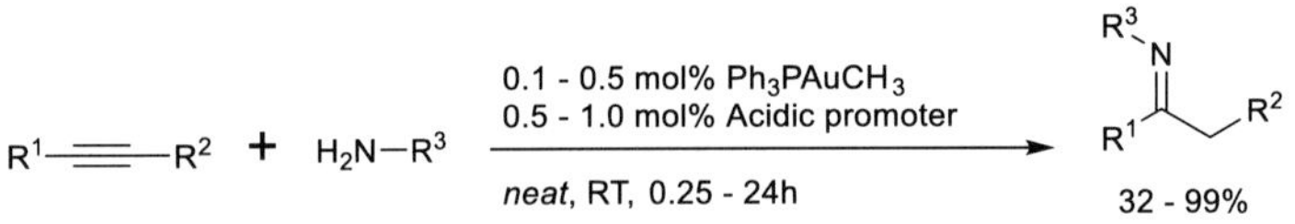

$$R^1-\!\!\!\equiv\!\!\!-R^2 + H_2N-R^3 \xrightarrow[\text{neat, RT, 0.25 - 24h}]{\substack{0.1 - 0.5 \text{ mol\% Ph}_3\text{PAuCH}_3 \\ 0.5 - 1.0 \text{ mol\% Acidic promoter}}} R^1\text{-C(=NR}^3\text{)-CH}_2R^2$$

32 - 99%

**Scheme 15.4** Intermolecular additions of nitrogen nuclephiles to alkynes deliver imines.

## 15.3.2 Oxygen Nucleophiles

Oxygen as a nucleophile, for example, in the form of water, alcohols, or carbonyl compounds, has also been extensively used. The basic principle is almost the same as with the hydroamination; the only difference is that in the case of triple bonds a second addition onto the resulting activated double bond is favoured, finally leading to acetals or ketones. This is most likely caused by the thermodynamically more stable acetals or ketones compared to the rather instable aminals or primary imines [10].

The first reaction was an intermolecular reaction: the addition of water or methanol to alkynes in order to form ketones or dimethoxy acetals. The water addition was first reported in 1975 by Thomas and coworkers but with low yields or high catalyst loadings [11]. Then in 1991 Fukuda and Utimoto described the addition of water (Scheme 15.5) and methanol to alkynes with much better yields and catalyst loadings. Similar to the intermolecular hydroamination nonterminal unsymmetrical alkynes caused selectivity problems [12].

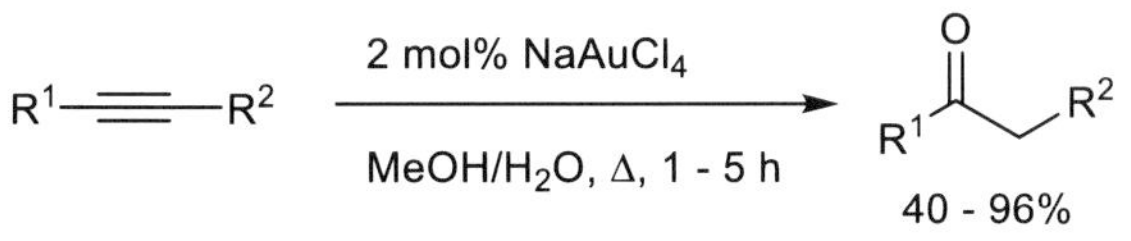

**Scheme 15.5** Water can serve as a nucleophile, too.

The first intramolecular example was published in 2000 by Hashmi et al. [13]. Initiated by the attack of a carbonyl oxygen onto a triple bond, symmetric furans are formed in quantitative yield after aromatization (Scheme 15.6) [13].

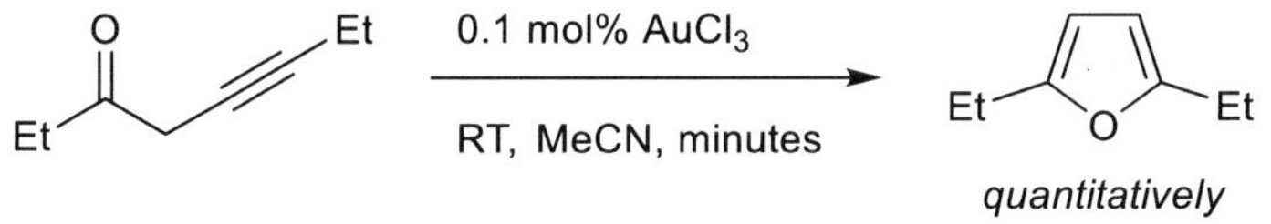

**Scheme 15.6** A carbonyl oxygen atom can serve as an intramolecular O-nucleophile, which opens access to a range of oxygen heterocycles.

Five years later Antoniotti, Genin, Michelet, and Genêt reported on the first intramolecular formation of a polycyclic acetal (Scheme 15.7) [14].

In this context, recently very high turnover numbers for the twofold intamolcular alcohol addition to alkynes, providing spiroketals, were reported. Up to 32 million turnovers could be reached (Scheme 15.8) [15].

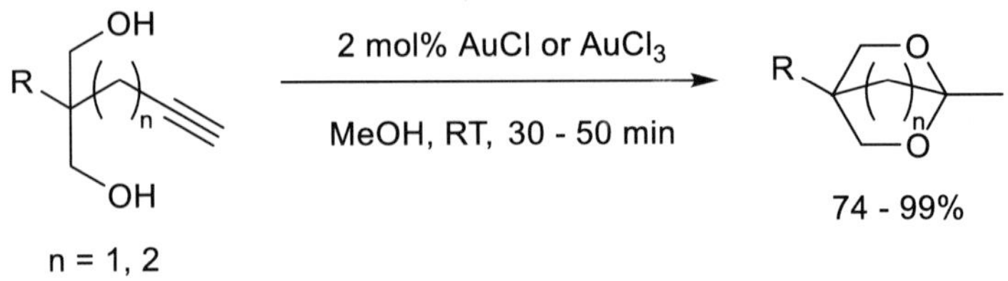

**Scheme 15.7** A twofold addition of alcohols to alkynes allows the synthesis of acetalic structures.

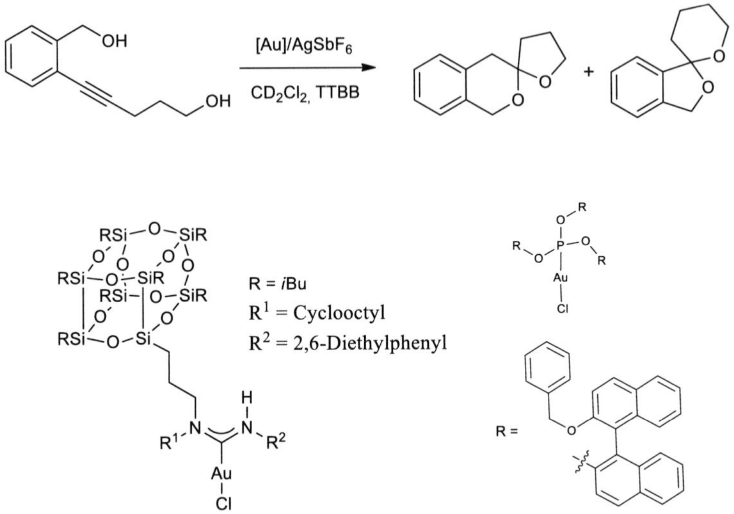

**Scheme 15.8** A catalyst basing on a nitrogen acyclic carbene ligand on gold(I) achieved many million turnovers for an intramolecular spicroketal formation.

### 15.3.3 Carbon Nucleophiles

Carbon nucleophiles can also be applied. If electron-rich arenes are used, gold-catalyzed hydroarylation of multiple bonds is also possible.

The first example is an intermolecular reaction published in 2003 by Reetz and Sommer [16]. An excess of activated arenes like mesitylene furnished phenylacetylene derivates in good yields (Scheme 15.9) [16].

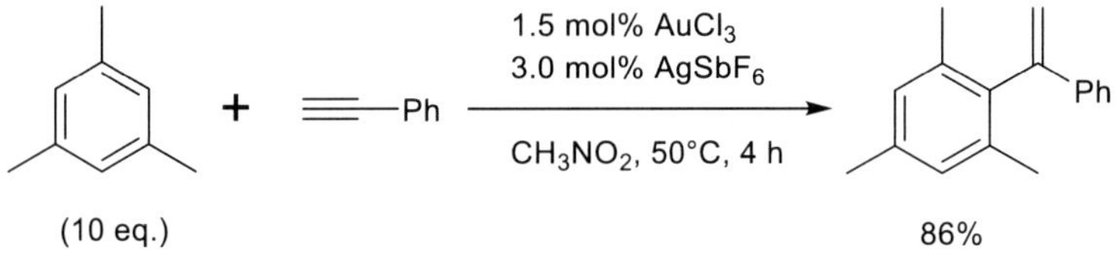

**Scheme 15.9** Electron-rich arenes allow efficient hydroarylation reactions.

An intramolecular example was first presented by Nevado and Echavarren in 2005, again with activated arenes. The reaction itself was also possible with platinum(II) chloride, but yields were only low to moderate (Scheme 15.10) [17].

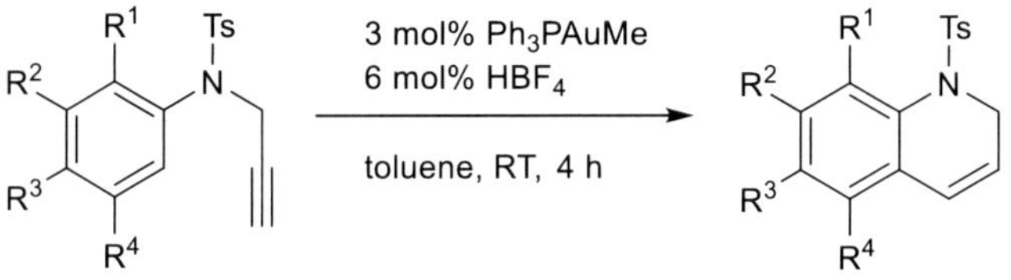

a: $R^1 = R^4 = $ OMe, $R^2 = R^3 = $ H; 71%
b: $R^1 = R^4 = $ H, $R^2 = R^3 = $ OMe; 71%
c: $R^1 = $ H, $R^2 = R^3 = R^4 = $ OMe; 92%

**Scheme 15.10** Intramolecular hydroarylation as a pathway to benzo-anellated heterocycles.

Although later described, the intramolecular hydroarylation was more often utilized in synthesis as selectivity problems are easy to control and complex annulated systems are obtained in this case.

## 15.4 Enyne Cyclizations

The cyclization of enynes is a very important, intensely researched area in the field of homogenous gold catalysis. The reaction principle is that the gold catalyst activates the alkyne as an electrophile, so the alkene can perform a nucleophilic attack on it. Then rearrangements take place, so the reaction itself can provide a board array of possible products; hence the prediction of the outcome of a certain reaction is rather difficult (Scheme 15.11). It can even vary further when a nucleophile like an alcohol is present, which intercepts intermediates of the reaction [18].

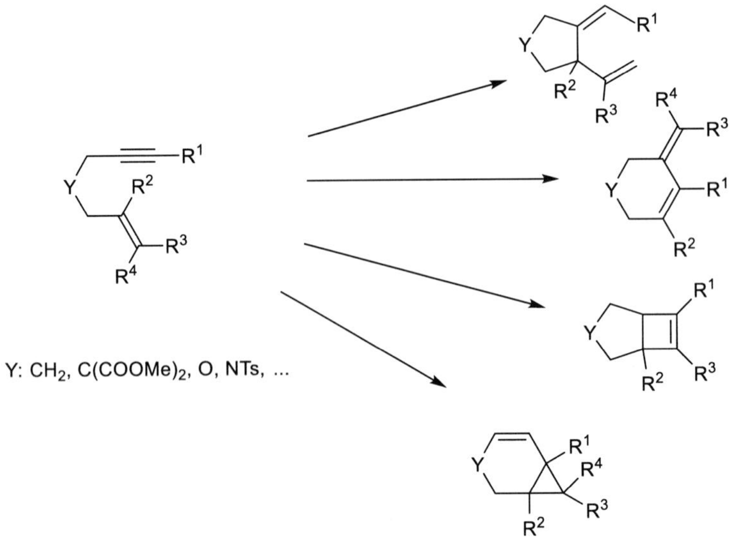

Y: CH$_2$, C(COOMe)$_2$, O, NTs, ...

**Scheme 15.11** Many different products can be formed in enyne cycloisomerization reactions.

The cyclization of furanynes to phenols reported by Hashmi, Frost, and Bats in 2000 is in principle the first gold-catalyzed enyne cyclization (Scheme 15.12) [19]. Despite the rather complex mechanism of this reaction it turned out to be highly efficient for a broad range of substrates and it was one of the intensely researched reactions during the beginning of the big interest in homogenous gold catalysis.

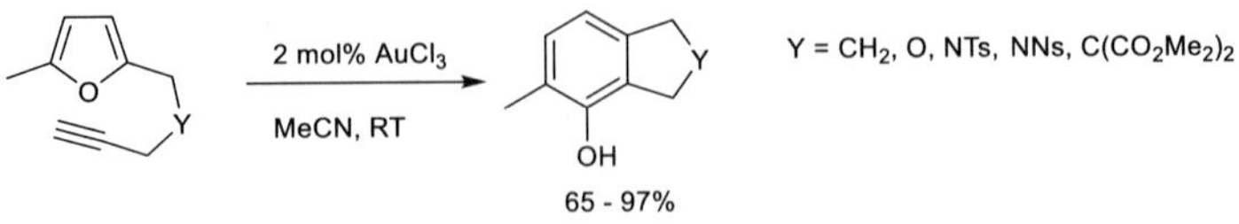

2 mol% AuCl$_3$
MeCN, RT
65 - 97%
Y = CH$_2$, O, NTs, NNs, C(CO$_2$Me$_2$)$_2$

**Scheme 15.12** The furanyne reaction as the first gold-catalyzed enyne-type reaction that was investigated.

The first gold-catalyzed enyne cyclization with an isolated double bond was published 2004 by Echavarren and coworkers (Scheme 15.13) [20]. As previously described the cycloisomerization is initiated by an activation of the alkyne, which enables the intramoleculary nucleophilic attack of the alkene. However, the entire mechanism is too complex to be explained in detail here. After rearrangement of the highly reactive cyclopropyl carbenoid intermediates (that can also be regarded as gold-stabilized carbocations) the final product is obtained. If methanol is employed as a solvent, methanol addition takes place, which delivers the addition product ZZ.

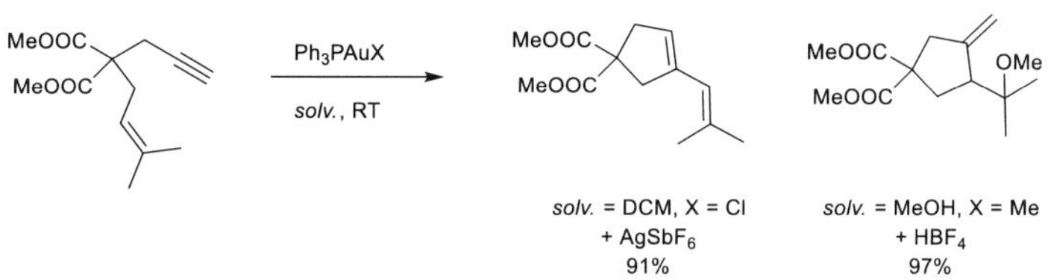

**Scheme 15.13** Often competing nucleophilic additions are observed in enyne cyclization reactions.

## 15.5 Gold Catalysis with Propargyl Esters and Related Compounds

Propargyl esters are an important recurring motive in gold catalysis. Depending on the substitution pattern of the applied starting materials, they can undergo either a 1,2-migration or a 1,3-migration of the acetate moiety (Scheme 15.14). The 1,2-migration is also referred to as a [2,3]-rearrangement because the newly formed bond between the two fragments is 2 and 3 atoms away from the broken bond. Therefore the 1,3-migration is a [3,3]-rearrangement. Often also the nomenclature of Baldwin's rules are used to describe the cyclizations, so the 1,2-migration is a 5 *exo* dig cyclization, while the 1,3 migration is a 6 *endo* dig cyclization. Here the number stands for the size of the newly formed ring, *exo* and *endo* tell whether the broken bond to form the ring (from the alkyne)

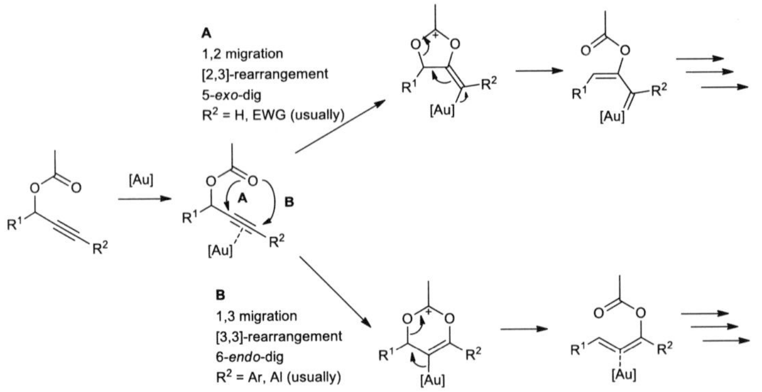

**Scheme 15.14** Two different modes compete in gold-catalyzed reactions that are based on propargylic esters.

is found outside (*exo*) or inside (*endo*) the newly formed ring, and "dig" stands for the geometry of the electrophile, which is in both cases, as it is an alkyne, diagonal.

Terminal alkynes and alkynes bearing electron-withdrawing groups usually undergo a 1,2-migration and then form possibly a gold carbenoid as an intermediate, which reacts further in different ways depending on the substrates. This methodology offers a convenient alternative to the common strategies that apply metal-catalyzed decomposition of diazo compounds as precursors for metal carbenoids.

Nonterminal alkynes without electron-withdrawing groups usually perform a 1,3-migration, which then leads to a gold-coordinated allene, which can then also proceed in different reactions. Depending on the substrates the formed allene can either react as a nucleophile or be activated by the gold catalyst and then serve as an electrophile [21].

### 15.5.1 *1,2-Migration*

The first gold catalysis with propargyl esters was reported in 2003 by Miki, Ohe, and Uemura. In this reaction the proposed carbenoid intermediate undergoes an intermolecular cyclopropanation with styrene (Scheme 15.15). However, for this transformation ruthenium proved to be a better catalyst [22].

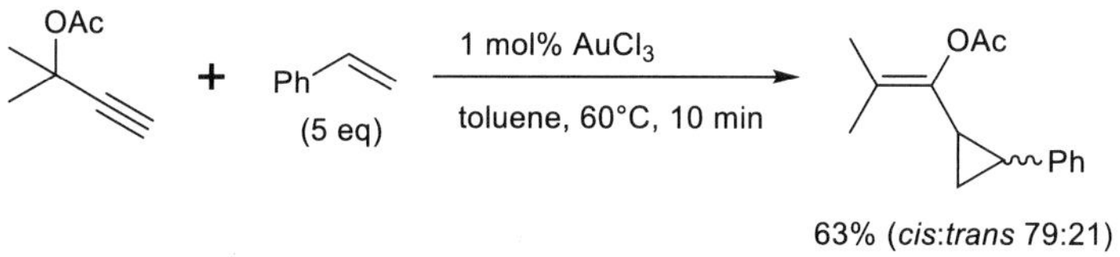

**Scheme 15.15** First use of propargylic esters for cyclopropanation reactions.

Soon after this publication Fürstner and coworkers published in 2004 an effective intramolecular cyclopropanation with a propargyl ester (Scheme 15.16). Though the alkyne had a phenyl substituent the substrate most probably underwent a 1,2-migration, showing that the two modes of cyclizations are not absolutely defined by the substituents of the alkyne [23].

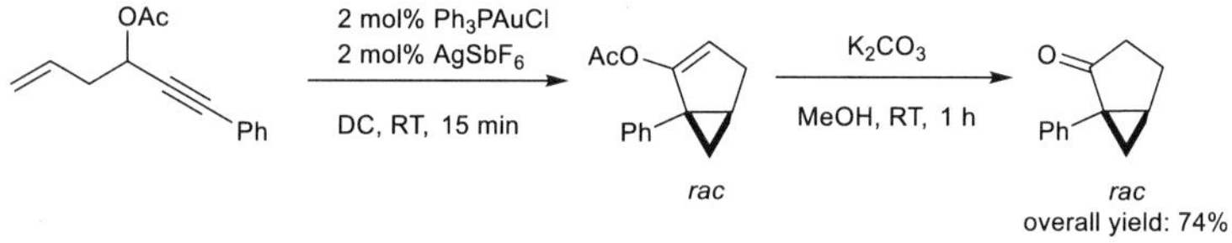

**Scheme 15.16** Terpene-type structures are readily accessible from gold-catalyzed cyclo-isomerizations of propargylic esters.

### 15.5.2 1,3-Migration

The first 1,3-migration was introduced by Zhang in 2005. Starting with indole-based substrates a 1,3-migration takes place and in a subsequent process the allene, which is activated by coordination to the gold, is intramoleculary attacked by the indole as a nucleophile to form a lactone (Scheme 15.17). The proposed gold intermediate then forms a rather unexpected four-membered ring with the indole [24].

Two years later Echavarren and coworkers reported the first reaction in which an allene generated by a 1,3-acetoxy migration was intermoleculary reacted. However, for the reaction with 1,3-dicarbonyl compounds the mechanistic pathway, and therefore also the product outcome, is strongly dependent on the substrates (Scheme 15.18). The function of the copper additive is not

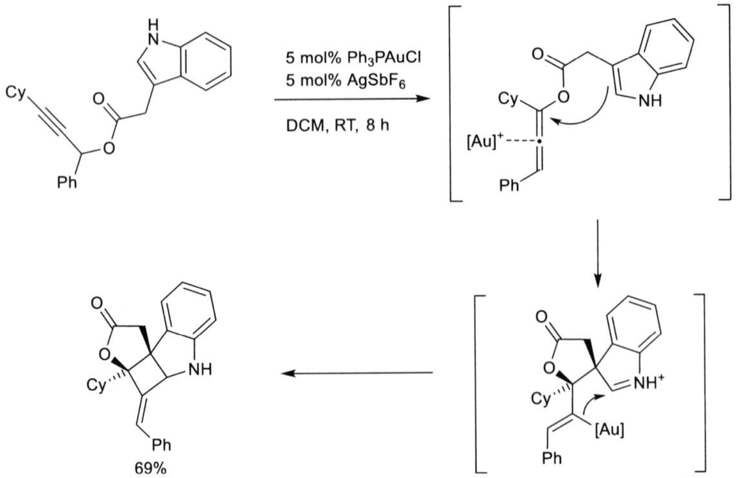

**Scheme 15.17** Even four-membered rings can be synthesized from propargylic esters.

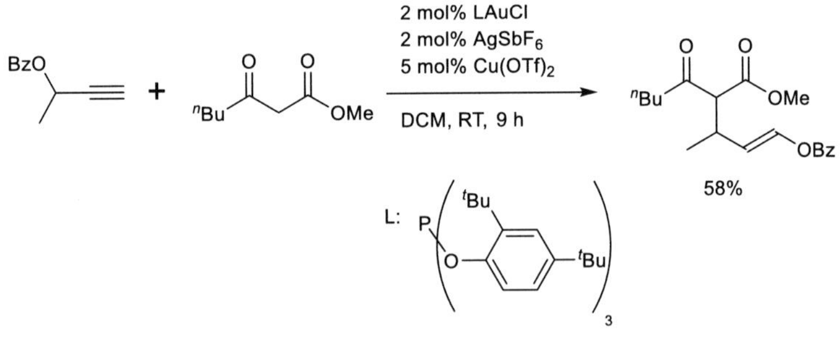

**Scheme 15.18** Intermolecular reactions with carbon nucleophiles are another aspect of propargylic esters.

completely clear; combined with gold it is increasing the yield, and without gold it promotes the substitution of the ester group instead [25].

### 15.5.3 Long-Range Migrations

With particular substrates a formal 1,6-migration could be achieved, which mechanistically is based on an initial 1,3-rearrangement, a cyclization, and a final 1,5-migration (Scheme 15.19) [26]. Isotope

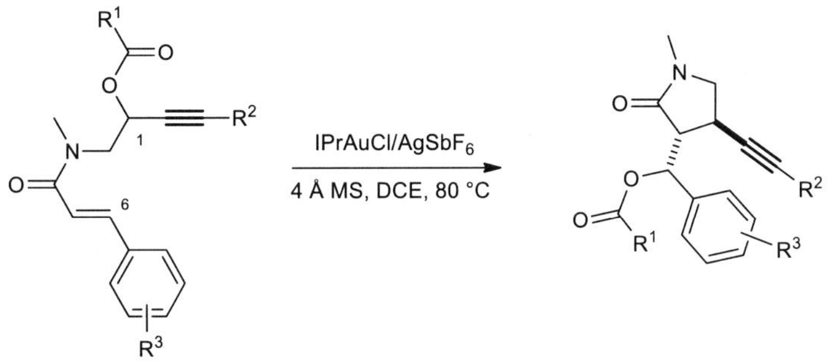

**Scheme 15.19** A rare example of a gold-catalyzed reaction of an alkyne-based substrate in which the C–C triple bond is restored in the course of the reaction.

labeling experiments and computational studies confirmed that reaction pathway. Furthermore, the reaction could be extended to carbonates or phosphates instead of the carboxylates [27]. In all cases in high diastereo-selectivity the same diastereomer was formed. This reaction is also remarkable because—different from most gold-catalyzed reactions—the alkyne survives the reaction. This is also the reason why the molecular sieves have to be added; otherwise a competing gold-catalyzed hydratation of the alkyne reduced the yields.

## 15.6 Gold-Catalyzed Oxidations of Alkynes

Gold can catalyze the oxidation of alkynes by oxygen/nitrogen transfer from 1,2-dipoles like *N*-oxides or sulfoxides (Scheme 15.20). In the proposed mechanism the dipole $Y^- – X^+$ consists of a nucleophile Y, which attacks the gold-activated alkyne, and a leaving group X, which subsequently is eliminated. Thereby a gold carbenoid is generated and the alkyne is oxidized. Before the discovery of this methodology, $\alpha$-oxo gold carbenoid intermediates were usually generated from $\alpha$-diazo carbonyl compounds.

Nowadays mostly *N*-oxides are used, but this reaction principle is not completely limited to this oxidant [28].

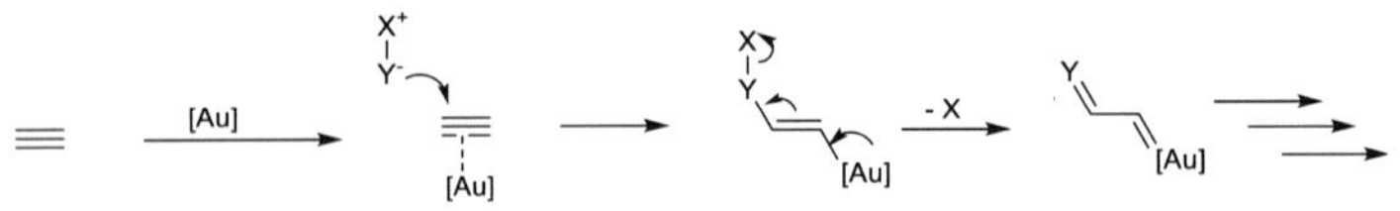

Scheme 15.20 Heteroatom transfer to alkynes provides an easy access to gold(I) carbenoids.

### 15.6.1 Sulfoxides

The first sulfoxide reactions following this reaction principle were simultaneously published by Li and Zhang [29] and Shapiro and Toste [30] in 2007 (Scheme 15.21). Both publications are describing almost the same reactions. In the chosen example from Li and Zhang, the oxygen of the sulfoxide attacks as a nucleophile after the alkyne is activated by the pyridinecarboxylate gold(III) catalyst [31]. The sulfur–oxygen bond breaks, and a thioether and an $\alpha$-oxo carbene are formed. The carbene inserts into the *ortho*-C–H bond of the arene and then forms the final product.

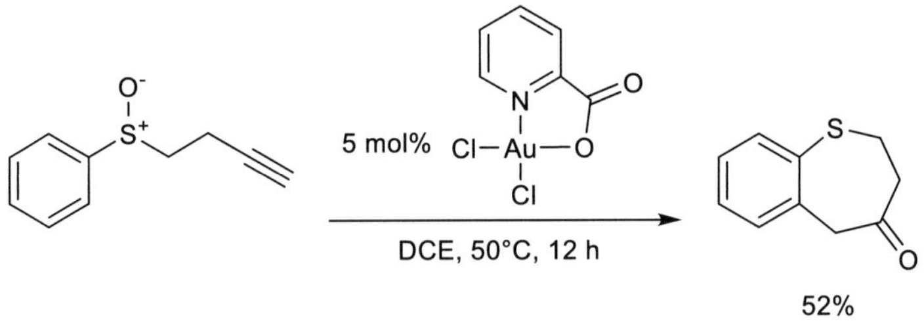

Scheme 15.21 Even intramolecular redox reactions, going along with oxygen transfer from an sulfoxide, have been investigated.

### 15.6.2 Amine Oxides

The first use of amine oxides in combination with gold catalysis was published in 2009 by Shin and coworkers. In this intramolecular reaction an isoindole was generated after the oxidation and gold carbenoid generation (Scheme 15.22) [32].

One year later Zhang and coworkers published the oxygen transfer from pyridine-*N*-oxides to alkynes catalyzed by gold. In this reaction the *N*-oxide intermolecularly transfers the oxygen onto

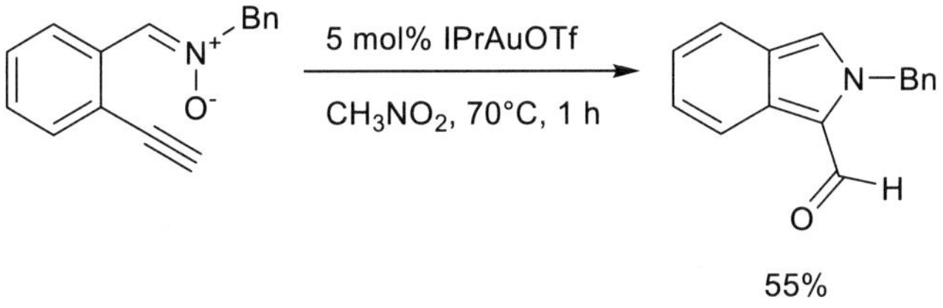

**Scheme 15.22** *N*-oxides are a preferred family of reagents in this field.

the substrate under generation of an oxo-gold carbenoid [33]. The carbene then inserts into the hydrogen–oxygen bond, resulting in the cyclized keto-ether (Scheme 15.23).

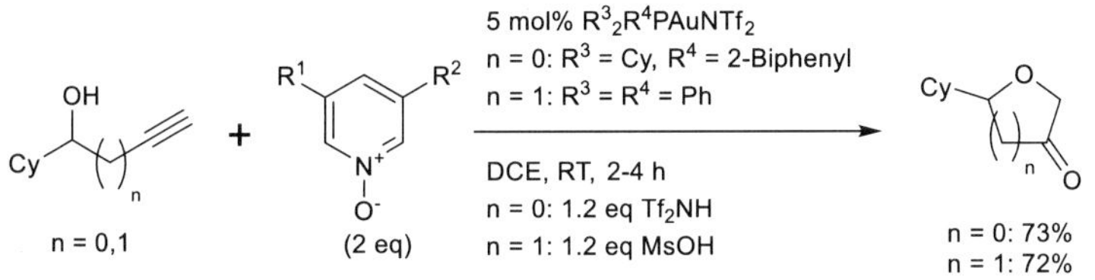

**Scheme 15.23** Pyridine-*N*-oxides represent the most versatile family of reagents for intermolecular oxygen transfer.

## 15.7 Oxidative Couplings with Gold

With the aid of an oxidant gold can efficiently catalyze oxidative carbon–carbon couplings. As a mechanism it has been proposed that gold(I) organyls are oxidized to gold(III) organyls. Then another organyl is transferred to the extended coordination sphere of gold(III), where finally reductive elimination can take place to form the product and regenerate the gold(I) species (Scheme 15.24) [34].

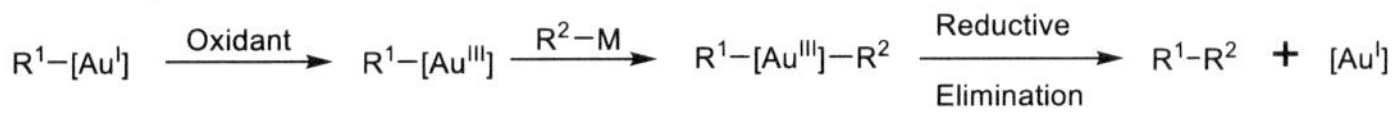

**Scheme 15.24** Only a few oxidants can initiate catalytic cycles involving oxidation of the gold catalyst.

The elemental step for a homocoupling of gold(I) organyls with a fluoro-oxidant was first reported by Hashmi, Ramamurthi, and Rominger in 2009 (Scheme 15.25) [35]. At that time such a C–C bond formation from organogold species was completely unexpected.

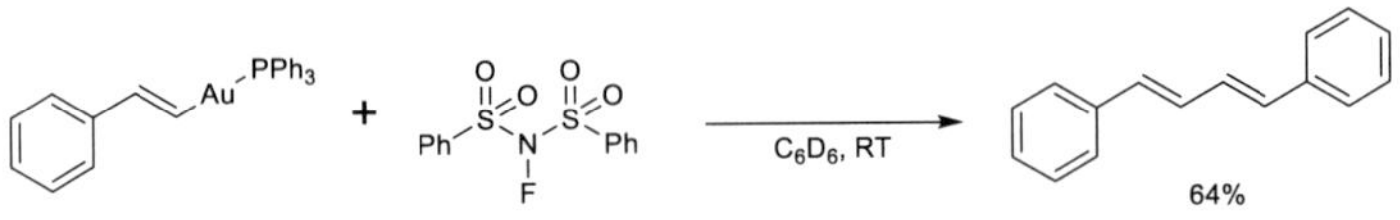

**Scheme 15.25** A F+ donor initiated unexpected C–C bond formation.

Shortly after that Zhang and coworkers published a catalytic homocoupling using 1-chloromethyl-4-fluoro-1,4-diazoniabicyclo[2.2.2]octane bis(tetrafluoroborate) (Selectfluor) as the oxidant (Scheme 15.26). The gold organyl was in situ prepared by a 1,3-migration of an acetate, as discussed in Section 15.5 of this chapter [36].

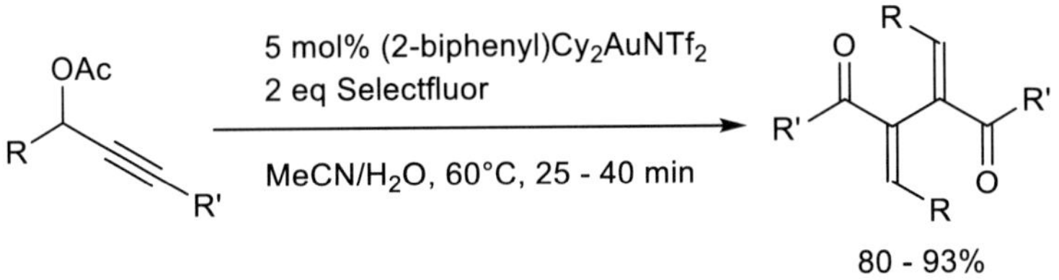

**Scheme 15.26** Selectfluor also proves to be efficient in catalytic conversions.

A few months later the same group reported on a Selectfluor-assisted gold-catalyzed cross-coupling reaction of the previously used substrates with aromatic boronic acids that are believed to transmetalate the organic fragment onto the intermediate gold (III) species (Scheme 15.27) [37].

In the same year Waser and coworkers showed alkynyl-substituted hypervalent iodine reagents can be coupled with indoles in gold-catalyzed conversions (Scheme 15.28) [38]. The mechanism is not fully understood yet, but it could be an oxidative coupling.

In 2010, Gouverneur and coworkers showed the possibility of an aromatic C–H activation combined with oxidative gold

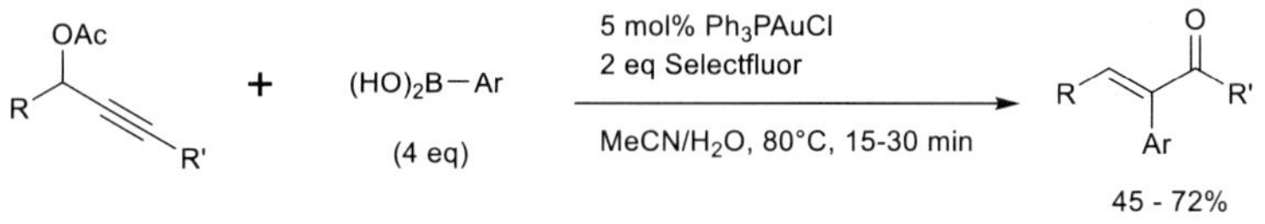

**Scheme 15.27** Instead of homodimerizations, with boronic acids also heterodimerizations were possible.

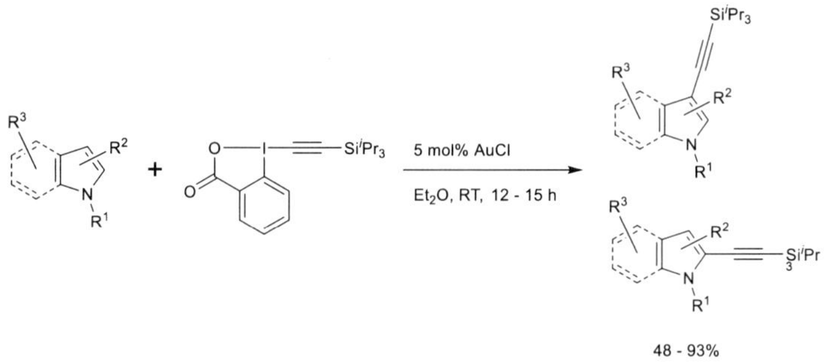

**Scheme 15.28** The nucleophilic indole core can be coupled to alkynes by the use of hypervalent iodocompounds.

catalysis (Scheme 15.29). The proposed gold intermediate probably underwent an oxidation, a C–H activation, and in the end a reductive elimination to form the product. This was only possible with an arene nearby, so an annulation at the other side of the lactone could also be achieved for $R^2 =$ Bn [39].

Haro and Nevado in the same year published a gold-catalyzed cross-coupling between activated arenes and alkynes, which could be called a double C–H activation. Probably the tendency of gold for coordinating alkynes and arenes made this possible [40].

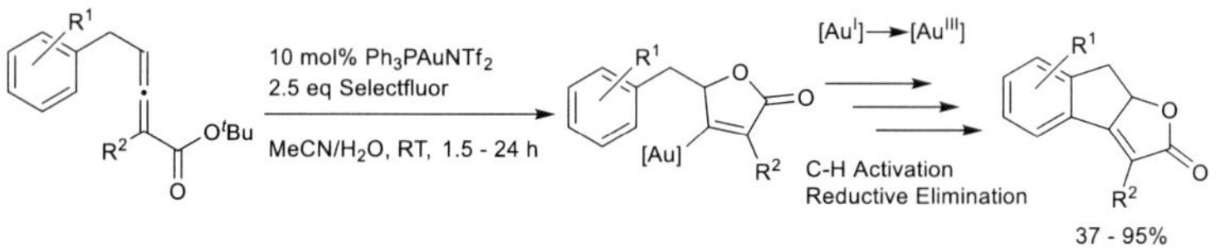

**Scheme 15.29** Hammond-type vinylgold(I) intermediates can be coupled to arenes [39].

One year later Ball, Lloyd-Jones, and Russell achieved a C–H activation arene-arene cross-coupling (Scheme 15.30). Trimethylsilyl (TMS)-substituted arenes were coupled with another arene at the position of a C–H bond. A combination of PhI(OAc)$_2$ (diacetoxyiodo)benzene) and camphersulfonic acid formed in situ the needed oxidant. Electron-deficient TMS-substituted arenes worked best, while no distinct trend could be observed for the C–H-activated arenes [41].

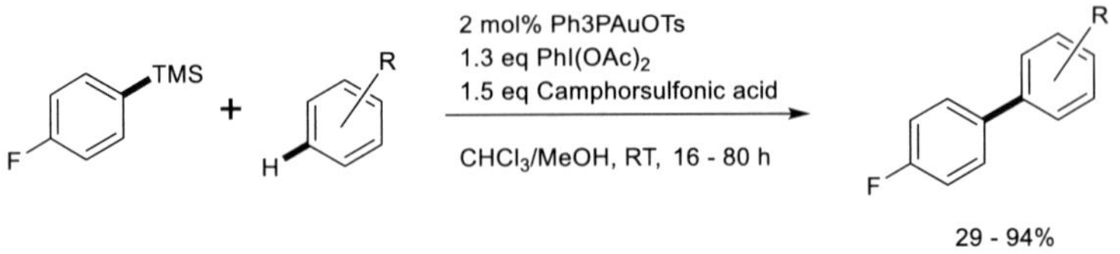

**Scheme 15.30** Intermolecular cross-coupling under oxidative conditions.

A different principle was introduced by Haro and Nevado, also in 2011, where two nucleophiles are added to a double bond (Scheme 15.31). According to the proposed mechanisms, gold enables first an intramolecular hydroamination, and while still be being bound to the substrate, it is oxidized by Selectfluor. Then either the nucleophile binds to gold(III) and a reductive elimination happens to form the product or the gold(III) acts as a leaving group (being itself reduced to gold(I)) to form a cationic aziridine, which then can be attacked by the nucleophile to form the product [42].

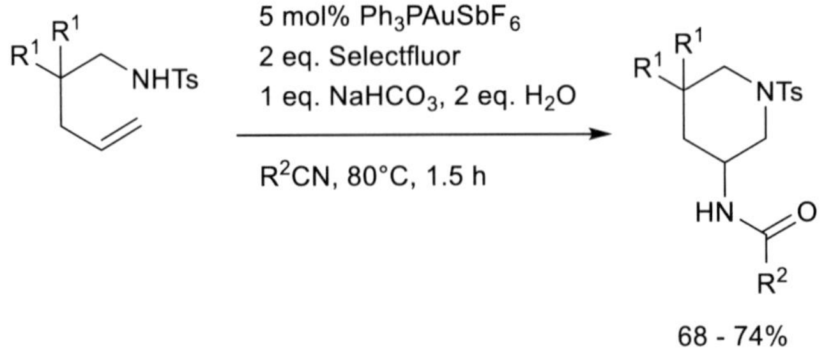

**Scheme 15.31** Under the oxidative conditions an alkene can add two nucleophiles.

But not just oxidative carbon–carbon couplings are possible; multiple gold-catalyzed oxidations have been reported as well. The first proof for homogenous oxidative gold catalysis (often the formation of active nanoparticles is discussed in this context) was published by Hashmi et al. in 2010 (Scheme 15.32). An esterification of an aldehyde with an alcohol was catalyzed by hydrogen tetrachloroaurate with 2-(2-methoxyphenyl)pyridine as a ligand and *tert*-butyl hydroperoxide as an oxidant. The absence of elemental gold and gold nanoparticles was proven by extended X-ray absorption fine structure (EXAFS) [43].

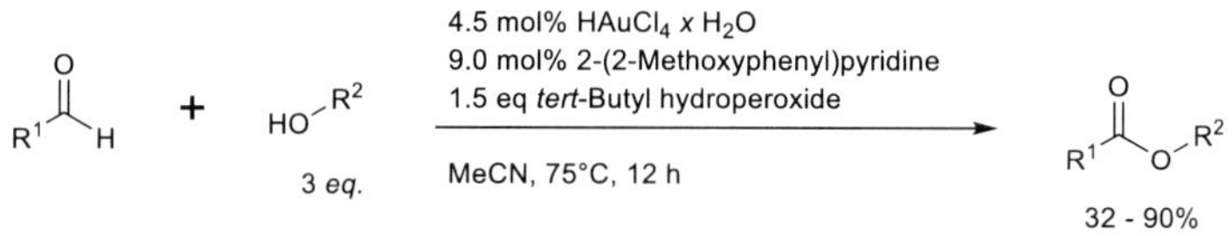

**Scheme 15.32** Selective oxidative esterification with a homogeneous, soluble catalyst.

## 15.8 Transmetalation/Cross-Coupling

In analogy to other organometallic compounds like Grignard compounds, organozinc compounds, organostannanes, and gold organyls can be used for palladium-catalyzed cross-coupling [44].

The first examples of such cross-couplings were simultaneously published 2009 by Blum and coworkers [45] and Hashmi et al. [45, 46] (Scheme 15.33).

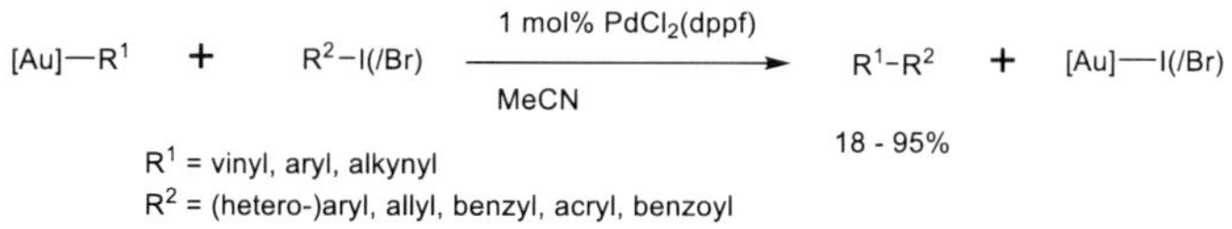

**Scheme 15.33** Palladium-catalyzed coupling of organogold compounds and organic halides.

A few months later Blum and coworkers published a transformation where palladium and gold both act as catalysts (Scheme

15.34). With regard to the mechanism, gold probably enables the nucleophilic attack of an ester onto an allene, which cyclizes to a gold-bound lactone under liberation of a palladium-stabilized allyl cation. Then a cross-coupling between the gold organyl and the allyl palladium species forms the product and regenerates both the catalysts [47].

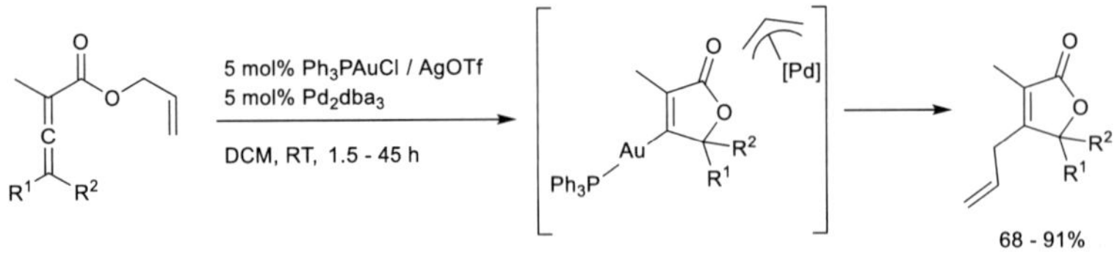

**Scheme 15.34** Both gold and palladium can be used in a catalytic manner at the same time.

## 15.9 Generation and Usage of Dipoles in Gold Catalysis

In gold catalysis (zwitterionic) dipoles can be generated as intermediates using the already described reaction modes. The intermediates can then undergo cycloadditions to form new, more complex products [28c, 48].

This strategy was applied as early as 2002 by Yamamoto and coworkers, where the triple bond of an *ortho*-alkynyl benzaldehyde was activated by gold. The intramolecular nucleophilic attack of the aldehyde forms a dipole intermediate (Scheme 15.35). This proposed pyrylium intermediate then reacts with another alkyne in a [4 + 2] cycloaddition to form the product [49].

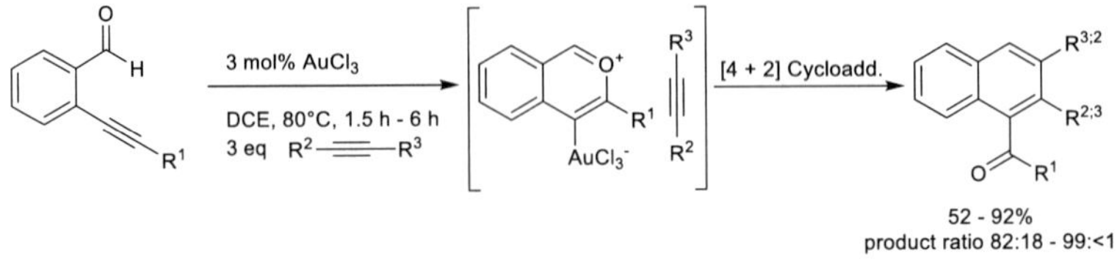

**Scheme 15.35** In situ generation of dipolar intermediates.

## 15.10 A³-Couplings

The substrates of A³-couplings are named for their three substrates starting with the letter A: **a**ldehydes, **a**mines, and **a**lkynes (Scheme 15.36). This reaction employs the gold catalyst to generate an alkynyl gold compound, which then reacts with an in situ–formed iminium cation to yield the product [50].

The first example of a gold-catalyzed version of this reaction was published by Wei and Li in 2003 [51].

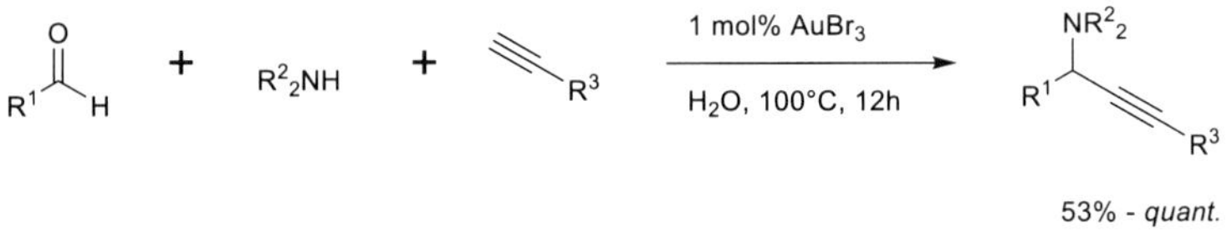

**Scheme 15.36** A three-component reaction, another example of the rare survival of a triple bond in a gold-catalyzed reaction.

## 15.11 Dual Activation

The gold-catalyzed cyclization of *ortho*-dialkynylarenes is often referred to as dual activation catalysis due to its mechanistic background. The reaction has great synthetic potential as it offers an easy access to diverse polycyclic scaffolds [52].

It was first reported independently by Zhang and coworkers and Hashmi et al. in the end of 2011 and the beginning of 2012 [53]. Both postulated dual activation of the substrate by two gold catalysts that enables the formation of a highly reactive gold vinylidene intermediate. The first evidence for a dual activation mechanism was reported earlier in 2008 by Toste and coworkers regarding cycloisomerization of alleneynes [54].

A follow-up paper from the Hashmi et al. group in 2012 was chosen as a representative as it is a rather simple example for this reaction mode (Scheme 15.37) [55]. Because the whole proposed mechanism is much more complicated, only the most important intermediates will be shown. While one gold center is $\sigma$-coordinated

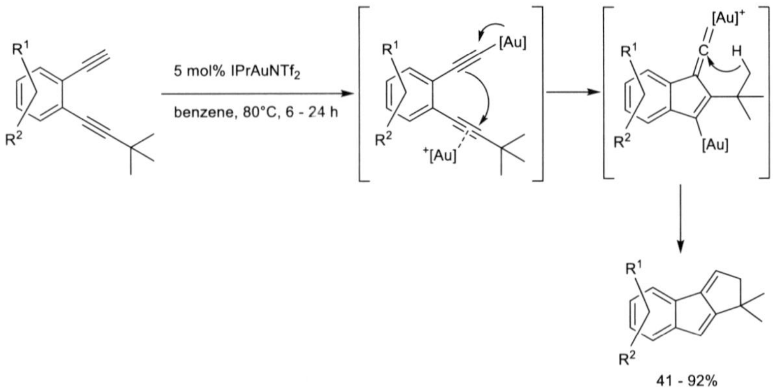

**Scheme 15.37** Dual activation of the substrate by two gold centers, and gold(I) vinylidene intermediates are characteristic for this family of reactions.

at the terminal alkyne to activate it as a nucleophile, another gold center is $\pi$-coordinated at the nonterminal alkyne to activate it as an electrophile. Cyclization takes place during which the rather uncommon vinylcarbene (short: vinylidene) gold intermediate is formed, which can then undergo an $sp^3$ C–H activation to cyclize again. While the previously vinylidene-bound gold center leaves as a cation, the other gold center transfers onto another terminal alkyne, generating a $\sigma$-activated substrate and cationic gold (catalyst transfer).

## 15.12 Gold Catalysis Combined with Organocatalysis

Due to its great functional group tolerance, gold catalysis is very well suited for the combination with the big emerging field of organocatalysis. The reaction mode of the applied organocatalyst differs widely, ranging from carbenes over amines and Brønsted acids to chiral catalysts that influence the substrate over H bonds. Therefore only two representative examples will be presented [56].

The first combination of organo- and gold catalysis was published in 2008 by Kirsch and coworkers. The organocatalyst was a secondary amine, which formed an enamine with the substrate to promote the cyclization with the gold-activated alkyne (Scheme 15.38) [57].

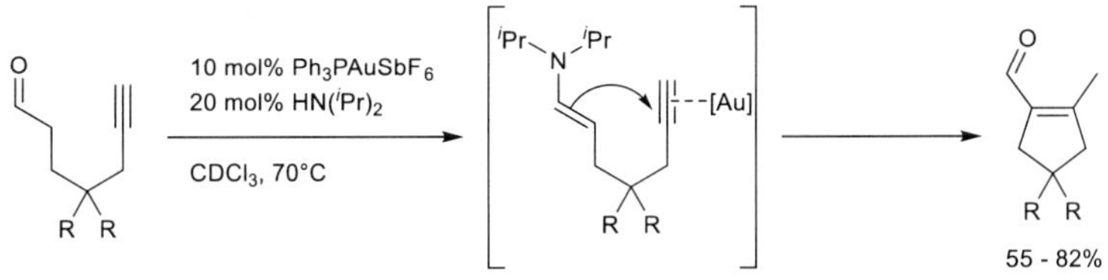

**Scheme 15.38** Kirsch's combination of organocatalysis and gold catalysis.

The first examples for a combination of a chiral Brønsted acid with a gold catalyst were published 2009 by the groups of Gong and coworkers [58], Dixon and coworkers [59], and Liu and Che [60]. Gong's publication shall serve as an example here. In this reaction the gold species first catalyzes an intramolecular hydroamination by activating the alkyne (Scheme 15.39). The formed enamine is then reduced to the chiral amine by the combination of the Hantzsch ester and the chiral phosphoric acid to form the product [58].

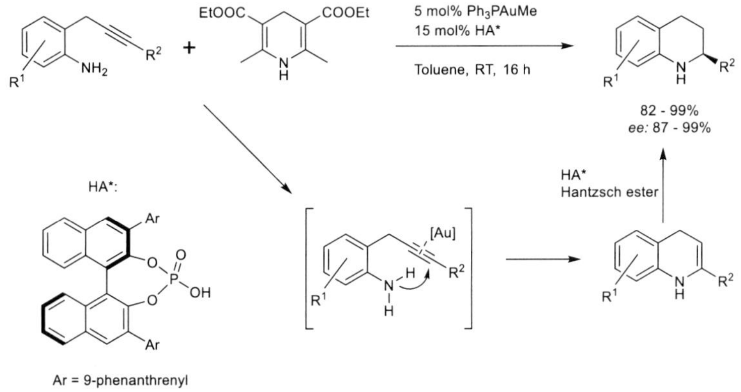

**Scheme 15.39** A combination of gold catalysis and chiral Bronsted acid catalysis.

## 15.13 Functionalizing Deauration

The elemental step of the protodeauration ($X^+ = H^+$), where the gold catalyst is released from the substrate, can be altered to functionalize the substrate on that position (Scheme 15.40). Often $I^+$ or $F^+$ sources are used, but the method is not limited to these electrophiles [34c].

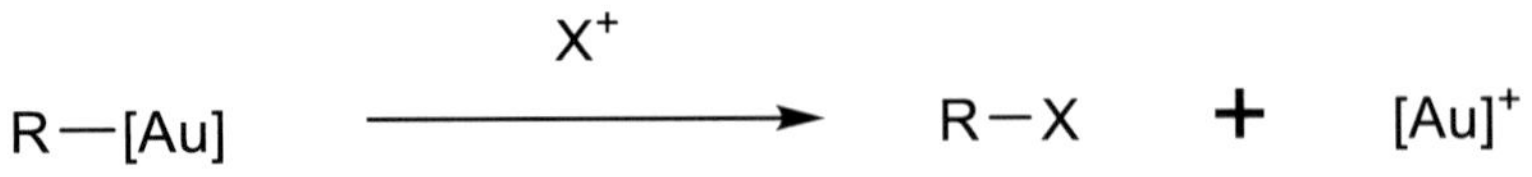

$$R-[Au] \xrightarrow{X^+} R-X \;+\; [Au]^+$$

**Scheme 15.40** Alternative electrophiles can be used to reliberate the cationic gold(I) catalyst from the organogold intermediates; this reaction is not specific for protons.

The first intramolecular example for this reaction mode was published in 2006 by Nakamura, Sato, and Yamamoto (Scheme 15.41). There the sulfur in a thioether acts as a nucleophile, generating a oxygen-stabilized carbocation, which acts as $X^+$ and substitutes the gold [61].

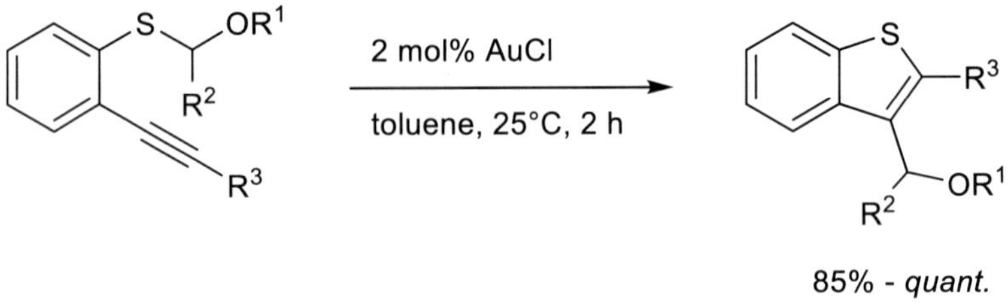

**Scheme 15.41** Carbon electrophiles for the formation of new C–C bonds.

One year later Kirsch et al. demonstrated that a functionalizing deauration was also possible in an intermolecular manner. N-iodosuccinimide was employed as an $I^+$ source, which substituted the gold center after an enyne cyclization (Scheme 15.42) [62].

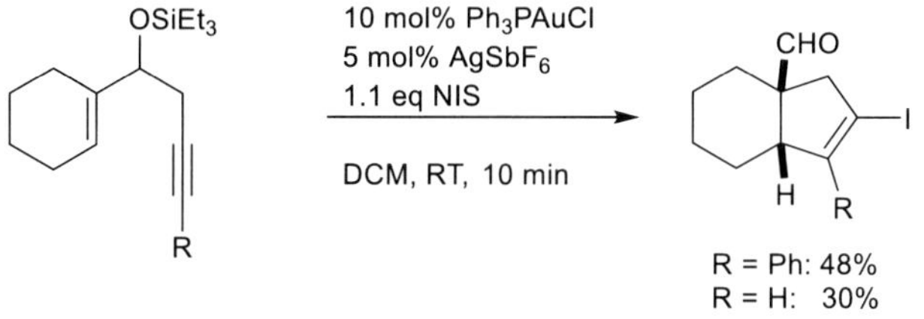

**Scheme 15.42** The vinylgold(I) intermediate can be intercepted with I⁺ donors.

## 15.14 Glycosylation via Gold Catalysis

In the field of carbohydrate chemistry, gold catalysis can facilitate the process of chemical glycosylation due to the rather mild reaction conditions and its selectivity. The principle is that the anomeric carbon is bound to a gold cyclizeable fragment. This gold-triggered cyclization liberates the positively charged sugar as an intermediate, which then reacts with another free hydroxyl group of a different sugar to form the product [63].

The first example was published by Hotha and Kashyap in 2006 on the basis of propargyl alcohol. Due to the gold activation, the propagyl alcohol cyclizes to methyleneoxirane, so another alcohol can attack as a nucleophile at the positively charged intermediate, forming the desired product (Scheme 15.43) [64].

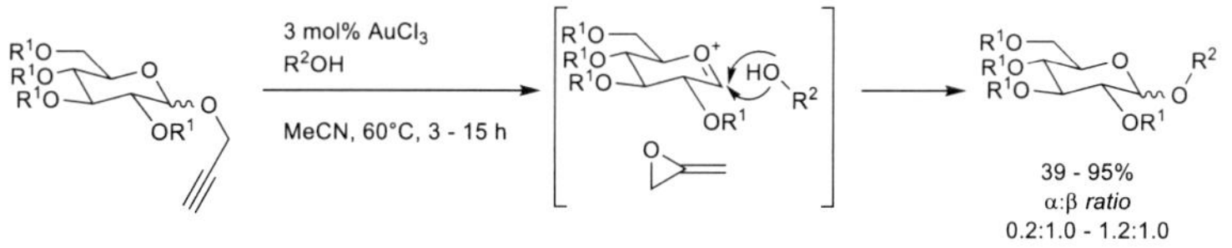

**Scheme 15.43** Gold-induced elimination generates carboxonium ions for gylcosidation reactions.

## 15.15 Ring Enlargements/Strained Substrates

As a Lewis acid, gold can initiate rearrangements of strained substrates, often resulting in ring enlargements. Since these processes are often driven by thermodynamics, this chemistry is very substrate specific. Due to its unique properties, gold-catalyzed rearrangements can yield different products in comparison to other Lewis acids [65].

It would go beyond the scope of this book to show all the possibilities for rearrangements of strained substrates; nevertheless two examples with good selectivity shall be shown.

Following the first example in 2004 [66], an example by Toste and coworkers, published in 2005, is a gold-catalyzed pinacol-like rearrangement (Scheme 15.44). Gold probably activates the alkyne as an electrophile, while the oxygen stabilizes a (partial) positive charge, so a ring expansion takes place and the product can form [67].

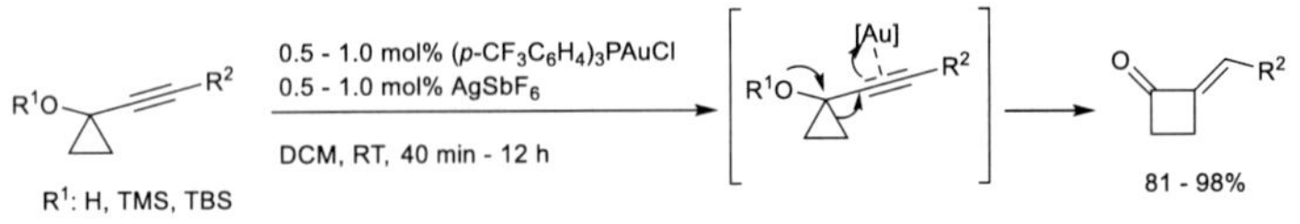

**Scheme 15.44** Ring expansion of a cyclopropylether or cyclo-alcohol to a cyclobutanone.

The second example by Shi, Liu, and Tang in 2006 shows that vinylcyclopropenes can act as electrophiles in the presence of gold (Scheme 15.45). The example shows a twofold gold-catalyzed hydroamination of such vinylcyclopropenes by a tosyl amine to yield pyrrolidines ZZ. Since the ring-opened, non-gold-coordinated intermediate ZZ was found if a lower reaction temperature was employed, it could be proposed that the ring opening happens first. It is then followed by intramolecular hydroamination [68].

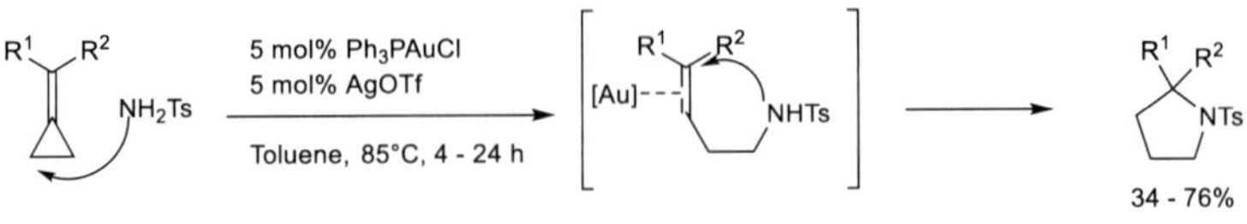

**Scheme 15.45** Ring expansion of an alkylidenecyclopropane to a pyrrolidine.

## 15.16 Dehydrative Gold Catalysis

In propagylic, allylic, and benzylic alcohols the hydroxyl group can be substituted by a nucleophile under gold catalysis (Scheme 15.46). In the general concept of this reaction, gold coordinates to the alkyne and the oxygen and then a gold hydroxide complex and a carbocation stabilized by conjugation is formed. If permitted by the substrate, this cation can isomerize, which is at the end attacked by a nucleophile to form the product, water, and regenerate the catalyst. Often this reaction mode is combined with other gold-catalyzed reactions like enyne cyclizations or nucleophilic attacks onto triple bonds [69].

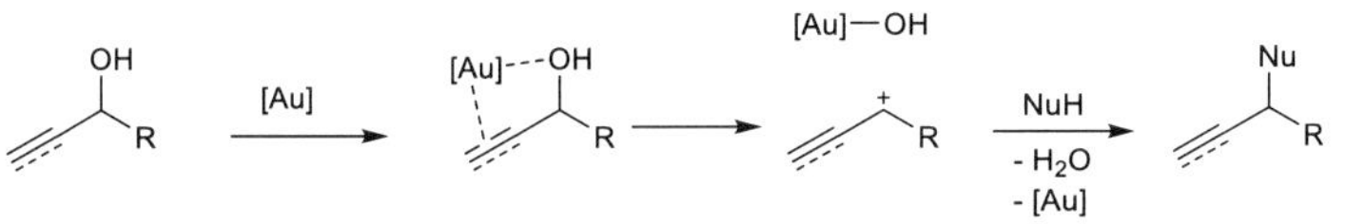

**Scheme 15.46** Activation of propargylic alcohols for nucleophilic additions.

The first example of such a reaction was published by Campagne and coworkers in 2005 (Scheme 15.47). In this reaction the hydroxyl group of propargylic alcohols was substituted by various nucleophiles (alcohols, thiols, and carbon nucleophiles) [70].

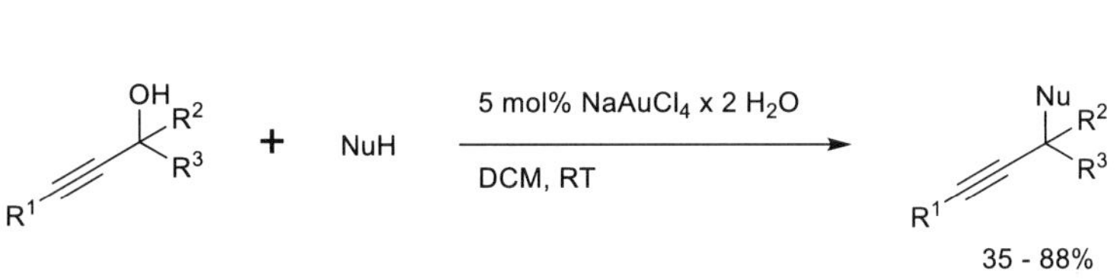

**Scheme 15.47** Campagne's first examples used gold(III) catalysts.

In the recent past also very successful allylic substitutions have been developed [71]. Even asymmetric reactions are possible in that way [72].

## 15.17 Conclusion

A manifold of different reactions have evolved from a simple basic reactivity pattern. The explosive growth of the field after the year 2000 has added many new and innovative methods with high synthetic efficiency to the sector of transition metal catalysis. At the moment it seems that this highly dynamic development will continue in the future, too.

## References

1. Obradors, C., and Echavarren, A.M. (2014). Intriguing mechanistic labyrinths in gold(I) catalysis, *Chem. Commun.*, **50**, pp. 16–28; Gulevich, A.V., Dudnik, A.S., Chernyak, N., and Gevorgyan, V. (2013). Transition metal-mediated synthesis of monocyclic aromatic heterocycles, *Chem. Rev.*, **113**, pp. 3084–3213; Rudolph, M., and Hashmi, A.S.K. (2012). Gold catalysis in total synthesis: an update, *Chem. Soc. Rev.*, **41**, pp. 2448–2462; Hashmi, A.S.K., and Toste, F.D. (2012). *Modern Gold Catalyzed Synthesis* (Wiley-VCH, Weinheim, Germany); Rudolph, M., and Hashmi, A.S.K. (2011). Heterocycles from gold catalysis, *Chem. Commun.*, **47**, pp. 6536–6544; Huang, H., Zhou, Y., and Liu, H. (2011). Recent advances in the gold-catalyzed additions to C-C multiple bonds, *Beilstein J. Org. Chem.*, **7**, pp. 897–936; Corma, A., Leyva-Pérez, A., and Sabater, M.J. (2011). Gold-catalyzed carbon-heteroatom bond-forming reactions, *Chem. Rev.*, **111**, pp. 1657–1712; Bandini, M. (2011). Gold-catalyzed decorations of arenes and heteroarenes with C-C multiple bonds, *Chem. Soc. Rev.*, **40**, pp. 1358–1367; Nevado, C. (2010). Gold catalysis: recent developments and future trends, *Chimia*, **64**, pp. 247–251; Das, A., Sohel, S.M.A., and Liu, R.-S. (2010). Carbo- and heterocyclisation of oxygen- and nitrogen-containing electrophiles by platinum, gold, silver and copper species. *Org. Biomol. Chem.*, **8**, pp. 960–979; Furstner, A. (2009). Gold and platinum catalysis: a convenient tool for generating molecular complexity, *Chem. Soc. Rev.*, **38**, pp. 3208–3221; Belmont, P., and Parker, E. (2009). Silver and gold catalysis for cycloisomerization reactions, *Eur. J. Org. Chem.*, pp. 6075–6089; Shen, H.C. (2008). Recent advances in syntheses of heterocycles and carbocycles via homogeneous gold catalysis. Part 1: heteroatom addition and hydroarylation reactions of alkynes, allenes, and alkenes, *Tetrahedron*, **64**, pp. 3885–3903; Shen, H.C. (2008). Recent advances in syntheses of carbocycles and

heterocycles via homogeneous gold catalysis, Part 2: cyclizations and cycloadditions, *Tetrahedron*, **64**, pp. 7847–7870; Li, Z., Brouwer, C., and He, C. (2008). Gold-catalyzed organic transformations, *Chem. Rev.*, **108**, pp. 3239–3265; Kirsch, S.F. (2008). Construction of heterocycles by the strategic use of alkyne $\pi$-activation in catalyzed cascade reactions, *Synthesis*, **20**, pp. 3183–3204; Hashmi, A.S.K., and Rudolph, M. (2008). Gold catalysis in total synthesis, *Chem. Soc. Rev.*, **37**, pp. 1766–1775; Arcadi, A. (2008). Alternative synthetic methods through new developments in catalysis by gold, *Chem. Rev.*, **108**, pp. 3266–3325; Hashmi, A.S.K. (2007). Gold-catalyzed organic reactions, *Chem. Rev.*, **107**, pp. 3180–3211; Fürstner, A., and Davies, P.W. (2007). Catalytic carbophilic activation: catalysis by platinum and gold $\pi$ acids, *Angew. Chem., Int. Ed.*, **46**, pp. 3410–3449; Hashmi, A.S.K., and Hutchings, G.J. (2006). Gold catalysis, *Angew. Chem., Int. Ed.*, **45**, pp. 7896–7936.

2. Ito, Y., Sawamura, M., and Hayashi, T. (1986). Catalytic asymmetric aldol reaction: reaction of aldehydes with isocyanoacetate catalyzed by a chiral ferrocenylphosphine-gold(I) complex, *J. Am. Chem. Soc.*, **108**, pp. 6405–6406.

3. Gulevich, A.V., Zhdanko, A.G., Orru, R.V.A., and Nenajdenko, V.G. (2010). Isocyanoacetate derivatives: synthesis, reactivity, and application, *Chem. Rev.*, **110**, pp. 5235–5331; Sawamura, M., and Ito, Y. (1992). Catalytic asymmetric synthesis by means of secondary interaction between chiral ligands and substrates, *Chem. Rev.*, **92**, pp. 857–871.

4. Widenhoefer, R.A. (2008). Recent developments in enantioselective gold(I) catalysis, *Chem. Eur. J.*, **14**, pp. 5382–5391; Bongers, N., and Krause, N. (2008). Golden opportunities in stereoselective catalysis, *Angew. Chem., Int. Ed.*, **47**, pp. 2178–2181.

5. Gu, P., Xu, Q., and Shi, M. (2014). Development and outlook of chiral carbene–gold(I) complexes catalyzed asymmetric reactions, *Tetrahedron Lett.*, **55**, pp. 577–584; Mourad, A.K. and Czekelius, C. (2013). Enantioselective functionalization of terminal alkynes by gold catalysis, *Synlett*, **24**, pp. 1459–1463; Patil, N.T. (2012). Chirality transfer and memory of chirality in gold-catalyzed reactions, *Chem. Asian J.*, **7**, pp. 2186–2194; Clavier, H., and Pellissier, H. (2012). Recent developments in enantioselective metal-catalyzed domino reactions, *Adv. Synth. Catal.*, **354**, pp. 3347–3403; Pradal, A., Toullec, P.Y., and Michelet, V. (2011). Recent developments in asymmetric catalysis in the presence of chiral gold complexes, *Synthesis*, **10**, pp. 1501–1514; Sengupta, S., and Shi, X. (2010). Recent advances in asymmetric gold catalysis, *ChemCatChem*, **2**, pp. 609–619.

6. Patil, N.T., Kavthe, R.D., and Shinde, V.S. (2012). Transition metal-catalyzed addition of C-, N- and O-nucleophiles to unactivated C-C multiple bonds, *Tetrahedron*, **68**, pp. 8079–8146; Hashmi, A.S.K., and Bührle, M. (2010). Gold-catalyzed addition of X–H bonds to C–C multiple bonds, *Aldrichim. Acta*, **43**, pp. 27–33; Krause, N., and Winter, C. (2011). Gold-catalyzed nucleophilic cyclization of functionalized allenes: a powerful access to carbo- and heterocycles, *Chem. Rev.*, **111**, pp. 1994–2009.
7. Widenhoefer, R.A., and Han, X. (2006). Gold-catalyzed hydroamination of C–C multiple bonds, *Eur. J. Org. Chem.*, pp. 4555–4563.
8. Fukuda, Y., Utimoto, K., and Nozaki, H. (1987). Preparation of 2,3,4,5-tetrahydropyridines from 5-alkynylamines under the catalytic action of Au(III), *Heterocycles*, **25**, pp. 297–300.
9. Mizushima, E., Hayashi, T., and Tanaka, M. (2003). Au(I)-catalyzed highly efficient intermolecular hydroamination of alkynes, *Org. Lett.*, **5**, pp. 3349–3352.
10. Muzart, J. (2008). Gold-catalysed reactions of alcohols: isomerisation, inter- and intramolecular reactions leading to C–C and C–heteroatom bonds, *Tetrahedron*, **64**, pp. 5815–5849; Liu, L., Xu, B., and Hammond, G.B. (2011). Construction of cyclic enones via gold-catalyzed oxygen transfer reactions, *Beilstein J. Org. Chem.*, **7**, pp. 606–614; Alcaide, B., Almendros, P., and Alonso, J.M. (2011). Gold catalyzed oxycyclizations of alkynols and alkyndiols, *Org. Biomol. Chem.*, **9**, pp. 4405–4416; Munoz, M.P. (2012). Transition metal-catalysed intermolecular reaction of allenes with oxygen nucleophiles: a perspective, *Org. Biomol. Chem.*, **10**, pp. 3584–3594.
11. Norman, R.O.C., Parr, W.J.E., and Thomas, C.B. (1976). The reactions of alkynes, cyclopropanes, and benzene derivatives with gold(III), *J. Chem. Soc., Perkin Trans.*, **1**, pp. 1983–1987.
12. Fukuda, Y., and Utimoto, K. (1991). Effective transformation of unactivated alkynes into ketones or acetals by means of Au(III) catalyst, *J. Org. Chem.*, **56**, pp. 3729–3731.
13. Hashmi, A.S.K., Schwarz, L., Choi, J.-H., and Frost, T.M. (2000). A new gold-catalyzed C–C bond formation, *Angew. Chem., Int. Ed.*, **39**, pp. 2285–2288.
14. Antoniotti, S., Genin, E., Michelet, V., and Genêt, J.-P. (2005). Highly efficient access to strained bicyclic ketals via gold-catalyzed cycloisomerization of bis-homopropargylic diols, *J. Am. Chem. Soc.*, **127**, pp. 9976–9977.

15. Blanco Jaimes, M.C., Böhling, C.R.N., Serrano-Becerra, J.M., and Hashmi, A.S.K. (2013). Highly active mononuclear NAC–gold(I) catalysts, *Angew. Chem., Int. Ed.*, **52**, pp. 7963–7966; Blanco Jaimes, M.C., Rominger, F., Pereira, M.M., Carrilho, R.M.B., Carabineiro, S.A.C. and Hashmi, A.S.K. (2014). Highly active phosphite gold(I) catalysts for intramolecular hydroalkoxylation, enyne cyclization and furanyne cyclization, *Chem. Commun.*, **50**, pp. 4937–4940.
16. Reetz, M. T., and Sommer, K. (2003). Gold-catalyzed hydroarylation of alkynes, *Eur. J. Org. Chem.*, **2003**, pp. 3485–3496.
17. Nevado, C., and Echavarren, A.M. (2005). Intramolecular hydroarylation of alkynes catalyzed by platinum or gold: mechanism and *endo* selectivity, *Chem. Eur. J.*, **11**, pp. 3155–3164.
18. Zhu, Z.-B., and Kirsch, S.F. (2013). Propargyl vinyl ethers as heteroatom-tethered enyne surrogates: diversity-oriented strategies for heterocycle synthesis, *Chem. Commun.*, **49**, pp. 2272–2283; Zhang, D.-H., Zhang, Z., and Shi, M. (2012). Transition metal-catalyzed carbocyclization of nitrogen and oxygen-tethered 1,*n*-enynes and diynes: synthesis of five ornynes and diynes: synthesis of five or six-membered heterocyclic compounds, *Chem. Commun.*, **48**, pp. 10271–10279; Hashmi, A.S.K. (2010). New pathways in the gold-catalyzed cycloisomerization of furanynes, *Pure Appl. Chem.*, **82**, pp. 1517–1528; Jiménez-Núñez, E., and Echavarren, A.M. (2008). Gold-catalyzed cycloisomerizations of enynes: a mechanistic perspective, *Chem. Rev.*, **108**, pp. 3326–3350.
19. Hashmi, A.S.K., Frost, T.M., and Bats, J.W. (2000). Highly selective gold-catalyzed arene synthesis, *J. Am. Chem. Soc.*, **122**, pp. 11553–11554.
20. Nieto-Oberhuber, C., Muñoz, M.P., Buñuel, E., Nevado, C., Cárdenas, D.J., and Echavarren, A.M. (2004). Cationic gold(I) complexes: highly alkynophilic catalysts for the *exo*- and *endo*-cyclization of enyne, *Angew. Chem., Int. Ed.*, **43**, pp. 2402–2406.
21. Marion, N., and Nolan, S.P. (2007). Propargylic esters in gold catalysis: access to diversity, *Angew. Chem., Int. Ed.*, **46**, pp. 2750–2752; Johansson, M.J., Gorin, D.J., Staben, S.T., and Toste, F.D. (2005). Gold(I)-catalyzed stereoselective olefin cyclopropanation, *J. Am. Chem. Soc.*, **127**, pp. 18002–18003; Fürstner, A., and Hannen, P. (2004). Carene terpenoids by gold-catalyzed cycloisomerization reactions, *Chem. Commun.*, **22**, pp. 2546–2547.
22. Miki, K., Ohe, K., and Uemura, S. (2003). Ruthenium-catalyzed cyclopropanation of alkenes using propargylic carboxylates as precursors of vinylcarbenoids, *J. Org. Chem.*, **68**, pp. 8505–8513.

23. Mamane, V., Gress, T., Krause, H., and Fürstner, A. (2004). Platinum- and gold-catalyzed cycloisomerization reactions of hydroxylated enynes, *J. Am. Chem. Soc.*, **126**, pp. 8654–8655.
24. Zhang, L. (2005). Tandem Au-catalyzed 3,3-rearrangement -[2+2] cycloadditions of propargylic esters: expeditious access to highly functionalized 2,3-indoline-fused cyclobutanes, *J. Am. Chem. Soc.*, **127**, pp. 16804–16805.
25. Amijs, C.H.M., López-Carrillo, V., and Echavarren, A.M. (2007). Gold-catalyzed addition of carbon nucleophiles to propargyl carboxylates, *Org. Lett.*, **9**, pp. 4021–4024.
26. Hashmi, A.S.K., Yang, W., Yu, Y., Hansmann, M.M., Rudolph, M., and Rominger, F. (2013). Gold-catalyzed formal 1,6-acyloxy migration leading to 3,4-disubstituted pyrrolidin-2-ones, *Angew. Chem., Int. Ed.*, **52**, pp. 1329–1332.
27. Yang, W., Yu, Y., Zhang, T., Hansmann, M.M., Pflästerer, D., and Hashmi, A.S.K. (2013). Gold-catalyzed highly diastereoselective synthesis of functionalized 3,4-disubstituted butyrolactams via phosphatyloxy or carbonate double migrations, *Adv. Synth. Catal.*, **355**, pp. 2037–2043.
28. (a) Zhang, L. (2014). A non-diazo approach to $\alpha$-oxo gold carbenes via gold-catalyzed alkyne oxidation, *Acc. Chem. Res.*, **47**, pp. 877–888; (b) Xiao, J., and Li, X. (2011). Gold $\alpha$-oxo carbenoids in catalysis: catalytic oxygen-atom transfer to alkynes, *Angew. Chem., Int. Ed.*, **50**, pp. 7226–7236; (c) López, F., and Mascareñas, J.L. (2011). Recent developments in gold-catalyzed cycloaddition reactions, *Beilstein J. Org. Chem.*, **7**, pp. 1075–1094.
29. Li, G., and Zhang, L. (2007). Gold-catalyzed intramolecular redox reaction of sulfinyl alkynes: efficient generation of $\alpha$-oxo gold carbenoids and application in insertion into R-CO bonds, *Angew. Chem., Int. Ed.*, **46**, pp. 5156–5159.
30. Shapiro, N.D., and Toste, F.D. (2007). Rearrangement of alkynyl sulfoxides catalyzed by gold(I) complexes, *J. Am. Chem. Soc.*, **129**, pp. 4160–4161.
31. Hashmi, A.S.K., Weyrauch, J.P., Rudolph, M., and Kurpejovic, E. (2004). Gold-Katalyse: die Vorteile von N- und N,O-Liganden, *Angew. Chem.*, **116**, pp. 6707–6709; (2004). Gold catalysis: the benefits of N and N,O ligands, *Angew. Chem., Int. Ed.*, **43**, pp. 6545–6547.
32. Yeom, H.-S., Lee, Y., Lee, J.-E., and Shin, S. (2009). Geometry-dependent divergence in the gold-catalyzed redox cascade cyclization of *o*-alkynylaryl ketoximes and nitrones leading to isoindoles, *Org. Biomol. Chem.*, **7**, pp. 4744–4752.

33. Ye, L., Cui, L., Zhang, G., and Zhang, L. (2010). Alkynes as equivalents of $\alpha$-diazo ketones in generating $\alpha$-oxo metal carbenes: a gold-catalyzed expedient synthesis of dihydrofuran-3-ones, *J. Am. Chem. Soc.*, **132**, pp. 3258–3259; Ye, L., He, W., and Zhang, L. (2010). Gold-catalyzed one-step practical synthesis of oxetan-3-ones from readily available propargylic alcohols, *J. Am. Chem. Soc.*, **132**, pp. 8550–8551.
34. (a) Wegner, H.A., and Auzias, M. (2011). Gold for C-C coupling reactions: a swiss-army-knife catalyst?, *Angew. Chem., Int. Ed.*, **50**, pp. 8236–8247; (b) Hopkinson, M.N., Gee, A.D., and Gouverneur, V. (2011). $Au^I/Au^{III}$ catalysis: an alternative approach for C-C oxidative coupling, *Chem. Eur. J.*, **17**, pp. 8248–8262; (c) Engle, K.M., Mei, T.-S., Wang, X., and Yu, J.-Q. (2011). Bystanding $F^+$ oxidants enable selective reductive elimination from high-valent metal centers in catalysis, *Angew. Chem., Int. Ed.*, **50**, pp. 1478–1491; (d) Garcia, P., Malacria, M., Aubert, C., Gandon, V., and Fensterbank, L. (2010). Gold-catalyzed cross-couplings: new opportunities for C-C bond formation, *ChemCatChem*, **2**, pp. 493–497.
35. Hashmi, A.S.K., Ramamurthi, T.D., and Rominger, F. (2009). Synthesis, structure and reactivity of organogold compounds of relevance to homogeneous gold catalysis, *J. Organomet. Chem.*, **694**, pp. 592–597.
36. Cui, L., Zhang, G., and Zhang, L. (2009). Homogeneous gold-catalyzed efficient oxidative dimerization of propargylic acetates, *Bioorg. Med. Chem. Lett.*, **19**, pp. 3884–3887.
37. Zhang, G., Peng, Y., Cui, L., and Zhang, L. (2009). Gold-catalyzed homogeneous oxidative cross-coupling reactions, *Angew. Chem., Int. Ed.*, **48**, pp. 3112–3115.
38. Brand, J.P., Charpentier, J., and Waser, J. (2009). Direct alkynylation of indole and pyrrole heterocycles, *Angew. Chem., Int. Ed.*, **48**, pp. 9346–9349.
39. Hopkinson, M.N., Tessier, A., Salisbury, A., Giuffredi, G.T., Combettes, L.E., Gee, A.D., and Gouverneur, V. (2010). Gold-catalyzed intramolecular oxidative cross-coupling of non-activated arenes, *Chem. Eur. J.*, **16**, pp. 4739–4743.
40. Haro, T. de and Nevado, C. (2010). Gold-catalyzed ethynylation of arenes, *J. Am. Chem. Soc.*, **132**, pp. 1512–1513.
41. Ball, L.T., Lloyd-Jones, G.C., and Russell, C.A. (2012). Gold-catalyzed direct arylation, *Science*, **337**, pp. 1644–1648.
42. Haro, T. de and Nevado, C. (2011). Flexible gold-catalyzed regioselective oxidative difunctionalization of unactivated alkenes, *Angew. Chem., Int. Ed.*, **50**, pp. 906–910.

43. Hashmi, A.S.K., Lothschütz, C., Ackermann, M., Doepp, R., Anantharaman, S., Marchetti, B., Bertagnolli, H., and Rominger, F. (2010). Compare to the homogeneous gold-catalyzed oxidative esterification reported recently, *Chem. Eur. J.*, **16**, pp. 8012–8019.
44. Pérez-Temprano, M.H., Casares, J.A., and Espinet, P. (2012). Bimetallic catalysis using transition and group 11 metals: an emerging tool for C–C coupling and other reactions, *Chem. Eur. J.*, **18**, pp. 1864–1884; Hirner, J.J., Shi, Y., and Blum, S.A. (2011). Organogold reactivity with palladium, nickel, and rhodium: transmetalation, cross-coupling, and dual catalysis, *Acc. Chem. Res.*, **44**, pp. 603–613.
45. Shi, Y., Ramgren, S.D., and Blum, S.A. (2009). Palladium-catalyzed carboauration of alkynes and palladium/gold cross-coupling, *Organometallics*, **28**, pp. 1275–1277; Hashmi, A.S.K., Lothschütz, C., Döpp, R., Rudolph, M., Ramamurthi, T.D., Rominger, F. (2009). Gold and palladium combined for cross-coupling, *Angew. Chem., Int. Ed.*, **48**, pp. 8243–8246.
46. Hashmi, A.S.K., Döpp, R., Lothschütz, C., Rudolph, M., Riedel, D., and Rominger, F. (2010). Scope and limitations of palladium-catalyzed cross-coupling reactions with organogold compounds, *Adv. Synth. Catal.*, **352**, pp. 1307–1314.
47. Shi, Y., Roth, K.E., Ramgren, S.D., and Blum, S.A. (2009). Catalyzed catalysis using carbophilic lewis acidic gold and Lewis basic palladium: synthesis of substituted butenolides and isocoumarins, *J. Am. Chem. Soc.*, **131**, pp. 18022–18023.
48. Garayalde, D., and Nevado, C. (2012). Gold-containing and gold-generated 1,n-dipoles as useful platforms toward cycloadditions and cyclizations, *ACS Catal.*, **2**, pp. 1462–1479.
49. Asao, N., Takahashi, K., Lee, S., Kasahara, T., and Yamamoto, Y. (2002). AuCl(3)-catalyzed benzannulation: synthesis of naphthyl ketone derivatives from o-alkynylbenzaldehydes with alkynes, *J. Am. Chem. Soc.*, **124**, pp. 12650–12651.
50. Li, C.-J. (2010). The development of catalytic nucleophilic additions of terminal alkynes in water, *Acc. Chem. Res.*, **43**, pp. 581–590.
51. Wei, C., and Li, C.-J. (2003). A highly efficient three-component coupling of aldehyde, alkyne, and amines via C–H activation catalyzed by gold in water, *J. Am. Chem. Soc.*, **125**, pp. 9584–9585.
52. Hashmi, A.S.K. (2014). Dual gold catalysis, *Acc. Chem. Res.*, **47**, pp. 864–846; Gómez-Suárez, A. and Nolan, S.P. (2012). Dinuclear gold catalysis: are two gold centers better than one?, *Angew. Chem., Int. Ed.*, **51**, pp. 8156–8159.

53. Hashmi, A.S.K., Braun, I., Rudolph, M., and Rominger, F. (2012). The role of gold acetylides as a selectivity trigger and the importance of *gem*-diaurated species in the gold-catalyzed hydroarylating-aromatization of arene-diynes, *Organometallics*, **31**, pp. 644–661; Ye, L., Wang, Y., Aue, D.H., and Zhang, L. (2012). Experimental and computational evidence for gold vinylidenes: generation from terminal alkynes via a bifurcation pathway and facile C–H insertions, *J. Am. Chem. Soc.*, **134**, pp. 31–34.
54. Cheong, P.H.-Y., Morganelli, P., Luzung, M.R., Houk, K.N., and Toste, F.D. (2008). Gold-catalyzed cycloisomerization of 1,5-allenynes via dual activation of an ene reaction, *J. Am. Chem. Soc.*, **130**, pp. 4517–4526.
55. Hashmi, A.S.K., Braun, I., Nösel, P., Schädlich, J., Wieteck, M., Rudolph, M., and Rominger, F. (2012). Simple gold-catalyzed synthesis of benzofulvenes: *gem*-diaurated species as "instant dual-activation" precatalysts, *Angew. Chem., Int. Ed.*, **51**, pp. 4456–4460.
56. Patil, N.T., Shinde, V.S., and Gajula, B. (2012). A one-pot catalysis: the strategic classification with some recent examples, *Org. Biomol. Chem.*, **10**, pp. 211–224; Loh, C.C.J., and Enders, D. (2012). Merging organocatalysis and gold catalysis: a critical evaluation of the underlying concepts, *Chem. Eur. J.*, **18**, pp. 10212–10225; Shao, Z., and Zhang, H. (2009). Combining transition metal catalysis and organocatalysis: a broad new concept for catalysis, *Chem. Soc. Rev.*, **38**, pp. 2745–2755.
57. Binder, J.T., Crone, B., Haug, T.T., Menz, H., and Kirsch, S.F. (2008). Direct carbocyclization of aldehydes with alkynes:? combining gold catalysis with aminocatalysis, *Org. Lett.*, **10**, pp. 1025–1028.
58. Han, Z.-Y., Xiao, H., Chen, X.-H., and Gong, L.-Z. (2009). Consecutive intramolecular hydroamination/asymmetric transfer hydrogenation under relay catalysis of an achiral gold complex/chiral Brønsted acid binary system, *J. Am. Chem. Soc.*, **131**, pp. 9182–9183.
59. Muratore, M.E., Holloway, C.A., Pilling, A.W., Storer, R.I., Trevitt, G., and Dixon, D.J. (2009). Enantioselective Brønsted acid-catalyzed N-acyliminium cyclization cascades, *J. Am. Chem. Soc.*, **131**, pp. 10796–10797.
60. Liu, X.-Y., and Che, C.-M. (2009). Highly enantioselective synthesis of chiral secondary amines by gold(I)/chiral Brønsted acid catalyzed tandem intermolecular hydroamination and transfer hydrogenation reactions, *Org. Lett.*, **11**, pp. 4204–4207.
61. Nakamura, I., Sato, T., and Yamamoto, Y. (2006). Gold-catalyzed intramolecular carbothiolation of alkynes: synthesis of 2,3-disubstituted benzothiophenes from ($\alpha$-alkoxy alkyl) (*ortho*-alkynyl phenyl) sulfides, *Angew. Chem., Int. Ed.*, **45**, pp. 4473–4475.

62. Kirsch, S.F., Binder, J.T., Crone, B., Duschek, A., Haug, T.T., Liébert, C., and Menz, H. (2007). Catalyzed tandem reaction of 3-silyloxy-1,5-enynes consisting of cyclization and pinacol pearrangement, *Angew. Chem., Int. Ed.*, **46**, pp. 2310–2313.

63. McKay, M.J., and Nguyen, H.M. (2012). Recent advances in transition metal-catalyzed glycosylation, *ACS Catal.*, **2**, pp. 1563–1595.

64. Hotha, S., and Kashyap, S. (2006). Propargyl glycosides as stable glycosyl donors: anomeric activation and glycoside syntheses, *J. Am. Chem. Soc.*, **128**, pp. 9620–9621.

65. Zhang, D.-H., Tang, X.-Y., and Shi, M. (2014). Gold-catalyzed tandem reactions of methylenecyclopropanes and vinylidenecyclopropanes, *Acc. Chem. Res.*, **47**, pp. 913–924; Lu, B.-L., Dai, L., and Shi, M. (2012). Strained small rings in gold-catalyzed rapid chemical transformations, *Chem. Soc. Rev.*, **41**, pp. 3318–3339; Miege, F., Meyer, C., and Cossy, J. (2011). When cyclopropenes meet gold catalysts, *Beilstein J. Org. Chem.*, **7**, pp. 717–734; Garayalde, D., and Nevado, C. (2011). Synthetic applications of gold-catalyzed ring expansions, *Beilstein J. Org. Chem.*, **7**, pp. 767–780.

66. Hashmi, A.S.K., and Sinha, P. (2004). Gold catalysis: mild conditions for the transformation of alkynyl epoxides to furans, *Adv. Synth. Catal.*, **346**, pp. 432–438.

67. Markham, J.P., Staben, S.T., and Toste, F.D. (2005). Gold(I)-catalyzed ring expansion of cyclopropanols and cyclobutanols, *J. Am. Chem. Soc.*, **127**, pp. 9708–9709.

68. Shi, M., Liu, L.-P., and Tang, J. (2006). Gold(I)-catalyzed domino ring-opening ring-closing hydroamination of methylenecyclopropanes (MCPs) with sulfonamides: facile preparation of pyrrolidine derivatives, *Org. Lett.*, **8**, pp. 4043–4046.

69. Biannic, B., and Aponick, A. (2011). Gold-catalyzed dehydrative transformations of unsaturated alcohols, *Eur. J. Org. Chem.*, **2011**, pp. 6605–6617.

70. Georgy, M., Boucard, V. and Campagne, J.-M. (2005). Gold(III)-catalyzed nucleophilic substitution of propargylic alcohols, *J. Am. Chem. Soc.*, **127**, pp. 14180–14181.

71. Mukherjee, P., and Widenhoefer, R.A. (2011). Gold(I)-catalyzed intramolecular amination of allylic alcohols with alkylamines, *Org. Lett.*, **13**, pp. 1334–1337.

72. Bandini, M., Monari, M., Romaniello, A., and Tragni, M. (2010). Gold-catalyzed direct activation of allylic alcohols in the stereoselective synthesis of functionalized 2-vinyl-morpholines, *Chem. Eur. J.*, **16**, pp. 14272–14277.

# Index

acetone  74–75, 77, 80, 91, 128, 268, 315–316, 325
acrolein  322, 422–423
activated carbon  15, 60–62, 126, 149, 155, 157–158
alcohol oxidation  352, 355, 361, 371, 375, 395
  aerobic  113–114, 118, 353–355, 360
  liquid-phase  158, 357, 375
alcohols, unsaturated  422
aldehydes  341–342, 352–354, 360, 390, 466, 483, 485
  $\beta$-unsaturated  420, 422, 424
alkanes  312–313
alkenes  5, 420–421, 467, 471, 473, 482
alkynes  74, 313, 421, 467–471, 473–475, 477–478, 481, 484–485, 487, 490–491
  gold-activated  477, 487
  gold-catalyzed oxidation of  477
  terminal  474, 486
allenes  467, 475, 484
allylic alcohols  353
alumina  3, 5, 7–9, 15–16, 19–20, 24, 59, 75, 82–84, 89–90, 92, 142, 322–323, 327, 358–359
ammonia  3, 8–10, 15, 21, 349
anion adsorption  6–9, 11, 14–15, 20, 25
arylboronic acids  390, 395, 397
ascorbic acid  41, 45, 54, 57

benzaldehyde  91, 355–356, 361, 366–372
benzyl alcohol  93, 158, 187, 324, 343, 354–356, 361, 365–366, 368–369, 371, 376
  aerobic oxidation of  115–116, 357, 359, 367, 369, 371
  liquid-phase oxidation of  354, 356, 358, 361, 368, 370, 375
  selective oxidation of  91, 370
  solvent-free oxidation of  366
bimetallic catalysts
  alloyed  160
  gold-based  156
bimetallic nanoparticles  156, 370, 374

capping ligands, chemisorbed  357
carbon monoxide  54, 259, 271
carbon nanofibers (CNFs)  143, 149–151, 153–154, 349, 351
carbon nanotubes (CNTs)  60–62, 136, 149
carbon nucleophiles  470, 476, 491
carbonate spectators  306
carbonates
  bidentate  242–243, 245
  hydrogen  243, 245
  sodium  15–16
catalysis
  gold-based  377
  transition metal  492

catalysts
  active 1–2, 130, 193, 305, 312, 348, 407, 466
  Au/CeO$_2$ 208–209, 215, 224, 232, 303–304, 315, 362, 397, 404
  Au/CeZrO$_4$ 208
  Au/MgO 266
  Au/SiO$_2$ 264, 328
  Au/TiO$_2$ 17, 155, 221, 271–272, 291, 315, 323–324, 351, 362
  Au/ZrO$_2$ 21, 269, 274–275
  calcined 306–307
  copper 13
  dried 3, 306
  gold-based 123, 144–145, 147, 149, 151, 153, 155–157, 159, 253–254, 256, 258, 260, 262, 272–274, 343–346
  gold-based trimetallic 375
  gold–palladium alloy 368, 373
  gold–platinum 363, 372
  gold–silver 374–375
  gold-supported 314, 323, 356
  iron 408
  Lindlar 422
  metal-free 136
  microgel-stabilized Au nanocluster 116
  microgel-supported 114
  monometallic 160
  monometallic platinum 372
  multimetallic 344
  MVS-derived Au–Pd bimetallic 90–91
  non-noble-metal 13
  palladium 392–393
  platinum 313, 343, 373
  quasi-homogeneous 113, 354
  recyclable 352, 372
  synthesized low-Na 177
  titania-supported 321, 323, 367
  transition metal 432
  trimetallic 377
  water–gas shift 208, 240
catalytic hydrodechlorination 417
catalytic oxidation 154, 190, 362
cationic gold clusters 234, 236, 238–239, 246
cationic gold species 296
chemical vapor deposition (CVD) 124–126, 128–129, 180, 324
chemisorption 257, 259–262, 264–267, 271, 274, 277–278, 418, 438
  hydrogen 259, 264, 418, 425
  methyl mercaptane 264
  oxygen 262
  pulse 258, 267
  selective 253–254, 256–258, 260, 262, 264, 266, 268, 270, 272, 274, 276, 278
  stoichiometry 259, 268, 277–278
  techniques 257, 259
chemoselectivity 415–416, 418, 420, 422, 424, 426, 428, 430, 432, 434, 436, 438, 440, 442
  enhanced 420
  hydrogenation 420
chlorobenzene 358, 425, 430
cinnamaldehyde 422
CNFs, see carbon nanofibers
CNTs, see carbon nanotubes
colloid immobilization 294, 296, 306, 325
colloidal method 342, 346, 349, 358, 366–367, 372–373
colloidal particles 46, 49
colloids 48–49, 51, 55, 58, 63, 100
core–shell microgels, thermosensitive 114–115
crotonaldehyde 371, 422
CVD, see chemical vapor deposition
cyclohexane 326, 328, 431
cyclohexane oxidation 326–328
cyclohexanol 326–328, 356, 427

cyclohexanone 326–328, 427

density functional theory (DFT) 194, 208, 322, 348, 420
deposition-precipitation 13–15, 17–26, 124, 176, 180, 269, 272–273, 276, 296, 299, 317, 324, 348–349, 436, 438
DFT, *see* density functional theory

EELS, *see* electron energy loss spectroscopy
EFTEM, *see* energy-filtered transmission electron microscopy
electron energy loss spectroscopy (EELS) 136, 140–141, 418
electron microscopy 191–192, 255
electron tomography 139, 142–143
electrophiles 471, 474, 486, 488, 490
electrostatic repulsion 48, 50, 148
electrostatic stabilization 48, 50, 52
electrostatic stabilizers 51, 55
electrosteric stabilization 48, 50–51
elemental gold 483
energy-filtered transmission electron microscopy (EFTEM) 137, 141–143
environmental transmission electron microscopy (ETEM) 136, 155, 160, 306
enyne cyclizations 471, 488, 491
ETEM, *see* environmental transmission electron microscopy
ethane 312–313, 315, 324

ethanol 56, 80, 128, 268, 321, 325, 353, 362, 431

fulminating gold 3–4
furfural 360–361

glucose 40, 326, 345–346, 373–374
glucose oxidation 5, 326, 345–346
glycerate 346–350, 352, 364, 367
glycerol 57, 154, 158, 194, 324, 326, 346–347, 349–352, 360, 364–365, 367, 372
  selective oxidation of 346, 349, 365
glycerol oxidation 130, 158, 326, 347–351, 375–376
gold
  bulk 144, 254, 264, 418
  catalytic activity of 82, 90, 254, 393–394, 402, 406
  cationic 178, 214, 302, 304, 486, 488
  metallic 11, 179, 226, 234
  nanosized 438
  oxidized 177–179, 184, 215, 226
  quantitative determination of 277
gold acetate 2, 5–6
  solubility of 6, 304
gold activation 489
gold anion adsorption 7–8
gold carbenoid 474, 477
gold catalysis 394–395
  asymmetric 466
  dehydrative 491
  oxidative 483
gold catalyst deactivation 195
gold catalysts 86, 206, 220, 232, 327, 342, 347–348, 377, 394, 397, 405, 409–410

active  123, 345, 408
alumina-supported  316
chemical deactivation/blocking of  193
impregnated  324
light-sensitive  191
monometallic  348, 372
nanoparticulate  292–293, 295, 297, 299, 301, 303
nanostructured  327
selective chemisorption on  262–263, 265
silica-based  12
titania-supported  315, 317
titanosilicate-supported  315
wet  302
X-ray photoelectron spectroscopy characterization of  171–172, 174, 176, 178, 180, 182, 184, 186, 188, 190, 192, 194, 196
XPS of  175, 177, 179, 181, 183
ZSM-5-supported  326
gold-catalyzed cross-coupling reaction  480
gold-catalyzed cyclo-isomerizations  475
gold-catalyzed enyne-type reaction  472
gold-catalyzed hydratation  477
gold-catalyzed hydroarylation  470
gold-catalyzed Sonogashira coupling  408, 410
gold colloids  41, 45, 47, 49, 325, 345
gold crystallites  125, 273
gold hydroxide  18, 491
gold nanoclusters  99–100, 102, 104, 106, 108, 110, 112–118
gold nanocrystals  56, 78, 80
gold nanoparticles, synthesis of  353, 355
gold nanorods  58
gold organyls  480, 483–484

gold precursors  1–2, 7, 21, 40, 45, 53, 105, 107, 118, 125, 127–128, 195, 325
gold salts  5, 54, 57
gold sols  40, 187
gold–palladium core–shell nanoparticles  370–371
gold–palladium nanoparticles  364, 367–368, 370–371, 375–376
gold–platinum nanoparticles  372–373

high-resolution transmission electron microscopy (HRTEM)  80, 136–137, 139, 148, 151–152, 157, 219, 236, 272, 274, 276, 350, 424
HRTEM, *see* high-resolution transmission electron microscopy
hydrocarbons  312, 423
hydrodechlorination  415
hydrogen peroxide  315, 319, 348, 361, 367, 376, 401
hydrogenation
  catalytic  417, 432
  chemoselective  434
  gas-phase  431, 438
  selective  5, 421–422, 424

iodobenzene  396–399, 402, 409

magnetron sputtering  129, 180–181
metal vapor synthesis (MVS)  74, 77, 79, 81, 83, 85, 87–89
metal vapors  74–76
metal–carbonyl bond  213
methane oxidation  312–314

methanol  56, 231, 313–314, 359–360, 362, 420, 431–432, 469, 473
methanol production  314, 420–421
methanol selectivity  314
methyl esters  359–360
microgel chains  102–103
microgel-stabilized Au nanocatalyst  113
microgel-stabilized Au nanoclusters  110, 112, 117–118
microgel stabilizers  104, 109, 117–119
microgel synthesis  101, 103, 119
microgels  99–118, 354
  sulfonated  105–106
monometallic gold  91, 372–373
MVS, see metal vapor synthesis

nanoparticles, Immobilization of  58–59, 61, 63
nitroarenes  434–436, 440
nitrobenzene  435, 439–441
noble metals  156, 254, 393, 408
nucleophiles  74, 467–469, 472, 474–475, 477–478, 482, 486, 488–489, 491

organic synthesis  390
organocatalysis  486–487
organogold compounds  483
organometallic chemistry  465
OSC, see oxygen storage capacity
oxidative coupling  313, 479–481
oxidative esterification  360–362
oxide
  base metal  286–287, 296, 298
  magnesium  24, 368

  propene  315–323
  propylene  315–317, 320–322
  zinc  4, 368
oxygen activation  225, 230, 233, 287
oxygen adsorption  259, 263
  static  263
oxygen-containing organic compounds, selective oxidation of  341–342
oxygen storage capacity (OSC)  207–208, 236, 305–306

$p$-chloronitrobenzene  433, 435
$p$-nitrophenol  116, 118–119
palladium  91–93, 156–159, 287, 312, 315, 317, 363–364, 366, 369–373, 376, 389–393, 395–396, 407–411, 439, 483–484
palladium-catalyzed carbon–carbon bond formation reactions  392
palladium-catalyzed carbon–carbon reactions  391
palladium-catalyzed coupling of organogold compounds  483
palladium-catalyzed reactions  391–392
phenols  425–427, 431, 472
phenylboronic acid  397–400
physical vapor deposition (PVD)  124, 128–130
point of zero charge (PZC)  6–8, 11, 15–16, 18–20, 24–25, 59
poly(vinyl alcohol) (PVA)  52, 55, 57, 59, 61–62, 143, 149, 152, 193, 351, 357
polyvinylpyrrolidone (PVP)  52, 55–56, 101, 357, 370, 394
propargylic alcohols  491
propargylic esters  474–476

propene 271, 315–322, 324
  oxidation of 123, 318, 321, 324–325
propene conversion 316, 319, 321–322
propene epoxidation 315, 317, 319–320, 322
propylene 316–318, 321–322, 324
propylene epoxidation 316
PVA, see poly(vinyl alcohol)
PVD, see physical vapor deposition
PVP, see polyvinylpyrrolidone
PZC, see point of zero charge

scanning transmission electron microscopy (STEM) 136–137, 139, 142, 160, 178, 191, 264, 366, 370
selective oxidation 312–314, 316, 318, 320, 322, 324, 326, 328, 364–365, 372, 374
SERS, see surface-enhanced Raman spectroscopy
$SiO_2$ 6, 8, 10, 61, 126, 130, 148, 192, 287, 296, 304, 317, 425, 431, 434
small alkanes 312–313, 315, 328
SMAs, see solvated metal atoms
SMSI, see strong metal support interaction
sodium citrate 41, 54, 105
sodium hydroxide 15, 325, 347
solid grinding 124, 127–128, 319–320, 325, 345
solvated metal atoms (SMAs) 73–78, 80, 82, 84, 86, 88–92, 124
Sonogashira coupling 402–403, 405, 407–409
spectroscopic characterization 172, 205, 227
STEM, see scanning transmission electron microscopy
steric stabilization 48–50, 52, 102
strong metal support interaction (SMSI) 126
sulfoxides 477–478
surface-enhanced Raman spectroscopy (SERS) 398

Tammann temperature 191, 193
TEM, see transmission electron microscopy
temperature-programmed desorption (TPD) 9, 240–241, 262
temperature-programmed reduction (TPR) 225, 233, 358, 422, 430, 436, 438
tetrachloroauric acid 1–5
titania 4–5, 7–11, 15–17, 19–20, 22, 90, 142, 263, 265, 317, 319, 322–323, 367–369
TMA, see trimethylamine
toluene 75, 77–78, 80, 93, 314, 365–366, 368, 370, 376, 425
TPD, see temperature-programmed desorption
TPR, see temperature-programmed reduction
transmetallation 407
transmission electron microscopy (TEM) 8, 12, 78, 80, 88, 135–144, 146, 156–158, 160–161, 171, 255–256, 260–264, 271–272, 277, 422
trimethylamine (TMA) 321

water–gas shift (WGS) 179, 193, 207, 271, 416
WGS, see water–gas shift

X-ray diffraction (XRD)  88, 137, 161, 191, 233, 256, 260–262, 274, 436
X-ray photoelectron spectroscopy (XPS)  88, 161, 172–180, 182, 184, 187, 191, 193–195, 325, 349, 357, 396, 403, 438, 440

XPS, *see* X-ray photoelectron spectroscopy
XRD, *see* X-ray diffraction

zirconia  8, 12, 15, 208, 233, 243, 245, 273